INTRODUCTION

Mathematics underpins everything. From a few axioms (basic statements accepted automatically), it is possible to derive all the Pure Mathematics contained within this book. When some modelling assumptions are included, techniques in Pure Mathematics can be used to explore Applied Mathematics, such as Statistics and Mechanics. A sound knowledge of Mathematics is necessary to understand how the world and the rest of the universe behave. For example, it is essential to the investigation of the Big Bang, the mapping of the human genome, understanding the environment and the financial markets.

The aim of this book is to help make your study of advanced mathematics interesting and successful. However, the content can be demanding and there is no need to worry if you do not understand it straight away. Discuss ideas with other students, and of course check with your teacher or tutor. Most important of all, keep asking questions.

This book covers the requirements of Year 2 A-level Mathematics.

Good luck and enjoy your study of Mathematics. We hope that this book will encourage you to study Mathematics further after you have completed your course.

Helen Ball

Kath Hipkiss

Michael Kent

Chris Pearce

FEATURES TO HELP YOU LEARN

Real-life context

Each chapter starts with a real life application of the mathematics you are learning in the chapter.

Learning objectives

A summary of the concepts, ideas and techniques that you will meet in the chapter.

Topic links

See how the material in the chapter relates to other chapters in the book.

Prior knowledge

See what mathematics you should know before you start the chapter, with some practice questions to check your understanding.

Key terms and glossary

Important words are written in **bold**. The words are defined in the glossary at the back of the book.

Stop and think

These boxes present you with probing questions and problems to help you to reflect on what you have been learning.

---

3 COORDINATE GEOMETRY: PARAMETRIC EQUATIONS

The concept of parametric equations may appear quite abstract at first, but they are very useful. Imagine a penalty being taken during a football match. The football will travel through three-dimensional space and consequently, at any moment, its position could be given by three coordinates (x, y, z). The path of the football could be expressed by equations in x, y and z. If you were to choose to define the motion in terms of a parameter, for example time t, then you can easily begin to see where the ball will be at any particular moment.

LEARNING OBJECTIVES

You will learn how to:

› understand and use the parametric equations of curves

› understand and use conversion between Cartesian and parametric forms

› use parametric equations in modelling in a variety of contexts.

TOPIC LINKS

The skills and techniques that you learn in this chapter will help you to solve problems in **Chapter 15 Kinematics** where the motion is defined using parametric equations.

PRIOR KNOWLEDGE

You should already know how to:

› manipulate polynomials algebraically, including:

  › expanding brackets and simplifying by collecting like terms

  › factorisation

  › using and manipulating surds, including rationalising the denominator.

› use and understand the graphs of functions

› sketch curves defined by simple equations, including quadratics, cubics, quartics and reciprocals

› interpret the algebraic solution of equations graphically

› use the intersection points of graphs to solve equations

› understand and use the coordinate geometry of the circle, including using the equation of a circle in the form $(x - a)^2 + (y - b)^2 = r^2$

Stop and think If you are asked to show that two lines, between two different pairs of coordinates, are parallel do you need to work out the equation of each line or is there a slightly simpler approach?

Helen Ball
Kath Hipkiss
Michael Kent
Chris Pearce

Mathematics

A Level Year 2

Book

Collins

ebook included

To access the eBook visit www.collins.co.uk/ebooks and follow the step-by-step instructions.

CONTENTS

CONTENTS

* Short answers are given in this book, with full worked solutions for all exercises, large data set activities, exam-style questions and extension questions available to teachers by emailing education@harpercollins.co.uk

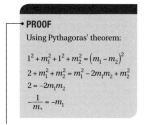

PROOF

Using Pythagoras' theorem:

$$1^2 + m_1^2 + 1^2 + m_2^2 = (m_1 - m_2)^2$$
$$2 + m_1^2 + m_2^2 = m_1^2 - 2m_1m_2 + m_2^2$$
$$2 = -2m_1m_2$$
$$-\frac{1}{m_2} = -m_1$$

TECHNOLOGY

Using a graphing software package, try finding the equations of lines passing through other points but that are parallel to the given equation.

KEY INFORMATION

You need to be able to use straight line models in a variety of contexts.

Proof

These boxes show proofs or describe what type of and how a proof is being applied.

Technology

These boxes show how you can use your calculator or other technologies such as graphing software to explore the mathematics.

Key information

These boxes highlight information that you need to pay attention to and learn, such as key formulae and learning points.

Explanations and examples

Each section begins with an explanation and one or more worked examples. Some show alternative solutions to get you thinking about different approaches to a problem.

Example 8

Find the equation of the line that is parallel to $3x - y + 7 = 0$ and passes through the point $(1, 8)$.

Write the equation of the line in the form $ax + by + c = 0$, where a, b and c are integers.

Solution

Rearrange the equation into the form $y = mx + c$.

$$y = 3x + 7$$

$m = 3$ and $(x_1, y_1) = (1, 8)$

Colour-coded questions

The questions in the exercises and the review questions are colour-coded (green, blue and red) to show you how difficult they are. Exercises start with more accessible (green) questions and then progress through intermediate (blue) to more challenging (red) questions. Red questions are particularly appropriate if you are studying for the full A-level.

(CM) **7** In each of the following cases, explain the mathematical manipulation steps required to show whether or not each pair of lines are perpendicular.

a $5x - 3y - 7 = 0$ and $-10y = 6x + 3$

b $9x - 10y - 7 = 0$ and $9y - 10x + 2 = 0$

(PS) (CM) **8** A quadrilateral is drawn with vertices at $(0, 6)$, $(2, 10)$, $(4, 4)$ and $(6, 8)$.

By considering gradients, work out which of the following the quadrilateral could possibly be and state your reasons.

rhombus   kite   parallelogram   square   trapezium   rectangle

Question-type indicators

Many questions in the book require you to think in different ways. Look for these icons:

(CM) Communicating mathematically – this requires an explanation, often in words. This is a key skill for Assessment Objective 2: Reason, interpret and communicate mathematically, and forms part of the overarching themes for AS and A-level Mathematics (OT1: Mathematical argument, language and proof).

(PF) Proof – you need to show how you arrived at the given result using logical steps. This is a key skill for Assessment Objective 2: Reason, interpret and communicate mathematically, and forms part of the overarching themes for AS and A-level Mathematics (OT1: Mathematical argument, language and proof).

(PS) Problem solving – you need to apply a multi-step process, often using concepts from different areas of mathematics. This is a key skill for Assessment Objective 3: Solve problems within mathematics and other contexts, and forms part of the overarching themes for AS and A-level Mathematics (OT2: Mathematical problem solving).

(M) Modelling – you will need to make one or more assumptions in order to proceed. This is a key skill for Assessment Objective 3: Solve problems within mathematics and other contexts, and forms part of the overarching themes for AS and A-level Mathematics (OT3: Mathematical modelling).

Using the large data set activity

These activities in the Statistics chapters can be used in groups or individually to explore your exam board's large data set, becoming familiar with the contexts and practising the statistical skills you've learnt.

Using the large data set 14.2

A statistician believes there is a relationship between two of the variables in the large data set.

a Select two variables from your large data set. Determine whether there is a connection between these variables.

b Would you expect this connection to be the same in different time periods or different locations? State your reasons carefully.

Summary of key points

At the end of each chapter, there is a summary of key formulae and learning points.

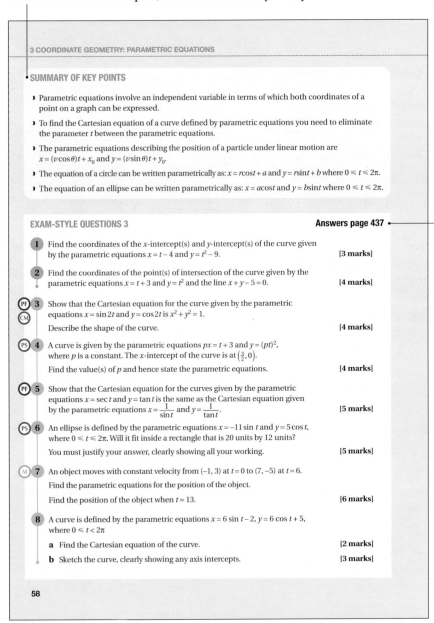

3 COORDINATE GEOMETRY: PARAMETRIC EQUATIONS

SUMMARY OF KEY POINTS

› Parametric equations involve an independent variable in terms of which both coordinates of a point on a graph can be expressed.

› To find the Cartesian equation of a curve defined by parametric equations you need to eliminate the parameter t between the parametric equations.

› The parametric equations describing the position of a particle under linear motion are $x = (v\cos\theta)t + x_0$ and $y = (v\sin\theta)t + y_0$.

› The equation of a circle can be written parametrically as: $x = r\cos t + a$ and $y = r\sin t + b$ where $0 \leqslant t \leqslant 2\pi$.

› The equation of an ellipse can be written parametrically as: $x = a\cos t$ and $y = b\sin t$ where $0 \leqslant t \leqslant 2\pi$.

EXAM-STYLE QUESTIONS 3   Answers page 437

1 Find the coordinates of the x-intercept(s) and y-intercept(s) of the curve given by the parametric equations $x = t - 4$ and $y = t^2 - 9$. **[3 marks]**

2 Find the coordinates of the point(s) of intersection of the curve given by the parametric equations $x = t + 3$ and $y = t^2$ and the line $x + y - 5 = 0$. **[4 marks]**

(PF) (CM) 3 Show that the Cartesian equation for the curve given by the parametric equations $x = \sin 2t$ and $y = \cos 2t$ is $x^2 + y^2 = 1$.

Describe the shape of the curve. **[4 marks]**

(PS) 4 A curve is given by the parametric equations $px = t + 3$ and $y = (pt)^2$, where p is a constant. The x-intercept of the curve is at $\left(\frac{3}{2}, 0\right)$.

Find the value(s) of p and hence state the parametric equations. **[4 marks]**

(PF) 5 Show that the Cartesian equation for the curves given by the parametric equations $x = \sec t$ and $y = \tan t$ is the same as the Cartesian equation given by the parametric equations $x = \dfrac{1}{\sin t}$ and $y = \dfrac{1}{\tan t}$. **[5 marks]**

(PS) 6 An ellipse is defined by the parametric equations $x = -11\sin t$ and $y = 5\cos t$, where $0 \leqslant t \leqslant 2\pi$. Will it fit inside a rectangle that is 20 units by 12 units?

You must justify your answer, clearly showing all your working. **[5 marks]**

(M) 7 An object moves with constant velocity from $(-1, 3)$ at $t = 0$ to $(7, -5)$ at $t = 6$.

Find the parametric equations for the position of the object.

Find the position of the object when $t = 13$. **[6 marks]**

8 A curve is defined by the parametric equations $x = 6\sin t - 2$, $y = 6\cos t + 5$, where $0 \leqslant t < 2\pi$

a Find the Cartesian equation of the curve. **[2 marks]**

b Sketch the curve, clearly showing any axis intercepts. **[3 marks]**

58

Exam-style questions

Practise what you have learnt throughout the chapter with questions written in examination style and allocated marks, progressing in order of difficulty.

At the end of the book you will find **Exam-style extension questions**, providing additional practice at A-level standard for each chapter.

1 ALGEBRA AND FUNCTIONS 1: FUNCTIONS

Do you play computer games? Even if you don't, you probably have a friend who does. How does a computer game work? The basis of a computer game is an algorithm: a list of operations that need to be performed, which incorporates the order in which they need to be performed. The computer game software is the codification of the algorithm. A function is just like an algorithm – it is a step-by-step process containing mathematical operations.

LEARNING OBJECTIVES

You will learn how to:

> understand and use the definition of a function

> understand and use composite functions and their graphs

> understand and use inverse functions and their graphs

> sketch and use the graphs of $y = |ax + b|$

> find the graphs of $y = |f(x)|$ and $y = f(|x|)$, given the graph of $y = f(x)$

> use functions in modelling, including consideration of limitations and refinements of the models, including exponential growth and decay.

TOPIC LINKS

The skills and techniques that you learn in this chapter will help you to solve problems involving trigonometric functions in **Chapter 5 Trigonometry**. They will also help you to understand the results of your differentiation in **Chapter 6 Differentiation** and **Chapter 7 Further differentiation**. The ability to use different types of functions will help you to understand and solve area problems in **Chapter 8 Integration** and mechanics problems in **Chapter 15 Kinematics**.

PRIOR KNOWLEDGE

You should already know how to:

> work with coordinates in all four quadrants

> recognise, sketch and interpret graphs of:

> > linear functions, quadratic functions, cubic functions, quartic functions, the reciprocal function $y = \frac{1}{x}$ with $x \neq 0$, $y = \frac{1}{x^2}$ with $x \neq 0$, exponential and logarithmic functions, the trigonometric functions (with x in degrees) $y = \sin x$, $y = \cos x$ and $y = \tan x$ for angles of any size, circles

> sketch transformations of a given function.

You should be able to complete the following questions correctly:

1 Sketch the graph of each of the following functions:

   a $y = x^2 - 3$ **b** $y = \sqrt{x}$ **c** $y = \dfrac{1}{x}$

   d $y = e^x$ **e** $y = \sin x$ **f** $y = \tan x$

2 If $f(x) = x^2$, sketch the graph of each of the following functions:

   a $f(2x)$ **b** $2f(x)$

   c $f(x + 2)$ **d** $f(x) + 2$

3 If $y = 2 - x$, find the value of y when:

   a $x = 0$ **b** $x = 1$

   c $x = -3$ **d** $x = \dfrac{1}{2}$

1.1 Definition of a function

A **set** of numbers can be transformed into a different set of numbers by applying a **mapping**. For example, if the input set is defined as {1, 2, 3, 4} and a mapping 'subtract 1' is applied, then the output set becomes {0, 1, 2, 3}. The input set of numbers is called the **domain** and the output set of numbers is called the **range**. Throughout this chapter, where the domain is not stated, it should be taken as the set of real numbers ($\mathbb{R}$).

A **function** is a special mapping such that every member of the domain (input set) is mapped to exactly one member of the range (output set). A function can be written as $f(x) =$ or $f: x \mapsto$. For example, $f(x) = x - 1$ or $f: x \mapsto x - 1$.

Consider the input set {−2, −1, 0, 1, 2}. If $f(x) = 3x - 1$ then the output set will be {−7, −4, −1, 2, 5}, with each member of the domain mapping exactly to one member of the range. This is a **one-to-one function**.

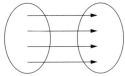

one-to-one function

Once again, consider the input set {−2, −1, 0, 1, 2}. If $f(x) = x^2$ then the output set will be {0, 1, 4}, with each member of the domain mapping to exactly one member of the range. However, in this case some members of the domain (eg −1 and 1) both map to the same member of the range (1) and so this is a **many-to-one function**.

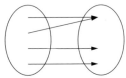

many-to-one function

> **KEY INFORMATION**
>
> The domain is the input set of numbers to a mapping or function.
>
> The range is the output set of numbers from a mapping or function.

> **KEY INFORMATION**
>
> In a function, every member of the domain is mapped to exactly one member of the range. It can be written as $f(x) =$ or $f: x \mapsto$.

> **KEY INFORMATION**
>
> In a one-to-one function, every member of the domain maps to exactly one member of the range.

> **KEY INFORMATION**
>
> In a many-to-one function, every member of the domain maps to exactly one member of the range and some members of the domain map to the same member of the range.

Consider the input set {1, 4, 9}. If $f(x) = \pm\sqrt{x}$ then the output set will be {−3, −2, −1, 1, 2, 3}, with each member of the domain mapping to more than one member of the range. This is a one-to-many relationship and consequently *not* a function.

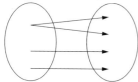

one-to-many relationship

Example 1

Which of the following relationships are functions on the set of real numbers? Clearly justify your answer. For those that are functions, state if the function is one-to-one or many-to-one.

a $f(x) = 2 - x$ **b** $g(x) = x^3$ **c** $h(x) = +\sqrt{x}$

d

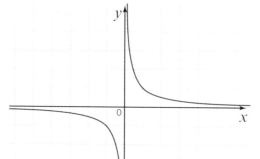

e

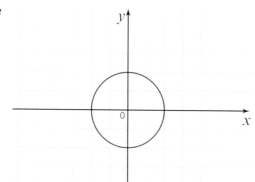

f

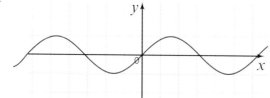

Solution

a $f(x) = 2 - x$ is a linear relationship whereby every input value is mapped to only one output value and every output value is mapped to only one input value. This is a one-to-one function.

b $g(x) = x^3$ is a cubic relationship whereby every input value is mapped to only one output value and every output value is mapped to only one input value. This is a one-to-one function.

c $h(x) = +\sqrt{x}$ is not a function because negative input values of x do not map to any output values because no negative number has a square root.

> If the domain was restricted to all positive real numbers, then $h(x) = +\sqrt{x}$ would be a function.

d This is not a function because an input value of $x = 0$ does not map to any output value at the **discontinuity** at $x = 0$.

e This is a circle in the form $x^2 + y^2 = r^2$ where an input value can map to two output values; consequently, this is *not* a function.

> If a vertical line from the domain on the x-axis can be drawn through a graph which intersects the curve more than once then the relationship is one-to-many and consequently *not* a function.

f This is a relationship whereby every input value is mapped to only one output value. As an output value may be mapped to more than one input value, this is a many-to-one function.

Example 2

A function $f(x) = 4x - 3$ has domain $\{x = -3, -2, -1, 0, 1, 2\}$.

a Find $f(2)$.

b Find the range.

c State if the function is one-to-one or many-to-one.

Solution

a $f(2)$ means find the output value when the input value is 2.

$$f(2) = (4)(2) - 3 = 5$$

b To work out the range, each input value in the domain needs to be substituted into the function.

$$f(-3) = (4)(-3) - 3 = -15$$
$$f(-2) = (4)(-2) - 3 = -11$$
$$f(-1) = (4)(-1) - 3 = -7$$
$$f(0) = (4)(0) - 3 = -3$$
$$f(1) = (4)(1) - 3 = 1$$
$$f(2) = (4)(2) - 3 = 5$$

So the range of $f(x)$ is $\{-15, -11, -7, -3, 1, 5\}$

c This is a linear relationship whereby every input maps to only one output value and every output value is related to only one input value. This is a one-to-one function.

Example 3

Given that g: $x \mapsto 2x^2 + 1$ with domain $\{x \in \mathbb{R}, -3 \leqslant x \leqslant 3\}$.

a Find g(−1).

b Sketch the graph of the function and state the range.

c Find the value of b where g$(b) = \dfrac{27}{2}$.

d State if the function is one-to-one or many-to-one.

Solution

a g$(-1) = (2)(-1)^2 + 1 = 3$

b g: $x \mapsto 2x^2 + 1$ with domain $\{x \in \mathbb{R}, -3 \leqslant x \leqslant 3\}$

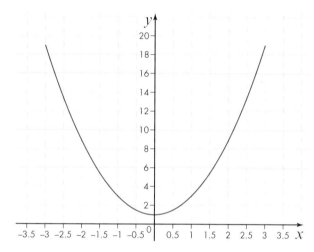

Range of g(x) is $1 \leqslant \text{g}(x) \leqslant 19$.

c Substitute the input and output values into the function and solve.

$$2b^2 + 1 = \frac{27}{2}$$

$$2b^2 = \frac{25}{2}$$

$$b^2 = \frac{25}{4}$$

$$b = \pm\frac{5}{2}$$

d This is a relationship whereby every input value is mapped to only one output value. As an output value may be mapped to more than one input value, this is a many-to-one function.

Example 4

The function $h(x)$ is defined as $h(x) = \begin{cases} 1 - x, & x < 2 \\ +\sqrt{x}, & x \geqslant 2 \end{cases}$

a Sketch the graph of the function and state the range.

b Find the values of c where $h(c) = 2$.

c State if the function is one-to-one or many-to-one.

Solution

a For $x < 2$ the function is linear. For $x \geqslant 2$ the function is a positive square root. The value of $h(2)$ needs to be calculated for both parts of the function.

$$h(2) = 1 - 2 = -1, \, x < 2$$

$$h(2) = +\sqrt{x} = 1.414, \, x \geqslant 2$$

We calculate $h(2)$ for this part of the function to define the sketch, even though $x = 2$ is not included in this part of $h(x)$

Using your knowledge from **Book 1, Chapter 3 Algebra and functions 3: Sketching curves**, sketch the two curves that comprise the function.

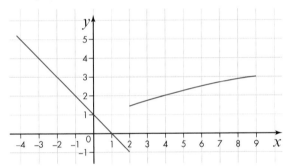

The range is $h(x) > -1$.

b Use your graph to find c where $h(c) = 2$.

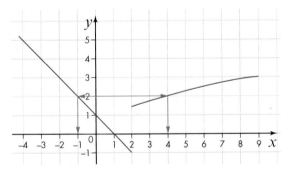

You can check by substituting the input and output values into each part of the function and solving.

$$1 - c = 2$$
$$c = -1$$

or

$$+\sqrt{c} = 2$$
$$c = 4$$

c This is a relationship whereby every input value is mapped to only one output value. As an output value may be mapped to more than one input value, this is a many-to-one function. •

Although it might appear it is many-valued for $x = 2$, the 2 separate conditions and the square root for x ensure it isn't.

Exercise 1.1A

Answers page 421

1 Which of the following relationships are functions on the set of real numbers? Clearly justify your answer. For those that are functions, state if the function is one-to-one or many-to-one.

a $f(x) = 5 - 3x$ **b** $g(x) = x^2 + x$ **c** $h(x) = \pm\sqrt{x + 1}$

d

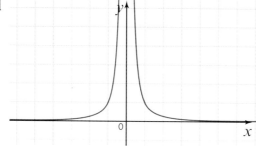

e

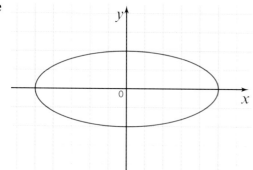

f

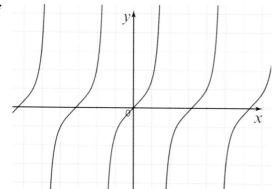

2 Given that $f(x) = 4x - 3$ with domain $\{x = -3, -2, -1, 0, 1, 2\}$:

    a find the range

    b state if the function is one-to-one or many-to-one.

3 Given that g: $x \mapsto 3 - 5x - 2x^2$ with domain $\{x \in \mathbb{R}, -5 \leqslant x \leqslant 5\}$:

 a find g(-2)

 b sketch the graph of the function and state the range

 c find the values of b where g(b) = 6

 d state if the function is one-to-one or many-to-one.

(PS) 4 Given that h(x) = $d - x$ and j(x) = $\dfrac{x+2}{4}$ where $x \in \mathbb{R}$ and d is a constant, find the value of d for which h(2) = j(2).

5 The function k(x) is defined as k(x) = $\begin{cases} x^2 - 2, & x < 0 \\ 2x - 3, & x \geqslant 0 \end{cases}$

 a Sketch the graph of the function and state the range.

 b Find the values of c where k(c) = 7.

 c State if the function is one-to-one or many-to-one.

(CM) 6 The following graph is a sketch of the curve given by the equation $y = \dfrac{1}{(x+a)^2}$.

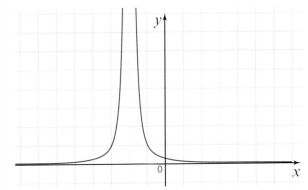

Suggest a possible domain, or domains that would allow the relationship to become a function. Clearly justify your answer.

(PS) 7 The function m(x) is defined as m(x) = $\begin{cases} -(3x + 2), & x \leqslant 1 \\ x^3, & x > 1 \end{cases}$

Find the value(s) of x such that m(x) = x.

1.2 Composite functions

You can apply one function more than once or combine two or more functions to make a new function. The new function is known as a **composite function**.

The notation for composite functions may be written in one of two ways: either fg(x) or, slightly more simply, fg. The order in which a composite function is written is significant. fg means 'do g first, then f', whereas gf means 'do f first, then g'.

> **KEY INFORMATION**
>
> fg(x) or fg means 'do g first, then f'.

Example 5

Given that $f(x) = x^2 - 1$ and $g(x) = 3 - x$, find the following:

a $fg(x)$ **b** $gf(x)$ **c** $gg(x)$

d $f^2(x)$ **e** $fg(2)$

f the values of a such that $gf(a) = -5$

Solution

a $fg(x)$ means 'do g first, then f'. This could also be written as $f(g(x))$.

$$= f(3 - x)$$

Wherever you see x in f, substitute $3 - x$.

$$= (3 - x)^2 - 1$$

$$= 9 - 6x + x^2 - 1$$

$$= 8 - 6x + x^2$$

b $gf(x)$ means 'do f first, then g'.

$$= g(x^2 - 1)$$

Wherever you see x in g, substitute $(x^2 - 1)$.

$$= 3 - (x^2 - 1)$$

$$= 3 - x^2 + 1$$

$$= 4 - x^2$$

c $gg(x)$ means 'do g, then g again'. $gg(x)$ could also be written as $g^2(x)$.

$$= g(3 - x)$$

Wherever you see x in g, substitute $3 - x$.

$$= 3 - (3 - x)$$

$$= 3 - 3 + x$$

$$= x$$

d $f^2(x)$ means 'do f, then f again'. $f^2(x)$ could also be written as $ff(x)$.

$$= f(x^2 - 1)$$

Wherever you see x in f, substitute $x^2 - 1$.

$$= (x^2 - 1)^2 - 1$$

$$= x^4 - 2x^2 + 1 - 1$$

$$= x^4 - 2x^2$$

e $fg(2)$ can be found in two ways.

$$g(2) = 3 - 2 = 1$$

Then substitute the value of g(2) into f.

$$f(1) = 1^2 - 1 = 0$$

Alternatively, you can find fg and then substitute in $x = 2$.

From **part a**, $fg(x) = 8 - 6x + x^2$

$$fg(2) = 8 - 12 + 4 = 0$$

f $gf(a) = -5$

From **part b**, $gf(x) = 4 - x^2$

$$gf(a) = 4 - a^2$$

$$gf(a) = -5$$

$$4 - a^2 = -5$$

$$a^2 = 9$$

$$a = \pm 3$$

Stop and think   Is it always true that fg ≠ gf?

Example 6

Given that $f(x) = \dfrac{1}{x+2}$, $x \neq -2$, and $g(x) = 2x^2 + 5$, find the following in terms of f and g:

a $\dfrac{2}{(x+2)^2} + 5$ **b** $\dfrac{1}{2x^2 + 7}$ **c** $8x^4 + 40x^2 + 55$

Solution

a As $f(x) = \dfrac{1}{x+2}$, then $\dfrac{2}{(x+2)^2} + 5$ can be rewritten.

$$= 2\left(\frac{1}{x+2}\right)^2 + 5$$

$$= 2(f(x))^2 + 5$$

But $g(x) = 2x^2 + 5$ so this can be rewritten.

$$= g\left(\frac{1}{x+2}\right)$$

$$= gf(x)$$

You could check this result by doing the process in reverse to see if $gf(x) = \dfrac{2}{(x+2)^2} + 5$.

b As $g(x) = 2x^2 + 5$, then $\dfrac{1}{2x^2 + 7}$ can be rewritten.

$$= \frac{1}{(2x^2 + 5) + 2}$$

$$= \frac{1}{g(x) + 2}$$

But $f(x) = \dfrac{1}{x+2}$ so this can be rewritten.

$$= f(2x^2 + 5)$$

$$= fg(x)$$

c As $g(x) = 2x^2 + 5$, then $8x^4 + 40x^2 + 55$ can be rewritten.

$$= 2(4x^4 + 20x^2 + 25) + 5$$

$$= 2(2x^2 + 5)^2 + 5$$

But $g(x) = 2x^2 + 5$ so this can be rewritten.

$$= g(2x^2 + 5)$$

$$= gg(x) \text{ or } g^2(x)$$

Exercise 1.2A

Answers page 421

1 Given that $f(x) = 4x + 3$ and $g(x) = x^2$, find the following:

    a $fg(1)$ **b** $gf(1)$

    c $gg(-2)$ **d** $f^2(-2)$

2 Given that $f(x) = \dfrac{1}{x}$, $x \neq 0$ and $g(x) = x^2 - 1$, find the following:

    a $fg(x)$ **b** $gf(x)$

    c $gg(x)$ **d** $f^2(x)$

(PS) 3 Given that $f(x) = x^2$ and $g(x) = x - 5$, find the values of a such that $fg(a) = 4$.

(PF) 4 The functions f and g are defined by:

    $f: x \mapsto 5x - 2$ $g: x \mapsto x^2 + 3$

    Show that $fg \neq gf$.

(PS) 5 The functions f and g are defined by

    $f: x \mapsto x + 1$ $g: x \mapsto x^2 - 7$

    Express the following in terms of f and g:

    a $x + 2$ **b** $x^2 + 2x - 6$

(PF) 6 Given that $f(x) = 2x - 3$, $g(x) = x^3$ and $h(x) = \dfrac{1}{x+1}$, show that $f^2hg = \dfrac{4}{x^3+1} - 6$

(PS) 7 The functions f and g are defined by

    $f: x \mapsto 2^x$ $g: x \mapsto x - 3$

    Find a value(s) of x such that $fg = gf$.

(PS) 8 Given that $f(x) = \sqrt{x}$, $g(x) = x + 7$ and $h(x) = \dfrac{1}{x}$, $x > 0$, express the following in terms of f, g and h:

    a $\sqrt{x + 7}$ **b** $x^{-\frac{1}{2}} + 7$ **c** $x^{\frac{1}{2}} + 7$

1.3 Inverse functions

The **inverse function** maps members of the range back to members of the domain. Consequently, inverse functions can only be found for one-to-one functions.

Inverse functions cannot be found for many-to-one functions because the inverse would be a one-to-many relationship, which is not a function.

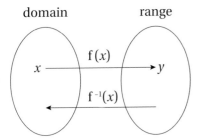

The inverse function

The notation used to represent the inverse of a function f(x) is f^{-1}(x).

If f(x) = x + 1, then f^{-1}(x) = x − 1.

What is the meaning of ff^{-1}(x)?

$$ff^{-1}(x) = f(x - 1)$$

$$= x - 1 + 1$$

$$= x$$

What is the meaning of f^{-1}f(x)?

$$f^{-1}f(x) = f^{-1}(x + 1)$$

$$= x + 1 - 1$$

$$= x$$

You can conclude that if f^{-1} exists, then ff^{-1}(x) = f^{-1}f(x) = x.

If f(x) = x + 1 then f^{-1}(x) = x − 1. The following graphs demonstrate this visually:

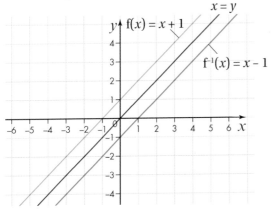

KEY INFORMATION

The inverse can only be found for one-to-one functions.

KEY INFORMATION

f^{-1}(x) represents the inverse of f(x).

KEY INFORMATION

ff^{-1}(x) = f^{-1}f(x) = x

KEY INFORMATION

y = f^{-1}(x) is the image of y = f(x) reflected in the line y = x.

Notice that $y = f^{-1}(x)$ is the **image** of $y = f(x)$ reflected in the line $y = x$.

Example 7

A function is defined as $f(x) = x^2 + 4$ where $\{x \in \mathbb{R}, x > 0\}$.

a Find $f^{-1}(x)$ and state its domain.

b Sketch the graphs of $f(x)$ and its inverse on the same pair of axes.

c Show that $ff^{-1}(x) = x$.

Solution

a To find the inverse you need to rearrange $f(x)$ to make x the subject.

Let $y = x^2 + 4$.

$$y - 4 = x^2$$
$$\sqrt{y - 4} = x$$

You need to state the inverse in terms of x.

$$f^{-1}(x) = \sqrt{x - 4} \text{ where } \{x \in \mathbb{R}, x > 4\}$$

b

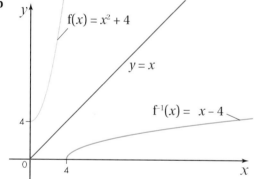

c $ff^{-1}(x) = f\left(\sqrt{x - 4}\right)$

$\qquad = \left(\sqrt{x - 4}\right)^2 + 4$

$\qquad = x - 4 + 4$

$\qquad = x$

Example 8

A function is defined as $f(x) = \dfrac{1}{x}$ where $\{x \in \mathbb{R}, x \neq 0\}$.

a Find $f^{-1}(x)$ and state its domain. What do you notice?

b Sketch the graphs of $f(x)$ and its inverse on the same pair of axes. What do you notice?

c Show that if $f(a) = b$ then $f^{-1}(b) = a$.

d Find the values of x such that $f(x) = f^{-1}(x)$.

Solution

a To find the inverse you need to rearrange $f(x)$ to make x the subject.

Let $y = \dfrac{1}{x}$.

$$xy = 1$$
$$x = \dfrac{1}{y}$$

You need to state the inverse in terms of x.

$$f^{-1}(x) = \dfrac{1}{x} \text{ where } \{x \in \mathbb{R}, x \neq 0\}$$
$$f(x) = f^{-1}(x)$$

This is known as a **self inverse**.

b

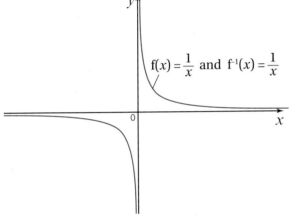

$$f(x) = \dfrac{1}{x} \text{ and } f^{-1}(x) = \dfrac{1}{x}$$

> **KEY INFORMATION**
>
> A self inverse is when $f(x) = f^{-1}(x)$.

As $f(x) = f^{-1}(x)$, their graphs are the same.

c $f(a) = b$

Substituting into $f(x)$ gives: $\dfrac{1}{a} = b$

Rearranging: $\dfrac{1}{b} = a$

which is the same as $f^{-1}(b) = a$

d As $f(x) = f^{-1}(x)$ all their values are the same, i.e. the solution set is $\{x \in \mathbb{R}, x \neq 0\}$.

Exercise 1.3A

Answers page 422

1 For each of the following functions, find $f^{-1}(x)$, state its domain and sketch the graphs of $f(x)$ and $f^{-1}(x)$ on the same pair of axes.

a $f(x) = 5 - 4x, \{x \in \mathbb{R}\}$

b $f(x) = (x - 3)^2, \{x \in \mathbb{R}, x \geqslant 3\}$

c $f(x) = x^3, \{x \in \mathbb{R}\}$

(PF) 2 Show that if $f(x) = 7x - 2$ then $f^{-1}(2) = \frac{4}{7}$.

(PS) 3 For each of the following functions, find $f(x)$, state its domain and sketch the graphs of $f(x)$ and $f^{-1}(x)$ on the same pair of axes.

    a $f^{-1}(x) = 5 - 4x$, $\{x \in \mathbb{R}\}$

    b $f^{-1}(x) = +\sqrt{x} + 4$, $\{x \in \mathbb{R} \ x \geqslant 0\}$

    c $f^{-1}(x) = e^x$, $\{x \in \mathbb{R}\}$

(PS) 4 Given that $f(x) = x^2 + 7x - 11$ where $\{x \in \mathbb{R}, x \geqslant -\frac{7}{2}\}$, find $f^{-1}(x)$. (Hint: complete the square).

(PS) 5 $f(x) = \dfrac{2}{x-1} + 5$ where $\{x \in \mathbb{R}, x > 1\}$.

    a Find $f^{-1}(x)$ and state its domain.

    b Sketch the graphs of $f(x)$ and $f^{-1}(x)$ on the same pair of axes.

    c Find the value of x such that $f(x) = f^{-1}(x)$.

(PF) 6 Given that $f(x) = 2x^2 + 5x - 13$, $x \geqslant -\dfrac{5}{4}$ show that $ff^{-1}(x) = x$.

1.4 The modulus of functions

The modulus of a number is its positive numerical value. For example, $|7| = 7$ and $|-7| = 7$. Generally, the **modulus** function is of the form $y = |f(x)|$.

For example if $f(x) = 2$ then $|f(x)| = 2$.

$|f(x)| = f(x)$ for $f(x) \geqslant 0$

For example if $f(x) = -2$ then $|f(x)| = -(-2) = 2$.

$|f(x)| = -f(x)$ for $f(x) < 0$

For example, consider the graphs of $y = x$ and $y = |x|$. You will see that any negative values from the function (negative y values) are changed to be positive by the modulus. This is a reflection in the x-axis.

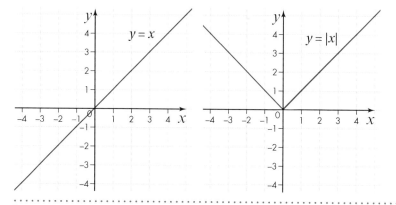

TECHNOLOGY

The modulus of a number is sometimes known as the **absolute value**. Consequently, on some calculators the button to find the modulus is labelled 'Mod', whereas on others it is labelled 'Abs'.

KEY INFORMATION

The modulus function is of the form $y = |f(x)|$.

› $|f(x)| = f(x)$ for $f(x) \geqslant 0$

› $|f(x)| = -f(x)$ for $f(x) < 0$

Example 9

Sketch the graph of $y = |x^2 - 2x - 3|$.

Solution

First, sketch the graph of $y = x^2 - 2x - 3$.

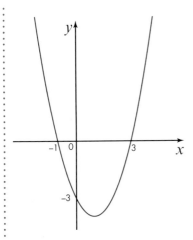

Any negative values from the function (negative y values) will be changed to positive by the modulus, a reflection in the x-axis.

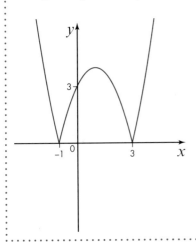

TECHNOLOGY

Using a graphic calculator or graphing software package, plot the graphs of $y = |f(x)|$ and $y = |-f(x)|$ where $f(x)$ is a quadratic function for which the discriminant is greater than zero. What do you notice?

Example 10

Sketch the graph of $y = |x|^2 - 2|x| - 3$.

Solution

First you need to sketch the graph of $y = x^2 - 2x - 3$ for $x \geq 0$.

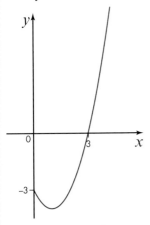

You then need to reflect the curve in the y-axis.

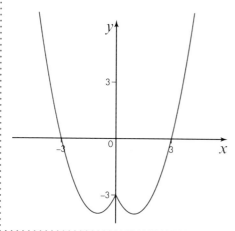

If you are not sure why the curve is reflected in the y-axis, create and compare the table of values for each of $y = x^2 - 2x - 3$ and $y = |x|^2 - 2|x| - 3$.

Stop and think What do you think the graph of $y = |x^2 - 2x| - 3$ will look like? Why?

Exercise 1.4A

Answers page 423

(PS) 1 Sketch the graph of each of the following functions on different pairs of axes. Clearly show the coordinates of any axis intercepts.

    a $y = |3x + 2|$ **b** $y = |x - 2|$ **c** $y = |x^2 - 4|$

    d $y = 3|x| + 2$ **e** $y = |x| - 2$ **f** $y = |x|^2 - 4$

(CM) 2 Which of the following is the graph of $y = \left|\dfrac{1}{x}\right|$? Clearly explain and justify your choice.

A

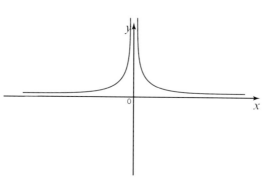

B

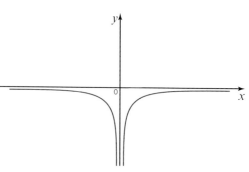

C

D

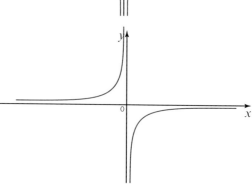

(PS) 3 Sketch the graph of each of the following functions on different pairs of axes. Clearly show the coordinates of any axis intercepts.

a $y = |7 - x|$ **b** $y = |x^3 - 1|$ **c** $y = |-x|$

d $y = 7 - |x|$ **e** $y = |x|^3 - 1$ **f** $y = -|x|$

(CM) 4 When the graphs of $y = |\sin x|$ and $y = \sin|x|$ for $-2\pi \leqslant x \leqslant 2\pi$ are sketched, are the curves symmetrical about the x-axis and/or the y-axis? Clearly explain and justify your answer.

(PS) 5 Sketch the graph of each of the following functions on different pairs of axes. Clearly show the coordinates of any axis intercepts.

a $y = |e^x|$ **b** $y = |\ln x|$ **c** $y = |\tan x|$

d $y = e^{|x|}$ **e** $y = \ln|x|$ **f** $y = \tan|x|$

You need to be able to use the graphs of functions containing moduli to solve both equations and inequalities. For example, you need to be able to use the graph of $y = |2x - 1|$ to solve the equation $|2x - 1| = x$ or the inequality $|2x - 1| > x$.

Example 11

Solve the equation $|2x - 1| = x$.

Solution

On the same pair of axes, sketch the graphs of $y = |2x - 1|$ and $y = x$.

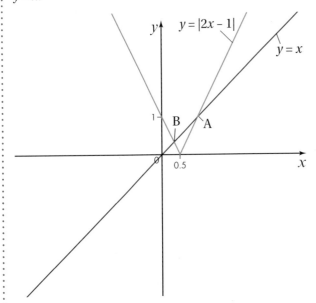

There are two points of intersection, A and B.

At A: $\qquad 2x - 1 = x$

Solving gives $\qquad x = 1$

At B: $\qquad -(2x - 1) = x$

Solving gives $-2x + 1 = x$

$$x = \frac{1}{3}$$

Example 12

Solve the inequality $|2x - 1| > x$.

Solution

On the same pair of axes sketch the graphs of $y = |2x - 1|$ and $y = x$.

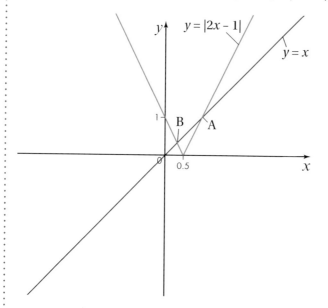

There are two points of intersection, A and B.

From Example 11: At A, $x = 1$ and at B, $x = \frac{1}{3}$

So, from the graphs, we can see the solution is $x > 1$ or $x < \frac{1}{3}$.

Exercise 1.4B

Answers page 425

1 **a** On the same pair of axes, sketch the graphs of $y = |3x - 4|$ and $y = 2$.

   b Use your graphs to solve $|3x - 4| = 2$.

2 **a** On the same pair of axes, sketch the graphs of $y = |x - 5|$ and $y = 2x$.

   b Use your graphs to solve $|x - 5| > 2x$.

(PS) **3** Solve the equation $|2x + 5| = |x - 1|$.

(PS) **4** Solve the inequality $|2^x - 3| < 2$.

(PS) **5** Solve $|x^2 - 3x - 4| = x$.

(PS) **6** Solve $|\sin x| \leq \frac{1}{2}$ for $-\frac{\pi}{2} \leq x \leq \frac{\pi}{2}$.

1.5 Transformations involving the modulus function

In **Book 1, Chapter 3 Algebra and functions 3: Sketching curves** you looked at how to apply different, individual transformations to curves. These were:

f($x + a$)	a horizontal translation of $-a$
f(x) $+ a$	a vertical translation of $+a$
f(ax)	a horizontal stretch with a scale factor of $\frac{1}{a}$
af(x)	a vertical stretch with a scale factor of a

You need to be able to apply these transformations to functions involving the modulus function.

Example 13

Starting with a sketch of $y = |x|$, show how it could be transformed to $y = |2x + 1| - 3$ through successive transformations, illustrating your results with sketches.

Solution

First let f(x) = $|x|$ and sketch the graph.

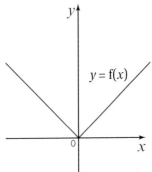

$f(2x) = |2x|$

Let $g(x) = |2x|$ and sketch the graph.

> This is a horizontal stretch with a scale factor of $\frac{1}{2}$.

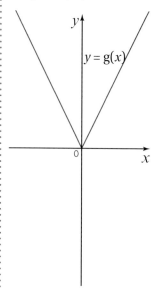

$g\left(x + \frac{1}{2}\right) = \left|2\left(x + \frac{1}{2}\right)\right|$

$2\left(x + \frac{1}{2}\right) = 2x + 1$

Let $h(x) = \left|2\left(x + \frac{1}{2}\right)\right| = |2x + 1|$ and sketch the graph.

> This is a horizontal translation of $-\frac{1}{2}$.

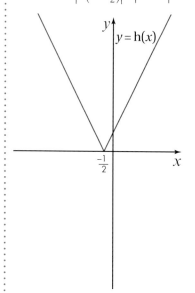

$h(x) - 3 = |2x + 1| - 3$

> This is a vertical translation of -3.

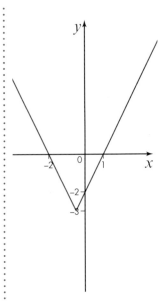

This is a sketch of $|2x + 1| - 3$ as required.

Exercise 1.5A

Answers page 426

1 Given that $f(x) = |x|$, $\{x \in \mathbb{R}\}$, sketch the graphs of the following functions:

    a $f(2x)$ **b** $f(x - 3)$ **c** $f(x) + 5$

(PS) 2 Given that $g(x) = |2x|$, match the sketches of graphs to the transformations.

    a $g(x + 1)$ **b** $g(2x)$

    c $g(x) - 1$ **d** $3g(x)$

A

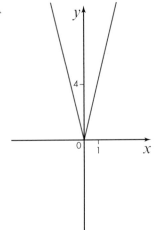

B

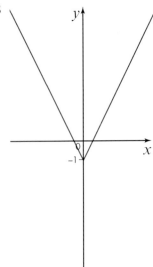

C

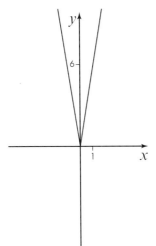

D

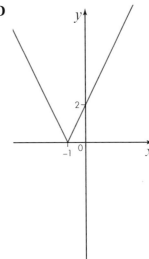

3 Given that $h(x) = |x^2 - 1|$, $\{x \in \mathbb{R}\}$, sketch the graphs of the following functions:

    a $h(2x)$ **b** $h(x - 3)$ **c** $h(x) + 5$

(PS) 4 Sketch the graph of $j(x) = a|x - 1| + b$ for $a > 0$ and $b > 0$. Clearly mark the coordinates of any axis intercepts.

5 Given that $k(x) = |x^2 - 3x - 4|$, $\{x \in \mathbb{R}\}$, sketch the graphs of the following functions:

    a $k(2x)$ **b** $k(x - 3)$ **c** $k(x) + 5$

(PS) 6 Given that $m(x) = |x^2 - 2x| + 1$, sketch the graph of $m(x - 2)$. Clearly mark the coordinates of any axis intercepts.

1.6 Functions in modelling

You need to be able to assess whether or not a function chosen to model a real-life situation is suitable and, where appropriate, suggest possible improvements.

Example 14

The following formula models how the value of a car varies in relation to the age of the car:

$$V = 15\,000 - 1500e^{\frac{t}{5}}$$

where V is the value of the car, in pounds (£), and t is the age of the car, in years.

a Find the value of the car when new, to the nearest pound.

b Find the value of the car after 5 years and after 10 years, to the nearest pound.

c Sketch the graph of V against t.

23

d Is this a suitable model for the value of the car over time? Clearly justify your answer.

e Comment on the suitability of the alternative formula

$$V = 13500e^{-\frac{t}{c}} \text{ where } c \text{ is a constant.}$$

Solution

a When the car is new, $t = 0$. Substitute this value into the formula.

$$V = 15\,000 - 1500e^0 = 13\,500$$

So the value of the car when new was £13 500.

b When the car is 5 years old, $t = 5$. Substitute this value into the formula.

$$V = 15\,000 - 1500e^1 = 10\,923$$

So the value of the 5-year-old car is £10 923.

When the car is 10 years old, $t = 10$. Substitute this value into the formula.

$$V = 15\,000 - 1500e^2 = 3916$$

So the value of the 10-year-old car is £3916.

c

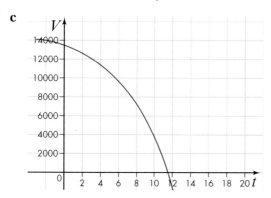

d One reason this is not a suitable model is because the value of the car can become negative, which in reality is not possible.

e This alternative formula is an improvement because the value of the car cannot become negative.

Example 15

The pressure exerted by a rocket on Earth is proportional to its mass divided by the area over which its weight is exerted. The rocket has four circular landing feet, the radius of which can be adjusted. The pressure exerted by the same rocket on another planet varies according to the following formula:

$$P = 2450\mathrm{e}^{-\frac{r}{5}}$$

where P is the pressure in $\mathrm{N\,m}^{-2}$ and r is the radius of the rocket feet.

a   Using the formula, determine the pressure exerted by the rocket feet on the planet just before it lands. Explain your answer.

b   After the rocket has landed, calculate the pressure exerted by the rocket if the radius of its feet is 1 m.

c   After the rocket has landed, find the radius, to the nearest metre, when the pressure exerted by the rocket first falls below $1000\,\mathrm{N\,m}^{-2}$.

d   Sketch the graph to show how the pressure varies in relation to the radius of the rocket feet. Is this a good model? Justify your answer.

Solution

a   Before it lands, the rocket feet are not exerting any pressure on the surface of the planet, so set $r = 0$. Substitute this value into the formula.

$$P = 2450\mathrm{e}^0 = 2450$$

Using the formula, the pressure exerted before the rocket lands is $2450\,\mathrm{N\,m}^{-2}$. However, the actual pressure would be zero as the rocket is not actually in contact with the surface of the planet.

b   When the radius of its feet is 1 m, then $r = 1$. Substitute this value into the formula.

$$P = 2450\mathrm{e}^{-\frac{1}{5}} = 2005.89$$

Using the formula, the pressure exerted by the rocket is $2005.89\,\mathrm{N\,m}^{-2}$.

c   In this case, we are given the value of P and can form an inequality which can be solved.

$$2450\mathrm{e}^{-\frac{r}{5}} < 1000$$

$$\mathrm{e}^{-\frac{r}{5}} < \frac{1000}{2450}$$

Take logs of both sides of the inequality.

$$-\frac{r}{5} < \ln\left(\frac{1000}{2450}\right)$$

Solving gives $r = 5\,\mathrm{m}$; this is the radius, to the nearest metre, when the pressure first falls below $1000\,\mathrm{N\,m}^{-2}$.

d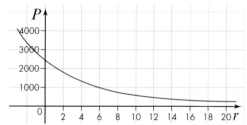

Overall the model is suitable for the scenario – except just before the rocket lands, when the model indicates the rocket exerts pressure, but in reality the rocket would not exert any pressure.

Exercise 1.6A

Answers page 427

1 The number of cells in a developing embryo is modelled by the equation

$$N = 2^t$$

where N is the number of cells and t is the age of the embryo in days.

a How many cells does the embryo have just before it starts to grow?

b Calculate how many cells the embryo has after 10 days and after 28 days.

c Sketch the graph of N against t.

CM 2 The value of a mobile phone can be modelled by the following formula:

$$V = 50 + 550e^{-\frac{t}{4}}$$

where V is the value of the phone in pounds (£) and t is the age of the phone in years.

a What is the price of the mobile phone when new?

b Calculate the value of the phone after 6 months, 2 years and 4 years.

c Sketch the graph of the value of the phone, V, against time t.

d Is this a suitable model for the value of the phone over time? Clearly justify your answer.

CM 3 The number of students achieving first degrees at UK universities can be modelled by the formula:

$$S = 200\,000 + 20\,000e^{\frac{t}{10}}$$

where S is the number of students and t is the number of years after 1994 (the year in which polytechnics became universities).

a How many students achieved their first degree at a UK university in 1994?

b How many students achieved their first degree at a UK university in 2000 and in 2010?

c In what year will the number of students achieving their first degree at a UK university first exceed 500 000 students, according to this model?

d In what year will the number of students achieving their first degree at a UK university first exceed 1 000 000 students, according to this model?

e Comment on the suitability of this model.

SUMMARY OF KEY POINTS

❯ A function is a special mapping such that every member of the domain is mapped to exactly one member of the range. It can be written as $f(x) =$ or $f: x \rightarrow$.

❯ The domain is the input set of numbers to a mapping or function. The range is the output set of numbers from a mapping or function.

❯ In a one-to-one function, every member of the domain maps to exactly one member of the range.

❯ In a many-to-one function, every member of the domain maps to exactly one member of the range and some members of the domain map to the same member of the range.

❯ The application of one function more than once, or the combination of two or more functions, makes a composite function. $fg(x)$ or fg means 'do g first, then f'.

❯ The inverse function maps members of the range back to members of the domain. The inverse can only be found for one-to-one functions.

❯ The notation used to represent the inverse of a function $f(x)$ is $f^{-1}(x)$.

   ❯ $f\,f^{-1}(x) = f^{-1}\,f(x) = x$

   ❯ $y = f^{-1}(x)$ is the image of $y = f(x)$ reflected in the line $y = x$.

❯ A self inverse is when $f(x) = f^{-1}(x)$.

❯ The modulus of a number is its positive numerical value. The modulus function is of the form $y = |f(x)|$.

   ❯ $|f(x)| = f(x)$ for $f(x) \geqslant 0$

   ❯ $|f(x)| = -f(x)$ for $f(x) < 0$

EXAM-STYLE QUESTIONS 1

Answers page 428

1 A function $f(x) = 8x - 5$ has domain $\{x = -3 \leqslant x \leqslant 3\}$.

  a Find $f(2)$.   **[1 mark]**

  b Represent the function in a mapping diagram showing input and output sets.   **[1 mark]**

  c Find the value of b where $f(b) = -13$.   **[1 mark]**

  d State if the function is a one-to-one or many-to-one.   **[1 mark]**

2 Given that $f(x) = 2x$ and $g(x) = x^2 - 4$, $\{x \in \mathbb{R},\ x \geqslant 0\}$, find the following:

  a $fg(x)$ **b** $gf(x)$ **c** $gg(x)$

  d $f^{-1}(x)$ **e** $g^{-1}(x)$ **[5 marks]**

CM 3 The value of a house can be modelled using the following formula:

$$V = 250000\,e^{\frac{t}{12}}$$

where V is the value of the house in pounds (£) and t is the age of the house in years.

   a What is the value of the house when it is newly built?   **[1 mark]**

   b What is the value of the house after 10 years?   **[1 mark]**

   c Sketch the graph of the value of the house, V, against time t.   **[2 marks]**

   d Is this a suitable for the value of the house over time? Clearly justify your answer.   **[1 mark]**

PS 4 **a** Given that $f(x) = |3x + 5|$ and $g(x) = 4x$, on the same pair of axes sketch the graphs of $f(x)$ and $g(x)$.   **[2 marks]**

   b Use your graphs to solve $|3x + 5| > 4x$.   **[1 mark]**

   c On the same pair of axes sketch the graphs of $2f(x)$ and $|g(x)|$.   **[2 marks]**

   d Use your graphs to solve $2f(x) > |g(x)|$.   **[1 mark]**

5 The function $k(x)$ is defined as $k(x) = \begin{cases} 3 - x^2, & x < 1 \\ 3x + 4, & x \geqslant 1 \end{cases}$

   a Sketch the graph of the function and state the range.   **[3 marks]**

   b Find the value of c where $k(c) = -6$.   **[1 mark]**

   c State if the function is one-to-one or many-to-one.   **[1 mark]**

PS 6 Given that $f(x) = x^2 + 5x - 13$, $\{x \in \mathbb{R} \ x > 0\}$ and $g(x) = \dfrac{1}{x}$, $\{x \in \mathbb{R} \ x > 0\}$, find:

   a $gf(x)$ **[1 mark]**

   b $g^2(x)$ **[1 mark]**

   c $f^{-1}(x)$ **[2 marks]**

PS 7 **a** Given that $f(x) = x^3 - 1$, solve the inequality $|f(x)| < 26$.   **[2 marks]**

   b Sketch the graphs of $f(x)$, $|f(x)|$ and $f(|x|)$ on the same pair of axes, clearly labelling each graph.   **[2 marks]**

   c Write down the transformation to reflect $|f(x)|$ in the x-axis.   **[1 mark]**

8 Given that $f(x) = 11 - 3x$, $\{x \in \mathbb{R}, x \geqslant 0\}$, find:

   a $f^{-1}(x)$ and state its domain   **[3 marks]**

   b the coordinates of the point of intersection of the graphs of $f(x)$ and $f^{-1}(x)$ **[2 marks]**

   c the transformation that maps $f(x)$ onto $f^{-1}(x)$.   **[1 mark]**

9 Given that f(x) = |x|, deduce a transformation to match each of the following graphs and write the expression for the function. In each case justify your choice of transformation.

a

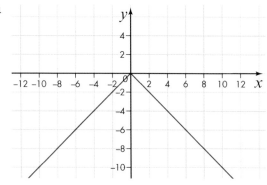

[1 mark]

b

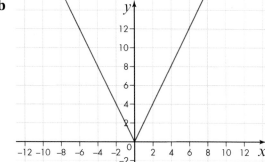

[1 mark]

c

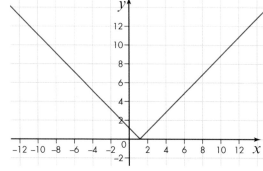

[1 mark]

d

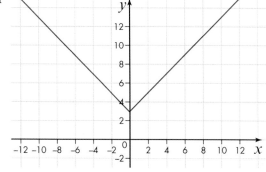

[1 mark]

10 The following graph is a sketch of the curve given by the equation $y = \dfrac{1}{a-x} + b$.

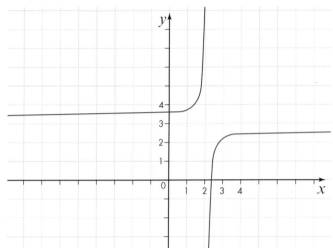

Suggest a possible domain or domains that would allow the relationship to become a function. Clearly justify your answer. **[4 marks]**

 11 Given that $f(x) = 2x^2 + 7x - 11$, $\{x \in \mathbb{R}\}$:

a find $f^{-1}(x)$ **[2 marks]**

b show that $ff^{-1}(x) = x$ **[3 marks]**

 12 Two scientists study the same population of breeding rabbits. Scientist A says that the population of rabbits can be modelled using the following formula:

$$P = 2^{\frac{t}{3}}$$

where P is the total number of rabbits and t is the age of the population of rabbits in months.

Scientist B says that a better formula for modelling the population of rabbits is:

$$P = 1000000 - 600000 \left(2^{-\frac{t}{3}} \right)$$

Discuss the merits of each model. Use clear mathematical language in your discussions. **[4 marks]**

 13 a Given that $f(x) = |\cos x|$, solve $f\left(\dfrac{x}{2}\right) \leq \dfrac{1}{2}$ for $-\dfrac{\pi}{2} \leq x \leq \dfrac{\pi}{2}$. **[3 marks]**

b On the same pair of axes sketch the graphs of $f(x)$ and $f\left(\dfrac{x}{2}\right)$ for $-\dfrac{\pi}{2} \leq x \leq \dfrac{\pi}{2}$. **[4 marks]**

14 Given that $f(x) = 4x + 3$, $\{x \in \mathbb{R}\}$ and $g(x) = \dfrac{2}{3x-3}$, $\{x \in \mathbb{R}, x \neq 1\}$, find:

a the domain of $fg(x)$ such that it is a function. **[2 marks]**

b the domain of $gf(x)$ such that it is a function. **[2 marks]**

c Hence, or otherwise, solve $fg(x) = gf(x)$, giving your answers to 3 significant figures. **[3 marks]**

2 ALGEBRA AND FUNCTIONS 2: PARTIAL FRACTIONS

In an overview of social media and technology in 2015, the PewResearchCenter reported that '92% of teens report going online daily – including 24% who say they go online "almost constantly"'. As a fraction, 92% is equivalent to $\frac{23}{25}$ and 24% is equivalent to $\frac{6}{25}$. By manipulating the fractions, how could you work out how many teens who go online daily aren't online "almost constantly"? You would simply subtract $\frac{6}{25}$ from $\frac{23}{25}$, which gives a fraction of $\frac{17}{25}$, or 68%.

If the number of teens going online "almost constantly" was 25% instead, the fraction calculation would have been a little more involved. As a fraction, 25% is equivalent to $\frac{1}{4}$. To subtract this from $\frac{23}{25}$ you need to find a common denominator – for example, $\frac{92}{100} - \frac{25}{100} = \frac{67}{100}$.

These are examples of fraction calculations using purely numeric values to work out an answer. You are going to look at writing algebraic fractions as partial fractions, which is a way of simplifying more complex functions.

LEARNING OBJECTIVES

You will learn how to:

› simplify rational expressions including by factoring, cancelling and algebraic division

› decompose rational functions into partial fractions.

TOPIC LINKS

The skills and techniques that you learn in this chapter will help you to solve problems where you need to integrate complex functions in **Chapter 8 Integration**.

PRIOR KNOWLEDGE

You should already know how to:

› manipulate polynomials algebraically

› perform algebraic division.

You should be able to complete the following questions correctly:

1 Use algebraic division to divide $x^3 + 6x^2 + 11x + 6$ by $(x + 2)$.

2 Use algebraic division to divide $x^3 + 6x^2 + 11x + 6$ by $(x - 3)$ and find the remainder.

3 Use algebraic division to divide $2x^3 - 3x^2 + 1$ by $(2x + 1)$.

2.1 Simplifying algebraic fractions

You should always write numeric fractions in their simplest form – this is no different for algebraic fractions. You can **simplify** fractions by combinations of factorising, cancelling and algebraic division. You are now going to look at factorising and cancelling, and in **Section 2.4** you will revisit algebraic division in the context of improper fractions.

Example 1

Simplify $\dfrac{2x+1}{3+5x-2x^2}$.

Solution

Attempt to factorise the denominator.

$$\frac{2x+1}{3+5x-2x^2} = \frac{2x+1}{(2x+1)(3-x)}$$

The numerator and denominator have a common factor, $2x+1$, which can be cancelled.

$$\frac{2x+1}{(2x+1)(3-x)} = \frac{1}{3-x}$$

> Before starting any manipulations, ask yourself whether there is anything obvious that can be cancelled.

Example 2

Simplify $\dfrac{2x^2-2x-40}{2x^2+5x+12}$.

Solution

Attempt to factorise the numerator and the denominator.

$$\frac{2x^2-2x-40}{2x^2+5x+12} = \frac{2(x+4)(x-5)}{(x+4)(2x-3)}$$

The numerator and denominator have a common factor, $x+4$, which can be cancelled.

$$\frac{2(x+4)(x-5)}{(x+4)(2x-3)} = \frac{2(x-5)}{2x-3}$$

Exercise 2.1A

Answers page 430

1 Simplify the following:

  a $\dfrac{5x+10}{x+2}$ **b** $\dfrac{x+3}{x^2+x-6}$ **c** $\dfrac{x+4}{x^2-16}$

2 Simplify the following:

  a $\dfrac{x^2-x^3}{1-x^2}$ **b** $\dfrac{4x-12}{(x+3)^2}$ **c** $\dfrac{x^2-7x}{x^2-49}$

3 Simplify the following:

a $\dfrac{x^3 - 8x}{x^2 - 4}$

b $\dfrac{3\left(\frac{1}{3}x^2 - \frac{4}{3}x - 4\right)}{\frac{1}{2}x^2 + \frac{7}{2}x + 5}$

c $\dfrac{2x^3 - 8x}{x^2 - 3x - 28}$

2.2 Partial fractions without repeated terms

In your GCSE course you simplified expressions like $\dfrac{1}{x-1} + \dfrac{2}{x+1}$.
What steps would you have needed to take?

First, you would have found a common denominator.

$$\frac{1}{x-1} + \frac{2}{x+1} = \frac{(x+1) + 2(x-1)}{(x-1)(x+1)}$$

Then you would have simplified the numerator and, maybe, expanded the brackets in the denominator.

$$= \frac{x + 1 + 2x - 2}{x^2 - 1}$$

Then you would have collected like terms in the numerator.

$$= \frac{3x - 1}{x^2 - 1}$$

When you split a fraction into **partial fractions**, you are doing the reverse of the above. So, to split $\dfrac{3x-1}{x^2-1}$ into partial fractions, you would have $\dfrac{3x-1}{x^2-1} \equiv \dfrac{1}{x-1} + \dfrac{2}{x+1}$.

The **identity** sign $\equiv$ has been used in the above example, rather than an equals sign, to show that $\dfrac{1}{x-1} + \dfrac{2}{x-1}$ is the same as $\dfrac{3x-1}{x^2-1}$ but just written in a different way.

An expression with two or more linear terms in the denominator,

for example $\dfrac{f(x)}{(x+p)(x+q)(x+r)}$, can be split into partial fractions

in the form $\dfrac{A}{x+p} + \dfrac{B}{x+q} + \dfrac{C}{x+r}$, where A, B and C are constants.

> **KEY INFORMATION**
>
> Partial fractions are two or more fractions into which a more complex fraction can be split.

> **KEY INFORMATION**
>
> $$\frac{f(x)}{(x+p)(x+q)(x+r)}$$
> $$\equiv \frac{A}{x+p} + \frac{B}{x+q} + \frac{C}{x+r}$$

Example 3

Split $\dfrac{2x - 13}{(x+1)(x-2)}$ into partial fractions using substitution.

Solution

Write the expression, using constants as the numerators, as partial fractions.

$$\frac{2x - 13}{(x+1)(x-2)} \equiv \frac{A}{x+1} + \frac{B}{x-2}$$

You now need to find the values of A and B. Add the two fractions on the right-hand side (RHS) of the identity by finding a common denominator.

$$\frac{2x - 13}{(x+1)(x-2)} \equiv \frac{A(x-2) + B(x+1)}{(x+1)(x-2)}$$

As the denominators on both sides of the identity are the same, the numerators are also the same.

$$2x - 13 \equiv A(x - 2) + B(x + 1)$$

Substitute $x = 2$.

$$2 \times 2 - 13 = B(2 + 1)$$

$$-9 = 3B$$

$$B = -3$$

$x = 2$ has been chosen so that you are left with an equation with only one unknown, which can then be found.

Substitute $x = -1$.

$$2 \times -1 - 13 = A(-1 - 2)$$

$$-15 = -3A$$

$$A = 5$$

State the original fraction as partial fractions.

$$\frac{2x - 13}{(x + 1)(x - 2)} \equiv \frac{5}{x + 1} - \frac{3}{x - 2}$$

If you are asked to 'split a fraction into partial fractions' or similar, you must state what the partial fractions are. If you are asked to 'find the values of A and B' or similar, it isn't necessary to state the partial fractions.

Example 4

Split $\dfrac{36 + 5x}{16 - x^2}$ into partial fractions by equating **coefficients**.

Solution

Write the expression, using constants as the numerators, as partial fractions.

$$\frac{36 + 5x}{16 - x^2} \equiv \frac{A}{4 - x} + \frac{B}{4 + x}$$

You now need to find the values of A and B. Add the two fractions on the RHS of the identity by finding a common denominator.

$$\frac{36 + 5x}{16 - x^2} \equiv \frac{A(4 + x) + B(4 - x)}{(4 - x)(4 + x)}$$

As the denominators on both sides of the identity are the same, the numerators are also the same.

$$36 + 5x \equiv A(4 + x) + B(4 - x)$$

By substituting different values of x into the identity, the values of constants A and B can be determined by finding and solving simultaneous equations.

Expand the brackets in the RHS of the identity.

$$36 + 5x \equiv 4A + Ax + 4B - Bx$$

Collect like terms.

$$36 + 5x \equiv 4(A + B) + x(A - B)$$

Equate the constant terms.

$$4A + 4B = 36 \qquad ①$$

Equate the coefficients of x.

$$A - B = 5 \qquad ②$$

Equations ① and ② now need to be solved simultaneously. Multiply ② by 4.

$$4A - 4B = 20 \qquad ③$$

Add ① and ③.

$$8A = 56$$

$$A = 7$$

Substitute into ②.

$$7 - B = 5$$

$$B = 2$$

State the original fraction as partial fractions.

$$\frac{36 + 5x}{16 - x^2} \equiv \frac{7}{4 - x} + \frac{2}{4 + x}$$

> You now have an equation where the constant elements of $36 + 5x \equiv 4(A + B) + x(A - B)$ have been equated and consequently are the same.

> You now have another equation where the coefficients of x in $36 + 5x \equiv 4(A + B) + x(A - B)$ have been equated and consequently are the same.

> You can choose either method – substitution or equating coefficients – to split a fraction into partial fractions, unless you are given specific instructions about which method to use.

Example 5

Split $\dfrac{3x^2 + 2x - 1}{(x - 1)(x - 2)(x - 3)}$ into partial fractions using substitution.

Solution

Write the expression, using constants as the numerators, as partial fractions.

$$\frac{3x^2 + 2x - 1}{(x - 1)(x - 2)(x - 3)} \equiv \frac{A}{x - 1} + \frac{B}{x - 2} + \frac{C}{x - 3}$$

You now need to find the values of A, B and C. Add the three fractions on the RHS of the identity by finding a common denominator.

$$\frac{3x^2 + 2x - 1}{(x - 1)(x - 2)(x - 3)}$$

$$\equiv \frac{A(x - 2)(x - 3) + B(x - 1)(x - 3) + C(x - 1)(x - 2)}{(x - 1)(x - 2)(x - 3)}$$

As the denominators on both sides of the identity are the same, the numerators are also the same.

$$3x^2 + 2x - 1 \equiv A(x - 2)(x - 3) + B(x - 1)(x - 3) + C(x - 1)(x - 2)$$

Substitute $x = 1$.

$$3 + 2 - 1 = A(-1)(-2)$$

$$A = 2$$

Substitute $x = 2$.

$$12 + 4 - 1 = B(1)(-1)$$
$$B = -15$$

Substitute $x = 3$.

$$27 + 6 - 1 = C(2)(1)$$
$$C = 16$$

State the original fraction as partial fractions.

$$\frac{3x^2 + 2x - 1}{(x-1)(x-2)(x-3)} \equiv \frac{2}{x-1} - \frac{15}{x-2} + \frac{16}{x-3}$$

Stop and think   Can you simplify $\dfrac{x^2 - 19x - 32}{x^3 - 7x - 6}$?

Exercise 2.2A

Answers page 430

1   Express each of the following as partial fractions:

   a $\dfrac{2x - 5}{(x+2)(x+3)}$ **b** $\dfrac{6x - 12}{(x-1)(x+5)}$ **c** $\dfrac{3 - 5x}{(x-3)(x-7)}$

   using: **i** substitution   **ii** equating coefficients.

(PF) 2   Show that $\dfrac{11 - 2x}{(x-2)(x+5)}$ can be written as $\dfrac{A}{x-2} + \dfrac{B}{x+5}$. State the values of A and B.

(PS) 3   Express each of the following as partial fractions:

   a $\dfrac{2 - 3x}{(2x+1)(3-x)}$ **b** $\dfrac{2x + 22}{x^2 + 2x}$ **c** $\dfrac{4x - 30}{x^2 - 8x + 15}$

(CM) 4   Express $\dfrac{1}{x^2 - 9}$ and $\dfrac{1}{x^2 - 16}$ as partial fractions. What do you notice?

5   Express each of the following as partial fractions:

   a $\dfrac{2x^2 - 4x + 8}{(x-1)(x-2)(x-3)}$ **b** $\dfrac{2 - 3x - 4x^2}{x(x-1)(1-2x)}$ **c** $\dfrac{6 - 6x - 5x^2}{(x-1)(x-2)(x+4)}$

(PF) 6   Show that $\dfrac{1}{x^2 - a^2}$ can be written as $\dfrac{1}{2a(x-a)} - \dfrac{1}{2a(x+a)}$, where $a \in \mathbb{R}$.

(CM) 7   Express $\dfrac{5 + 3x - x^2}{-x^3 + 3x^2 + 4x - 12}$ as partial fractions. Provide a detailed commentary for each step in your working, ensuring that you clearly explain and justify any decisions you make.

2.3 Partial fractions with repeated terms

An expression that has repeated linear terms in the denominator, for example $\dfrac{f(x)}{(x+p)(x+q)^2}$, can be split into partial fractions in the form $\dfrac{A}{x+p} + \dfrac{B}{x+q} + \dfrac{C}{(x+q)^2}$, where A, B and C are constants.

KEY INFORMATION

$$\frac{f(x)}{(x+p)(x+q)^2}$$
$$\equiv \frac{A}{x+p} + \frac{B}{x+q} + \frac{C}{(x+q)^2}$$

Example 6

Split $\dfrac{3x^2 + 2x + 2}{(x-2)(x-3)^2}$ into partial fractions.

Solution

Write the expression, using constants as the numerators, as partial fractions.

$$\frac{3x^2 + 2x + 2}{(x-2)(x-3)^2} \equiv \frac{A}{x-2} + \frac{B}{x-3} + \frac{C}{(x-3)^2}$$

You now need to find the values of A, B and C. Add the three fractions on the RHS of the identity by finding a common denominator.

$$\frac{3x^2 + 2x + 2}{(x-2)(x-3)^2} \equiv \frac{A(x-3)^2 + B(x-2)(x-3) + C(x-2)}{(x-2)(x-3)^2}$$

As the denominators on both sides of the identity are the same, the numerators are also the same.

$$3x^2 + 2x + 2 \equiv A(x-3)^2 + B(x-2)(x-3) + C(x-2)$$

Expand the brackets in the RHS of the identity.

$$3x^2 + 2x + 2 \equiv Ax^2 - 6Ax + 9A + Bx^2 - 5Bx + 6B + Cx - 2C$$

Collect like terms.

$$3x^2 + 2x + 2 \equiv x^2(A + B) + x(-6A - 5B + C) + 9A + 6B - 2C$$

Equate the constant terms.

$$9A + 6B - 2C = 2 \qquad \qquad ①$$

Equate the coefficients of x.

$$-6A - 5B + C = 2 \qquad \qquad ②$$

Equate the coefficients of x^2.

$$A + B = 3 \qquad \qquad ③$$

Rearrange ③ to make B the subject.

$$B = 3 - A$$

Substitute B into ① and ②.

$$9A + 6(3 - A) - 2C = 2$$

$$3A - 2C = -16 \qquad \qquad ④$$

$$-6A - 5(3 - A) + C = 2$$

$$-A + C = 17 \qquad \qquad ⑤$$

Equations ④ and ⑤ now need to be solved simultaneously. Multiply ⑤ by 3.

$$-3A + 3C = 51 \qquad \qquad ⑥$$

Add ④ and ⑥.

$$C = 35$$

Substitute into ⑤.

$$A = 18$$

Substitute into ③.

$$B = -15$$

State the original fraction as partial fractions.

$$\frac{3x^2 + 2x + 2}{(x-2)(x-3)^2} \equiv \frac{18}{x-2} - \frac{15}{x-3} + \frac{35}{(x-3)^2}$$

Exercise 2.3A

Answers page 431

(PF) **1** Show that $\dfrac{x^2 + 8x + 4}{x^2(x-2)}$ can be written as $\dfrac{x^2 + 8x + 4}{x^2(x-2)} = \dfrac{A}{x} + \dfrac{B}{x^2} + \dfrac{C}{x-2}$. State the values of A, B and C.

(PS) **2** Given that $\dfrac{5x}{(x-3)^2} \equiv \dfrac{p}{x-3} + \dfrac{3p}{(x-3)^2}$, find the value of p.

3 Express $\dfrac{7x - 3}{x^2 - 8x + 16}$ as partial fractions.

(PS) **4** Identify and correct the mistakes in the following attempt to split $\dfrac{2x^2 - x - 6}{x^3 + 4x^2 + 4x}$ into partial fractions:

$$\frac{2x^2 - x - 6}{x^3 + 4x^2 + 4x} = \frac{2x^2 - x - 6}{x(x+2)^2}$$

$$\equiv \frac{A}{x} + \frac{B}{x+2} + \frac{C}{(x+2)^2}$$

$$\equiv \frac{A(x+2)^2 + B(x)(x+2) + C(x)}{x(x+2)^2}$$

$$2x^2 - x - 6 \equiv A(x+2)^2 + B(x)(x+2) + C(x)$$

Substitute $x = 0$.

$$-6 = 4A$$

$$A = -\frac{3}{2}$$

Substitute $x = -2$.

$$8 + 2 - 6 = -2C$$

$$C = -2$$

Substitute $x = 1$.

$$2 - 1 - 6 = \left(-\frac{3}{2}\right)(9) + B(1)(3) + (-2)(1)$$

$$B = \frac{7}{2}$$

$$\frac{2x^2 - x - 6}{x^3 + 4x^2 + 4x} \equiv -\frac{3}{x} + \frac{7}{x+2} + \frac{2}{(x+2)^2}$$

5 Can $\dfrac{1}{(x+1)(x-2)^2}$ be split into partial fractions? You must clearly justify your answer.

 6 Write down each step to express $\dfrac{2x^2+6x+5}{(x-2)^3}$ as partial fractions.

 7 By choosing suitable examples to support your arguments, discuss the advantages and disadvantages of the two different methods to split a rational function with linear terms in its denominator into partial fractions.

2.4 Improper fractions

An **improper fraction** that contains no algebraic terms is one where the numerator is greater than or equal to the denominator, for example $\frac{7}{4}$. An algebraic fraction is improper when the **degree** (highest power) of the numerator is greater than or equal to the degree of the denominator. For example, in the algebraic fraction $\dfrac{x^3}{(x-2)^2}$ the degree of the numerator is 3 and the degree of the denominator is 2. Therefore it is an improper fraction.

If you are asked to split an improper fraction into partial fractions, you first need to convert the fraction into a **mixed number**. In a mixed number that contains no algebraic terms, the fraction part is a **proper fraction** where the numerator is less than the denominator, for example $1\frac{1}{3}$. In a mixed number that contains algebraic terms, the fraction part is proper when the degree of the numerator is less than the degree of the denominator. For example, in the fraction part of $3+\dfrac{(x-2)^2}{x^3}$ the degree of the numerator is 2 and the degree of the denominator is 3. Therefore it is a proper fraction. Once the improper fraction has been converted to a mixed number, the proper fraction part can then be split into partial fractions.

> **KEY INFORMATION**
>
> An improper fraction containing algebraic terms is one where the degree of the numerator is greater than or equal to the degree of the denominator.

> **KEY INFORMATION**
>
> To split an improper fraction into partial fractions you need first to convert the fraction into a mixed number and then to split the proper fraction part of the mixed number into partial fractions.

Example 7

Express $\dfrac{2x^3-5x^2+x}{x^2-1}$ as partial fractions.

Solution

Divide the numerator by the denominator.

$$
\begin{array}{r}
2x-5 \\
x^2-1 \overline{\smash{\big)}\ 2x^3-5x^2+\ x} \\
\underline{2x^3-2x} \\
-5x^2+3x \\
\underline{-5x^2+5} \\
3x-5
\end{array}
$$

You can now write the improper fraction as a mixed number.

$$\frac{2x^3 - 5x^2 + x}{x^2 - 1} \equiv 2x - 5 + \frac{3x - 5}{x^2 - 1}$$

Split the proper fraction part of the mixed number into partial fractions.

$$\frac{3x - 5}{x^2 - 1} \equiv \frac{A}{x - 1} + \frac{B}{x + 1}$$

You now need to find the values of A and B. Add the two fractions on the RHS of the identity by finding a common denominator.

$$\frac{3x - 5}{x^2 - 1} \equiv \frac{A(x + 1) + B(x - 1)}{(x - 1)(x + 1)}$$

As the denominators on both sides of the identity are the same, the numerators are also the same.

$$3x - 5 \equiv A(x + 1) + B(x - 1)$$

Substitute $x = 1$.

$$A = -1$$

Substitute $x = -1$.

$$B = 4$$

> When you start with an improper fraction, remember to write out the whole solution and not just the proper fraction expressed as partial fractions.

State the original fraction as partial fractions.

$$\frac{2x^3 - 5x^2 + x}{x^2 - 1} \equiv 2x - 5 - \frac{1}{x - 1} + \frac{4}{x + 1}$$

Exercise 2.4A

Answers page 432

 1 Categorise each of the following as either a proper fraction, an improper fraction or a mixed number. State a reason for each of your categorisations.

a $\dfrac{4 - 7x}{(1 - 3x)(x - 2)}$ **b** $\dfrac{10x^2 - 7x + 3}{(2 - 5x)(x + 2)}$ **c** $7 + \dfrac{3x^2 - 4x + 1}{(x - 1)(x + 2)(x - 2)}$

d $\dfrac{x^3 + 6}{(x + 3)^2}$ **e** $\dfrac{7x^2 - 6x}{x^2 - 9}$ **f** $\dfrac{4x - 9}{(3x - 2)(2x + 5)}$

g $3 + \dfrac{5x - 7}{(5 - 4x)^2}$ **h** $\dfrac{x^4 - x^3 + x^2}{(x + 2)^3}$ **i** $\dfrac{2x^2 + 3x - 11}{(2x - 1)^3}$

2 **a** By dividing the denominator into the numerator, convert $\dfrac{6x^2}{x^2 - 3x + 2}$ from an improper fraction into a mixed number.

    b Factorise $x^2 - 3x + 2$ and hence split the proper fraction part of the mixed number found in **part a** into partial fractions.

(PF) 3 Show that $\dfrac{x^3+6}{(x+3)^2}$ can be written as $Ax+B+\dfrac{C}{x+3}+\dfrac{D}{(x+3)^2}$.

State the values of A, B, C and D.

4 Express $\dfrac{10x^2-7x+3}{(2-5x)(x+2)}$ as partial fractions.

(PS) 5 Find the coefficient of $\dfrac{1}{x+3}$ when $\dfrac{7x^2-6x}{x^2-9}$ is split into partial fractions.

(PF) 6 Show that $\dfrac{x^4-x^3+x^2}{(x+2)^3}$ can be written as $Ax+B+\dfrac{C}{x+2}+\dfrac{D}{(x+2)^2}+\dfrac{E}{(x+2)^3}$.

Find the sum of A, B, C, D and E.

2.5 Using partial fractions in differentiation, integration and series expansion

Splitting an expression into partial fractions can make differentiation calculations easier.

For example, if you were asked to find the first and second

derivatives of $f(x)=\dfrac{2x^2+4x-13}{(x+3)(x+2)^2}$ it is much easier to first express

$f(x)$ as partial fractions and subsequently differentiate the fractions individually.

Example 8

Express $f(x)=\dfrac{2x^2+4x-13}{(x+3)(x+2)^2}$ as partial fractions and hence find

the first and second derivatives.

Solution

Write the expression, using constants as the numerators, as partial fractions.

$$\frac{2x^2+4x-13}{(x+3)(x+2)^2} \equiv \frac{A}{x+3}+\frac{B}{x+2}+\frac{C}{(x+2)^2}$$

You now need to find the values of A, B and C. Add the three fractions on the RHS of the identity by finding a common denominator.

$$\frac{2x^2+4x-13}{(x+3)(x+2)^2} \equiv \frac{A(x+2)^2+B(x+2)(x+3)+C(x+3)}{(x+3)(x+2)^2}$$

As the denominators on both sides of the identity are the same, the numerators are also the same.

$$2x^2+4x-13 \equiv A(x+2)^2+B(x+2)(x+3)+C(x+3)$$

Substitute $x=-2$.

$$8-8-13 = C(1)$$

$$C=-13$$

Substitute $x = -3$.

$$18 - 12 - 13 = A(-1)(-1)$$

$$A = -7$$

Substitute $x = 0$.

$$-13 = (-7)(2)(2) + (B)(2)(3) + (-13)(3)$$

$$B = 9$$

State the original fraction as partial fractions.

$$\frac{2x^2 + 4x - 13}{(x+3)(x+2)^2} \equiv -\frac{7}{x+3} + \frac{9}{x+2} - \frac{13}{(x+2)^2}$$

To find the first derivative, each partial fraction simply needs to be differentiated in turn.

$$f'(x) = \frac{7}{(x+3)^2} - \frac{9}{(x+2)^2} + \frac{26}{(x+2)^3}$$

> The first term here is the first derivative of $-\frac{7}{(x+3)}$, the second term is the first derivative of $\frac{9}{(x+2)}$, and so on. Remember, when you differentiate you need to multiply by the 'old' power and subsequently reduce the power by 1.

To find the second derivative, each partial fraction again simply needs to be differentiated in turn.

$$f''(x) = -\frac{14}{(x+3)^3} + \frac{18}{(x+2)^3} - \frac{78}{(x+2)^4}$$

> The first term here is the second derivative of $-\frac{7}{(x+3)}$, the second term is the second derivative of $\frac{9}{x+2}$, and so on.

Stop and think If the expression hadn't been split into partial fractions, what steps would have been required to find the first and second derivatives?

In **Chapter 7 Further differentiation** you will learn to differentiate functions such as this in a different way, using the quotient rule.

Some expressions can only be integrated once they have been split into partial fractions. For example, if you were asked to find the integral $\int \frac{2x^2 + 4x - 13}{(x+3)(x+2)^2} \, dx$, this can only be done if the expression $\frac{2x^2 + 4x - 13}{(x+3)(x+2)^2}$, is first redefined as partial fractions. You will explore this in detail in **Chapter 8 Integration**.

You may be asked to find the binomial expansion, or another type of expansion of a function, but the format of the function does not readily lend itself to the expansion method requested.

For example, if you were asked to find the binomial expansion of $f(x) = \frac{2x^2 + 4x - 13}{(x+3)(x+2)^2}$, it is necessary to first express f(x) as partial fractions and subsequently find the binomial expansion of the fractions individually. You will encounter this again in **Chapter 4 Sequences and series**.

SUMMARY OF KEY POINTS

> Partial fractions are two or more fractions into which a more complex fraction can be split.

> An expression with two or more linear terms in the denominator, for example $\dfrac{f(x)}{(x+p)(x+q)(x+r)}$, can be split into partial fractions in the form $\dfrac{A}{x+p} + \dfrac{B}{x+q} + \dfrac{C}{x+r}$, where A, B and C are constants.

> An expression that has repeated linear terms in the denominator, for example $\dfrac{f(x)}{(x+p)(x+q)^2}$, can be split into partial fractions in the form $\dfrac{A}{x+p} + \dfrac{B}{x+q} + \dfrac{C}{(x+q)^2}$, where A, B and C are constants.

> An improper fraction that contains algebraic terms is one where the degree of the numerator is greater than or equal to the degree of the denominator.

> To split an improper fraction into partial fractions, you first need to convert the fraction into a mixed number and then split the proper fraction part of the mixed number into partial fractions.

EXAM-STYLE QUESTIONS 2

Answers page 432

1 Simplify $\dfrac{x+3}{x+9}$. [1 mark]

(PF) 2 Show that $\dfrac{2x^2}{(x-1)(x-2)}$ is an improper fraction that can be expressed as partial fractions. [3 marks]

(PS) (PF) 3 Show that the product of two fractions, one with an odd-number denominator and the other with an even-number denominator, can be written as the sum of two fractions with the same denominators as the original fractions in the product. The even and odd numbers are consecutive numbers. [4 marks]

4 Express $\dfrac{11x-5}{(x-3)(3x-2)}$ in the form $\dfrac{A}{x-3} + \dfrac{B}{3x-2}$. [3 marks]

(PF) 5 Show that $\dfrac{18x-11}{(x-3)(x-2)}$ can be expressed in the form $\dfrac{A}{x-3} + \dfrac{B}{x-2}$. [3 marks]

6 Express $\dfrac{6x+7}{(x+1)^2}$ in the form $\dfrac{A}{x+1} + \dfrac{B}{(x+1)^2}$. [3 marks]

(PF) 7 Express $f(x) = \dfrac{9}{(x+1)(x+2)}$ as partial fractions.

Show that the first derivative of $f(x)$ is $-\dfrac{9}{(x+1)^2} + \dfrac{9}{(x+2)^2}$. [5 marks]

(PF) 8 Show that $\dfrac{5-9x-2x^2}{6x^2+27x-15}$ simplifies to $-\dfrac{1}{3}$. [3 marks]

9 Express $\dfrac{42}{4x^2-9}$ as partial fractions. [3 marks]

(PF) 10 Show that when $\dfrac{5x^3}{(x-1)(x-2)^2}$ is written in the form $A + \dfrac{B}{x-1} + \dfrac{C}{x-2} + \dfrac{D}{(x-2)^2}$, then A, B, C and D are all multiples of 5. [6 marks]

11 Express $\dfrac{30x}{(x-1)(x^2-16)}$ as partial fractions. [4 marks]

(CM) 12 Consider the fraction $\dfrac{1}{ax^2 + bx + c}$. If this fraction were to be expressed as partial fractions, what form would they take if:

   a the discriminant of the denominator was zero   **[2 marks]**

   b the discriminant of the denominator was greater than zero   **[2 marks]**

   c the discriminant was less than zero?   **[2 marks]**

(PS) 13 Given that $\dfrac{7x-5}{x-a\,\,x-3} \equiv -\dfrac{9}{x-a} + \dfrac{16}{x-3}$, find the value of a.   **[4 marks]**

(CM) 14 Express the fraction part of $f(x) = x + \dfrac{3x^2 - 4x + 1}{(x-1)(x+2)(x-2)}$ as partial fractions and hence find the first and second derivatives of $f(x)$. What might it be easy to forget to do in this question, and why?   **[6 marks]**

(PS) 15 Express $f(x) = \dfrac{2}{x^3 + 3x^2}$ as partial fractions.   **[5 marks]**

(PS) 16 **a** Express $f(x) = \dfrac{17x}{6x^2 + 13x - 5}$ as partial fractions.   **[3 marks]**

   b Hence, or otherwise, find the gradient of the curve when $x = 1$.   **[4 marks]**

17 Express $\dfrac{1}{(x+4)^2 - 9}$ as partial fractions.   **[4 marks]**

(PS) 18 Simplify $\dfrac{3x^2 + 5x - 2}{\frac{x}{2} + 1}$.   **[3 marks]**

(PF) (CM) 19 Show that $\dfrac{3x^2 + 2x + 2}{(x-2)(x-3)^2}$ cannot be split into partial fractions in the form $\dfrac{A}{x-2} + \dfrac{B}{x-3} + \dfrac{C}{x-3}$. Clearly state the reasoning behind your conclusion.   **[6 marks]**

20 Express $\dfrac{16x^2 + 29x + 7}{(x+4)(2x+1)^2}$ in the form $\dfrac{A}{x+4} + \dfrac{B}{2x+1} + \dfrac{C}{(2x+1)^2}$.   **[6 marks]**

(CM) 21 Can the expression $\dfrac{4x^5 - 3x + 2}{(x^2 - 4)(2x+1)}$ be split into partial fractions? You must justify your answer. If it can, find the coefficient of $(x+2)^{-1}$.   **[6 marks]**

(PF) 22 Show that $\dfrac{27x^4 - 9x^2 + 5}{x(2-3x)^2}$ can be written as $Ax + B + \dfrac{C}{x} + \dfrac{D}{2-3x} + \dfrac{E}{(2-3x)^2}$.

Find the sum of A, B, C, D and E.   **[6 marks]**

(PS) 23 Express $\dfrac{x^4 - x^3 + x^2 + 1}{(x-2)^3}$ as partial fractions and hence find the gradient of the curve when $x = 6$.   **[7 marks]**

(PS) (PF) 24 Given that $f(x) = 2x^3 + 11x^2 + 10x - 8$, show that $\dfrac{8}{f(x)}$ can be expressed as partial fractions.   **[6 marks]**

(PF) 25 Show that $f(x) = \dfrac{1}{c^2 - x^2}$ can be written as $f(x) = \dfrac{1}{2c(x+c)} - \dfrac{1}{2c(x-c)}$.   **[5 marks]**

(PF) 26 Show that $f(x) = \dfrac{x^2 + 5x + 4}{(x-1)(x-4)(x+3)}$ can be written in the form

$f(x) = \dfrac{A}{x-1} + \dfrac{B}{x-4} + \dfrac{C}{x+3}$ and that the sum of A, B and C is 1.   **[7 marks]**

3 COORDINATE GEOMETRY: PARAMETRIC EQUATIONS

The concept of parametric equations may appear quite abstract at first, but they are very useful. Imagine a penalty being taken during a football match. The football will travel through three-dimensional space and consequently, at any moment, its position could be given by three coordinates (x, y, z). The path of the football could be expressed by equations in x, y and z. If you were to choose to define the motion in terms of a parameter, for example time t, then you can easily begin to see where the ball will be at any particular moment.

LEARNING OBJECTIVES

You will learn how to:

› understand and use the parametric equations of curves

› understand and use conversion between Cartesian and parametric forms

› use parametric equations in modelling in a variety of contexts.

TOPIC LINKS

The skills and techniques that you learn in this chapter will help you to solve problems in **Chapter 15 Kinematics** where the motion is defined using parametric equations.

PRIOR KNOWLEDGE

You should already know how to:

› manipulate polynomials algebraically, including:

  › expanding brackets and simplifying by collecting like terms

  › factorisation

  › using and manipulating surds, including rationalising the denominator.

› use and understand the graphs of functions

› sketch curves defined by simple equations, including quadratics, cubics, quartics and reciprocals

› interpret the algebraic solution of equations graphically

› use the intersection points of graphs to solve equations

› understand and use the coordinate geometry of the circle, including using the equation of a circle in the form $(x - a)^2 + (y - b)^2 = r^2$

> ❱ understand and use composite functions
> ❱ understand and use inverse functions and their graphs
> ❱ understand and use trigonometric identities.

You should be able to complete the following questions correctly:

1 Show that $\cos^2 x + \sin^2 x = 1$.

2 Given that $f(x) = x^2 - 9$ find $f^{-1}(x)$.

3 Given that $f(x) = x^3 - 1$ and $g(x) = 2x$ find $fg(x)$.

4 Given that $f(x) = \dfrac{1}{(x^2 + 7)}$ find $f^{-1}(x)$.

5 Given that $f(x) = x^2 + 7x - 11$ find $f^{-1}(x)$.

3.1 Parametric equations of curves

For equations that are expressed parametrically (that are functions of a parameter), for example $x = f(t)$ and $y = g(t)$, you need to be able to determine the coordinates of the points on the curve separately and subsequently draw the curve.

For example, the path of the ball during a particular penalty kick could be given by $x = t + 2$ and $y = 4 - t^2$, where both x and y are functions of a parameter, time t. If a **domain** is specified for t, such as $-2 \leqslant t \leqslant 2$, then you need to work out the domains of both x and y.

KEY INFORMATION

Parametric equations involve an independent variable, a parameter, in terms of which both coordinates of a point on a graph can be expressed.

| Stop and think | The penalty spot on a football pitch is 11 metres from the goal. Assuming the goalkeeper doesn't stop it, would a ball kicked from the penalty spot following the above equations result in a goal? |

Example 1

Draw the curve given by the parametric equations $x = t + 2$ and $y = 4 - t^2$, for $-2 \leqslant t \leqslant 2$. Where does the curve meet the x-axis?

KEY INFORMATION

The domain of a function is the set of input values for the function.

Solution

You need to draw a table to show the values of t, x and y.

t	-2	-1	0	1	2
$x = t + 2$					
$y = 4 - t^2$					

For each value of t you need to work out the corresponding values of x and y.

The domain of t has been specified as $-2 \leqslant t \leqslant 2$, so values of t outside of this range should not be used. Only values of x and y that have been determined for values of t in this range should be used.

When $t = -2$:

$$x = -2 + 2 = 0$$

$$y = 4 - (-2)^2 = 0$$

When $t = -1$:

$$x = -1 + 2 = 1$$

$$y = 4 - (-1)^2 = 3$$

And so on until the table is complete:

t	-2	-1	0	1	2
$x = t + 2$	0	1	2	3	4
$y = 4 - t^2$	0	3	4	3	0

You then need to draw a pair of axes, plot the points and draw the graph through the points.

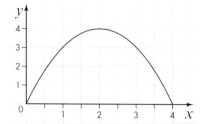

The curve meets the x-axis at $(0, 0)$ and $(4, 0)$.

The curve has been plotted only for the values in the table which are for the specified domain of t. No other values should be plotted.

In this example it was necessary to draw the curve for the given parametric equations. If this hadn't been required, how else could you have worked out the coordinates of the points where the curve meets the x-axis?

When the curve meets the x-axis, the y-coordinate will be zero.

$$y = 4 - t^2$$

$$0 = 4 - t^2$$

Rearrange.

$$t^2 = 4$$

$$t = \pm 2$$

Substitute into $x = t + 2$.

When $t = -2$, $x = 0$

When $t = 2$, $x = 4$

The curve meets the x-axis at $(0, 0)$ and $(4, 0)$.

Example 2

A curve is given by the parametric equations $x = t + 3$ and $y = t^2$. The line $y = 3x - 11$ intersects the curve.

Find the coordinates of the point(s) of intersection.

Solution

$y = 3x - 11$

Substitute the given values of x and y, in terms of t, into this equation.

$$t^2 = 3(t + 3) - 11$$

Expand and simplify.

$$t^2 - 3t + 2 = 0$$

Solve for t

$$(t - 1)(t - 2) = 0$$

$$t = 1 \text{ or } t = 2$$

It is easy to make the mistake of stopping at this point because an equation has been solved, but the problem is only partially completed.

When $t = 1$:

$$x = 1 + 3 = 4$$

$$y = (1)^2 = 1$$

The coordinates of one point of intersection are (4, 1).

t has two values, so there will be two points of intersection.

When $t = 2$:

$$x = 2 + 3 = 5$$

$$y = (2)^2 = 4$$

The coordinates of the other point of intersection are (5, 4).

Example 3

A curve is given by the parametric equations $x = \dfrac{t}{c}$ and $y = c(t^2 - 1)$, where c is a constant. The y-intercept of the curve is at $(0, -3)$. Find the value of c and hence state the parametric equations.

Solution

At the y-intercept, $x = 0$.

$$0 = \frac{t}{c}$$

$$t = 0$$

Substitute $t = 0$ and $y = -3$ (the y-coordinate of the y-intercept) into the parametric equation for y.

$$-3 = c(0 - 1)$$

Rearrange.

$$t = 0$$

$$c = 3$$

The parametric equations are $x = \dfrac{t}{3}$ and $y = 3(t^2 - 1)$.

Exercise 3.1A

Answers page 435

1 Draw the curve given by the parametric equations $x = t + 3$ and $y = t^2$, for $-4 \leqslant t \leqslant 4$.

Use the graph to write down the coordinates of the y-intercept.

2 Draw the curve given by the parametric equations $x = t + 2$ and $y = t^2 - 1$, for $-4 \leqslant t \leqslant 4$.

Use the graph to write down the coordinates of the turning point.

3 Find the coordinates of the x-intercept(s) and y-intercept(s) for each of the curves given by the following parametric equations:

a $x = t + 2$ and $y = 3t$ **b** $x = t - 1$ and $y = t^2$ **c** $x = \dfrac{t}{2}$ and $y = t^2 - 4$

4 Find the coordinates of the point(s) of intersection of the curve given by the parametric equations $x = t + 3$ and $y = t^2$ and the line $x + y - 5 = 0$.

5 Copy and complete the following table of values, to 2 decimal places, for the curve given by the parametric equations $x = \sin t + 1$ and $y = \cos t - 2$, for $0 \leqslant t \leqslant 2\pi$.

t	0	$\dfrac{\pi}{6}$	$\dfrac{\pi}{3}$	$\dfrac{\pi}{2}$	$\dfrac{2\pi}{3}$	$\dfrac{5\pi}{6}$	π	$\dfrac{7\pi}{6}$	$\dfrac{4\pi}{3}$	$\dfrac{3\pi}{2}$	$\dfrac{5\pi}{3}$	$\dfrac{11\pi}{6}$	2π
x													
y													

Draw a graph of the curve. What is the shape of the curve?

What do you notice about the curve and the numeric values in the parametric equations?

6 Find the coordinates of the x-intercept(s) and y-intercept(s) for each of the curves given by the following parametric equations:

a $x = t - 1$ and $y = t^4 - 2$ **b** $x = \dfrac{1}{t}$ and $y = 2t^2 - 3$ **c** $x = t^2$ and $y = t^2 - 3t + 2$

7 Find the coordinates of the point(s) of intersection of the curve given by the parametric equations $x = \dfrac{1}{t}$ and $y = 2t^2 - 3$ and the curve $x^2 - y - 2 = 0$.

(PS) 8 A curve is given by the parametric equations $x = pt - 1$ and $y = \dfrac{p}{t}$, where p is a constant.

The y-intercept of the curve is at $(0, 9)$.

Find the value(s) of p and hence state the parametric equations.

9 Find the coordinates of the x-intercept(s) and y-intercept(s) of the curves given by the parametric equations $x = 2\sin t + 1$ and $y = \cos t$, for $0 \leqslant t \leqslant 2\pi$.

(PS) 10 A curve is given by the parametric equations $x = qt + 2$ and $y = t^3 + q$, where q is a constant.

The y-intercept of the curve is at $\left(0, \dfrac{617}{25}\right)$.

Find the value(s) of q and hence state the parametric equations.

3.2 Converting between Cartesian and parametric forms

Although a curve may be defined by parametric equations it is possible to convert the parametric equations into a **Cartesian equation**, eliminating the parameter.

The **Cartesian coordinate** system was invented by René Descartes (1596–1650). It is a system that allows each point to be specified uniquely by a pair of coordinates, for example (1, 2). A Cartesian equation is one in which the variables are the Cartesian coordinates of a point on a line, curve or surface, for example $y = 2x + 3$.

Example 4

Find the Cartesian equation for the curve given by the parametric equations $x = t + 2$ and $y = 4 - t^2$. Where does the curve meet the x-axis?

Solution

In **Example 1**, the curve given by the parametric equations $x = t + 2$ and $y = 4 - t^2$, for $-2 \leqslant t \leqslant 2$, was drawn. The shape of the curve was a parabola so the Cartesian equation should be a quadratic.

As one of the parametric equations is linear and the other is quadratic, t will be eliminated by substitution.

$$x = t + 2$$

Rearrange to make t the subject.

$$x - 2 = t$$

Substitute for t in the parametric equation for y.

$$y = 4 - (x - 2)^2$$

Expand and simplify.

$$y = 4x - x^2$$

When the curve meets the x-axis, the y-coordinate will be zero.

$$4x - x^2 = 0$$

Factorise.

$$x(4 - x) = 0$$

$$x = 0 \text{ or } x = 4$$

The curve meets the x-axis at (0, 0) and (4, 0).

> **KEY INFORMATION**
>
> To find the Cartesian equation of a curve defined by parametric equations, you need to eliminate the parameter t between the parametric equations.

Example 5

Draw the curve given by the parametric equations $x = \sin t + 1$ and $y = \cos t - 2$, for $0 \leqslant t \leqslant 2\pi$.

Solution

To eliminate t from this pair of parametric equations, a trigonometric identity involving sine and cosine is needed.

$$\sin^2 t + \cos^2 t = 1$$

Rearrange the two parametric equations to make $\sin t$ and $\cos t$ the subjects respectively.

$$x = \sin t + 1$$

$$x - 1 = \sin t$$

So $\sin^2 t = (x - 1)^2$

$$y = \cos t - 2$$

$$y + 2 = \cos t$$

So $\cos^2 t = (y + 2)^2$

Substitute for $\sin^2 t$ and $\cos^2 t$ in the identity.

$$(x - 1)^2 + (y + 2)^2 = 1$$

Sketch the curve (which you weren't asked to do, but it does aid understanding).

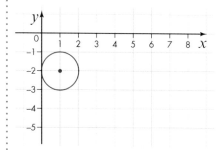

So the curve is a circle with centre (1, –2) and radius 1. The coordinates of the centre of the circle are the same as the values added to $\sin t$ and $\cos t$ in the original parametric equations.

TECHNOLOGY

Using a graphing software package, investigate the shapes of each of the curves given by the following pairs of parametric equations. In each case, vary the values of a and b:

> $x = \sin t \pm a$ and $y = \cos t \pm b$

> $x = \sin at$ and $y = \cos bt$

What shapes do you see? How do the values of a and b affect the shapes?

A curve given by parametric equations in the form $x = a\sin t + b$ and $y = a\cos t + c$ is a circle.

Stop and think   What would be the shape of the curve given by the parametric equations $x = \sin 2t$ and $y = \cos t$, for $0 \leqslant t \leqslant 2\pi$? Can you explain why?

Exercise 3.2A

Answers page 436

1 Find the Cartesian equation for each of the curves given by the following parametric equations:

a $x = t - 1$ and $y = t^2$

b $x = t + 2$ and $y = t^2 - 1$

c $x = \dfrac{t}{2}$ and $y = t^2 - 4$

To which family of curves does each of these curves belong: linear, quadratic, cubic, quartic, reciprocal or none of these?

PF **2** Show that the Cartesian equation for the curve given by the parametric equations $x = t + 2$ and $y = 3t$ can be written in the form $ax + by + c = 0$. State the values of a, b and c.

PS **3** A circle is given by the parametric equations $x = \sin t$ and $y = \cos t$. Find the radius and the coordinates of the centre of the circle.

4 Find the Cartesian equation for each of the curves given by the following parametric equations:

   a $x = t - 1$ and $y = t^4 - 2$

   b $x = \dfrac{1}{t}$ and $y = 2t^2 - 3$

   c $x = t^2$ and $y = t^2 - 3t + 2$

   To which family of curves does each of these curves belong: linear, quadratic, cubic, quartic, reciprocal or none of these?

5 Identify and correct the mistake(s) in the following workings to find the Cartesian equation for the curve given by the parametric equations $x = 4\sin t - 3$ and $y = \dfrac{1}{(5\sin t)}$.

$$x = 4\sin t - 3$$

$$4(x - 3) = \sin t$$

$$y = \frac{\operatorname{cosec} t}{5}$$

$$5y = \operatorname{cosec} t$$

But $\operatorname{cosec} t = \dfrac{1}{\cos t}$

$$5y = \frac{1}{\cos t}$$

$$\cos t = \frac{1}{5y}$$

Also $\sin^2 t + \cos^2 t = 1$

Therefore $(4x - 12)^2 + \left(\dfrac{1}{5y}\right)^2 = 1$

3.3 Problems involving parametric equations

You have already seen that a curve given by parametric equations in the form $x = a\sin t + b$ and $y = a\cos t + c$ is a circle. You are now going to use the parametric equations for other shapes to understand and solve problems.

Example 6

The graph shows an ellipse with parametric equations $x = 3\sin t$ and $y = 7\cos t$, where $0 \leqslant t \leqslant 2\pi$. a is the length of the minor (shortest) axis and b is the length of the major (longest) axis. Find the value of a, b and $a + b$.

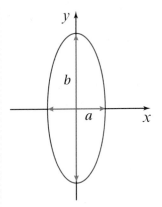

Solution

You need to find the coordinates of the x and y-intercepts so you can work out the lengths of a and b.

At the x-intercepts, $y = 0$.

$7\cos t = 0$

$$t = \frac{\pi}{2}, \frac{3\pi}{2}$$

Substitute $t = \frac{\pi}{2}$ into the parametric equation for x.

$x = 3$

Substitute $t = \frac{3\pi}{2}$ into the parametric equation for x.

$x = -3$

So $a = 6$.

At the y-intercepts, $x = 0$.

$3\sin t = 0$

$$t = 0, \pi$$

Substitute $t = 0$ into the parametric equation for y.

$y = 7$

Substitute $t = \pi$ into the parametric equation for y.

$y = -7$

So $b = 14$.

Therefore $a + b = 20$.

Example 7

A tie company has designed a bow tie. The shape of the bow tie is a curve, as shown in the graph, with parametric equations $x = 6\sin t$ and $y = \frac{5}{2}\sin 2t$, where $0 \leqslant t \leqslant 2\pi$. The tie company wants to package the bow tie in a rectangular box which is $2\,\text{cm}$ deep. If a and b are the measurements of the bow tie in cm, what is the minimum volume of the box?

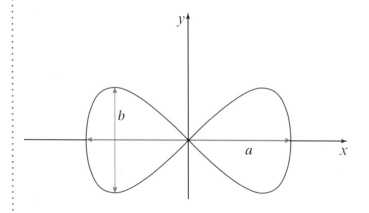

TECHNOLOGY

Using a graphing software package, investigate the shapes of each of the curves given by the following pairs of parametric equations. Vary the values of a and b:

➤ $x = a\sin t$ and $y = b\cos t$

➤ $x = a\cos t$ and $y = b\sin t$

What shapes do you see? How do the values of a and b affect the shapes?

Solution

You need to find the coordinates of the x-intercepts so you can work out the length of a.

For the x-intercepts, $y = 0$.

$$\frac{5}{2}\sin 2t = 0$$

$$t = 0, \frac{\pi}{2}$$

Substitute $t = 0$ into the parametric equation for x.

$$x = 0$$

Substitute $t = \frac{\pi}{2}$ into the parametric equation for x.

$$x = 6$$

So $a = 12$.

To work out the value of b you first need to work out the y-coordinates of the maximum and minimum points.

$$y = \frac{5}{2}\sin 2t$$

$$\frac{dy}{dt} = \frac{5}{4}\cos 2t$$

At both maximum and minimum points, the gradient $= 0$.

$$\frac{5}{4}\cos 2t = 0$$

$$2t = \frac{\pi}{2}, \frac{3\pi}{2}$$

Substitute $2t = \frac{\pi}{2}$ into the parametric equation for y.

$$y = \frac{5}{2}$$

Substitute $2t = \frac{3\pi}{2}$ into the parametric equation for y.

$$y = -\frac{5}{2}$$

So $b = 5$.

The minimum volume of the bow tie box is $5 \times 12 \times 2 = 120\,\text{cm}^3$.

Stop and think

The 'bow tie' curve was given by the parametric equations $x = a\sin t$ and $y = b\sin 2t$. What effect does changing the values of a and b have on the curve? How would you change the equations so the bow tie was vertical rather than horizontal?

In **Book 1, Chapter 15 Kinematics** you learnt about the motion of a particle in a straight line. Consider an object starting at (x_0, y_0) when $t = 0$. How can you represent the position of the object at any time t? Displacement is the product of velocity and time, with velocity having two components as shown in the diagram.

Consequently the parametric equations that represent the position of the object at any time t are:

$x = (v\cos\theta)t + x_0$

$y = (v\sin\theta)t + y_0$

where (x_0, y_0) is the position of the object when $t = 0$.

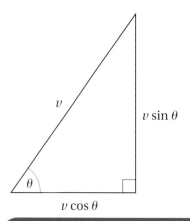

Example 8

An object moves with constant velocity from $(1, 8)$ at $t = 0$ to $(6, 20)$ at $t = 5$. Find the parametric equations for the position of the object.

Solution

$x = (v\cos\theta)t + x_0$

$y = (v\sin\theta)t + y_0$

Substitute $(1, 8)$ for (x_0, y_0) and $(6, 20)$ for (x, y).

$6 = (v\cos\theta)t + 1$

$20 = (v\sin\theta)t + 8$

Simplify.

$(v\cos\theta)t = 5$ ①

$(v\sin\theta)t = 12$ ②

Divide $(v\sin\theta)t$ by $(v\cos\theta)t$.

$$\frac{(v\sin\theta)t}{(v\cos\theta)t} = \frac{12}{5}$$

$$\tan\theta = \frac{12}{5}$$

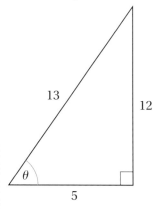

$\sin\theta = \dfrac{12}{13}$

$\cos\theta = \dfrac{5}{13}$

Substitute $\cos\theta = \dfrac{5}{13}$ and $t = 5$ (given) into ① to work out v.

$5\left(\dfrac{5}{13}v\right) = 5$ ①

$v = \dfrac{13}{5}$

> **KEY INFORMATION**
>
> The parametric equations describing the position of a particle under linear motion are:
>
> $x = (v\cos\theta)t + x_0$ and
> $y = (v\sin\theta)t + y_0$

> You should recognise this as a Pythagorean triple.

Substitute the values of v, $\sin\theta$, $\cos\theta$ and (x_0, y_0) into the original parametric equations.

$x = (v\cos\theta)t + x_0$

$x = \left(\dfrac{13}{5} \times \dfrac{5}{13}\right)t + 1$

$x = t + 1$

$y = (v\sin\theta)t + y_0$

$y = \left(\dfrac{13}{5} \times \dfrac{12}{13}\right)t + 8$

$y = \dfrac{12}{5}t + 8$

Exercise 3.3A

Answers page 436

(PS) 1 A circle is defined by the parametric equations $x = 11\cos 3t$ and $y = 11\sin 3t$, where $0 \leqslant t \leqslant 2\pi$.

Find the diameter of the circle.

2 An ellipse is defined by the parametric equations $x = 7\cos t$ and $y = 3\sin t$, where $0 \leqslant t \leqslant 2\pi$. a is the length of the major axis and b is the length of the minor axis.

Find the values of a and b.

(PS) (CM) 3 Which of the following graphs is for the curve defined by the parametric equations $x = t + 2t^2$ and $y = 5t$? You must provide justification for your answer.

A
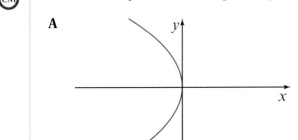

B
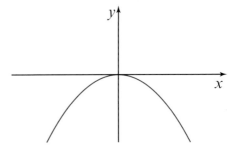

C
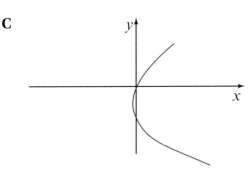

D
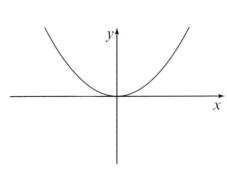

(M) 4 An object moves with constant velocity from $(1, 5)$ at $t = 0$ to $(4, 9)$ at $t = 3$.

Find the parametric equations for the position of the object.

5 A spectacles company has a new sunglasses frame. The shape of the frame is a curve, as shown in the graph, with parametric equations $x = 4\sin t$ and $y = \frac{3}{2}\sin 2t$, where $0 \leqslant t \leqslant 2\pi$.

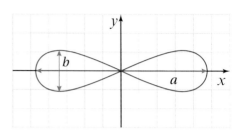

By finding a and b, demonstrate a mathematical relationship between a and b, and the parametric equations.

What would be the parametric equations for a frame that is twice as long and twice as tall?

 6 An object moves with constant velocity from (2, 7) at $t = 0$ to (5, 17) at $t = 4$.

Find the parametric equations for the position of the object.

 7 A child walks with constant velocity towards a favourite toy, as shown in the graph. It takes the child 21 seconds walk to point A.

Find the parametric equations for the position of the child at time t.

What is the velocity of the child?

Where should the toy be placed so that the child reaches it after exactly one minute?

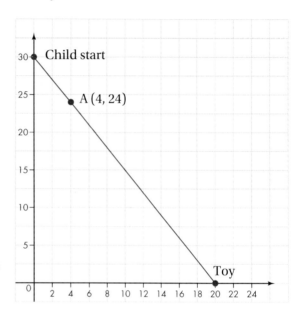

 8 **Example 6** shows how to find the sum of the minor and major axes of an ellipse with parametric equations $x = 3\sin t$ and $y = 7\cos t$, where $0 \leqslant t \leqslant 2\pi$.

How could this method be modified to find the sum of the minor and major axes of an ellipse with parametric equations $x = 5 + 2\sin t$ and $y = 6 + 5\cos t$, where $0 \leqslant t \leqslant 2\pi$?

Demonstrate that your solution works. (Hint: consider the location of the centre of the ellipse in each case).

 9 A graphic design company has developed a new character style for the numeral 8. The shape of the new character is a curve as shown in the graph, with parametric equations $x = 1 - 3\sin 2t$ and $y = 3 + 5\sin t$, where $0 \leqslant t \leqslant 2\pi$. The graphic design company wants to know the height of the new character.

What is the height of the numeral 8?

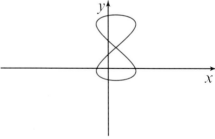

 10 An object moves with constant velocity, and the parametric equations for the position of the object are $x = 9t + 11$ and $y = 10 - 41t$.

What is the velocity of the object?

On what bearing, to the nearest degree, is it moving?

SUMMARY OF KEY POINTS

> Parametric equations involve an independent variable in terms of which both coordinates of a point on a graph can be expressed.

> To find the Cartesian equation of a curve defined by parametric equations you need to eliminate the parameter t between the parametric equations.

> The parametric equations describing the position of a particle under linear motion are $x = (v\cos\theta)t + x_0$ and $y = (v\sin\theta)t + y_0$.

> The equation of a circle can be written parametrically as: $x = r\cos t + a$ and $y = r\sin t + b$ where $0 \leqslant t \leqslant 2\pi$.

> The equation of an ellipse can be written parametrically as: $x = a\cos t$ and $y = b\sin t$ where $0 \leqslant t \leqslant 2\pi$.

EXAM-STYLE QUESTIONS 3   **Answers page 437**

1 Find the coordinates of the x-intercept(s) and y-intercept(s) of the curve given by the parametric equations $x = t - 4$ and $y = t^2 - 9$.   **[3 marks]**

2 Find the coordinates of the point(s) of intersection of the curve given by the parametric equations $x = t + 3$ and $y = t^2$ and the line $x + y - 5 = 0$.   **[4 marks]**

(PF) (CM) **3** Show that the Cartesian equation for the curve given by the parametric equations $x = \sin 2t$ and $y = \cos 2t$ is $x^2 + y^2 = 1$.

Describe the shape of the curve.   **[4 marks]**

(PS) **4** A curve is given by the parametric equations $px = t + 3$ and $y = (pt)^2$, where p is a constant. The x-intercept of the curve is at $\left(\frac{3}{2}, 0\right)$.

Find the value(s) of p and hence state the parametric equations.   **[4 marks]**

(PF) **5** Show that the Cartesian equation for the curves given by the parametric equations $x = \sec t$ and $y = \tan t$ is the same as the Cartesian equation given by the parametric equations $x = \dfrac{1}{\sin t}$ and $y = \dfrac{1}{\tan t}$.   **[5 marks]**

(PS) **6** An ellipse is defined by the parametric equations $x = -11 \sin t$ and $y = 5 \cos t$, where $0 \leqslant t \leqslant 2\pi$. Will it fit inside a rectangle that is 20 units by 12 units?

You must justify your answer, clearly showing all your working.   **[5 marks]**

(M) **7** An object moves with constant velocity from $(-1, 3)$ at $t = 0$ to $(7, -5)$ at $t = 6$.

Find the parametric equations for the position of the object.

Find the position of the object when $t = 13$.   **[6 marks]**

8 A curve is defined by the parametric equations $x = 6 \sin t - 2$, $y = 6 \cos t + 5$, where $0 \leqslant t < 2\pi$

a Find the Cartesian equation of the curve.   **[2 marks]**

b Sketch the curve, clearly showing any axis intercepts.   **[3 marks]**

9 **a** The parametric equations for the curve shown are $x = \cos t$ and $y = \sin 2t$, where $0 \leqslant t < 2\pi$.

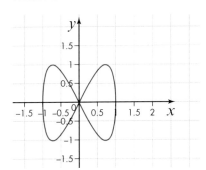

Find the Cartesian equation of the curve. **[3 marks]**

b Find the value(s) of t where the curve intersects the x-axis at -1 and 1.

PS 10 An ellipse is defined by the parametric equations $x = 7\cos t + \frac{3}{2}$ and $y = 3\sin t - 1$, where $0 \leqslant t \leqslant 2\pi$. a is the length of the major axis and b is the length of the minor axis.

What is the product of a and b? **[6 marks]**

M PS 11 An object moves with constant velocity, and the parametric equations for the position of the object are $x = 24t + 5$ and $y = 7t - 8$.

What is the velocity of the object and on what bearing, to the nearest degree, is it moving? **[7 marks]**

CM PS 12 Which of the following graphs is for the curve defined by the parametric equations $x = 3\cos^2 t$ and $y = 7\cos^2 t$? You must provide justification for your answer.

A

B

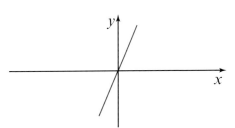

C

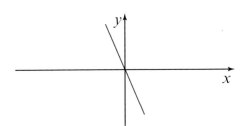

D

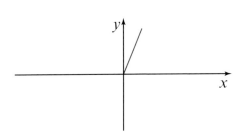

[4 marks]

13 Given that $\sin^2 t \cos^2 t = 1$ and $0 \leqslant t < 2\pi$:

    a define parametric equations for a circle centre $(2, -3)$ and radius 5 units   **[2 marks]**

    b use the parametric equations to determine the coordinates of the x- and y-intercepts. Give answers to 1 decimal place.   **[4 marks]**

 14 **a** A curve has parametric equations $x = \dfrac{1}{(s-1)^2}$ and $y = s + 1$, where $s \neq 1$.

    Find the Cartesian equation of the curve, y in terms of x.   **[4 marks]**

    b Does the curve intersect the y-axis? You must provide mathematical justification for your answer.   **[2 marks]**

 15 Show that the Cartesian equation of the curve defined by the parametric equations $x = \dfrac{t}{1 + 2t}$ and $y = \dfrac{t-1}{t+1}$, where $t \neq -1$ or $-\frac{1}{2}$, can be written as $y = \dfrac{1 - 3x}{x - 1}$.   **[6 marks]**

16 Find the Cartesian equation of the curve defined by the parametric equations $x = \cos t$ and $y = \cos 2t$, where $0 \leqslant t < 2\pi$. Sketch the curve.   **[5 marks]**

4 SEQUENCES AND SERIES

In 1967 Benoit Mandelbrot published a paper titled 'How long is the coast of Britain?' His surprising answer was that it is impossible to say. The closer you look, the more 'rough' the coast seems to be and the longer your answer becomes.

Mandelbrot is best known for his work on fractals, in particular the discovery of the Mandelbrot Set.

The set is generated from looking at the behaviour of sequences generated by a simple term-to-term rule. Like the coastline example, the boundary of the set gets more complex as you zoom in on any part of it. If you look closely at the image on the right, you can see sections which look like copies of other sections but on a smaller scale. This is one of the properties of fractal objects.

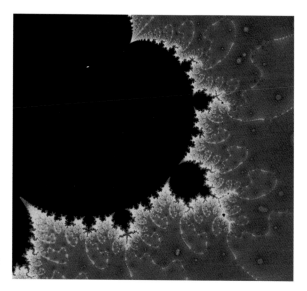

Mandelbrot's work helps to explain the behaviour of systems that deviate suddenly from what is expected, where small variations can have very large effects. It helps us to understand, for example, fluctuations on the stock market or why weather forecasting is so difficult.

LEARNING OBJECTIVES

You will learn how to:

› use functions to generate sequences

› describe the properties of different sequences

› interpret and use sigma notation

› identify and work with arithmetic sequences and their associated series

› identify and work with geometric sequences and their associated series

› extend the range of applications of binomial expansions.

TOPIC LINKS

At GCSE level you encountered simple sequences and term-to-term rules. You will also need to use skills from **Chapter 2 Algebra and functions 2: Partial fractions** to evaluate a function for different values of the variable. In this chapter you will learn notation that you will be able to use in **Chapter 14 Statistical hypothesis testing**.

PRIOR KNOWLEDGE

You should already know how to:

- › use a simple rule to find the terms of a sequence
- › work out the coefficients in a binomial expansion.

You should be able to complete the following questions correctly:

1 The nth triangle number is $\frac{1}{2}n(n+1)$.

    a Find the 5th triangle number.

    b Show that 45 is a triangle number but 54 is not.

2 Find the coefficient of x^3 in the binomial expansion of $(2+x)^6$

3 Sam earns £24 000 a year.
He expects to get a 5% pay rise each year.
How much does he expect to be earning in five years' time?

4.1 Types of sequences

A lot of new building is planned for a town.

The population is expected to rise by 10% per year for the next few years.

Suppose the population now is p_0 thousand people.

In one year's time it will be p_1, in two years' time it will be p_2, and so on.

The multiplier for an increase of 10% is 1.1.

Therefore $p_1 = 1.1p_0$, $p_2 = 1.1p_1$, $p_3 = 1.1p_2$, and so on.

We can write a **recurrence relation**

$$p_{n+1} = 1.1p_n$$

where n can be 0, 1, 2, 3, … or any non-negative integer.

Once you know p_0 you can then find all the subsequent values.

Here are some examples for three different values of p_0. Values are rounded to 1 decimal place.

> $100\% + 10\% = 110\% = 1.1$

p_0	p_1	p_2	p_3	p_4	p_5
5.0	5.5	6.1	6.7	7.3	8.1
10.0	11.0	12.1	13.3	14.6	16.1
25.0	27.5	30.3	33.3	36.6	40.3

Notice that once you have a value for p_0 you can then generate a **sequence** of numbers which give the population, in thousands, in subsequent years.

TECHNOLOGY

You should be able to generate a table like this easily in a spreadsheet. Most calculators allow you to enter a starting value and quickly generate further terms. Make sure you know how to do this.

Stop and think   This might be a good model of how the population will grow for a few years. Why is it not a good model for a longer period of time?

The recurrence relation can be used to find the next number in the sequence if you know the current one.

In this example, you could also write a specific formula for p_n in terms of n.

If $p_0 = 5$, then $p_n = 5 \times 1.1^n$.

If $p_0 = 10$, then $p_n = 10 \times 1.1^n$.

This is an example of a **divergent sequence**. As n gets larger, p_n gets larger without an upper limit.

You can write this as $n \to \infty$, $p_n \to \infty$.

A more realistic model of population growth is the recurrence relation

$$p_{n+1} = 1.1p_n - 0.005p_n^{\,2}$$

You can see why if you look at an example.

Look again at the specific example where $p_0 = 10$.

p_0	p_1	p_2	p_3	p_4	p_5	p_6	p_7	p_8	p_9
10.0	10.5	11.0	11.5	12.0	12.5	12.9	13.4	13.8	14.3

If you compare it with the values in the appropriate row in the first table, you will see that the growth becomes less rapid over time. This is more realistic than a population which keeps growing at a constant rate.

Here are some values for larger values of n:

n	10	20	30	40	50
p_n	14.7	17.7	19.1	19.7	19.9

In this case, the values form a **convergent sequence**.

As the value of n increases, the value of p_n approaches a limit of 20.

You can write that as $n \to \infty$, $p_n \to 20$ or $\lim\limits_{n \to \infty} p_n = 20$.

For a third type of sequence, imagine a wheel with a radius of 1 m, rotating once every eight seconds. A point on the rim of the wheel is initially level with the hub. If x_n metres is the distance above or below the hub after n seconds, then

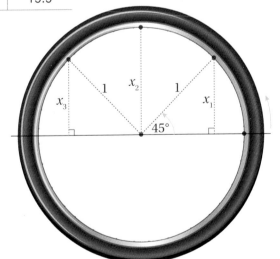

$x_0 = 0$

$x_1 = \sin 45° = 0.707$

$x_2 = \sin 90° = 1$

$x_3 = \sin 135° = 0.707$

The general term is $x_n = \sin 45n°$ and the sequence is

$0, 0.707, 1, 0.707, 0, -0.707, -1, -0.707, 0, 0.707, 1, 0.707, \ldots$

The values repeat every eight terms. This is a **periodic sequence** of **order** 8.

Here are three more sequences:

> Sequence 1: $x_{n+1} = \frac{1}{2}x_n + 3$ $x_1 = 10$

The sequence is $10, 8, 7, 6.5, 6.25, 6.125, \ldots$

> Sequence 2: $x_{n+1} = \frac{1}{2}x_n + 3$ $x_1 = 0$

The sequence is $0, 3, 4.5, 5.25, 5.625, 5.8125, \ldots$

> Sequence 3: $x_{n+1} = \frac{30}{x_n + 1}$ $x_1 = 8$

The sequence is $8, 4.75, 7.3158, 5.1007, 6.8815, 5.3595, 6.5975, 5.5472, \ldots$

All three sequences converge to 6 and you can write $\lim_{x \to \infty} x_n = 6$.

The first is a **decreasing sequence** because $x_{n+1} > x_n$ for all values of n.

The second is an **increasing sequence** because $x_{n+1} < x_n$ for all values of n.

The third is an **oscillating sequence**.

> The first term is usually denoted with a subscript of 0 or 1 (e.g. x_0 or x_1), whichever is more convenient.

TECHNOLOGY

Make sure you can generate these sequences with a calculator or a spreadsheet.

KEY INFORMATION

Sequences can be divergent, convergent or periodic.

They can also be increasing, decreasing or oscillating.

Example 1

Describe each of these sequences.

a $a_{n+1} = 3a_n - 10$ $a_1 = 2$
b $a_{n+1} = \frac{a_n}{3} + 10$ $a_1 = 2$
c $a_{n+1} = \frac{20}{a_n}$ $a_1 = 2$

Solution

a The sequence starts $2, -4, -22, -76, -238, \ldots$
It is decreasing and diverging.

b The sequence starts $2, 10.6667, 13.5556, 14.5185, 14.8395, \ldots$
It is increasing and converging to 15.

c The sequence starts $2, 10, 2, 10, \ldots$
The sequence is periodic of order 2.

Exercise 4.1A   **Answers page 438**

1 A sequence is defined by the recurrence relation $x_{n+1} = x_n + 2n + 1$.

   a If $x_1 = 1$, work out the value of x_4.

   b If $x_1 = 1$, find an expression for x_n in terms of n.

   c If $x_1 = 2$, find an expression for x_n in terms of n.

2 The triangle numbers are defined by the function $x_n = \frac{1}{2}n(n+1)$, integer $n \geqslant 1$.

  a Find the 10th and the 20th triangle numbers.

  b Find a recurrence relation satisfied by the triangle numbers.

(PF) **3** A sequence is defined by the recurrence relation $x_{n+1} = \sqrt[3]{10 - x_n}$ $\quad x_1 = 6$.

  a Show that the sequence converges to a limit.

  b Show that the limit satisfies the equation $x = \sqrt[3]{10 - x}$.

(PF) **4** A sequence is defined by the recurrence relation $t_n = \tan 60n$, integer $n \geqslant 0$.

  Show that the sequence is periodic and find its order.

(M) (CM) **5** Albert has a balance of £1000 owing on his credit card.
  Each month, 2% of the balance is added to the debt and he pays off £50.
  The amount he owes after n months is £x_n.

  a Explain why $x_0 = 1000$ and $x_{n+1} = 1.02x_n - 50$.

  b Find the amount he owes after his 4th monthly payment.

  c A formula for the amount he owes is $x_n = 2500 - 1500 \times 1.02^n$.

  Show that it will take 26 months to pay off the debt.

  d What would happen if he paid off only £15 a month instead of £50?

(PS) **6** A sequence is defined by $x_{n+1} = \frac{1}{2}\left(x_n + \frac{2}{x_n}\right)$.

  a If $x_1 = 1$, find the value of x_5.

  b If $x_1 = 2$, find the value of x_5.

  c Both the sequences in **parts a** and **b** converge to the same limit, which you may recognise.

  What is the significance of that number?

  d Another sequence is defined by $x_{n+1} = \frac{1}{2}\left(x_n + \frac{3}{x_n}\right)$, $x_1 = 2$.

  Find the limit of this sequence.

  e Is the limit to the sequence in **part d** the same if you change the value of x_1?

  f A sequence is defined by $x_{n+1} = \frac{1}{2}\left(x_n + \frac{a}{x_n}\right)$, where a is a positive number.

  Predict a limit for this sequence and test your prediction with particular values of a.

(PS) **7** This question is about the recurrence relation $x_{n+1} = \frac{12}{(x_n - 1)} + 1$, integer $n \geqslant 1$.

  a Describe the sequence if $x_1 = 3$.

  b Describe the behaviour of the sequence if $x_1 = 5$.

  c Describe the behaviour of the sequence if $x_1 = -4$.

  d Describe the behaviour of the sequence if $x_1 = -5$.

(PS) **8** This question is about the sequence defined by $x_{n+1} = x_n^2 - x_n - 1$, integer $n \geqslant 1$.

  a Describe the behaviour of the sequence if $x_1 = 2$.

  b Describe the behaviour if $x_1 > 2$.

  c What happens if $x_1 = 0.5$?

(M) Modelling   (PS) Problem solving   (PF) Proof   (CM) Communicating mathematically   **65**

4.2 Sigma notation

The sequence of square numbers has the definition $s_n = n^2$, $n = 1, 2, 3, \ldots$

The terms are 1, 4, 9, 16, …

The sum of the first three terms is $1 + 4 + 9 = 14$.

You can write that using **sigma notation** as $\displaystyle\sum_{n=1}^{3} n^2 = 14$.

Σ is the Greek capital letter sigma.

Read this as 'the sum from n equals 1 to 3 of n^2'. It means that you add together those three terms, with n equal to 1, then 2, then 3.

Check that $\displaystyle\sum_{n=1}^{2} n^2 = 5$ and $\displaystyle\sum_{n=1}^{6} n^2 = 91$.

Any letter can be used, it does not have to be n.

$\displaystyle\sum_{i=1}^{3} i^2$ is identical to $\displaystyle\sum_{n=1}^{3} n^2$.

For example, $\displaystyle\sum_{i=3}^{6} i(i+2) = 3 \times 5 + 4 \times 6 + 5 \times 7 + 6 \times 8 = 122$.

You can sometimes write expressions involving Σ in different ways.

Example 2

$\displaystyle\sum_{i=1}^{20} i = 210$ and $\displaystyle\sum_{i=1}^{20} i^2 = 2870$.

Use these facts to find

a $\displaystyle\sum_{i=1}^{20} 5i$ **b** $\displaystyle\sum_{i=1}^{20} (i^2 - 1)$ **c** $\displaystyle\sum_{i=1}^{20} i(i+4)$

Solution

a $\displaystyle\sum_{i=1}^{20} 5i = 5 \times 1 + 5 \times 2 + \ldots + 5 \times 20$

Take out 5 as a factor.

$\displaystyle\sum_{i=1}^{20} 5i = 5 \times \sum_{i=1}^{20} i = 5 \times 210 = 1050$

b $\displaystyle\sum_{i=1}^{20} (i^2 - 1) = 1^2 - 1 + 2^2 - 1 + \ldots + 20^2 - 1$

Reordering, it is clear that $\displaystyle\sum_{i=1}^{20} (i^2 - 1) = \sum_{i=1}^{20} i^2 - \sum_{i=1}^{20} 1$

$= 2870 - 20 = 2850$

c Multiply out the bracket.

$\displaystyle\sum_{i=1}^{20} i(i+4) = \sum_{i=1}^{20} (i^2 + 4i)$

Write that as two separate sums.

$\displaystyle\sum_{i=1}^{20} 1 = 20$ and more generally

$\displaystyle\sum_{i=1}^{n} 1 = n$.

$$= \sum_{i=1}^{20} i^2 + \sum_{i=1}^{20} 4i$$

$$= \sum_{i=1}^{20} i^2 + 4 \sum_{i=1}^{20} i$$

$$= 2870 + 4 \times 210$$

$$= 3710$$

Exercise 4.2A

Answers page 438

1 Work out:

a $= \sum_{i=1}^{4}(3i + 2)$ **b** $\sum_{i=1}^{4}\left(i^2 + i\right)$ **c** $\sum_{i=1}^{4}(i-1)(i+4)$

2 Work out $\sum_{k=1}^{5}\dfrac{30}{k}$.

3 Work out $\sum_{n=1}^{4}\cos 90n$.

(PS) 4 $\sum_{i=1}^{n} i^2 = \frac{1}{6}n(n+1)(2n+1)$

Use this fact to find:

a the sum of the first six square numbers

b the sum of all the square numbers between 100 and 1000.

(PS) 5 The sum of the first 50 positive integers is $\sum_{r=1}^{50} r = 1275$.

Use this fact to find:

a the sum of the first 50 multiples of 3

b the sum of the first 50 odd numbers.

6 Find $\sum_{r=1}^{20}\left\{(r+1)^3 - r^3\right\}$

7 Write, using Σ notation:

a $1 + 4 + 7 + \dots + 100$ **b** $3 + 6 + 11 + 18 + \dots + 83 + 102$

c $\dfrac{2}{3} + \dfrac{4}{5} + \dfrac{8}{7} + \dfrac{16}{9} + \dots$ (20 terms)

(PF) 8 The ith triangle number, t_i, is given by $\frac{1}{2}i(i+1)$.

a Show that the 100th triangle number is 5050. **b** Work out $\sum_{i=1}^{5} t_i$.

c $\sum_{i=1}^{n} i = \frac{1}{2}n(n+1)$ and $\sum_{i=1}^{n} i^2 = \frac{1}{6}n(n+1)(2n+1)$

Use these facts to show that the sum of the first n triangle numbers, $\sum_{i=1}^{n} t_i = \frac{1}{6}n(n+1)(n+2)$.

4.3 Arithmetic sequences and series

An **arithmetic sequence** is one in which there is a common difference between successive terms.

Here are some examples:

13, 16, 19, 22, 25, ...   First term = 13, common difference = 3

100, 89, 78, 67, 56, ...   First term = 100, common difference = −11

−35.5, −35.3, −35.1, −34.9, ...   First term = −35.5, common difference = 0.2

When you say the *adjective* 'arithmetic' aloud, the stress is on the third syllable, 'met'.

The difference is positive for an increasing sequence and negative for a decreasing one.

You can write a general arithmetic sequence as:

$a, a + d, a + 2d, a + 3d, \ldots$

The nth term of an arithmetic sequence is $a + (n − 1)d$,

For example, the first sequence above is 13, 16, 22, 25, ...

The 100th term is $13 + 99 \times 3 = 310$.

The second sequence is 100, 89, 78, 67, 56, ...

The 100th term is $100 − 99 \times 11 = −989$.

The sum of a sequence is called a **series**.

Here is a series of the first 50 positive integers:

$S = 1 + 2 + 3 + 4 + \ldots + 48 + 49 + 50$

An easy way to find this sum is to write it backwards.

$S = 50 + 49 + 48 + 47 + \ldots + 3 + 2 + 1$

Add the two series together term by term.

$2S = (1 + 50) + (2 + 49) + (3 + 48) + \ldots + (48 + 3) + (49 + 2) + (50 + 1)$

You can see that there are 50 pairs and each pair adds up to 51.

$2S = 50 \times 51$

Therefore $S = \frac{1}{2} \times 50 \times 51 = 1275$.

You can do this for any arithmetic sequence.

This shows that if you know the first term, a, and the last term, l, and there are n terms, then the sum of the series is $\frac{1}{2}n(a + l)$.

However, it is more usual just to know the first term and the common difference.

Suppose S_n is the sum of the first n terms of the arithmetic sequence with the first term a and the common difference d.

$$S_n = \sum_{i=1}^{n} \{a + (i-1)d\}$$

$$S_n = a + (a+d) + (a+2d) + \ldots + \{a + (n-2)d\} + \{a + (n-1)d\}$$

Write the terms in reverse order.

$$S_n = \{a + (n-1)d\} + \{a + (n-2d)\} + \ldots + (a+2d) + (a+d) + a$$

Add.

$$2S_n = \{2a + (n-1)d\} + \{2a + (n-1)d\} + \ldots$$

There are n terms, each equal to $\{2a + (n-1)d\}$, so

$$2S_n = n\{2a + (n-1)d\}$$

Therefore $S_n = \dfrac{n}{2}\{2a + (n-1)d\}$.

> **KEY INFORMATION**
>
> You should be able to prove the sum formula for an arithmetic sequence in this way.

> **KEY INFORMATION**
>
> › An arithmetic sequence has a first term a and a common difference d.
>
> › The nth term is $a + (n-1)d$.
>
> › The sum of the first n terms, S_n, is $\dfrac{n}{2}\{2a + (n-1)d\}$.

Example 3

On her birthday Ada's Uncle Isaac gives her some money.

Each birthday after the first one he gives her £5 more than he gave her on her previous birthday.

On her seventh birthday he gave her £76.

After she has received her 21st birthday present, she notices that the total amount in pounds she has been given is equal to the year in which she is 21.

What year was Ada born?

Solution

Suppose her uncle gives her £x_n on her nth birthday.

The amounts he gives her form an arithmetic sequence with $d = 5$.

$x_7 = 76$

Use the formula $x_n = a + (n-1)d$ with $n = 7$.

$76 = a + 6 \times 5$

$a = 76 - 30 = 46$

Now use $S_n = \dfrac{n}{2}\{2a + (n-1)d\}$ with $n = 21$.

$S_{21} = \dfrac{21}{2}\{2 \times 46 + 20 \times 5\} = 10.5 \times 192 = 2016$

She was 21 in 2016.

She was born in $2016 - 21 = 1995$.

Exercise 4.3A   Answers page 439

1 An arithmetic sequence starts 20, 27, 34, 41, …

    a Find the 10th term.

    b Find the sum of the first 10 terms.

    c Find the largest term that is less than 200.

2 Find the sum of each of these arithmetic series:

    a $20 + 25 + 30 + 35 + …$ (10 terms)

    b $120 + 111 + 102 + 93 + …$ (12 terms)

    c $9 + 11.5 + 14 + 16.5 + …$ (15 terms)

3 These are the first three rows of a pattern of squares:

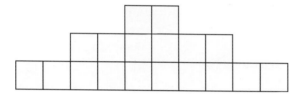

    There are two squares in the first row and six squares in the second row.

    Find the total number of squares in 20 rows.

4 On Sunday, 15 mm of snow falls.

    On each subsequent day, 6 mm more snow falls than on the day before.

    a How much snow falls on Saturday?

    b Find the total snowfall for the week.

5 Lucy is saving money.
In January she saves £50.
In each subsequent month she saves £5 more than the month before.

    How much will she save in a month in two years' time?

(PS) 6 The 5th term of an arithmetic sequence is 27.6 and the 10th term is 24.1.
Find the 20th term.

(PF) 7 **a** Find the sum of the first ten multiples of 6.

    b Show that the sum of the first n multiples of 6 is $3n(n + 1)$.

(PF) 8 **a** Prove that the sum of the first n odd numbers is n^2.

    b Find an expression for the sum of the first n even numbers.

(PS) 9 The first term of an arithmetic series is 10 and the common difference between terms is 4.
The sum of the series is 792.

    How many terms are there in the series?

4.4 Geometric sequences and series

One Sunday Sam says:

'I received 10 emails last Sunday. On each day after that, I received twice as many as the day before.'

How many emails did Sam claim she received last week?

There were 10 on Monday, 20 on Tuesday, 40 on Wednesday, and so on.

The total is $10 + 20 + 40 + 80 + 160 + 320 + 640 = 1270$

The numbers 10, 20, 40, 80 form a **geometric sequence**.

In a geometric sequence, each term is a common multiple of the previous term. In this example the multiple is 2.

In general, a geometric sequence is $a, ar, ar^2, ar^3, \ldots$

and the nth term is ar^{n-1}.

Suppose S_n is the sum of the first n terms of a geometric sequence.

$$S_n = a + ar + ar^2 + \ldots + ar^{n-2} + ar^{n-1}$$

Multiply every term by r.

$$rS_n = ar + ar^2 + \ldots + ar^{n-1} + ar^n$$

Subtract S_n from rS_n and all the terms cancel except two.

$$rS_n - S_n = ar^n - a$$

Factorise.

$$(r-1)S_n = a(r^n - 1)$$

Divide by $r - 1$.

$$S_n = \frac{a(r^n - 1)}{r - 1}$$

Applying this to the sum above, $10 + 20 + 40 + \ldots + 640$, for which $n = 7$, $a = 10$ and $r = 2$,

$S_7 = \dfrac{10 \times (2^7 - 1)}{2 - 1} = 1270$ as before.

The value of r can also be smaller than 1, or negative.

Here are some examples:

$a = 10$, $r = 0.8$:   10, 8, 6.4, 5.12, 4.096, ...

$a = 4$, $r = -3$:   4 −12, 36, −108, 324, −972, ...

$a = 24$, $r = -\dfrac{1}{4}$:   24, −6, 1.5, −0.375, ...

> **KEY INFORMATION**
>
> You should be able to prove the sum formula for a geometric sequence in this way.

When $r < 1$, the numerator and the denominator of the formula for S_n are both negative.

In that case you might prefer to use the formula as $S_n = \dfrac{a(1 - r^n)}{1 - r}$. It gives the same answer.

Suppose $|r| < 1$.

Then, as n gets larger, r^n gets closer to 0.

As $n \to \infty$, $r^n \to 0$ and therefore $1 - r^n \to 1$,

As $n \to \infty$, $S_n \to \dfrac{a}{1 - r}$.

This limit is called the **sum to infinity** and you can write $S_\infty = \dfrac{a}{1 - r}$.

As an example, look at this sequence: $24, -6, 1\frac{1}{2}, -\frac{3}{8}, \frac{3}{32}, \ldots$

$a = 24$ and $r = -\frac{1}{4}$

$S_4 = 19\frac{1}{8} = 19.125$

$S_5 = 19\frac{7}{32} = 19.219$ (to 3 d.p.)

$S_6 = 19\frac{25}{128} = 19.195$ (to 3 d.p.)

$S_\infty = \dfrac{a}{1 - r} = \dfrac{24}{1 - \left(-\frac{1}{4}\right)} = \dfrac{24}{1\frac{1}{4}} = 19\frac{1}{5} = 19.2$

You can see that the value of S_n is getting closer to 19.2.

> $|x|$ is called the **absolute value** of x. It is a number equal to x, ignoring any negative sign.
>
> So $|3.5| = 3.5$ and $|-2.3| = 2.3$
>
> $|r| < 1$ is a shorter way of writing $-1 < x < 1$.

KEY INFORMATION

> ▸ The first term of a geometric progression is a and the common ratio is r.
>
> ▸ The nth term is ar^{n-1}.
>
> ▸ The sum of the first n terms, S_n, is $\dfrac{a(r^n - 1)}{r - 1}$ or $\dfrac{a(1 - r^n)}{1 - r}$.
>
> ▸ If $|r| < 1$, the series converges and the sum to infinity, S_∞, is $\dfrac{a}{1 - r}$.
>
> ▸ If a sequence can't be bounded by any integer, it is known as a divergent geometric sequence.

Example 4

Elton hires a piano.

He pays £100 in the first month.

In each subsequent month he pays 80% of what he paid in the previous month.

a How much does he pay in the first year?

b How long will it be until he has paid £480?

c Show that, however long he hires the piano for, he will never pay more than £500.

Solution

a The amounts he pays each month, in pounds, are

$$100, \ 100 \times 0.8 = 80, \ 80 \times 0.8 = 64, \ \ldots$$

This is a geometric sequence with $a = 100$ and $r = 0.8$.

The total paid in the first year (12 months) is

$$S_{12} = \frac{100(1 - 0.8^{12})}{1 - 0.8} = 465.64$$

He pays £465.64 in the first year.

b The total paid after n months is £480.

$$S_n = \frac{100(1 - 0.8^n)}{1 - 0.8} = 480$$

Rearrange.

$1 - 0.8^n = \dfrac{480 \times 0.2}{100}$

$1 - 0.8^n = 0.96$

$0.8^n = 0.04$

Take logarithms of both sides.

$n \ln 0.8 = \ln 0.04$

$n = \dfrac{\ln 0.04}{\ln 0.8} = 14.43$

The value of n must be a whole number, so he will have paid £480 after 15 months.

c $S_\infty = \dfrac{a}{1 - r} = \dfrac{100}{1 - 0.8} = 500$

The limit for the sum is £500. Elton will never pay more than this.

Exercise 4.4A

Answers page 439

1 Here is the start of a geometric sequence: 5, 10, 20, 40, …

Find the 15th term.

2 Find the sum of each of these geometric series:

a $2 + 6 + 18 + 54 + …$ (10 terms)

b $100 + 90 + 81 + …$ (8 terms)

c $4 - 8 + 16 - 32 + …$ (12 terms)

d $20 + 22 + 24.2 + 26.62 + …$ (20 terms)

3 Jon is starting to get a pension from his job.
This year he will receive £12 000.
In each of the following years the pension will increase by 3%.

Find the total he will receive over the next ten years.

4 Find the sum to infinity of each of these geometric series:

a $1 + \dfrac{1}{2} + \dfrac{1}{4} + \dfrac{1}{8} + …$

b $1 + \dfrac{1}{3} + \dfrac{1}{9} + \dfrac{1}{27} + …$

c $1 - \dfrac{1}{3} + \dfrac{1}{9} - \dfrac{1}{27} + …$

d $1 + \dfrac{3}{4} + \dfrac{9}{16} + \dfrac{27}{64} + …$

e $1 - \dfrac{2}{3} + \dfrac{4}{9} - \dfrac{8}{27} + …$

5 Paul is given 1p on his first birthday.

On each subsequent birthday he is given twice as much as he was given on his previous birthday.

a How much will he receive on his 21st birthday?

b How much will he have received altogether after his 21st birthday?

c Why is it impossible to continue this process until his 65th birthday?

6 Mia has a plan to save some money.
In the first year she will save £2000.
In each subsequent year she will save 25% more than the previous year.

a How much will she save in the third year?

b How long will it take her to save £50 000?

(PS) 7 A square is divided into four smaller squares and the top left-hand quarter is shaded:

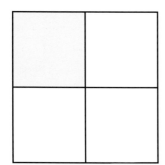

This process is repeated with the bottom right-hand square:

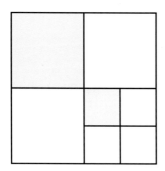

a What fraction of the whole square is shaded now?

b The process is repeated with the small bottom right-hand square.

What fraction of the whole square is shaded now?

c Imagine this process could be repeated an infinite number of times.

What fraction of the whole square would be shaded?

(M) 8 Jess is going for a walk.

She walks 400 m north.

Then she walks 200 m south.

Then she walks 100 m north.

Then she walks 50 m south.

She continues in this way.

a How far is she from her starting point when she finishes?

b How far does she walk altogether?

(PF) 9 $0.454545\ldots = 0.45 + 0.0045 + 0.000045 + \ldots$

Use this fact to write $0.454545\ldots$ as a fraction.

(PS) 10 The first term of a geometric sequence is 80.

The sum to infinity of the sequence is 200.

How many terms must be added until the total is more than 199?

4.5 Binomial expansions

You encountered **binomial expansions** in **Book 1, Chapter 1 Algebra and functions 1: Manipulating algebraic expressions**.

$$(a + bx)^n = a^n + na^{n-1}bx + \frac{n(n-1)}{2}a^{n-2}(bx)^2 + \ldots + (bx)^n$$

where n is a positive integer

Put $a = b = 1$ and it becomes

$$(1 + x)^n = 1 + nx + \frac{n(n-1)}{2}x^2 + \frac{n(n-1)(n-2)}{3!}x^3 + \ldots \quad \textcircled{1}$$

This is correct if n is a positive integer.

For example, if $n = 7$ it becomes $(1 + x)^7 = 1 + 7x + 21x^2 + \ldots + x^7$.

The series terminates if n is a positive integer.

In fact, you can put any rational number into formula $\textcircled{1}$ and it gives an infinite series.

Here is a general result:

$$(1 + x)^n = 1 + nx + \frac{n(n-1)}{2}x^2 + \frac{n(n-1)(n-2)}{3!}x^3 + \ldots$$

if n is a rational number.

The series will converge if $|x| < 1$.

For example, if $n = -1$ you get the series

$$(1 + x)^{-1} = 1 + (-1)x + \frac{(-1)(-1-1)}{2}x^2 + \frac{(-1)(-1-1)(-1-2)}{3!}x^3 + \ldots, \quad |x| < 1$$

> The coefficient of x^3 is
> $$\frac{\frac{1}{2} \times \left(-\frac{1}{2}\right) \times \left(-\frac{3}{2}\right)}{6} = \frac{3}{8 \times 6} = \frac{1}{16}$$

Tidy up the minus signs.

$$\frac{1}{1 + x} = 1 - x + \frac{1 \times 2}{2}x^2 - \frac{1 \times 2 \times 3}{3!}x^3 + \ldots$$

All the **coefficients** cancel to 1.

$$\frac{1}{1 + x} = 1 - x + x^2 - x^3 + \ldots$$

The terms in the series on the right form a geometric series with $a = 1$ and $r = -x$, which converges if $|x| < 1$.

As another example, try $n = \frac{1}{2}$. This gives the series

$$(1 + x)^{\frac{1}{2}} = 1 + \frac{1}{2}x + \frac{\frac{1}{2}\left(\frac{1}{2} - 1\right)}{2}x^2 + \frac{\frac{1}{2}\left(\frac{1}{2} - 1\right)\left(\frac{1}{2} - 2\right)}{3!}x^3 + \ldots$$

$$\sqrt{1 + x} = 1 + \frac{1}{2}x - \frac{1}{8}x^2 + \frac{1}{16}x^3 + \ldots, \quad |x| < 1$$

> **Stop and think** Show that the next term in the binomial series for $\sqrt{1 + x}$ is $-\frac{5}{128}x^4$.

KEY INFORMATION

$$(1 + x)^n = 1 + nx + \frac{n(n-1)}{2}x^2 + \frac{n(n-1)(n-2)}{3!}x^3 + \ldots$$

if n is a rational number and $|x| < 1$.

This is the general binomial expansion.

The original form of the binomial expansion (where n is a positive integer) is valid for any value of x, and it produces a finite sequence of terms ($n + 1$ of them).

This new form gives an infinite sequence that will converge if $|x| < 1$.

Calculators do not 'know' square roots or any other function. They work them out by using series approximations of the type you have found here. That is one reason why series approximations are so useful.

Here are a few binomial approximations.

You have already seen $\dfrac{1}{1+x} = 1 - x + x^2 - x^3 + \ldots \; |x| < 1$

If you replace x with $-x$ you get $\dfrac{1}{1-x} = 1 + x + x^2 + x^3 + \ldots$

Putting $n = -\frac{1}{2}$ gives $\dfrac{1}{\sqrt{1+x}} = 1 - \frac{1}{2}x + \frac{3}{8}x^2 - \frac{5}{16}x^3 - \ldots$

> Always give the coefficients as exact fractions rather than approximate decimals.

Example 5

a Work out a binomial series for $\dfrac{1}{\sqrt[3]{1-x}}$.

Give the terms up to x^3 and state the range of values of x for which it is valid.

b Use your series to find an approximate value for $\dfrac{1}{\sqrt[3]{7.2}}$.

c Use your answer to **part a** to find a binomial series for

$\dfrac{1}{\sqrt[3]{8-x}}$ and state the range of values for x for which it is valid.

Solution

a $\dfrac{1}{\sqrt[3]{1-x}} = (1-x)^{-\frac{1}{3}}$

$$= 1 + \left(-\frac{1}{3}\right) \times (-x) + \frac{\left(-\frac{1}{3}\right) \times \left(-\frac{4}{3}\right)}{2}(-x)^2$$

$$+ \frac{\left(-\frac{1}{3}\right) \times \left(-\frac{4}{3}\right) \times \left(-\frac{7}{3}\right)}{6}(-x)^3 + \ldots$$

$$= 1 + \frac{1}{3}x + \frac{2}{9}x^2 + \frac{14}{81}x^3 + \ldots$$

The series is valid if $|-x| < 1$, which is the same as $|x| < 1$.

b $\dfrac{1}{\sqrt[3]{7.2}} = \dfrac{1}{\sqrt[3]{8-0.8}}$

To use a binomial expansion you must have 1 rather than 8, so take out 8 as a factor.

$$\frac{1}{\sqrt[3]{8-0.8}} = \frac{1}{\sqrt[3]{8(1-0.1)}}$$

$$= \frac{1}{2\sqrt[3]{1-0.1}}$$

$$= \frac{1}{2} \times \frac{1}{\sqrt[3]{1-0.1}}$$

Use the expansion from **part a** with $x = 0.1$.

$$\frac{1}{\sqrt[3]{7.2}} \approx \frac{1}{2}\left(1 + \frac{1}{3} \times 0.1 + \frac{2}{9} \times 0.01 + \frac{14}{81} \times 0.001\right)$$

$$= 0.517864 \text{ to } 6 \text{ d.p.}$$

Note that this is close to the exact answer of 0.517872 to 6 d.p.

c The expression in the cube root sign must start with 1 to form a binomial expansion.

Write $\dfrac{1}{\sqrt[3]{8-x}}$ as $\dfrac{1}{\sqrt[3]{8\left(1-\frac{1}{8}x\right)}} = \dfrac{1}{2}\left(1-\dfrac{1}{8}x\right)^{-\frac{1}{3}}$

Replace x in the series in **part a** by $\frac{1}{8}x$.

$$\dfrac{1}{\sqrt[3]{8-x}} = \dfrac{1}{2}\left\{1 + \dfrac{1}{3}\times\dfrac{1}{8}x + \dfrac{2}{9}\times\dfrac{1}{64}x^2 + \dfrac{14}{81}\times\dfrac{1}{512}x^3 + \ldots\right\}$$

$$= \dfrac{1}{2} + \dfrac{1}{48}x + \dfrac{1}{576}x^2 + \dfrac{7}{41472}x^3 + \ldots$$

The series is valid when $\left|\frac{1}{8}x\right| < 1$, which means that $|x| < 8$.

Exercise 4.5A

Answers page 439

1 **a** Work out the first three terms in the binomial expansion of $\dfrac{1}{\left(1+x\right)^2}$.

b Work out the first three terms in the binomial expansion of $\dfrac{1}{\left(1-x\right)^2}$.

2 **a** Work out the coefficient of x^3 in the binomial expansion of $(1+x)^{\frac{5}{2}}$.

b Work out the coefficient of x^4 in the binomial expansion of $(1+x)^{\frac{5}{2}}$.

3 **a** Work out the first three terms in the binomial expansion of $\sqrt[3]{1+x}$.

b Write down the range of values of x for which the expansion is valid.

c Use the expansion to find an approximation to $\sqrt[3]{1.1}$.

Round your answer to 3 decimal places.

4 **a** Work out the binomial expansion for $\sqrt{1+x}$ as far as the term in x^3.

b Find the binomial expansion for $\sqrt{1+2x}$ as far as the term in x^3.

c Find the range of values of x for which the expansion in **part b** is valid.

5 **a** Work out a cubic approximation to $\dfrac{1}{1+x}$ for small x.

b Use your answer to **part a** to find an approximation to $\dfrac{1}{1+x^2}$ for small x.

6 **a** Find the binomial expansion of $(1+y)^{-2}$ as far as the term in y^3.

b Use your answer to **part a** to find an expansion for $(2+3x)^{-2}$ and state the range of values of x for which your expansion is valid.

7 Find the coefficient of x^3 in the binomial expansion of $\dfrac{1}{\sqrt{1-2x}}$.

PS **8** Use a binomial approximation to find a value of $4.01^{\frac{3}{2}}$ correct to 5 decimal places.

Show your method.

SUMMARY OF KEY POINTS

> Sequences can be defined by a term-to-term rule or by a formula for the nth term.

> Sequences can be increasing or decreasing or periodic.

> The order of a periodic sequence is the number of repeating terms.

> Sequences can be convergent to a limit or divergent.

> In an arithmetic sequence the difference between successive terms is constant. If the first term is a and the common difference is d, then the nth term is $a + (n-1)d$ and the sum of the first n terms, S_n, is $\frac{n}{2}\{2a + (n-1)d\}$.

> In a geometric sequence the ratio of successive terms is constant. If the first term is a and the common ratio is r, then the nth term is ar^{n-1} and the sum of the first n terms, S_n, is $\frac{a(r^n - 1)}{r - 1}$ or $\frac{a(1 - r^n)}{1 - r}$.

> If $|r| < 1$ in a geometric sequence, then the series converges and the sum to infinity, S_∞, is $\frac{a}{1 - r}$.

> The binomial expansion of $(1 + x)^n$ is $1 + nx + \frac{n(n-1)}{2}x^2 + \frac{n(n-1)(n-2)}{3!}x^3 + \dots$ if n is a rational number and $|x| < 1$.

EXAM-STYLE QUESTIONS 4   Answers page 440

1 A sequence is defined by $x_{n+1} = x_n + 2n + 1$ $x_1 = -1$.

    a Find x_4.   **[2 marks]**

    b Find a formula for x_n in terms of n.   **[2 marks]**

2 A sequence is defined by $x_{n+1} = \frac{1}{4}x_n + 12$, $x_1 = 8$.

    Show that the series is convergent and find the limit.   **[3 marks]**

3 Here is the start of a sequence: 14, 16.8, …

    a If this is an arithmetic sequence, find the 20th term.   **[2 marks]**

    b If this is a geometric sequence, find the 20th term.   **[2 marks]**

(M) 4 Tom starts a new job. His employer offers him £1200 for the first month and says he can have an extra £70 a month each month if he does well.

    a If Tom does well, how much can he expect to earn in his first 12 months? **[3 marks]**

    b Tom says he would prefer to have an increase of 5% each month.

        Show that in this case he will earn less in the second month but more in the whole year. **[4 marks]**

5 Find the sum to infinity of $4 - \frac{8}{3} + \frac{16}{9} - \frac{32}{27} + \dots$ **[4 marks]**

(M) 6 At the start of each month Nina puts £400 into a savings account where it earns 0.5% per month interest.

    Work out how much she has in her savings account at the end of 12 months. **[5 marks]**

7 An arithmetic series starts $20 + 24 + \ldots$

There are n terms in the series. The sum of the series is 504.

  a Show that $n^2 + 9n = 252$.   **[5 marks]**

  b Find the number of terms in the series.   **[2 marks]**

8 **a** Find a series expansion for $\dfrac{1}{1-x}$.   **[3 marks]**

  b Use the result to find a series expansion for $\dfrac{1}{2-x}$ and state the range of values of x for which it is valid.   **[4 marks]**

(PS) 9 $x_{n+1} = 2x_n^{\,2}$

Find the positive range of values of x_1 for which this sequence is convergent.   **[3 marks]**

(PS) 10 **a** Find a cubic approximation to $\sqrt{1-2x}$.   **[2 marks]**

  b The cubic approximation is used to find an approximation to $\sqrt{24.5}$.

    A calculator gives the value of this square root as $4.949\,747\,468$.

    How accurate is the cubic approximation?   **[2 marks]**

(PF) 11 **a** Show that the sum of the first n odd numbers is n^2.   **[3 marks]**

  b Find a formula for the sum of the first n multiples of 2.   **[3 marks]**

12 **a** Find a binomial approximation to $\sqrt{4-8x}$ as far as the term in x^3.   **[6 marks]**

  b State the range of values of x for which the binomial expansion is valid.   **[1 mark]**

(PS) 13 A square spiral is constructed on a coordinate grid as follows.

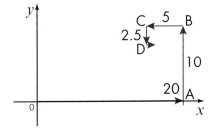

The first line, OA, is 20 units long.
The second line, AB, is 10 units long and parallel to the y-axis.
The third line, BC, is 5 units long and parallel to the x-axis.

This process continues to a final limit point.

  a Find the total length of the spiral.   **[3 marks]**

  b Find the coordinates of the end point of the spiral.   **[5 marks]**

14 A geometric series begins $10 + 15 + 22.5 + \ldots$

Work out the number of terms that are required until the sum is more than $10\,000\,000$.   **[6 marks]**

(PS) 15 A geometric series starts $40 + 32 + 14.4 + \ldots$

How many terms are needed before the total is closer than 0.1 to the sum to infinity?   **[5 marks]**

5 TRIGONOMETRY

At the beginning of the 20th century, doctors started using X-rays to look inside the human body. This revolutionary technique enabled them to identify problems and diseases without the need for invasive surgery. X-rays are used today in CAT (computerised axial tomography) scans, producing detailed images such as these, which show 'slices' through a brain.

From the 1960s onwards other methods of imaging, such as MRI (magnetic resonance imaging) have been developed.

The object being scanned produces a disrupted wave pattern, and getting from that to an image of the object which caused the disruption requires powerful mathematical techniques. These techniques involve the use of trigonometrical functions.

There is no doubt that these techniques save lives, and not just in a medical context. The same techniques are used in the detection of tripwires for landmines, so that the landmines can be found and made safe.

LEARNING OBJECTIVES

You will learn how to:

› use radians as well as degrees to measure angles

› prove and use formulae that show the connections between different trigonometric ratios

› define new trigonometric ratios alongside the sine, cosine and tangent ratios

› use trigonometric ratios in practical situations.

TOPIC LINKS

For this chapter you must know the definitions of sine, cosine and tangent for any angle, and you should know what the graphs of these functions look like – see **Book 1, Chapter 6 Trigonometry**. You need to be able to manipulate algebraic expressions and solve equations, using the techniques in **Book 1, Chapter 2 Algebra and functions 2: Equations and inequalitites**.

PRIOR KNOWLEDGE

You should already know how to:

- › sketch graphs of simple trigonometric functions

- › express the relationship between the sine, cosine and tangent of an angle

- › solve linear and quadratic equations

- › use a calculator accurately when using trigonometric ratios.

You should be able to complete the following questions correctly:

1 Sketch the graph of $y = 3\sin 2x$, $0 \leqslant x \leqslant 360°$.

2 Solve the equation $\cos(x + 20°) = 0.4$, $-180° \leqslant x \leqslant 180°$.

3 Prove that $\sin^2 x = (1 + \cos x)(1 - \cos x)$.

5.1 Radians

Up to now you have measured angles in degrees. In some contexts it is more convenient to use different units: **radians**.

One radian is the angle subtended at the centre of a circle by an **arc** with a length equal to the radius of the circle.

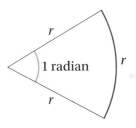

The angle between two straight lines from a point to a section of a curve is called a *subtended angle*. In this case, the curve is part of a circle and the lines are radii.

Since the circumference of a circle is $2\pi r$, this means that a whole turn is 2π radians.

$$2\pi \text{ radians} = 360° \text{ so } 1 \text{ radian} = \frac{360°}{2\pi} = 57.3° \text{ to 1 d.p.}$$

Angles in radians are often given as multiples of π. For example:

π radians $= 180°$

$\dfrac{\pi}{2}$ radians $= 90°$

$\dfrac{\pi}{4}$ radians $= 45°$

$\dfrac{\pi}{3}$ radians $= 60°$

KEY INFORMATION

π radians $= 180°$

1 radian $= \dfrac{180°}{\pi}$

Example 1

a Write $240°$ in radians.

b Change 0.83 radians to degrees.

Solution

a $240 = \frac{2}{3}$ of 360 so $240° = \frac{2}{3} \times 2\pi = \frac{4}{3}\pi$ radians.

b 1 radian $= \dfrac{180°}{\pi}$ so 0.83 radians $= 0.83 \times \dfrac{180°}{\pi} = 47.6°$ to 1 d.p.

Leave π in your answer.

Arc length and area of a sector

The angle at the centre of this **sector** of a circle is θ radians, the radius is r and the length of the arc is l.

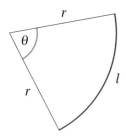

> If there is no degree symbol (°), then the angle is in radians.

A whole turn is 2π radians, so the length of the arc, l, is given by

$$l = \frac{\theta}{2\pi} \times \text{the circumference of the circle}$$

But the circumference of a circle is $2\pi r$, so

$$l = \frac{\theta}{2\pi} \times 2\pi r = r\theta$$

Similarly, the area of the sector, A, is given by

$$A = \frac{\theta}{2\pi} \times \text{the area of the circle}$$

The area of a circle is πr^2, so

$$A = \frac{\theta}{2\pi} \times \pi r^2 = \frac{1}{2}r^2\theta$$

The formulae for the length of an arc and the area of a sector are very simple if the angle is measured in radians.

KEY INFORMATION

For a sector of a circle of angle θ, the length of the arc, $l = r\theta$ and the area, $A = \frac{1}{2}r^2\theta$.

Example 2

A wedge of cheese is in the shape of a sector of a circle.

The radius is 25 cm and the angle is 0.3 radians.

Work out:

a the perimeter of the sector

b the area of the sector.

25 cm

0.3

25 cm

Solution

a The length of the arc is $25 \times 0.3 = 7.5$ cm.

The perimeter is the arc + 2 radii = $7.5 + 2 \times 25 = 57.5$ cm

b The area is $\frac{1}{2} \times 25^2 \times 0.3 = 93.75$ cm^2

Exercise 5.1A

Answers page 441

1 Give the following angles in degrees. Give your answers to 1 decimal place.

    a 0.5 radians   **b** 1.2 radians   **c** 0.1 radians   **d** 4 radians   **e** 5.5 radians

2 Give the exact value in radians of each of the following angles.

    a 30°   **b** 150°   **c** 270°   **d** 315°   **e** 720°

3 These angles are in radians. Change them to degrees.

a $\dfrac{\pi}{5}$ **b** $\dfrac{7\pi}{6}$ **c** $\dfrac{13\pi}{8}$ **d** 3.5π **e** 0.3π

4 Find the arc length of each of the following sectors.

a

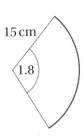

15 cm

1.8

b

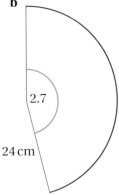

2.7

24 cm

c

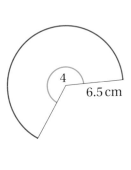

4

6.5 cm

5 Find the perimeters of the sectors in **question 4**.

6 Find the areas of the sectors in **question 4**.

7 A sector has been removed from this circle.

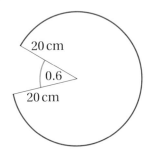

20 cm

0.6

20 cm

Calculate the area of the sector that remains.

8 This shape is the result of removing a sector from a larger sector.

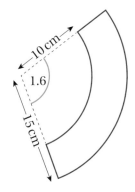

10 cm arc

1.6

15 cm

a Calculate the area of the shape.

b Calculate the perimeter of the shape.

(PS) 9 The area of a sector is 500 cm². The length of the arc of the sector is 20 cm.

Find:

a the radius of the sector   **b** the angle of the sector.

(PS) 10 The length of the arc of a sector in centimetres is numerically equal to the area of the sector in square metres. What, if anything, can you say about:

a the radius of the sector   **b** the angle of the sector?

5.2 Trigonometric ratios

You can find the trigonometric ratios (sine, cosine and tangent) for angles measured in radians.

Your calculator can work in radians or degrees.

You must learn the exact values of the trigonometric ratios for some angles.

To start with, these values are straightforward.

> Make sure you know how to calculate with trigonometric ratios on your calculator for angles in radians.

Angle in degrees	0°	90°	180°
Angle in radians	0	$\frac{\pi}{2}$	π
Sine	0	1	0
Cosine	1	0	−1

This is an isosceles right-angled triangle. The shorter sides are 1 unit.

By Pythagoras' theorem, the hypotenuse is $\sqrt{1^2 + 1^2} = \sqrt{2}$.

The angles are $\frac{\pi}{2}$ (90°) and $\frac{\pi}{4}$ (45°).

The diagram shows that

$$\sin\frac{\pi}{4} = \frac{1}{\sqrt{2}} \qquad \cos\frac{\pi}{4} = \frac{1}{\sqrt{2}} \qquad \tan\frac{\pi}{4} = 1$$

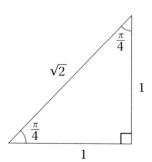

This is an equilateral triangle with a side of length 2, divided in half with a vertical line.

By Pythagoras' theorem, the height is $\sqrt{2^2 - 1^2} = \sqrt{3}$.

The angles are $\frac{\pi}{3}$ (60°) and $\frac{\pi}{6}$ (30°).

The diagram shows that

$$\sin\frac{\pi}{3} = \frac{\sqrt{3}}{2} \qquad \cos\frac{\pi}{3} = \frac{1}{2} \qquad \tan\frac{\pi}{3} = \sqrt{3}$$

and $\qquad \sin\frac{\pi}{6} = \frac{1}{2} \qquad \cos\frac{\pi}{6} = \frac{\sqrt{3}}{2} \qquad \tan\frac{\pi}{6} = \frac{1}{\sqrt{3}}$

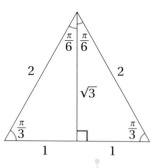

You can use these results to find the sines, cosines and tangents of angles bigger than $\frac{\pi}{2}$ or 90°. You can use the symmetries of the graphs of these functions help you.

> Memorise the triangles and sketch them to find the ratios when you need them.

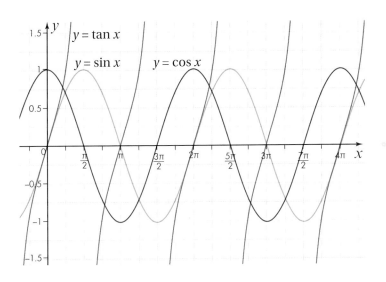

Notice that the values of x are in radians.

Example 3

Find the exact value of:

a $\sin\dfrac{5\pi}{4}$ **b** $\cos\dfrac{11\pi}{6}$ **c** $\tan\left(-\dfrac{\pi}{4}\right)$

Solution

a $\sin\dfrac{5\pi}{4}$ (or $\sin 225°$) $= -\sin\dfrac{\pi}{4}$ (or $-\sin 45°$) $= -\dfrac{1}{\sqrt{2}}$

b $\cos\dfrac{11\pi}{6}$ (or $\cos 330°$) $= \cos\dfrac{\pi}{6}$ (or $\cos 30°$) $= \dfrac{\sqrt{3}}{2}$

c $\tan\left(-\dfrac{\pi}{4}\right) = -\tan\dfrac{\pi}{4} = -1$

Make sure that you understand what the question is asking for and are using your calculator appropriately to give the required answer, e.g. do not truncate numbers in mid-calculation.

Exercise 5.2A

Answers page 441

1 Find the exact value of:

    a $\sin 30°$ **b** $\cos 30°$ **c** $\tan 30°$

2 Find the exact value of:

    a $\sin 60°$ **b** $\cos 60°$ **c** $\tan 60°$

3 Find the exact value of:

    a $\sin\dfrac{3\pi}{4}$ **b** $\cos\dfrac{3\pi}{4}$ **c** $\tan\dfrac{3\pi}{4}$

4 Find the exact value of:

    a $\sin\dfrac{2\pi}{3}$ **b** $\cos\dfrac{2\pi}{3}$ **c** $\tan\dfrac{2\pi}{3}$

5 Solve the following equations:

    a $2\sin\theta = \sqrt{3}$ $0 < \theta < 2\pi$ **b** $\tan\theta = \dfrac{\sqrt{3}}{3}$ $0 < \theta < 2\pi$

(PS) 6 Calculate $\sin^2\dfrac{\pi}{4} + \sin^2\dfrac{\pi}{3} + \cos^2\dfrac{\pi}{4}$.

5.3 Sketching graphs of trigonometric functions using radians

From your GCSE course and **Book 1, Chapter 6 Trigonometry**, you already know about the graphs of trigonometric functions. The only difference when you use radians is the values on the x-axis.

Example 4

Sketch the graph of:

a $f(x) = \sin(x - 0.2\pi)$ **b** $g(x) = 3\sin 2x$, where $0 \leqslant x \leqslant 2\pi$

Solution

a $y = \sin(x - 0.2\pi)$ will be a translation of $y = \sin x$ by $\begin{bmatrix} 0.2\pi \\ 0 \end{bmatrix}$

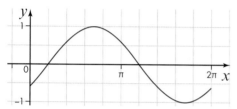

Notice that it crosses the x-axis at 0.2π and has a maximum value at $\frac{\pi}{2} + 0.2\pi = 0.7\pi$.

b $y = 3\sin 2x$ will be a stretch of $y = \sin x$ by 3 from the x-axis and by $\frac{1}{2}$ from the y-axis.

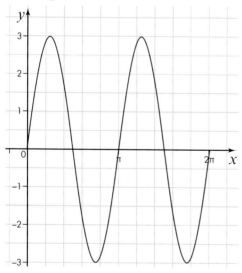

There are two complete cycles between 0 and 2π.

In **Example 4** you looked at the function $f(x) = \sin(x - 0.2\pi)$.

The variable x can take any value but the value of the function is always between -1 and 1.

The set of possible values of x is the **domain** of the function.

The domain of f(x) is all the real numbers, although in the example the domain was restricted to $0 \leqslant x \leqslant 2\pi$.

The set of possible values is called the **range** of the function.

The range of f(x) is $-1 \leqslant$ f(x) $\leqslant 1$.

In **part b** of **Example 4**, the range of g(x) is $-3 \leqslant$ g(x) $\leqslant 3$.

In **Example 4** you sketched the graph of g(x) = $3 \sin 2x$.

The **amplitude** of this graph is 3. The values can be up to 3 away from the middle value.

The **period** is π. The graph repeats every π units on the x-axis.

More generally, the graph of $y = a \sin \omega x$ has an amplitude of a and a period of $\frac{2\pi}{\omega}$.

If the variable is in degrees rather than radians, the amplitude is the same but the period is different. The graph of $y = a \sin \omega x°$ has an amplitude of a and a period of $\frac{360°}{\omega}$.

> Before you start sketching a graph, check whether you are using degrees or radians.

Exercise 5.3A

Answers page 441

1 **a** Find the period and the amplitude of $y = 4 \sin 2x$.

   b Sketch the graph of $y = 4 \sin 2x$ for $0 \leqslant x \leqslant 2\pi$.

2 **a** Find the period and the amplitude of $y = 10 \cos 4x$.

   b Sketch the graph of $y = 10 \cos 4x$ for $0 \leqslant x \leqslant \frac{\pi}{2}$.

3 **a** Find the period and the amplitude of $y = 2 \sin 0.5x$.

   b Sketch the graph of $y = 2 \sin 0.5x$ for $0 \leqslant x \leqslant 2\pi$.

   c On the same axes sketch the graph of $y = 1 + 2 \sin 0.5x$ for $0 \leqslant x \leqslant 2\pi$.

   d Describe the translation that maps $y = 2 \sin 0.5x$ onto $y = 1 + 2 \sin 0.5x$.

4 Find the equation of this graph.

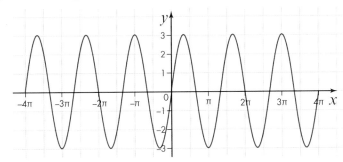

5 The equation of a graph is $y = 10 \sin \left(x + \frac{\pi}{4} \right)$.

Sketch the graph for the domain $-\pi \leqslant x \leqslant \pi$.

5.4 Practical problems

Graphs that look like a sine wave can arise in practical situations.

Example 5

A disc is rotating. The height (y cm) of a point on the edge above its lowest point after t seconds is given by the formula $y = 4 + 4\cos 0.2t$.

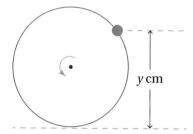

a Find the diameter of the disc.

b Find the time for one rotation of the disc.

c Sketch the graph of $y = 4 + 4\cos 0.2t$ for one rotation of the disc.

d Find the first and second times when the point is 7 cm above the lowest point.

Solution

a Since the point is on the edge of the disc, the amplitude of $4 + 4\cos 0.2t$ is equal to the radius.

The amplitude is 4, so the diameter is 8 cm.

b The period is the time for one rotation.

The units are radians so the period is $\frac{2\pi}{0.2} = 10\pi$, or 31.4 seconds to 3 s.f.

c

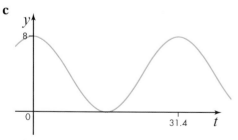

d When the height is 7 cm

$$4 + 4\cos 0.2t = 7$$

$$4\cos 0.2t = 3$$

> The first 4 in the formula translates the graph vertically but it does not change the amplitude or the period.

$$\cos 0.2t = 0.75$$

$$0.2t = \cos^{-1} 0.75 = 0.7227 \text{ or } 2\pi - 0.7227$$

$$t = 3.6 \text{ or } 27.8 \text{ to } 1 \text{ d.p.}$$

The first two times are after 3.6 and 27.8s.

Looking at the sketch graph, you can see that these are reasonable results.

> These are the first two positive values.

Exercise 5.4A

Answers page 442

 1 A mass on the end of a string is swinging backwards and forwards.

The distance (x cm) of the mass from its equilibrium position after t seconds is given by the formula $x = 5 \sin 2.1t$.

a Show that the time for one complete swing is 3.0 s to 1 d.p.

b Sketch the graph of $x = 5 \sin 2.1t$ for the interval $0 \leqslant t \leqslant 6$.

c Find the furthest distance of the mass from its equilibrium position.

2 This diagram illustrates a sound wave, approximately the note middle C. The horizontal axis represents time in seconds.

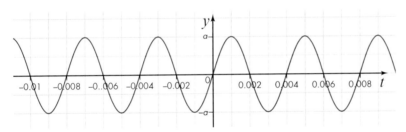

a Write down the period of the wave.

b Work out the number of cycles every second.

c The equation of the curve is $y = a \sin \omega t$ where a is the amplitude and ω is a constant.

Work out the value of ω.

3 This is the London Eye, a large rotating wheel.

You start at the bottom and your height (h m) after t minutes is given by the formula

$$h = 67.5 - 67.5\cos(12t)°$$

You travel round the circle once.

a Find your height after 10 minutes.

b Work out the diameter of the wheel.

c Work out how long the ride takes.

d Sketch a graph to show how your height changes during the ride.

e Work out for how many minutes you are over 100 m high.

M 4 To test the shock absorber of a vehicle, the spring is subjected to repeated compressions and extensions.

One end of the spring is fixed. The distance (y cm) of the moving end of the spring from its equilibrium position after t seconds is given by the formula $y = 1.5\sin 4\pi t$.

a Work out for the number of times the spring is compressed and extended each second.

b Sketch a graph of $y = 1.5\sin 4\pi t$ for one cycle of compression and extension.

c Work out for how long in each cycle the moving end of the spring is more than 1 cm from its equilibrium position.

M 5 The depth of the water (h m) in a harbour t hours after midnight is given by the formula

$$h = 2.1 + 1.6\sin 30(t + 1)°$$

a Find the depth of the water at midnight.

b Show that there is a high tide at 2 a.m.

c Find the time of the first low tide after midnight.

d Sketch a graph to show how the depth of the water varies over 24 hours.

e A boat needs 1.5 m of water to enter the harbour.

Work out at what times during the day the boat can safely do this.

5.5 Small angle approximations

There are three lines on this graph: $y = \sin x$, $y = \tan x$ and $y = x$.

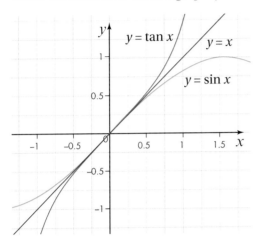

You can see that if x is small then these three lines are very close together.

Stop and think   Would all three lines still be close together if the x-axis was in degrees rather than radians?

What would the difference be?

Here is a table of values, given to 4 decimal places.

θ	0	0.05	0.07	0.10	0.15
$\sin \theta$	0	0.0500	0.0699	0.0998	0.1494
$\tan \theta$	0	0.0500	0.0701	0.1003	0.1511

The values are close even if θ is as large as 0.15 (about 9°).

Stop and think   Is the approximation still a good one if θ is a small negative value? How do you know?

When θ is small, $\theta \approx \sin \theta \approx \tan \theta$.

What about an approximation to $\cos x$?

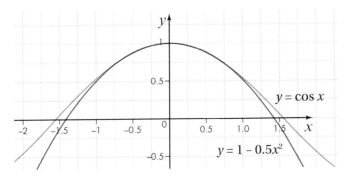

This graph shows that the quadratic curve $y = 1 - 0.5x^2$ is close to $y = \cos x$ if x is small.

When θ is small, $\cos\theta \approx 1 - \dfrac{\theta^2}{2}$.

> **KEY INFORMATION**
>
> If θ is small, $\sin\theta \approx \theta$,
>
> $\cos\theta \approx 1 - \dfrac{\theta^2}{2}$ and $\tan\theta \approx \theta$.

Example 6

Find an approximate value for $\dfrac{1 - \cos 2x}{\sin^2 3x}$ when x is small.

Solution

$\cos 2x \approx 1 - \dfrac{(2x)^2}{2} = 1 - \dfrac{4x^2}{2} = 1 - 2x^2$ and $\sin^2 3x \approx (3x)^2 = 9x^2$

So $\dfrac{1 - \cos 2x}{\sin^2 3x} \approx \dfrac{1 - (1 - 2x^2)}{9x^2} = \dfrac{2x^2}{9x^2} = \dfrac{2}{9}$

Exercise 5.5A

Answers page 442

1 **a** Work out the error in using 0.1 as an approximation to $\sin 0.1$.

 b Write the error as a percentage of the correct value.

(PF) **2** Show that if θ is small, then $1 - \theta^2$ is an approximation to $\cos^2\theta$.

(PS) **3** **a** Find an approximation to $\cos 2x$ when x is small.

(CM) **b** Find an approximation to $1 - 2\sin^2 x$ when x is small.

 c What conclusion can you draw from your answers to **parts a** and **b**?

4 Find an approximation to $\dfrac{1 - \cos x}{\sin^2 2x}$ when x is small.

(PF) **5** You know that $\sin^2\theta + \cos^2\theta = 1$ for all values of θ.

(CM) **a** Show that when you use small angle approximations for $\sin\theta$ and $\cos\theta$ then

$$\sin^2\theta + \cos^2\theta = 1 + \frac{1}{4}\theta^4$$

 b What can you say about the error in the formula in **part a** when $|\theta| < 0.5$?

(PS) **6** The hypotenuse of this right-angled triangle is 1 unit and one of the angles is x radians.

(PF)

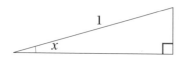

 a Show that if x is small, the area of the triangle is approximately $\dfrac{1}{2}x - \dfrac{1}{4}x^3$.

 b Calculate the error as a percentage of the correct value in using this formula if the angle is $5°$.

7 Find an approximation to $\sin 2x \cos x - \sin x \cos 2x$ when x is small.

8 Find an approximation to $\dfrac{\cos 4x - 1}{\tan 2x}$ when x is small.

5.6 Addition and subtraction formulae

Is there a connection between $\sin A$, $\sin B$ and $\sin (A + B)$?

Stop and think Try values for A and B to check that $\sin (A + B) \neq \sin A + \sin B$.

Look at the diagram on the right.

The length of OS is 1 and angle SOP is $A + B$.

From triangle OSP you should be able to see that

$$PS = OS \times \sin (A + B) = \sin (A + B) \qquad \textcircled{1}$$

and $\quad OP = OS \times \cos (A + B) = \cos (A + B) \qquad \textcircled{2}$

You now need to find other expressions for PS and OP.

From triangle ORS you can see that

$$SR = OS \times \sin A = \sin A$$

and $\quad OR = OS \times \cos A = \cos A$

These are marked on the diagram.

From triangle OQR you can see that

$$RQ = OR \times \sin B = \cos A \sin B \qquad \textcircled{3}$$

and $\quad OQ = OR \times \cos B = \cos A \cos B \qquad \textcircled{4}$

Now look at triangle STR, in which angle $RST = B$.

$$ST = SR \times \cos B = \sin A \cos B \qquad \textcircled{5}$$

$$RT = SR \times \sin B = \sin A \sin B \qquad \textcircled{6}$$

Finally, you can see from the diagram that $PT = RQ$ and $TR = PQ$.

So $\quad PS = ST + PT = ST + RQ$

Combine $\textcircled{1}$, $\textcircled{3}$ and $\textcircled{5}$.

$$\sin (A + B) = \sin A \cos B + \cos A \sin B \qquad \textcircled{7}$$

Also, $OP = OQ - PQ = OQ - TR$.

Combine $\textcircled{2}$, $\textcircled{4}$ and $\textcircled{6}$.

$$\cos (A + B) = \cos A \cos B - \sin A \sin B \qquad \textcircled{8}$$

The **addition formulae** $\textcircled{7}$ and $\textcircled{8}$ are important. They show how the sine and cosine of the sum of two angles are related to the sines and cosines of the separate angles.

They are valid whether the angles are measured in degrees or in radians.

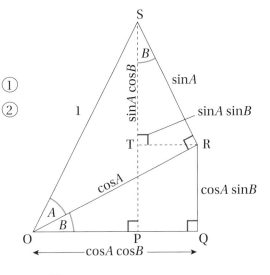

PROOF

The idea of this proof is to find different expressions for the sides of triangle SOP and equate them. The explanation shows how all the lengths marked on the diagram are derived.

Example 7

Find an *exact* expression for $\sin 75°$.

Solution

$\sin 75° = \sin (45° + 30°)$

Use the addition formula.

$$\sin 75° = \sin 45° \cos 30° + \cos 45° \sin 30°$$

$$= \frac{1}{\sqrt{2}} \times \frac{\sqrt{3}}{2} + \frac{1}{\sqrt{2}} \times \frac{1}{2}$$

$$= \frac{\sqrt{3} + 1}{\sqrt{2}}$$

You can derive other formulae from these two formulae.

$$\sin (A + B) = \sin A \cos B + \cos A \sin B$$

If you replace B with $-B$ you get

$$\sin (A - B) = \sin A \cos (-B) + \cos A \sin (-B)$$

But $\cos (-B) = \cos B$ and $\sin (-B) = -\sin B$, so

$$\sin (A - B) = \sin A \cos B - \cos A \sin B$$

In a similar way you can show that

$$\cos (A - B) = \cos A \cos B + \sin A \sin B$$

> Remember the symmetry of the sine and cosine graphs.

Stop and think   Derive the formula for $\cos (A - B)$ by replacing B with $-B$ in the formula for $\cos (A + B)$.

You can use these formulae to find formulae for $\tan (A + B)$ and $\tan (A - B)$.

$$\tan (A + B) = \frac{\sin (A + B)}{\cos (A + B)}$$

$$= \frac{\sin A \cos B + \cos A \sin B}{\cos A \cos B - \sin A \sin B}$$

Divide every term in the numerator and denominator by $\cos A \cos B$ and use the facts that

$$\frac{\sin A}{\cos A} = \tan A \quad \text{and} \quad \frac{\sin B}{\cos B} = \tan B$$

The formula simplifies to

$$\tan (A + B) = \frac{\tan A + \tan B}{1 - \tan A \tan B}$$

In a similar way you can prove that

$$\tan (A - B) = \frac{\tan A - \tan B}{1 + \tan A \tan B}$$

KEY INFORMATION

- $\sin (A \pm B) = \sin A \cos B \pm \cos A \sin B$

- $\cos (A \pm B) = \cos A \cos B \mp \sin A \sin B$

- $\tan (A \pm B) = \dfrac{\tan A \pm \tan B}{1 \mp \tan A \tan B}$

Be careful to use the correct signs.

These formulae are included in the formulae booklet.

> **Stop and think** Show how to deduce the formula for tan $(A - B)$ from the appropriate sine and cosine formulae.

Example 8

Simplify $\cos\left(A + \dfrac{\pi}{4}\right) - \cos\left(A - \dfrac{\pi}{4}\right)$.

Solution

Use the formulae for $\cos(A + B)$ and $\cos(A - B)$.

$$\cos\left(A + \frac{\pi}{4}\right) - \cos\left(A - \frac{\pi}{4}\right)$$

$$= \cos A \cos\frac{\pi}{4} - \sin A \sin\frac{\pi}{4} - \left(\cos A \cos\frac{\pi}{4} + \sin A \sin\frac{\pi}{4}\right)$$

$$= -2\sin A \sin\frac{\pi}{4}$$

$$= -2 \times \frac{1}{\sqrt{2}}\sin A$$

$$= -\sqrt{2}\sin A$$

Exercise 5.6A　　　　　　　　　　　　　　　　　　　　**Answers page 443**

1　Show that $\cos 75° = \dfrac{\sqrt{3} - 1}{2\sqrt{2}}$.

2　Find the exact value of $\tan 15°$.

3　Simplify $\sin 2A \cos 3A + \sin 3A \cos 2A$.

4　Find the exact value of $\cos 10° \cos 20° - \sin 10° \sin 20°$.

(PF) 5　Prove that $\sin A + \cos A = \sqrt{2}\sin\left(A + \dfrac{\pi}{4}\right)$.

6　$A + B = \dfrac{\pi}{2}$

　　Show that $\tan A \tan B = 1$.

(PF) 7　**a**　Prove that $\tan A + \tan B = \dfrac{\sin(A + B)}{\cos A \cos B}$.

　　b　Find a similar expression for $\tan A - \tan B$.

(PF) 8　**a**　Show that $\sin(A + B) + \sin(A - B) = 2\sin A \cos B$.

　　b　Hence show that $\sin x + \sin y = 2\sin\left(\dfrac{x + y}{2}\right)\cos\left(\dfrac{x - y}{2}\right)$.

(PF) 9　$A + B = \dfrac{\pi}{4}$

　　a　Prove that $\tan A = \dfrac{1 - \tan B}{1 + \tan B}$.

　　b　Find a similar formula for $\tan B$.

Double angle formulae

If you replace B with A in the addition formulae you get the following **double angle formulae**:

$$\sin 2A = 2 \sin A \cos A$$

$$\cos 2A = \cos^2 A - \sin^2 A$$

$$\tan 2A = \frac{2 \tan A}{1 - \tan^2 A}$$

> **Stop and think** Show how to derive each of these from the addition formulae.

The formula for $\cos 2A$ can be written in different ways.

Remember the formula derived from Pythagoras' theorem:

$$\sin^2 A + \cos^2 A = 1$$

If you rearrange this as $\sin^2 A = 1 - \cos^2 A$ and substitute into the formula for $\cos 2A$, you get

$$\cos 2A = \cos^2 A - (1 - \cos^2 A)$$

$$= 2 \cos^2 A - 1$$

Alternatively, substitute $\cos^2 A = 1 - \sin^2 A$ and get

$$\cos 2A = (1 - \sin^2 A) - \sin^2 A = 1 - 2 \sin^2 A$$

These formulae are examples of **identities** which are true for all values of the variable. We sometimes use the symbol $\equiv$ instead of $=$ to emphasise this. It means 'is identical to'. By contrast, in an equation the two expressions only have the same value for one or more particular values of the variable.

> **KEY INFORMATION**
>
> ❯ $\sin 2A = 2 \sin A \cos A$
>
> ❯ $\cos 2A = \cos^2 A - \sin^2 A$
> $\quad = 2\cos^2 A - 1 = 1 - 2\sin^2 A$
>
> ❯ $\tan 2A = \dfrac{2 \tan A}{1 - \tan^2 A}$
>
> You need to remember these formulae or be able to derive them from the addition and subtraction formulae.

Example 9

Prove that $\cos 3A \equiv 4\cos^3 A - 3\cos A$.

Solution

$$\cos 3A = \cos (2A + A)$$

$$= \cos 2A \cos A - \sin 2A \sin A$$

$$= (2\cos^2 A - 1)\cos A - 2\sin A \cos A \sin A$$

$$= 2\cos^3 A - \cos A - 2\cos A \sin^2 A$$

$$= 2\cos^3 A - \cos A - 2\cos A(1 - \cos^2 A)$$

$$= 2\cos^3 A - \cos A - 2\cos A + 2\cos^3 A$$

$$= 4\cos^3 A - 3\cos A$$

Writing it this way, you can use the addition formula.

Using the double angle formulae.

Using $\sin^2 A + \cos^2 A = 1$

Exercise 5.6B Answers page 443

(PF) 1 Show that $\frac{1}{2}\sin x = \sin\frac{x}{2}\cos\frac{x}{2}$.

(PF) 2 **a** Show that $\cos^2\theta = \frac{1}{2}(1+\cos 2\theta)$.

        b Show that $\sin^2\theta = \frac{1}{2}(1-\cos 2\theta)$.

(PF) 3 Show that $\cos x = \frac{1}{2}(\sin x \sin 2x + 2\cos^3 x)$.

(PF) 4 Show that $\cos 2\theta + 2\cos\theta + 1 = 2\cos\theta\,(\cos\theta + 1)$.

(PF) 5 **a** If $x = \frac{\pi}{8}$, show that $2\tan x = 1 - \tan^2 x$.

        b Solve the quadratic equation in **part a** to find the exact value of $\tan\frac{\pi}{8}$.

(PF) 6 Prove that $\sin 3A \equiv 3\sin A - 4\sin^3 A$.

5.7 Expressions of the form $a\cos\theta + b\sin\theta$

An expression of the form $a\cos\theta + b\sin\theta$ can be written in the form of a single sinusoidal wave.

For example, here is a graph of $y = 5\cos\theta + 2\sin\theta$:

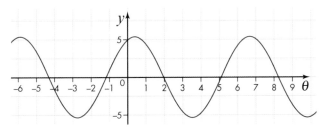

It looks like a sine wave.

Assume the equation can be written in the form $y = r\sin(\theta + \alpha)$, where r and α are numbers to be found.

$$5\cos\theta + 2\sin\theta \equiv r\sin(\theta + \alpha)$$

Use the addition formula.

> The sign $\equiv$ means that the two expressions are identical.

$$5\cos\theta + 2\sin\theta \equiv r(\sin\theta\cos\alpha + \cos\theta\sin\alpha)$$

$$5\cos\theta + 2\sin\theta \equiv r\sin\alpha\cos\theta + r\cos\alpha\sin\theta$$

If these are to be identical, then the coefficients of $\sin\theta$ and $\cos\theta$ must be the same.

So you want to find r and α so that

$$r\sin\alpha = 5 \quad \text{and} \quad r\cos\alpha = 2$$

Divide one by the other.

$$\frac{r\sin\alpha}{r\cos\alpha} = \frac{5}{2}$$

> The rs cancel and $\frac{\sin\alpha}{\cos\alpha} = \tan\alpha$

$$\tan\alpha = 2.5 \text{ so } \alpha = 1.19$$

To find r, square both equations and add:

$$r^2 \sin^2 \alpha + r^2 \cos^2 \alpha = 25 + 4$$

$$r^2 = 29 \quad \text{so} \quad r = \sqrt{29} = 5.385$$

Using $\sin^2 \alpha + \cos^2 \alpha = 1$

Put these values in the original identity.

$$5 \cos \theta + 2 \sin \theta \equiv 5.385 \sin (\theta + 1.19)$$

If you look at the graph, you can check that the amplitude is 5.385 and the graph is a translation of $y = 5.385 \sin \theta$ by $\begin{bmatrix} -1.19 \\ 0 \end{bmatrix}$.

You can generalise this method.

If $\quad a \cos \theta + b \sin \theta \equiv r \sin (\theta + \alpha)$

expand $\sin (\theta + \alpha)$ and equate coefficients.

$$\alpha = \tan^{-1} \frac{a}{b} \quad \text{and} \quad r = \sqrt{a^2 + b^2}$$

Example 10

Solve the equation $5 \cos \theta + 2 \sin \theta = 4$, $0 < \theta < 2\pi$.

Solution

First you must write $5 \cos \theta + 2 \sin \theta$ in the form $r \sin (\theta + \alpha)$.

This has already been done, so the equation becomes

$$5.385 \sin (\theta + 1.19) = 4$$

$$\sin (\theta + 1.19) = 0.7428$$

$$\sin^{-1} 0.7428 = 0.84$$

$$\text{so } \theta + 1.19 = 0.84 \text{ or } \pi - 0.84 \text{ or } 2\pi + 0.84 \text{ or } \ldots$$

So $\theta = -0.35$ or 1.11 or 5.94.

The first value is out of range so the two answers are $\theta = 1.11$ or 5.94.

You can use the graph to check that these look correct.

$r \sin (\theta + \alpha)$ is not the only way that $a \cos \theta + b \sin \theta$ can be rewritten.

$r \sin (\theta - \alpha)$, $r \cos (\theta + \alpha)$ and $r \cos (\theta - \alpha)$ could all be used instead.

Whichever form is used, follow the same procedure – expand the sum or difference and then equate coefficients.

Exercise 5.7A

Answers page 443

1 a Write $\cos \theta + \sqrt{3} \sin \theta$ in the form $r \sin (\theta + \alpha)$.

   b Sketch the graph of $y = \cos \theta + \sqrt{3} \sin \theta$, $0 \leqslant \theta \leqslant 2\pi$.

2 Write $3 \sin \theta + 4 \cos \theta$ in the form $r \sin (\theta + \alpha)$.

3 **a** Write $\cos\theta + \sin\theta$ in the form $r\sin(\theta + \alpha)$.

  b Sketch the graph of $y = \cos\theta + \sin\theta$, $0 \leqslant \theta \leqslant 2\pi$.

  c Write $\cos\theta - \sin\theta$ in the form $r\cos(\theta + \alpha)$.

  d Sketch the graph of $y = \cos\theta - \sin\theta$, $0 \leqslant \theta \leqslant 2\pi$.

4 Write $8\sin x - 6\cos x$ in the form $r\sin(x - \alpha)$.

5 **a** Write $12\sin\theta + 5\cos\theta$ in the form $r\sin(\theta + \alpha)$.

  b Find the amplitude of the graph of $y = 12\sin\theta + 5\cos\theta$.

  c A stretch from the x-axis followed by a translation will map the graph of $y = \sin\theta$ onto the graph of $y = 12\sin\theta + 5\cos\theta$.

  Find the vector to describe the translation.

  d Find the smallest positive solution of the equation $12\sin\theta + 5\cos\theta = 8$.

6 **a** Write $10\cos\theta - 12\sin\theta$ in the form $r\cos(\theta + \alpha)$.

  b Solve the equation $10\cos\theta - 12\sin\theta = 5$, $0 \leqslant \theta \leqslant 2\pi$.

(M) (PF) **7** A moving sculpture consists of two metal rods, OR and RP, of length 3 m and 4 m in a vertical plane. Angle ORP is a right angle.

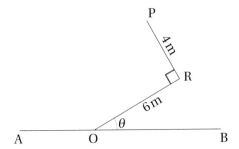

The rods can rotate about O. AOB is a straight line on the ground beneath the sculpture and angle ROB = θ radians.

  a Show that the height of P above the ground, in metres, is $4\cos\theta + 6\sin\theta$.

  b Find the maximum height of P above the ground.

  c Find the two possible values of θ for which P is 7 m above the ground.

(PF) **8** **a** Solve the equation $0.5\sin\theta + 0.4\cos\theta = 0.2$, $0 \leqslant \theta \leqslant 2\pi$.

  b Prove that the equation $0.5\sin\theta + 0.4\cos\theta = 0.7$ has no solution.

5.8 More trigonometric ratios

There are three more trigonometric ratios which can be useful. They are defined in terms of sine, cosine and tangent.

'Cosecant' can be abbreviated as cosec or as csc. Calculators and computer software often use the abbreviation csc.

The **secant** is defined as $\sec\theta = \dfrac{1}{\cos\theta}$.

The **cosecant** is defined as $\operatorname{cosec}\theta = \dfrac{1}{\sin\theta}$.

The **cotangent** is defined as $\cot\theta = \dfrac{1}{\tan\theta}$.

The graphs of the three functions (green), together with the graphs of the functions from which they are derived (red), are shown on the next page.

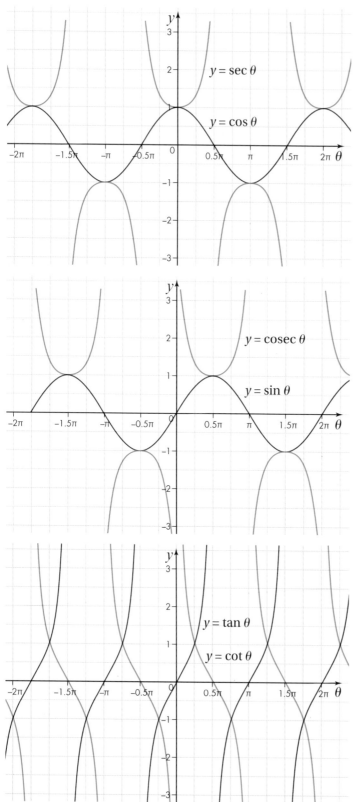

The new functions are not defined if $\sin\theta$, $\cos\theta$ or $\tan\theta = 0$, and values of θ where this is true are excluded from the domain. You can see that the graphs are **discontinuous** at these values.

Example 11

a Find the exact value of $\sec\frac{\pi}{6}$.

b Describe the domain of the function $f(x) = 1 + \sec x$.

c Describe the range of the function $f(x) = 1 + \sec x$.

Solution

a $\sec\frac{\pi}{6} = \dfrac{1}{\cos\frac{\pi}{6}} = \dfrac{1}{\frac{\sqrt{3}}{2}} = \dfrac{2}{\sqrt{3}}$

b Exclude values of x for which $\cos x = 0$.

That is, $x \neq \pm\dfrac{\pi}{2}, \pm\dfrac{3\pi}{2}, \pm\dfrac{5\pi}{2}$, etc.

c The range for $\cos x$ is $-1 \leqslant \cos x \leqslant 1$.

Since $\sec x = \dfrac{1}{\cos x}$ it follows that the range for $\sec\ x$ is $\sec x \geqslant 1$ and $\sec x \leqslant -1$.

The range for $1 + \sec x$ is $1 + \sec x \geqslant 2$ and $1 + \sec x \leqslant 0$.

Here is a familiar formula: $\sin^2\theta + \cos^2\theta = 1$.

Divide each term by $\cos^2\theta$.

$$\frac{\sin^2\theta}{\cos^2\theta} + \frac{\cos^2\theta}{\cos^2\theta} = \frac{1}{\cos^2\theta}$$

$$\left(\frac{\sin\theta}{\cos\theta}\right)^2 + 1 = \left(\frac{1}{\cos\theta}\right)^2$$

$$\tan^2\theta + 1 = \sec^2\theta \qquad \text{①}$$

Alternatively, you could divide each term by $\sin^2\theta$.

$$\frac{\sin^2\theta}{\sin^2\theta} + \frac{\cos^2\theta}{\sin^2\theta} = \frac{1}{\sin^2\theta}$$

$$1 + \cot^2\theta = \operatorname{cosec}^2\theta \qquad \text{②}$$

Equations ① and ② can be useful in rearranging expressions.

> **KEY INFORMATION**
> - $\sec^2\theta = 1 + \tan^2\theta$
> - $\operatorname{cosec}^2\theta = 1 + \cot^2\theta$

Example 12

Solve the equation $\tan x + \sec^2 x = 1$ $\qquad -\pi \leqslant x \leqslant \pi$.

Solution

Substitute $\tan^2 x + 1$ for $\sec^2 x$.

$$\tan x + \tan^2 x + 1 = 1$$

$$\tan x + \tan^2 x = 0$$

Factorise.

$$\tan x \, (1 + \tan x) = 0$$

Either $\tan x = 0$ or $1 + \tan x = 0$.

If $\tan x = 0$, $x = 0$, $-\pi$ or π.

If $1 + \tan x = 0$ then $\tan x = -1$ and $x = \dfrac{3\pi}{4}$ or $-\dfrac{\pi}{4}$.

There are five solutions in the given range: $x = 0$, $-\pi$, π, $\dfrac{3\pi}{4}$ or $-\dfrac{\pi}{4}$.

Exercise 5.8A

Answers page 444

1 Find the exact value of the following:

     a $\sec\dfrac{\pi}{4}$ **b** $\cot\dfrac{\pi}{6}$ **c** $\csc\dfrac{\pi}{3}$ **d** $\csc\dfrac{2\pi}{3}$

2 Copy this table and insert the exact values.

θ	$\sin\theta$	$\cos\theta$	$\tan\theta$	$\sec\theta$	$\csc\theta$	$\cot\theta$
$\dfrac{\pi}{6}$			$\dfrac{1}{\sqrt{3}}$			
$\dfrac{\pi}{3}$			$\sqrt{3}$			

3   **a** Solve the equation $\csc x = 4$, $0 \leqslant x \leqslant 2\pi$.

     b Solve the equation $\sec x = -3$, $0 \leqslant x \leqslant 2\pi$.

4 Find the exact value of:

     a $\sec 60°$ **b** $\csc 120°$ **c** $\cot 210°$

5   **a** Sketch the graph of the function $f(x) = \csc 2x$, $0 \leqslant x \leqslant 2\pi$.

     b Describe the range of the function.

6 Describe the range of each of these functions:

     a $f(x) = 0.5\sec x$ **b** $f(x) = 0.5 + \sec x$ **c** $f(x) = \sec(x + 0.5)$

(PF) 7 Prove that $\tan\theta + \cot\theta \equiv \sec\theta \csc\theta$.

(PF) 8 Prove that $\csc 2x = \dfrac{1}{2}\csc x \sec x$.

(PF) 9 Prove that $\cot 2x = \dfrac{\cot^2 x - 1}{2\cot x}$.

10 Solve the equation $2\sin x + 1 = \csc x$, $0 \leqslant x \leqslant 2\pi$.

     Give your solutions as multiples of π.

5.9 Inverse trigonometric functions

Suppose $y = \sin x$.

The inverse function is the **arcsine**, $f(x) = \arcsin x$.

The graph of $y = \arcsin x$ looks like this:

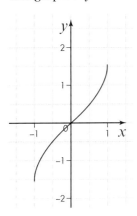

This is sometimes written as $\sin^{-1} x$, particularly on calculators.

Be careful: $\sin^{-1} x$ is *not* the same as $(\sin x)^{-1}$ or $\dfrac{1}{\sin x}$.

The domain for this graph is $-1 \leqslant x \leqslant 1$ because the sine function can only take a value between 1 and -1.

The range for the function is $-\dfrac{\pi}{2} \leqslant \arcsin x \leqslant \dfrac{\pi}{2}$ and the line does not go beyond these y values. These are called the **principal values** of the function.

A calculator will always tell you this value.

This diagram shows $y = \arcsin x$, $y = \sin x$ and $y = x$ on the same axes.

If the angle is measured in degrees, then the range is $-90° \leqslant \arcsin x \leqslant 90°$ and the scale on the y-axis will be changed accordingly. The shape of the graph will be the same.

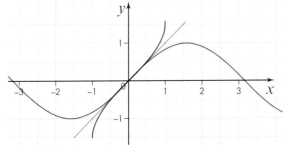

You can see that $y = \arcsin x$ is a reflection of part of $y = \sin x$ in the diagonal line $y = x$.

Here is a graph of the **arccosine** function, $f(x) = \arccos x$:

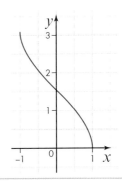

Stop and think   What are the domain and range for this graph?

Here is a graph of the **arctangent** function $f(x) = \arctan x$:

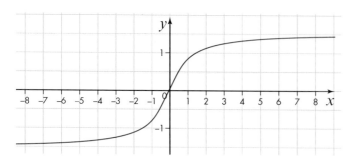

This graph is different from the previous ones because the domain is the set of all real numbers, the whole number line.

> **Stop and think**
>
> Why is the domain for $f(x) = \arctan x$ any real number?
>
> What is the range for $f(x) = \arctan x$?

Exercise 5.9A

Answers page 444

1 **a** Write down the domain of the function $f(x) = \arccos x$.

    b Find the domain of the function $g(x) = \arccos 2x$.

    c Find the domain of the function $h(x) = \arccos (x - 5)$.

2 **a** Sketch the graph of the function $f(x) = 3\arcsin 2x$.

    b Write down the range of the function.

3 Find the point of intersection of the curves $y = \arcsin x$ and $y = \arccos x$.

PS **CM** **4** **a** Draw the graph of $y = \arcsin x + \arccos x$.

    b Explain the shape of the graph in **part a**.

5 Solve the equation $\arcsin x^2 = 0.5$.

6 **a** Find the domain of the function $f(x) = \arcsin \dfrac{1}{x}$, $x \neq 0$.

    b Find the range of the function.

SUMMARY OF KEY POINTS

❯ π radians = 180°

❯ Arc length of a sector, $l = r\theta$ and area, $A = \dfrac{1}{2}r^2\theta$

❯ Small angle approximations: $\sin\theta \approx \tan\theta \approx \theta \qquad \cos\theta \approx 1 - \dfrac{\theta^2}{2}$

❯ $\sin(A \pm B) = \sin A \cos B \pm \cos A \sin B$

❯ $\cos(A \pm B) = \cos A \cos B \mp \sin A \sin B$

❯ $\tan(A \pm B) = \dfrac{\tan A \pm \tan B}{1 \mp \tan A \tan B}$

❯ $\sin 2A = 2\sin A \cos A$

❯ $\cos 2A = \cos^2 A - \sin^2 A$ or $2\cos^2 A - 1$ or $1 - 2\sin^2 A$

❯ $\tan 2A = \dfrac{2\tan A}{1 - \tan^2 A}$

❯ $a\cos\theta + b\sin\theta$ can be written in the form $r\sin(\theta \pm \alpha)$ or $r\cos(\theta \pm \alpha)$.

❯ $\sec\theta = \dfrac{1}{\cos\theta} \qquad \operatorname{cosec}\theta = \dfrac{1}{\sin\theta} \qquad \cot\theta = \dfrac{1}{\tan\theta}$

❯ $\tan^2\theta + 1 = \sec^2\theta$

❯ $1 + \cot^2\theta = \operatorname{cosec}^2\theta$

EXAM-STYLE QUESTIONS 5

Answers page 445

1 Find the exact value of 210° in radians. **[1 mark]**

2 The diagram shows the dimensions of a sector of a circle.

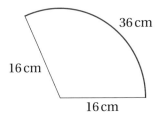

Calculate the area of the sector. **[3 marks]**

3 Show that $\tan 75° = \dfrac{\sqrt{3}+1}{\sqrt{3}-1}$. **[3 marks]**

(PF) 4 a Prove that $\sin\theta = 2\cos\dfrac{\theta}{2}\sin\dfrac{\theta}{2}$. **[2 marks]**

b Hence prove that $\sin\theta = 4\cos\dfrac{\theta}{2}\cos\dfrac{\theta}{4}\sin\dfrac{\theta}{4}$. **[2 marks]**

5 a Sketch the graph of $y = 2 + \arccos 2x$. **[3 marks]**

b Find the domain and range of $f(x) = 2 + \arccos 2x$. **[2 marks]**

6 Look at this question and solution.

Solve the equation $2\sin^2\theta = \sin\theta$ $0 \le \theta \le \pi$.

The following working for solving the equation is incorrect.

    Divide through by $\sin\theta$. $2\sin\theta = 1$

             Divide by 2. $\sin\theta = 0.5$

$$\theta = \frac{\pi}{6}$$

Write out a correct solution.

 7 A particle is oscillating between two points, A and B.

The distance, x metres, from the midpoint, M, after t seconds is given by the formula

$x = 5\sin 0.25\pi t$

a Find the length of the line AB.   **[1 mark]**

b Find the time the particle takes to go from A to B and then back to A.   **[2 marks]**

c Work out the first time and the second time that the particle is half way between M and A.   **[4 marks]**

 8 Prove that $\dfrac{\cos 2\theta}{\cos^2\theta} + \tan^2\theta = 1$ **[5 marks]**

9 The equilateral triangle and the sector have the same area.

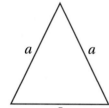 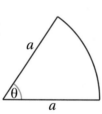

Show that the perimeter of the sector is approximately 4.5% less than the perimeter of the triangle.   **[4 marks]**

 10 Solve this equation, giving your answer in degrees.

$\cos^4\theta = 2\sin^2\theta\cos^2\theta$, $0° \le \theta \le 360°$.   **[3 marks]**

11 θ is an acute angle and $\cos\theta = \dfrac{4}{5}$.

Show that $\cos\left(\theta + \dfrac{\pi}{4}\right) = \dfrac{\sqrt{2}}{10}$.   **[6 marks]**

12 $10\sin\theta + 14\cos\theta \equiv r\sin(\theta + \alpha)$

a Work out the values of r and α.   **[5 marks]**

b Solve the equation $10\sin\theta + 14\cos\theta = 15$, $0 \le \theta \le 2\pi$.   **[4 marks]**

13 **a** Prove that $\cos^2 2A = 4\cos^4 A - 4\cos^2 A + 1$.   **[4 marks]**

b Prove that $\cos 4A = 8\cos^4 A - 8\cos^2 A + 1$.   **[3 marks]**

14 **a** Solve the equation $\tan x + \cot x = 5$, $0 \le x \le \pi$.   **[2 marks]**

b Find the values of k for which the equation $\tan x + \cot x = k$ has a solution.   **[2 marks]**

15 Solve the equation $6\sin^2 x + \cos x = 5$, where x is in radians and $0 \le x \le 2\pi$.   **[5 marks]**

6 DIFFERENTIATION

A business involved in making and selling particular products will have different types of costs. Fixed costs include things such as the rent of business premises and the cost of machinery and vehicles. Marginal costs include the cost of the raw materials used in manufacturing a product or the fuel used to produce it.

Profit will vary with the number of products sold and that in turn is affected by many things, particularly the selling price and the profit margin on each item.

Economists use mathematical models and equations to analyse the connections between these variables, and this can help to make sure that the company makes a profit and is able to plan for expansion.

This graph shows how the long-run average cost (LRAC) varies with the quantity of a product that is produced.

By differentiating the equation for this curve you can determine how varying the quantity produced will affect costs.

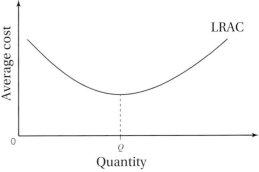

The minimum point (where the gradient is zero) shows the quantity Q that will give the lowest possible average cost.

LEARNING OBJECTIVES

You will learn how to:

- ❯ identify all the types of stationary point on a curve
- ❯ use the chain rule to differentiate a range of functions
- ❯ differentiate exponential and logarithmic functions
- ❯ differentiate sine and cosine functions.

TOPIC LINKS

For this chapter you should be able to differentiate linear combinations of powers of a variable. From **Book 1, Chapter 7 Exponentials and logarithms** you should be familiar with exponential functions and their graphs, including those involving the constant e, and be able to manipulate functions involving logarithms to base e. From **Chapter 5 Trigonometry** you should know about using radians to measure angles and small angle approximations to sines and cosines.

PRIOR KNOWLEDGE

You should already know how to:

› identify maximum and minimum points on a graph

› work out any power of a variable, including fractional and negative indices

› sketch graphs of $y = e^{kx}$ and $y = \ln x$

› use radian measure for angles

› use small angle approximations for $\sin \theta$ and $\cos \theta$.

You should be able to complete the following questions correctly:

1 Find the stationary points of the graph of $y = x^3 - 6x^2$ and identify whether they are maximum or minimum points.

2 Show that if α is small then $\cos 2\alpha \approx 1 - 2\alpha^2$.

3 Differentiate with respect to x:

   a $2x(x^2 - 1)$ **b** $\dfrac{x^2 - 1}{2x}$ **c** $\dfrac{1}{\sqrt{x}}$

6.1 Turning points

This is a graph of $y = x^5 - 15x^3$

At a **stationary point**, the gradient of the curve is zero; that is, $\dfrac{dy}{dx} = 0$.

To find the stationary points, differentiate the equation.

$$\frac{dy}{dx} = 5x^4 - 45x^2$$

When $\dfrac{dy}{dx} = 0$, $5x^4 - 45x^2 = 0$.

Divide by 5 and factorise.

$$x^4 - 9x^2 = 0$$

$$x^2(x^2 - 9) = 0$$

$$x^2(x - 3)(x + 3) = 0 \text{ so } x = 0, 3 \text{ or } -3$$

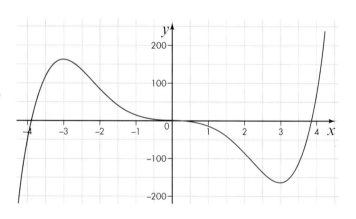

You can see on the graph that at these points the gradient is 0.

These stationary points are $(0, 0)$, $(3, -162)$ and $(-3, 162)$.

$(-3, 162)$ is a **maximum point**. The y-coordinate at this point is less than the y-coordinates at nearby points. The gradient of the curve changes sign from positive to negative as you go past that point.

$(3, -162)$ is a **minimum point**. The y-coordinate at this point is less than the y-coordinates at nearby points. At this point, the gradient changes from negative to positive.

$(0, 0)$ is a **point of inflection**. The gradient at neighbouring points on either side is negative.

> A section of a curve that has increasing gradient (positive second derivative) is known as concave upwards, and a section of a curve that has decreasing gradient (negative second derivative) is known as concave downwards.

An alternative to judging the type of stationary point by eye is to use the **second derivative**.

$$\frac{d^2y}{dx^2} = 20x^3 - 90x$$

When $x = -3$, $\frac{d^2y}{dx^2} = -270 < 0$, which confirms this is a maximum point.

When $x = 3$, $\frac{d^2y}{dx^2} = 270 > 0$, which confirms this is a minimum point.

When $x = 0$, $\frac{d^2y}{dx^2} = 0$. At a point of inflection the second derivative is always 0. However, this is not conclusive and the point could be a maximum, a minimum or a point of inflection. Look at the value of the gradient on either side of the point to decide which it is.

There are two types of stationary points that are points of inflection, illustrated by A and B on the upper graph.

The lower graph shows the curve $y = x^4$.

The origin is clearly a minimum point but $\frac{d^2y}{dx^2} = 12x^2$, which is 0 when $x = 0$.

Here is a summary:

> At the point of inflection, the curve changes from concave upwards to concave downwards (or vice versa), so the second derivative at this point is zero.

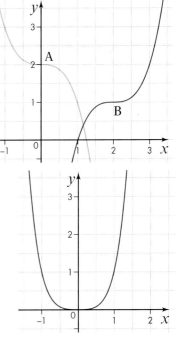

Value of $\dfrac{d^2y}{dx^2}$	negative	positive	zero
Type of stationary point	maximum point	minimum point	cannot tell

Exercise 6.1A

Answers page 446

1. Find the stationary point of the graph of $y = 16 + 10x - x^2$ and state, with a reason, whether it is a maximum or a minimum point.

2. **a** Show that the graph of $y = x^3 - 3x^2 + 3x - 20$ has just one stationary point.

   b Find its coordinates and show that it is a point of inflection.

3. Find and describe the stationary points for the graph of $f(x) = x^4 - 2x^2$.

4. **a** Show that the graph of $y = x^3$ has a point of inflection at the origin.

   b Describe any stationary points for the curve $y = (x - 10)^3 + 20$. Justify your answer.

(PF) 5. $y = ax^2 + bx + c$, $a \neq 0$, is a curve.

   Show that it must have a stationary point, which can be a maximum or a minimum but not a point of inflection.

(PS) (PF) 6. The sides of a cuboid are x cm, x cm and $(9 - 2x)$ cm.

   a Show that the total length of all the edges is 36 cm.

   b Show that the volume is a maximum when the shape is a cube.

   c Show that the total surface area is also a maximum when the shape is a cube.

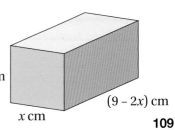

x cm

$(9 - 2x)$ cm

x cm

6.2 The chain rule

If you need to differentiate a function such as $y = (2x + 3)^3$ and find $\dfrac{dy}{dx}$, one way is to multiply out the bracket and then differentiate each term. That will take a lot of work. Is there a better way?

You can break the function into two parts by writing $y = u^3$ where $u = 2x + 3$.

Both of these are easy to differentiate: $\dfrac{dy}{du} = 3u^2$ and $\dfrac{du}{dx} = 2$.

How does this help? Remember that to differentiate from first principles you increase the variable by a small amount.

Suppose δx is a small increase in x – then both u and y will change. There will be a small change, δu, in u, and a small change, δy, in y.

The derivatives are $\dfrac{dy}{dx} = \lim\limits_{\delta x \to 0} \dfrac{\delta y}{\delta x}$, $\dfrac{dy}{du} = \lim\limits_{\delta u \to 0} \dfrac{\delta y}{\delta u}$ and $\dfrac{du}{dx} = \lim\limits_{\delta x \to 0} \dfrac{\delta u}{\delta x}$

Now we can write $\dfrac{\delta y}{\delta u} = \dfrac{\delta y}{\delta u} \times \dfrac{\delta u}{\delta x}$

Then $\qquad \lim\limits_{\delta x \to 0} \dfrac{\delta y}{\delta x} = \lim\limits_{\delta x \to 0} \dfrac{\delta y}{\delta u} \times \dfrac{\delta u}{\delta x} = \lim\limits_{\delta u \to 0} \dfrac{\delta y}{\delta u} \times \lim\limits_{\delta x \to 0} \dfrac{\delta u}{\delta x}$

So $\qquad\qquad \dfrac{dy}{dx} = \dfrac{dy}{du} \times \dfrac{du}{dx}$

> This is just the ordinary rule for multiplying fractions, since δx, δu and δy are just numbers and the δu cancels.

Using this formula in our example:

$$\dfrac{dy}{du} = 3u^2 \quad \text{and} \quad \dfrac{du}{dx} = 2 \quad \text{so} \quad \dfrac{dy}{dx} = 3u^2 \times 2 = 6u^2$$

The final step is to replace u with $2x + 3$, which gives

$$\dfrac{dy}{dx} = 6(2x + 3)^2$$

With practice you can use this method without actually writing down u.

Think of it is as $y = (\text{bracket})^3$ where the bracket stands for $2x + 3$.

Think of it as:	Write it as:
$y = (\text{bracket})^3$	$y = (2x + 3)^3$
$\dfrac{dy}{dx} = \dfrac{dy}{d(\text{bracket})} \times \dfrac{d(\text{bracket})}{dx}$	$\dfrac{dy}{dx} = 3(2x + 3)^2 \times 2$
	$= 6(2x + 3)^2$

This method is called the **chain rule** and it greatly increases the number of functions that you can differentiate.

> **KEY INFORMATION**
>
> If $y = f(u)$ and $u = g(x)$ then
> $$\dfrac{dy}{dx} = \dfrac{dy}{du} \times \dfrac{du}{dx}$$

Example 1

Differentiate $y = \dfrac{4}{x^2 + 1}$.

Solution

Method 1

Write it as $y = 4(x^2 + 1)^{-1} = 4u^{-1}$ where $u = x^2 + 1$.

Then $\dfrac{dy}{du} = 4 \times -1 \times u^{-2} = -\dfrac{4}{u^2}$ and $\dfrac{du}{dx} = 2x$.

Multiply these.

$$\frac{dy}{dx} = -\frac{4}{u^2} \times 2x = -\frac{8x}{u^2} = -\frac{8x}{(x^2+1)^2}$$

Method 2

If you do not use u explicitly, think of it as $y = 4(\text{bracket})^{-1}$

so that $\dfrac{dy}{dx} = 4 \times -1 \times (\text{bracket})^{-2} \times \dfrac{d(\text{bracket})}{dx}$

and write $\dfrac{dy}{dx} = -4(x^2+1)^{-2} \times 2x = \dfrac{-8x}{(x^2+1)^2}$ as before.

Use whichever method you are more comfortable with.

You will probably want to introduce u at first and you should find you can do without it as you become more confident.

Stop and think   Would you have been able to differentiate the function in **Example 1** if you had not known about the chain rule?

Exercise 6.2A

Answers page 446

1 Find $\dfrac{dy}{dx}$ in the following cases:

  a $y = (4x+2)^2$ **b** $y = (4x+2)^3$ **c** $y = (4x+2)^5$

2 Differentiate with respect to x:

  a $(8-x)^{10}$ **b** $(1+3x^2)^{10}$ **c** $(6x-3x^2)^{10}$

3 Find $f'(x)$ in the following cases:

  a $f(x) = \sqrt{x-3}$ **b** $f(x) = \sqrt{2x-3}$ **c** $f(x) = \sqrt{x^2-3}$

4 The equation of this curve is $y = \dfrac{20}{2+x}$, $x > 0$.

  a Find the gradient at $(2, 5)$.

  b Find the coordinates of the point where the gradient is -0.2.

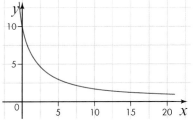

5 The equation of this curve is $y = \dfrac{600}{x^2+50}$.

  a Show that the point $(10, 4)$ is on the curve.

  b Find $\dfrac{dy}{dx}$.

  c Find the equation of the tangent at $(10, 4)$.

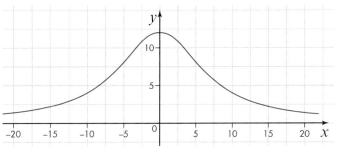

6.3 Differentiating e^{kx}

Here is a proof that if $y = e^x$ then $\dfrac{dy}{dx} = e^x$.

Suppose x increases by a small amount, δx, and y changes by
$\delta y = e^{x+\delta x} - e^x$.

$$\frac{\delta y}{\delta x} = \frac{e^{x+\delta x} - e^x}{\delta x} = \frac{e^x(e^{\delta x} - 1)}{\delta x}$$

$$\frac{dy}{dx} = \lim_{\delta x \to 0} \frac{\delta y}{\delta x} = \lim_{\delta x \to 0} \frac{e^x(e^{\delta x} - 1)}{\delta x} = e^x \lim_{\delta x \to 0} \frac{e^{\delta x} - 1}{\delta x}$$

So what is $\lim\limits_{\delta x \to 0} \dfrac{e^{\delta x} - 1}{\delta x}$?

Here is a table of values, rounded to 4 decimal places:

δx	0.1	0.01	0.001
$\dfrac{e^{\delta x} - 1}{\delta x}$	1.0517	1.0050	1.0005

> **PROOF**
> Factorising the expression for $\dfrac{\delta y}{\delta x}$ leads you to consider a simpler limit.

It is reasonable to assume that $\lim\limits_{\delta x \to 0} \dfrac{e^{\delta x} - 1}{\delta x} = 1$.

Therefore $\dfrac{dy}{dx} = e^{\delta x} \lim\limits_{\delta x \to 0} \dfrac{e^{\delta x} - 1}{\delta x} = e^x$

This means that for any point on the graph of $y = e^x$, the gradient is just the y-coordinate.

If $y = e^{kx}$ where k is a constant, then you can find $\dfrac{dy}{dx}$ by using the chain rule.

If $u = kx$ and $y = e^u$ then

$$\frac{dy}{dx} = \frac{dy}{du} \times \frac{du}{dx} = e^u \times k = ke^{kx}$$

> **KEY INFORMATION**
> If $y = e^x$ then $\dfrac{dy}{dx} = e^x$, or if $f(x) = e^x$ then $f'(x) = e^x$.

> **KEY INFORMATION**
> If $y = e^{kx}$ then $\dfrac{dy}{dx} = ke^{kx}$, or if $f(x) = e^{kx}$ then $f'(x) = ke^{kx}$.

If you multiply the function by a constant, the derivative is multiplied by the constant in the usual way.

If $y = ce^{kx}$, then $\dfrac{dy}{dx} = cke^{kx}$.

Example 2

Find the gradient of the curve $y = 10e^{-0.3x}$ at the point where $x = 5$.

Solution

If $y = 10e^{-0.3x}$ then $\dfrac{dy}{dx} = -0.3 \times 10e^{-0.3x} = -3e^{-0.3x}$.

When $x = 5$, $\dfrac{dy}{dx} = -3e^{-0.3 \times 5} = -3e^{-1.5} = -0.669$ to 3 d.p.

Sometimes exponential functions are written in the form a^{kx}, where a is not e but a different number. **Example 3** shows you how to differentiate such an expression.

Example 3

The population of a city, y millions, in x years' time is predicted to be given by the formula

$$y = 5 \times 1.02^x$$

Find the expected rate of increase in 4 years' time.

Solution

You need to find the value of $\dfrac{dy}{dx}$ when $x = 4$.

If $1.02 = e^a$ then $a = \ln 1.02$ and so $1.02 = e^{\ln 1.02}$.

That means $y = 5 \times (e^{\ln 1.02})^x = 5e^{(\ln 1.02)x}$

Therefore $\dfrac{dy}{dx} = 5(\ln 1.02)e^{(\ln 1.02)x} = 5(\ln 1.02)1.02^x$

When $x = 4$, $\dfrac{dy}{dx} = 5(\ln 1.02)1.02^4 = 0.107$ to 3 d.p.

The expected rate of increase is 107 000 to 3 s.f.

This uses the usual rule for combining indices.

The units are millions per year.

Here is the general method.

If you want to differentiate $= a^{kx}$, use the fact that $a = e^{\ln a}$ and write the expression as a power of e.

If $\qquad y = a^{kx} = (e^{\ln a})^{kx} = e^{(\ln a)kx}$

then $\qquad \dfrac{dy}{dx} = k(\ln a)e^{(\ln a)kx} = k(\ln a)a^{kx}$

KEY INFORMATION

If $y = x^{kx}$ then $\dfrac{dy}{dx} = k(\ln a)\, a^{kx}$.

Exercise 6.3A

Answers page 446

1 Find $\dfrac{dy}{dx}$ in the following cases:

  a $y = e^{2x}$ **b** $y = e^{-x}$ **c** $y = e^{0.4x}$ **d** $y = e^{4x+2}$

2 Work out $f'(x)$ in the following cases:

  a $f(x) = 4e^{0.5x}$ **b** $f(x) = 100e^{-0.1x}$ **c** $f(x) = 50e^{2x-10}$

3 **a** Write 1.5^x in the form e^{kx} where k is a constant.

  b If $y = 1.5^x$, find $\dfrac{dy}{dx}$.

4 $f(x) = 4e^{2x}$

  Find:

  a $f'(x)$ **b** $f''(x)$

5 The equation of this curve is $y = 4e^{-0.5x^2}$.

  a Find $\dfrac{dy}{dx}$.

  b Find the gradient at the point where the x-coordinate is -2.

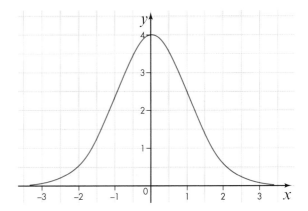

(M) **6** A saver invests £5000 at an annual rate of 3% compound interest.

    a Show that after t years the value is £5000 × 1.03^t.

    b Write 5000 × 1.03^t in the form 5000e^{at}.

    c Find the rate at which the savings are increasing when $t = 3$.

(PS) **7** $f(x) = e^{2x} + e^{-2x}$

    Show that $f''(x) = 4f(x)$.

(PS) **8** The equation of a curve is $y = e^x + 4e^{-x}$.

    Show that the curve has a minimum point and find its coordinates.

6.4 Differentiating in ax

On the right is a graph of $y = \ln x$.

On the same axes is a graph of $y = e^x$.

Because finding e to a power and finding ln are inverse operations, the graph of $y = \ln x$ is a reflection of the graph of $y = e^x$ in the line $y = x$.

Suppose you want to find the gradient of the curve $y = \ln x$ at the point A.

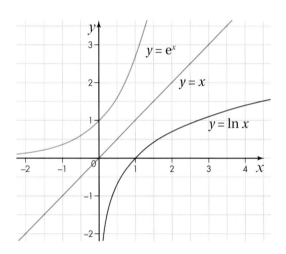

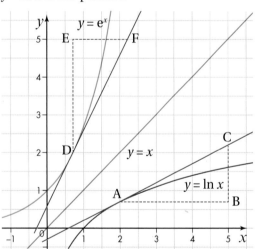

The tangent at A has been drawn; its gradient is $\dfrac{CB}{AB}$.

DEF is a reflection of ABC in the line $y = x$ so $\dfrac{CB}{AB} = \dfrac{FE}{DE}$.

But $\dfrac{FE}{DE} = \dfrac{1}{\frac{DE}{FE}} = \dfrac{1}{y\text{-coordinate of D}}$ because $\dfrac{DE}{FE}$ is the gradient of $y = e^x$ at the point D, which is just e^x or the y-coordinate.

Because of the reflection, the y-coordinate of D is the x-coordinate of A.

This gives the result that if $y = \ln x$ then $\dfrac{dy}{dx} = \dfrac{1}{x}$.

> **KEY INFORMATION**
>
> If $y = \ln x$ then $\dfrac{dy}{dx} = \dfrac{1}{x}$

Example 4

Differentiate: **a** $\ln(2x+3)$ **b** $\ln x^2$

Solution

a Use the chain rule:

If $y = \ln(2x+3)$ then $\dfrac{dy}{dx} = \dfrac{1}{2x+3} \times 2 = \dfrac{2}{2x+3}$

b Either use the chain rule:

If $y = \ln x^2$ then $\dfrac{dy}{dx} = \dfrac{1}{x^2} \times 2x = \dfrac{2x}{x^2} = \dfrac{2}{x}$

Or use the properties of logarithms:

$\ln \quad x^2 = 2\ln x$

So $\dfrac{dy}{dx} = 2 \times \dfrac{1}{x} = \dfrac{2}{x}$, which is the same result.

Using the fact that $\ln a^k = k \ln a$.

Example 5

Find the equation of the tangent to the curve $y = \ln(x+e)$ at the point (0, 1).

Solution

Use the chain rule to find $\dfrac{dy}{dx}$.

$$y = \ln u \quad \text{and} \quad u = x + e$$

$$\frac{dy}{du} = \frac{1}{u} \quad \text{and} \quad \frac{du}{dx} = 1$$

$$\frac{dy}{dx} = \frac{dy}{du} \times \frac{du}{dx} = \frac{1}{u} \times 1 = \frac{1}{x+e}$$

At (0, 1), $x = 0$ so $\dfrac{dy}{dx} = \dfrac{1}{e}$

The equation of the tangent is $y - 1 = \dfrac{1}{e}(x - 0)$

$$y = \frac{x}{e} + 1$$

Exercise 6.4A

Answers page 447

1 Differentiate with respect to x:

   a $\ln 3x$　　　**b** $\ln x^3$　　　**c** $\ln(x^3 + 2)$

2 Find $\dfrac{dy}{dx}$ in the following cases:

   a $y = 4\ln x$　　　**b** $y = \ln 4x$　　　**c** $y = \ln x^4$

3 **a** On the same axes, sketch the graphs of $y = \ln x$ and $y = \ln 2x$.

   b Find the vector for the translation that maps the graph of $y = \ln x$ onto $y = \ln 2x$.

   c $f(x) = \ln kx$ where k is a positive constant, $x > 0$.

      Show that $f'(x) = \dfrac{1}{x}$.

   d How are your answers to **parts b** and **c** related?

4 This is a graph of the curve $y = \ln(x + 5)$:

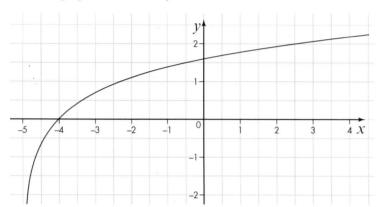

a Find the equation of the tangent to the curve at the point where it crosses the x-axis.

b Find the equation of the tangent to the curve at the point where it crosses the y-axis.

(PS) **5** a Show that the point $(e, 1)$ is on the graph of the curve $y = \ln x$.

b Show that the tangent to the curve $y = \ln x$ at $(e, 1)$ passes through the origin.

c Show that the normal to the curve $y = \ln x$ at $(e, 1)$ crosses the x-axis at $e + \dfrac{1}{e}$.

(PF) **6** a Find $\dfrac{dy}{dx}$ when $y = \ln\dfrac{10}{x}$.

b Find $\dfrac{dy}{dx}$ when $y = \ln\dfrac{10}{x^2}$.

c Generalise your results of **parts a** and **b** to find $\dfrac{dy}{dx}$ when $y = \ln\dfrac{10}{x^n}$ where n is a positive integer.

(PS) **7** $y = \log_{10} x$

Show that $\dfrac{dy}{dx} = \dfrac{1}{x \ln 10}$.

6.5 Differentiating sin x and cos x from first principles

In this section you will find the derivatives of $\sin x$ and $\cos x$.

Suppose $y = \sin x$.

If δx is a small increase in the value of x and δy is the corresponding change in y, then

$$\frac{dy}{dx} = \lim_{\delta x \to 0} \frac{\delta y}{\delta x}$$

$$\delta y = \sin(x + \delta x) - \sin x$$

so $\dfrac{\delta y}{\delta x} = \dfrac{\sin(x + \delta x) - \sin x}{\delta x}$

Use the **addition formula**, $\sin(x + \delta x) = \sin x \cos \delta x + \cos x \sin \delta x$.

$$\frac{\delta y}{\delta x} = \frac{\sin x \cos \delta x + \cos x \sin \delta x - \sin x}{\delta x}$$

See **Chapter 5 Trigonometry** for more on the addition formulae.

Factorise the numerator.

$$\frac{\delta y}{\delta x} = \frac{\cos x \sin \delta x + \sin x (\cos \delta x - 1)}{\delta x}$$

Write the expression as two separate fractions.

$$\frac{\delta y}{\delta x} = \cos x \left(\frac{\sin \delta x}{\delta x} \right) + \sin x \left(\frac{\cos \delta x - 1}{\delta x} \right)$$

$\frac{dy}{dx}$ is the limit of this as $\delta x \to 0$.

$$\frac{dy}{dx} = \lim_{\delta x \to 0} \frac{\delta y}{\delta x} = \lim_{\delta x \to 0} \left\{ \cos x \left(\frac{\sin \delta x}{\delta x} \right) + \sin x \left(\frac{\cos \delta x - 1}{\delta x} \right) \right\}$$

$$= \cos x \lim_{\delta x \to 0} \frac{\sin \delta x}{\delta x} + \sin x \lim_{\delta x \to 0} \frac{\cos \delta x - 1}{\delta x}$$

From **Chapter 5 Trigonometry** you should remember that if θ is small then $\sin \theta \approx \theta$ and $\cos \theta \approx 1 - \frac{1}{2}\theta^2$.

Use these to give the approximations $\sin \delta x \approx \delta x$ and $\cos \delta x \approx 1 - \frac{1}{2}(\delta x)^2$.

$$\frac{\sin \delta x}{\delta x} \approx 1 \quad \text{and} \quad \frac{\cos \delta x - 1}{\delta x} \approx \frac{1}{2}\delta x$$

$$\lim_{\delta x \to 0} \frac{\sin \delta x}{\delta x} = 1 \quad \text{and} \quad \lim_{\delta x \to 0} \frac{\cos \delta x - 1}{\delta x} \approx \lim_{\delta x \to 0} \frac{1}{2}\delta x = 0$$

Hence the result is $\frac{dy}{dx} = \cos x$.

This gives the very simple result that if $y = \sin x$ then $\frac{dy}{dx} = \cos x$.

Stop and think   Would this result still be true if the angle was in degrees rather than in radians?

You can differentiate $y = \cos x$ from first principles in a similar way.

$$\frac{\delta y}{\delta x} = \frac{\cos(x + \delta x) - \cos x}{\delta x}$$

$$= \frac{\cos x \cos \delta x - \sin x \sin \delta x - \cos x}{\delta x}$$

$$= \frac{\cos x (\cos \delta x - 1) - \sin x \sin \delta x}{\delta x}$$

$$\frac{dy}{dx} = \cos x \lim_{\delta x \to 0} \frac{\cos \delta x - 1}{\delta x} - \sin x \lim_{\delta x \to 0} \frac{\sin \delta x}{\delta x}$$

Which means that $\frac{dy}{dx} = -\sin x$.

So if $y = \cos x$, $\frac{dy}{dx} = -\sin x$.

KEY INFORMATION

› If $y = \sin x$, then $\frac{dy}{dx} = \cos x$.

› If $y = \cos x$, then $\frac{dy}{dx} = -\sin x$.

Example 6

Differentiate with respect to x:

a $2\sin 4x$ **b** $\cos 3x^2$

Solution

Use the chain rule in both cases.

a $\dfrac{dy}{dx} = 2\cos 4x \times 4 = 8\cos 4x$

b $\dfrac{dy}{dx} = -\sin 3x^2 \times 6x = -6x\sin 3x^2$

Exercise 6.5A

Answers page 447

1 Find $\dfrac{dy}{dx}$ in the following cases:

    a $y = \sin 2x$ **b** $y = \cos(5x - 2)$ **c** $y = \sin(x^2 + 1)$

2 Differentiate with respect to x:

    a $10\sin 0.5x$ **b** $\sin 3x + \cos 6x$ **c** $\cos(x^2 - 3x - 4)$

3 $f(x) = \sin 4x + \cos 4x$

    Show that $f''(x) + 16f(x) = 0$

(PF) **4** **a** Show that the derivative of $\sin^2 x$ is $\sin 2x$.

    b Find the derivative of $\cos^2 x$.

    c Explain the relationship between your answers to **parts a** and **b**.

5 This is a graph of $y = e^{\sin x}$.

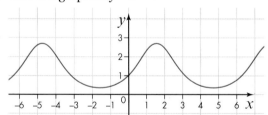

    a Find the gradient at the point $(0, 1)$.

    b Prove that the graph has a stationary point at $\left(\dfrac{\pi}{2}, e\right)$.

(M) **6** A bob on the end of a long string is a pendulum making small oscillations.

    The displacement, y metres, of the bob after t seconds is given by $y = 0.1\sin 2.4t$.

    a Find the speed of the bob when $t = 1$.

    b Find the position of the bob when the speed is zero.

    c Find the position of the bob when the acceleration is zero.

SUMMARY OF KEY POINTS

› If f'(x) = 0 and f''(x) > 0 then the curve has a minimum point. If f'(x) = 0 and f''(x) < 0 then the curve has a maximum point.

› The chain rule: if $y = f(u)$ and $u = g(x)$ then $\dfrac{dy}{dx} = \dfrac{dy}{du} \times \dfrac{du}{dx}$.

› If $y = e^{kx}$ then $\dfrac{dy}{dx} = ke^{kx}$.

› If $y = \ln x$ then $\dfrac{dy}{dx} = \dfrac{1}{x}$.

› If $y = \sin x$ then $\dfrac{dy}{dx} = \cos x$; if $y = \cos x$ then $\dfrac{dy}{dx} = -\sin x$.

EXAM-STYLE QUESTIONS 6

Answers page 448

1 The equation of a curve is $y = (x - 10)^3 + 15$.

  a Find $\dfrac{dy}{dx}$ and $\dfrac{d^2y}{dx^2}$.   **[3 marks]**

  b Show that the curve has a point of inflection and find its coordinates.   **[3 marks]**

2 $y = \sqrt{1 + x^2}$

  Show that $\dfrac{dy}{dx} = \dfrac{x}{y}$.   **[4 marks]**

3 $f(x) = \dfrac{1}{\cos x}$

  Show that $f'(x) = \dfrac{\tan x}{\cos x}$.   **[4 marks]**

4 Differentiate with respect to x:

  a $y = e^{-2x}$ **[2 marks]**

  b $y = e^{-x^2}$ **[2 marks]**

5 The equation of a curve is $y = \ln x^2$, $x > 0$.

  Find the coordinates of the point on the curve where the gradient is 0.5.   **[5 marks]**

CM **6** Here are attempts by two students to differentiate $\sin(x + 2)$:

Student A	Student B
$y = \sin(x + 2)$	$y = \sin(x + 2)$
$\quad = \sin x + \sin 2$	Use the addition formula.
So	$y = \sin x \cos 2 + \cos x \sin 2$
$\dfrac{dy}{dx} = \cos x + 0$	So
$\quad = \cos x$	$\dfrac{dy}{dx} = \cos x \cos 2 - \sin x \sin 2$

Is either student correct? Give a reason for your answer.   **[4 marks]**

(M) **7** The area covered by a colony of bacteria on a flat surface is $10\,\text{mm}^2$.

The area $y\,\text{mm}^2$, in t hours' time is modelled by the formula $y = 10 \times 2^{1.5t}$.

   a Show that the area will double in size in 40 minutes. **[2 marks]**

   b Find the area in 3 hours' time. **[2 marks]**

   c Find the rate at which the area will be increasing in 3 hours' time. **[6 marks]**

   d Why might the model no longer be valid after several hours? **[1 mark]**

8 The equation of this curve is $y = 0.1e^x + e^{-0.5x}$.

The curve has a minimum point. Find its coordinates. **[7 marks]**

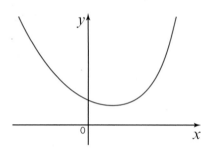

9 The equation of a curve is $y = 2^{0.5x + 3}$.

Find the equation of the tangent to the curve where it crosses the y-axis. **[4 marks]**

(CM) **10** The equation of a curve is $y = x^3 + 3x^2 + 4x + 5$.

   a Show that the curve has no stationary points. **[2 marks]**

   b The curve has a point of inflection.

   Find the equation of the tangent at this point. **[2 marks]**

11 The equation of a curve is $y = ae^{kx}$ where a and k are constants.

The point $(10, 20)$ is on the curve and the tangent at that point passes through $(0, -5)$ on the y-axis.

Find the values of a and k. **[4 marks]**

(PF) **12** $y = \sin^3 x$

   a Show that $\dfrac{dy}{dx} = 3\cos x - 3\cos^3 x$. **[4 marks]**

   b Show that $\dfrac{d^2y}{dx^2} = 6\sin x - 9\sin^3 x$. **[4 marks]**

13 The equation of a curve is $y = 10\cos 5(x - 20)°$.

Find the maximum and minimum values of the gradient of this curve. **[4 marks]**

14 The equation of this curve is $y = e^{-\frac{1}{2}x^2}$

The point P has x-coordinate a.

Show that the tangent at P crosses the x-axis at $(a + \dfrac{1}{a}, 0)$. **[4 marks]**

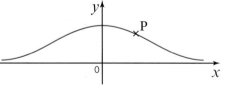

(PF) **15** $y = \sin 2x$

Prove from first principles that $\dfrac{dy}{dx} = 2\cos 2x$. **[5 marks]**

7 FURTHER DIFFERENTIATION

The diagram shows a mass oscillating on the end of a vibrating spring. To make it clearer, the positions of the mass at different times have been shown side by side, although of course the top of the spring is fixed and does not move.

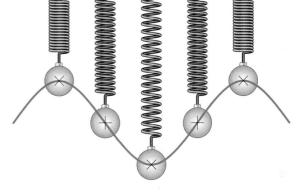

If the mass is not moving up or down, it is at an equilibrium point. When it moves, it is subject to a force towards the equilibrium point. The force is proportional to the displacement from that point.

The force produces an acceleration. If the displacement is x, then the acceleration is $\dfrac{d^2x}{dt^2}$ and you can write $\dfrac{d^2x}{dt^2} = -kx$, where the minus sign shows that the acceleration is in the opposite direction to the displacement, and k is a constant that depends upon the particular circumstances.

This is an example of a differential equation. It is called that because it has a derivative as part of it.

The solution of a differential equation is not a number but a function. In this case, it is a function of x in terms of t.

Differential equations are very common in all branches of science and engineering and in other areas of study too, such as economics. They are the natural mathematical way to describe variables that are changing continuously over time.

This particular example is called simple harmonic motion. You can check that a solution for this equation is $x = \sin \sqrt{k}\, t$. (Differentiate it twice.) This is a sine wave, as the mass moves up and down.

LEARNING OBJECTIVES

You will learn how to:

› differentiate products and quotients of functions

› differentiate trigonometric functions

› find a derivative when a curve is defined parametrically

› find a derivative when a curve is defined implicitly

› construct differential equations.

TOPIC LINKS

For this chapter you should be able to differentiate simple exponential and trigonometric expressions, as seen in **Chapter 6 Differentiation**. You should be able to manipulate trigonometric expressions (**Chapter 5 Trigonometry**) and other algebraic expressions. You should be able to sketch a curve defined parametrically, as seen in **Chapter 3 Coordinate geometry: Parametric equations**.

PRIOR KNOWLEDGE

You should already know how to:

> differentiate $\sin kx$, $\cos kx$ and e^{kx}

> find points on a curve defined parametrically

> use trigonometric identities such as the double angle formulae.

You should be able to complete the following questions correctly:

1 Find $\dfrac{dy}{dx}$ in the following cases:

   a $y = e^{2x} - e^{-2x}$

   b $y = 3\sin 4x + 5\cos 2x$

   c $y = \ln 2x$

2 A curve is defined parametrically by $x = t^2$ and $y = 2t$.

   Find where the curve crosses the line $x = 9$.

3 Show that $\dfrac{\tan^2 A}{\sec A + 1} + 1 = \sec A$.

7.1 The product rule

You know how to differentiate the functions $e^{-0.2x}$ and $\sin 4x$. Suppose you multiply them together to get $e^{-0.2x}\sin 4x$. How can you differentiate a product of two different functions?

Suppose $y = uv$, where u and v are both functions of x, and that δx is a small increase in x that produces changes in y, u and v of δy, δu and δv respectively.

This means that y changes to $y + \delta y$, u changes to $u + \delta u$ and v changes to $v + \delta v$.

$$\frac{dy}{dx} = \lim_{\delta x \to 0} \frac{\delta y}{\delta x}$$

$$\frac{\delta y}{\delta x} = \frac{(u + \delta u)(v + \delta v) - uv}{\delta x}$$

$$= \frac{uv + u\delta v + v\delta u + \delta u\delta v - uv}{\delta x}$$

$$= \frac{u\delta v + v\delta u + \delta u\delta v}{\delta x}$$

$$= u\frac{\delta v}{\delta x} + v\frac{\delta u}{\delta x} + \delta u\frac{\delta v}{\delta x}$$

$$\frac{dy}{dx} = \lim_{\delta x \to 0} \frac{\delta y}{\delta x}$$

$$= \lim_{\delta x \to 0} u\frac{\delta v}{\delta x} + \lim_{\delta x \to 0} v\frac{\delta u}{\delta x} + \lim_{\delta x \to 0} \delta u\frac{\delta v}{\delta x}t$$

The last term is $\displaystyle\lim_{\delta x \to 0} \delta u\frac{\delta v}{\delta x} = \lim_{\delta x \to 0} \delta u \times \lim_{\delta x \to 0} \frac{\delta v}{\delta x} = 0 \times \frac{dv}{dx} = 0.$

So $\qquad \displaystyle\frac{dy}{dx} = \lim_{\delta x \to 0} u\frac{\delta v}{\delta x} + \lim_{\delta x \to 0} v\frac{\delta u}{\delta x}$

$$\frac{dy}{dx} = u\frac{dv}{dx} + v\frac{du}{dx}$$

This is the **product rule** for differentiation.

> **KEY INFORMATION**
>
> If $y = uv$ where u and v are functions of x, then
> $$\frac{dy}{dx} = u\frac{dv}{dx} + v\frac{du}{dx}.$$

Example 1

This graph shows a damped sine wave. The amplitude is gradually decreasing.

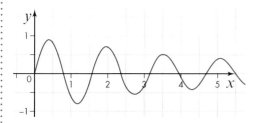

The equation of the curve is $y = e^{-0.2x}\sin 4x$, $x > 0$.

a Find $\dfrac{dy}{dx}$.

b Find the x-coordinate of the first stationary point.

Solution

a $\quad u = e^{-0.2x} \qquad$ and $\quad v = \sin 4x$

$\quad \dfrac{du}{dx} = -0.2e^{-0.2x} \quad$ and $\quad \dfrac{dv}{dx} = 4\cos 4x$

$\qquad \dfrac{dy}{dx} = u\dfrac{dv}{dx} + v\dfrac{du}{dx}$

$\qquad\qquad = e^{-0.2x} \times 4\cos 4x + \sin 4x \times -0.2e^{-0.2x}$

$\qquad\qquad = e^{-0.2x}(4\cos 4x - 0.2\sin 4x)$

b At a stationary point, $\dfrac{dy}{dx} = 0$.

$\qquad\qquad e^{-0.2x}(4\cos 4x - 0.2\sin 4x) = 0$

$\qquad\qquad\qquad 4\cos 4x - 0.2\sin 4x = 0$

You can cancel $e^{-0.2x}$ because it is always positive and cannot be zero.

Rearrange.

$$0.2\sin 4x = 4\cos 4x$$

Divide by 0.2.

$$\sin 4x = 20\cos 4x$$

Divide by $\cos 4x$.

$$\tan 4x = 20$$

$$\frac{\sin 4x}{\cos 4x} = \tan 4x$$

$$4x = \tan^{-1} 20$$

$$= 1.521$$

$$x = 0.380$$

The x-coordinate of the first stationary point is 0.380.

Exercise 7.1A

Answers page 449

1 Find $\dfrac{dy}{dx}$ in the following cases:

    a $y = x \sin x$ **b** $y = x \cos x$ **c** $y = x^2 \sin 2x$

2 Differentiate with respect to x:

    a $x e^x$ **b** $(x + 1) e^{-x}$ **c** $x^2 e^{2x}$

(PS) 3 $y = \sin x \cos x$

    a Find $\dfrac{dy}{dx}$ by using the product rule.

    b Find $\dfrac{dy}{dx}$ by using an identity for $\sin 2x$.

    c Show that your answers to **parts a** and **b** are identical.

4 $y = e^x \sin x$

    a Work out $\dfrac{dy}{dx}$.   **b** Work out $\dfrac{d^2 y}{dx^2}$.

5 The equation of a curve is $y = x \ln x$, $x > 0$.

    a Find $\dfrac{dy}{dx}$.   **b** Find the gradient at $(1, 0)$.

    c Find the coordinates of the point where the gradient is 2.

    d Find the coordinates of the stationary point.

(PS) 6 **a** The equation of a curve is $y = x e^{-x}$.

    Show that the curve has one stationary point and find its coordinates.

    b The equation of a second curve is $y = x^2 e^{-x}$.

    Show that this curve has two stationary points and find their coordinates.

7 This is a sketch of the graph of $y = 10x^2 e^{-x^2}$.

    Find the coordinates of the stationary points.

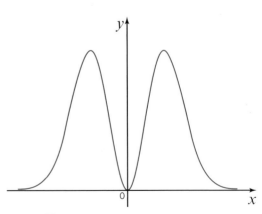

(M) Modelling   (PS) Problem solving   (PF) Proof   (CM) Communicating mathematically

PS **8** The equation of a curve is $y = e^{-0.2x} \sin 2x$, $x \geqslant 0$.

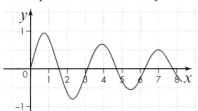

Show that the curve has stationary points at $x = \dfrac{n\pi}{2} + 0.736$ where n is 0, 1, 2, 3, ...

M **9** This graph shows the speed of a car ($y\,\text{m s}^{-1}$) over a 10-second interval.

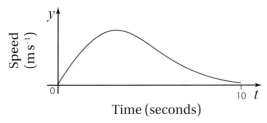

The speed at time t s is given by the formula $y = 15t\,e^{-0.05t^2}$, $0 \leqslant t \leqslant 10$.

Find the maximum speed of the car.

7.2 The quotient rule

You now know how to differentiate the product of two functions of x.

What if y is the *quotient* of two functions of x, that is $y = \dfrac{u}{v}$?

Use the same notation as in **Section 7.1**.

$$\delta y = \frac{u + \delta u}{v + \delta v} - \frac{u}{v}$$

Write this as a single fraction.

$$\delta y = \frac{(u + \delta u)v - u(v + \delta v)}{(v + \delta v)v}$$

$$= \frac{uv + v\delta u - uv - u\delta v}{v^2 + v\delta v}$$

$$= \frac{v\delta u - u\delta v}{v^2 + v\delta v}$$

$$\frac{\delta y}{\delta x} = \frac{v\dfrac{\delta u}{\delta x} - u\dfrac{\delta v}{\delta x}}{v^2 + v\delta v}$$

$$\frac{dy}{dx} = \lim_{\delta x \to 0} \frac{\delta y}{\delta x} = \frac{\lim\limits_{\delta x \to 0} v\dfrac{\delta u}{\delta x} - \lim\limits_{\delta x \to 0} u\dfrac{\delta v}{\delta x}}{\lim\limits_{\delta x \to 0}(v^2 + v\delta v)}$$

$$\frac{dy}{dx} = \frac{v\dfrac{du}{dx} - u\dfrac{dv}{dx}}{v^2}$$

This is the **quotient rule** for differentiation.

> **KEY INFORMATION**
>
> If $y = \dfrac{u}{v}$ then $\dfrac{dy}{dx} = \dfrac{v\dfrac{du}{dx} - u\dfrac{dv}{dx}}{v^2}$.

Example 2

Find $\dfrac{dy}{dx}$ when $y = \dfrac{1-x^2}{1+x^2}$.

Solution

$$u = 1 - x^2 \quad \text{and} \quad v = 1 + x^2$$

$$\frac{du}{dx} = -2x \quad \text{and} \quad \frac{dv}{dx} = 2x$$

$$\frac{dy}{dx} = \frac{v\dfrac{du}{dx} - u\dfrac{dv}{dx}}{v^2}$$

$$= \frac{(1+x^2) \times (-2x) - (1-x^2) \times 2x}{(1+x^2)^2}$$

$$= \frac{-2x - 2x^3 - 2x + 2x^3}{(1+x^2)^2}$$

$$= \frac{-4x}{(1+x^2)^2}$$

Be very careful with the signs. It is easy to make a mistake.

Exercise 7.2A

Answers page 449

1 Find $\dfrac{dy}{dx}$ in the following cases:

a $y = \dfrac{x}{x+1}$ b $y = \dfrac{x+1}{x^2+1}$ c $y = \dfrac{x^3}{2x-1}$

2 Differentiate with respect to x:

a $\dfrac{\sin x}{x}$ b $\dfrac{x+1}{\cos x}$ c $\dfrac{x^2}{\sin 2x}$

PS **3** $y = e^{-2x}\sin x$

a Use the product rule to find $\dfrac{dy}{dx}$.

b Show that you get the same answer if you write $y = \dfrac{\sin x}{e^{2x}}$ and use the quotient rule.

4 $y = \dfrac{\ln x}{x}, \qquad x > 0$

a Find $\dfrac{dy}{dx}$.   b Find $\dfrac{d^2y}{dx^2}$.

c Find the coordinates of the stationary point of the graph of $y = \dfrac{\ln x}{x}$ when $x > 0$, and show that it is a maximum point.

5 $y = \dfrac{1 - e^{-x}}{1 + e^{-x}}$

a Find $\dfrac{dy}{dx}$.   b Find the gradient of the graph of y at the origin.

c Show that the graph of y has no stationary points.

6 a Show that the derivative of $\dfrac{x}{\sqrt{x^2+1}}$ is $\dfrac{1}{\sqrt[3]{x^2+1}}$.   b Differentiate $\dfrac{\sqrt{x^2+1}}{x}$.

7 $y = \tan x$

Show that $\dfrac{dy}{dx} = \dfrac{1}{\cos^2 x}$.

7.3 Differentiating trigonometric functions

First, some useful notation:

$\frac{d}{dx}(x^3 - 2x)$ means 'the derivative of $x^3 - 2x$ with respect to x'.

So $\frac{d}{dx}(x^3 - 2x) = 3x^2 - 2$ and $\frac{d}{dx}(e^{-0.5x}) = -0.5e^{-0.5x}$

You know how to differentiate the sine and cosine functions.

$\frac{d}{dx}(\sin x) = \cos x$ and $\frac{d}{dx}(\cos x) = -\sin x$

What about other trigonometric functions, such as the tangent?

Differentiating tan x

$\tan x = \frac{\sin x}{\cos x}$ so you can differentiate this with the quotient rule.

$$u = \sin x \quad \text{and} \quad v = \cos x$$

$$\frac{du}{dx} = \cos x \quad \text{and} \quad \frac{dv}{dx} = -\sin x$$

$$\frac{dy}{dx} = \frac{v\frac{du}{dx} - u\frac{dv}{dx}}{v^2}$$

$$= \frac{\cos x \times \cos x - \sin x \times (-\sin x)}{\cos^2 x}$$

$$= \frac{\cos^2 x + \sin^2 x}{\cos^2 x}$$

$$= \frac{1}{\cos^2 x}$$

$\cos^2 x + \sin^2 x = 1$

But you already know that $\sec x = \frac{1}{\cos x}$.

Therefore $\frac{d}{dx}(\tan x) = \sec^2 x$.

Differentiating cot x, sec x and cosec x

In a similar way you can find $\frac{d}{dx}(\cot x)$.

$$\cot x = \frac{\cos x}{\sin x}$$

$$\frac{d}{dx}(\cot x) = \frac{\sin x \times (-\sin x) - \cos x \times \cos x}{\sin^2 x}$$

$$= \frac{-\sin^2 x - \cos^2 x}{\sin^2 x}$$

$$= \frac{-1}{\sin^2 x}$$

Therefore $\frac{d}{dx}(\cot x) = -\mathrm{cosec}^2 x$.

To find $\dfrac{d}{dx}(\sec x)$, write it as $\sec x = \dfrac{1}{\cos x} = (\cos x)^{-1}$ and use the chain rule.

$\sec x = \dfrac{1}{\cos x}$

$$\dfrac{d}{dx}(\sec x) = \dfrac{d}{dx}\left((\cos x)^{-1}\right)$$

$$= -1 \times (\cos x)^{-2} \times (-\sin x)$$

$$= \dfrac{\sin x}{\cos^2 x}$$

$$= \dfrac{1}{\cos x} \times \dfrac{\sin x}{\cos x}$$

Therefore $\dfrac{d}{dx}(\sec x) = \sec x \tan x$.

Similarly, $\dfrac{d}{dx}(\operatorname{cosec} x) = \dfrac{d}{dx}\left((\sin x)^{-1}\right)$

$$= -1 \times (\sin x)^{-2} \times (\cos x)$$

$$= -\dfrac{\cos x}{\sin^2 x}$$

$$= -\dfrac{1}{\sin x} \times \dfrac{\cos x}{\sin x}$$

Therefore $\dfrac{d}{dx}\left(\operatorname{cosec} x\right) = -\operatorname{cosec} x \cot x$.

Differentiating inverse trigonometric functions

Finally, what about the inverse trigonometric functions?

If $y = \arcsin x$ you can rewrite that as $x = \sin y$.

If you think of x, as a function of y, then $\dfrac{dx}{dy} = \cos y$.

What is the connection between $\dfrac{dy}{dx}$ and $\dfrac{dx}{dy}$?

$$\dfrac{dy}{dx} = \lim_{\delta x \to 0} \dfrac{\delta y}{\delta x} \qquad \text{and} \qquad \dfrac{dx}{dy} = \lim_{\delta y \to 0} \dfrac{\delta x}{\delta y}$$

But $\quad \dfrac{\delta x}{\delta y} \times \dfrac{\delta y}{\delta x} = 1 \qquad$ so $\qquad \dfrac{\delta x}{\delta y} = \dfrac{1}{\dfrac{\delta y}{\delta x}}$

If a and b are non-zero numbers, then $\dfrac{a}{b} \times \dfrac{b}{a} = 1$.

and so $\dfrac{dx}{dy}$ is just the reciprocal of $\dfrac{dy}{dx}$.

So if $y = \arcsin x$ and $x = \sin y$, then $\dfrac{dy}{dx} = \dfrac{1}{\cos y}$.

However, you want the answer in terms of x, not y.

Use the fact that $\cos^2 y + \sin^2 y = 1$.

$$\cos^2 y + x^2 = 1$$

Rearrange $\qquad\qquad \cos^2 y = 1 - x^2$

$$\cos y = \pm\sqrt{1 - x^2}$$

Therefore $\dfrac{d}{dx}(\arcsin x)$ is either $\dfrac{1}{\sqrt{1-x^2}}$ or $-\dfrac{1}{\sqrt{1-x^2}}$.

Here is the graph of $y = \arcsin x$.

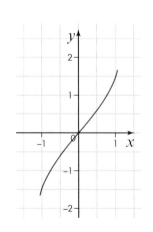

The gradient is always positive, so take the positive square root.

Therefore $\dfrac{d}{dx}(\arcsin x) = \dfrac{1}{\sqrt{1-x^2}}$.

There is a similar result for $\arccos x$.

If $y = \arccos x$, then $x = \cos y$ and $\dfrac{dy}{dx} = -\dfrac{1}{\sin y}$.

Use the fact that $\cos^2 y + \sin^2 y = 1$.

In this case,

$$x^2 + \sin^2 y = 1$$

Rearrange. $\sin^2 y = 1 - x^2$

$$\sin x = \pm\sqrt{1-x^2}$$

Therefore $\dfrac{d}{dx}(\arccos x)$ is either $\dfrac{1}{\sqrt{1-x^2}}$ or $-\dfrac{1}{\sqrt{1-x^2}}$.

Here is the graph of $y = \arccos x$.

The gradient is always negative, so take the negative square root.

Therefore $\dfrac{d}{dx}(\arccos x) = -\dfrac{1}{\sqrt{1-x^2}}$.

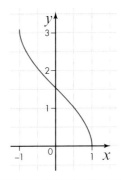

Stop and think You can see that the results for $\arcsin x$ and $\arccos x$ are very similar.

What does this tell you about the graphs of $y = \arcsin x$ and $y = \arccos x$?

The final result in this section is $\dfrac{d}{dx}(\arctan x)$.

If $y = \arctan x$, then $x = \tan y$.

$$\dfrac{dx}{dy} = \sec^2 y$$

$$\dfrac{dy}{dx} = \dfrac{1}{\sec^2 y}$$

Use the fact that $\sec^2 y = 1 + \tan^2 y$.

$$\sec^2 y = 1 + x^2$$

Therefore $\dfrac{d}{dx}(\arctan x) = \dfrac{1}{1+x^2}$.

Stop and think Here is a reminder of the graph of $y = \arctan x$.

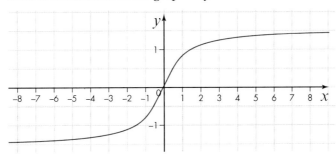

Does the result give values that match the shape of the graph?

What is the maximum gradient?

Example 3

Differentiate with respect to x: **a** $\sec 4x$ **b** $\sec^2 4x$

Solution

a Use the chain rule and the fact that the derivative of $\sec \theta$ is $\sec \theta \tan \theta$.

$$\frac{d}{dx}(\sec 4x) = (\sec 4x \tan 4x) \times 4$$

$$= 4\sec 4x \tan 4x$$

b Use the chain rule and the result from **part a**.

$$\frac{d}{dx}(\sec^2 4x) = 2\sec 4x \times 4\sec 4x \tan 4x$$

$$= 8\sec^2 4x \tan 4x$$

Exercise 7.3A

Answers page 450

1 Differentiate with respect to x:

   a $\tan 3x$ **b** $\cot 3x$ **c** $\tan(2x+3)$

2 Find $\dfrac{dy}{dx}$ in the following cases:

   a $y = \sec 2x$ **b** $y = \sec^2 x$ **c** $y = \sec^2 2x$

3 Differentiate with respect to x:

   a $\arcsin 2x$ **b** $\arcsin(x^2)$ **c** $\arctan \frac{1}{2} x$

4 Differentiate with respect to x:

   a $x \cot x$ **b** $x^2 \operatorname{cosec} x$ **c** $\dfrac{\sec x}{x}$

5 The equation of a curve is $y = \sec x$.

   Find the equation of the tangent to the curve at the point where $x = \frac{\pi}{3}$.

6 $y = \sec x$

   Work out $\dfrac{d^2 y}{dx^2}$.

(PS) 7 $y = \arctan 2x$

   Find the coordinates of the points where the gradient is 1.

(PS) 8 **a** Differentiate $\tan^2 x$.   **b** Differentiate $\sec^2 x$.

   c Explain the connection between your answers to **parts a** and **b**.

7.4 Differentiating parametric equations

For curves that are described parametrically, x and y are both written as functions of a parameter, t. You can use these to write $\dfrac{dy}{dx}$ as a function of the parameter.

If δx is a small change in x, there will be corresponding small changes in δy and δt.

KEY INFORMATION

Here is a summary of the results of this section:

> $\dfrac{d}{dx}(\tan x) = \sec^2 x$

> $\dfrac{d}{dx}(\cot x) = -\operatorname{cosec}^2 x$

> $\dfrac{d}{dx}(\sec x) = \sec x \tan x$

> $\dfrac{d}{dx}(\operatorname{cosec} x) = -\operatorname{cosec} x \cot x$

> $\dfrac{d}{dx}(\arcsin x) = \dfrac{1}{\sqrt{1 - x^2}}$

> $\dfrac{d}{dx}(\arccos x) = -\dfrac{1}{\sqrt{1 - x^2}}$

> $\dfrac{d}{dx}(\arctan x) = \dfrac{1}{1 + x^2}$

You should try to memorise these. Look for patterns that will help you to remember them.

$$\frac{\mathrm{d}y}{\mathrm{d}x} = \lim_{\delta x \to 0} \frac{\delta y}{\delta x}$$

$$\frac{\delta y}{\delta x} = \frac{\delta y}{\delta t} \div \frac{\delta x}{\delta t}$$

$$\lim_{\delta x \to 0} \frac{\delta y}{\delta x} = \lim_{\delta x \to 0} \frac{\delta y}{\delta t} \div \lim_{\delta x \to 0} \frac{\delta x}{\delta t}$$

But as $\delta x \to 0$, then $\delta t \to 0$.

So $\quad \lim_{\delta t \to 0} \frac{\delta y}{\delta x} = \lim_{\delta t \to 0} \frac{\delta y}{\delta t} \div \lim_{\delta t \to 0} \frac{\delta x}{\delta t}$

$$\frac{\mathrm{d}y}{\mathrm{d}x} = \frac{\mathrm{d}y}{\mathrm{d}t} \div \frac{\mathrm{d}x}{\mathrm{d}t}$$

This follows from the arithmetic rule that dividing by a fraction is the same as multiplying by the reciprocal.
$$\frac{a}{c} \div \frac{b}{c} = \frac{a}{c} \times \frac{c}{b} = \frac{a}{b}$$

KEY INFORMATION

If x and y are defined in terms of a parameter t, then
$$\frac{\mathrm{d}y}{\mathrm{d}x} = \frac{\mathrm{d}y}{\mathrm{d}t} \div \frac{\mathrm{d}x}{\mathrm{d}t}$$

Example 4

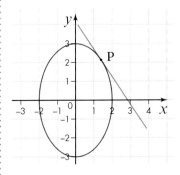

The equation of this ellipse is given parametrically as

$x = 2 \sin t$ and $y = 3 \cos t, \qquad 0 \leqslant t \leqslant 2\pi$

A tangent has been drawn at the point P, where $t = \frac{\pi}{4}$.

Find the equation of the tangent at P.

Solution

Differentiate x with respect to t.

$$x = 2 \sin t$$

$$\frac{\mathrm{d}x}{\mathrm{d}t} = 2 \cos t$$

Differentiate y with respect to t.

$$y = 3 \cos t$$

$$\frac{\mathrm{d}y}{\mathrm{d}t} = -3 \sin t$$

$$\frac{\mathrm{d}y}{\mathrm{d}x} = \frac{\mathrm{d}y}{\mathrm{d}t} \div \frac{\mathrm{d}x}{\mathrm{d}t}$$

$$= -3 \sin t \div 2 \cos t$$

$$= -\frac{3}{2} \frac{\sin t}{\cos t}$$

$$= -\frac{3}{2} \tan t$$

When $t = \frac{\pi}{4}$:

$$x = 2\sin\frac{\pi}{4} = 2 \times \frac{1}{\sqrt{2}} = \frac{2}{\sqrt{2}}$$

$$y = 3\cos\frac{\pi}{4} = 3 \times \frac{1}{\sqrt{2}} = \frac{3}{\sqrt{2}}$$

and $\frac{dy}{dx} = -\frac{3}{2}\tan\frac{\pi}{4} = -\frac{3}{2} \times 1 = -\frac{3}{2}$

The equation of the tangent at P is $y - \frac{3}{\sqrt{2}} = -\frac{3}{2}\left(x - \frac{2}{\sqrt{2}}\right)$

Multiply by 2. $\qquad\qquad 2y - \frac{3 \times 2}{\sqrt{2}} = -3x + \frac{3 \times 2}{\sqrt{2}}$

Rearrange. $\qquad\qquad 2y + 3x = 6\sqrt{2}$ •-------- Use the fact that $\frac{2}{\sqrt{2}} = \sqrt{2}$.

Exercise 7.4A

Answers page 450

1 Find $\frac{dy}{dx}$ as a function of t for each of the curves given by the following parametric equations:

   a $x = t - 1$ and $y = t^2$ **b** $x = t - 1$ and $y = t^4 - 2$ **c** $x = t^2$ and $y = t^2 - 3t + 2$

(PF) **2** A curve is given by the parametric equations $x = t + 2$ and $y = t^2 - 1$.

   Show that the equation of the tangent to the curve when $t = 1$ is given by $2x - y - 6 = 0$.

(PF) **3** A curve is given by the parametric equations $x = \frac{t}{2}$ and $y = t^2 - 4$.

   Show that the equation of the normal to the curve when $t = -1$ is given by $8y - 2x + 23 = 0$.

4 Find $\frac{dy}{dx}$ as a function of t for each of the curves given by the following parametric equations:

   a $x = 4\sin t$ and $y = 5\cos t$ **b** $x = 2\cos t$ and $y = 3\cos 2t$ **c** $x = \tan 2t$ and $y = \sin 2t$

(PF) **5** A curve is given by the parametric equations $x = 3\cos t + 1$ and $y = 7\sin t - 4$.

   Show that when $t = \frac{\pi}{4}$ the exact value of the gradient is $-\frac{7}{3}$.

(PS) **6** Identify and correct the mistakes in the following workings to find the gradient of the normal when $t = \frac{\pi}{3}$ for the curve given by the parametric equations $x = \sin 2t$ and $y = \cos 2t$.

$$x = \sin 2t$$
$$\frac{dx}{dt} = \frac{1}{2}\cos 2t$$
$$y = \cos 2t$$
$$\frac{dy}{dt} = -\frac{1}{2}\sin 2t$$
$$\frac{dy}{dx} = -\frac{1}{2}\sin 2t \div \frac{1}{2}\cos 2t$$
$$\qquad = -\tan 2t$$

When $t = \frac{\pi}{3}$

$$\frac{dy}{dx} = -\tan\frac{2\pi}{3} = -\sqrt{3}$$

Gradient of the normal is $-\sqrt{3}$.

(PS) 7 The parametric equations of this curve are $x = 2\sin 2t$ and $y = 3\cos t$, $0 \leqslant t \leqslant 2\pi$.

Find the gradient of each branch of the curve at the origin.

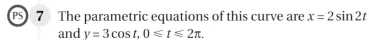

(PF) 8 A curve is given by the parametric equations $x = 2(t - \sin t)$ and $y = 2(1 - \cos t)$, where $0 \leqslant t \leqslant 2\pi$.

Show that the gradients of the tangent and normal are $1 + \sqrt{2}$ and $1 - \sqrt{2}$ respectively when $t = \frac{\pi}{4}$.

(PS) 9 A curve is given by the parametric equations $x = a\cos t$ and $y = b\sin t$, where $0 \leqslant t \leqslant 2\pi$.

In terms of a and b, find the y-intercept of the tangent to the curve when $t = \frac{\pi}{4}$.

(M) (PS) 10 This graph shows the cross-section of a speed bump. Lengths are in centimetres and the x-axis represents ground level.

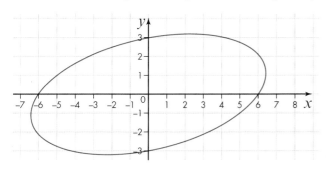

The curve of the speed bump is given parametrically by $x = 10\sin t$ and $y = 2 + 2\cos 2t$, $-\frac{\pi}{2} \leqslant t \leqslant \frac{\pi}{2}$.

a Find the gradient of the speed hump in terms of t.

b Find the value of t at the highest point on the speed bump.

c Find the angle of slope of the speed bump where $x = 9$ to the nearest degree.

7.5 Implicit equations

This curve is an ellipse. The equation is $x^2 + 4y^2 - 2xy = 36$.

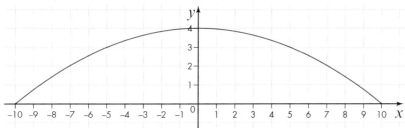

The equation $x^2 + 4y^2 - 2xy = 36$ is an **implicit equation**. That means that it shows a relationship between coordinates of points on the curve but it is not written in an explicit form of the type $y = f(x)$.

Stop and think   Why is it difficult to write the equation for y explicitly in this case?

To find a point on the curve you can substitute a value for x (or for y) and then solve the resulting equation to find y (or x).

For example, if you substitute $x = 6$ in the equation it becomes

$$36 + 4y^2 - 12y = 36$$

Rearrange.

$$4y^2 - 12y = 0$$

Divide by 4 and factorise.

$$y^2 - 3y = 0$$

$$y(y - 3) = 0$$

Therefore $y = 0$ or 3.

Two points on the curve are $(6, 0)$ and $(6, 3)$.

To find $\dfrac{dy}{dx}$ you can differentiate both sides of the equation.

$$\frac{d}{dx}(x^2 + 4y^2 - 2xy) = \frac{d}{dx}(36)$$

Differentiate each term on the left separately.

$$\frac{d}{dx}(x^2) + \frac{d}{dx}(4y^2) - \frac{d}{dx}(2xy) = \frac{d}{dx}(36)$$

The first and last terms are straightforward.

$$2x + \frac{d}{dx}(4y^2) - \frac{d}{dx}(2xy) = 0 \qquad \qquad ①$$

To differentiate $4y^2$, use the chain rule.

$$\frac{d}{dx}(4y^2) = \frac{d}{dy}(4y^2) \times \frac{dy}{dx}$$

$$= 8y\frac{dy}{dx}$$

To differentiate $2xy$, use the product rule.

$$\frac{d}{dx}(2xy) = 2\left\{\frac{d}{dx}(x) \times y + x\frac{dy}{dx}\right\}$$

$$= 2\left(y + x\frac{dy}{dx}\right)$$

Equation ① becomes

$$2x + 8y\frac{dy}{dx} - 2y - 2x\frac{dy}{dx} = 0 \qquad ②$$

Rearrange to make $\frac{dy}{dx}$ the subject.

$$8y\frac{dy}{dx} - 2x\frac{dy}{dx} = 2y - 2x$$

Divide by 2 and factorise.

$$4y\frac{dy}{dx} - x\frac{dy}{dx} = y - x$$

$$(4y - x)\frac{dy}{dx} = y - x$$

$$\frac{dy}{dx} = \frac{y - x}{4y - x} \qquad ③$$

So the gradient of the curve at $(6, 3)$ is $\frac{3 - 6}{4 \times 3 - 6} = \frac{-3}{6} = -\frac{1}{2}$

and the gradient of the curve at $(6, 0)$ is $\frac{0 - 6}{4 \times 0 - 6} = \frac{-6}{-6} = 1$.

The expression for $\frac{dy}{dx}$ includes both x and y.

Stop and think   Look at the graph and check that results for the gradients are reasonable.

Use the symmetry of the graph to write down the gradient at two other points.

You do not always need to rearrange equation ② to find $\frac{dy}{dx}$ in the explicit form of equation ③.

You could find the gradient at $(6, 3)$ by substituting the coordinates into equation ②, that is

$$12 + 24\frac{dy}{dx} - 6 - 12\frac{dy}{dx} = 0$$

and solving that equation.

Example 5
Find the stationary points on the curve $x^2 + 4y^2 - 2xy = 36$.

Solution
This is the curve explored above.

You know that $2x + 8y\frac{dy}{dx} - 2y - 2x\frac{dy}{dx} = 0$

At a stationary point, $\frac{dy}{dx} = 0$.

Therefore $2x - 2y = 0$ and $x = y$.

Substitute x for y in the equation of the curve.

$$x^2 + 4x^2 - 2x^2 = 36$$

$$3x^2 = 36$$

$$x^2 = 12$$

$$x = \sqrt{12}$$

Since $\pm x = y$, the stationary points are at $\left(\sqrt{12}, \sqrt{12}\right)$ and $\left(-\sqrt{12}, -\sqrt{12}\right)$.

Answers page 451

Exercise 7.5A

1 The equation of a curve is $2x^2 + y^2 = 34$.

   a Find an expression for $\dfrac{dy}{dx}$.

   b Find the gradient at $(3, 4)$.

   c Find the gradient at $(3, -4)$.

2 The equation of this curve is $x^2 y = 4(2 - y)$.

   a Show that $(2, 1)$ is on the curve.

   b Find the gradient at $(2, 1)$.

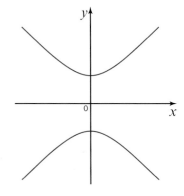

3 The equation of a curve is $x^2 + 4xy + y^2 = 25$.

   a Show that $(5, 0)$ and $(0, 5)$ are on the curve.

   b Find the gradient of the curve at $(5, 0)$.

   c Show that $\dfrac{dy}{dx} = -\dfrac{x + 2y}{2x + y}$.

PS **4** The equation of this curve is $y^2 - x^2 = 16$.

   a Show that $(3, 5)$ is on the curve.

   b Find the equation of the tangent to the curve at $(3, 5)$.

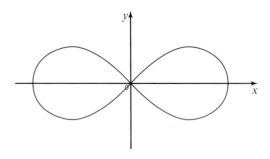

PS **5** The equation of this curve is $(x^2 + y^2)^2 = 50(x^2 - y^2)$.

Show that stationary points lie on a circle, centre the origin, with a radius of 5.

6 An implicit equation for this curve is $x^2 - y^2 = 1$.

a Find an expression for $\dfrac{dy}{dx}$.

b The curve can be written parametrically as $x = \sec t$ and $y = \tan t$, where $0 \leqslant t \leqslant 2\pi$. Use this fact to find $\dfrac{dy}{dx}$ in terms of t.

c Show that your two expressions for $\dfrac{dy}{dx}$ are equivalent.

d Find the points on the curve where the gradient is equal to 2.

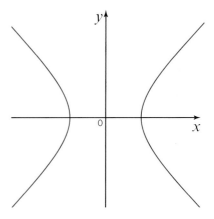

7.6 Constructing differential equations

If you make a hot drink it does not stay hot – the temperature gradually falls. Newton's law of cooling states that the rate at which the temperature falls is proportional to the difference between the temperature of the drink and room temperature, the temperature of the surroundings.

If the temperature of the drink after t minutes is $y\,°C$ and room temperature is a constant $r\,°C$, then you can write $\dfrac{dy}{dt} \propto y - r$.

You can change this to an equation by introducing a constant, c, and writing

$$\frac{dy}{dt} = -c(y - r)$$

The – sign indicates that y is decreasing with time, and c will be positive.

The value of c will depend on the particular situation. If the drink is in a wide shallow cup with a large surface area it will cool more quickly than if it is in an insulated mug. The value of c will be larger for the shallow cup than it is for the insulated mug.

This is an example of a **differential equation**. Differential equations are very common in all branches of science and engineering, where they can illustrate the relationship between variables that change over time or space.

The solution of a differential equation is a formula that gives one variable in terms of another. For our hot drink equation, a solution will be a formula that gives y as a function of t.

Finding the solution can be very difficult. However, testing whether a particular function is a solution is relatively easy.

Example 6

A cup of coffee is made and left to cool down. The temperature after t minutes is $y\,°C$ and the room temperature is $20\,°C$.

a Show that the temperature satisfies the differential equation $\dfrac{dy}{dt} = -c(y - 20)$, $c > 0$.

b Show that $y = ae^{-ct} + 20$, where a is a constant, is a solution to the differential equation.

c The coffee is initially at $100\,°C$ and after 5 minutes the temperature is $70\,°C$. Work out the values of a and c.

Solution

a Most of this has already been done at the start of **Section 7.6**.

$$\frac{dy}{dt} \propto y - 20 \quad \text{and so} \quad \frac{dy}{dt} = -c(y - 20),$$

where the − sign shows that the coffee is cooling.

b If $y = ae^{-ct} + 20$ you need to show that this satisfies the differential equation.

$$\frac{dy}{dt} = -cae^{-ct}$$

and $-c(y - 20) = -c(ae^{-ct} + 20 - 20)$

$$= -cae^{-ct}$$
$$= \frac{dy}{dt}$$

This shows that $y = ae^{-ct} + 20$ is a solution.

c When $t = 0$, $y = 100$. Substitute these values.

$$100 = a + 20$$

Therefore $a = 80$ and $y = 80e^{-ct} + 20$.

When $t = 5$, $y = 70$. Substitute these values.

$$70 = 80e^{-5c} + 20$$

Rearrange.

$$50 = 80e^{-5c}$$
$$e^{-5c} = \frac{50}{80} = 0.625$$
$$-5c = \ln 0.625$$
$$c = \frac{\ln 0.625}{-5} = 0.094$$

The solution is $y = 80e^{-0.094t} + 20$.

You were not asked to draw the graph, but it looks like this:

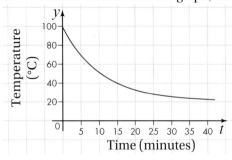

Stop and think

Use the graph to describe how the temperature of the coffee is changing over time.

When is the temperature changing most quickly? What is the temperature after a long time?

Is this how you would expect the temperature of a cup of coffee to change?

Exercise 7.6A **Answers page 451**

1 The gradient at any point on the graph of $y = f(x)$ is proportional to the square of the x-coordinate.

  a Write down a differential equation to express this relationship.

  b Show that $y = x^3 + 4$ is a solution of your differential equation.

2 The gradient at any point on the graph of $y = f(x)$ is inversely proportional to the y- coordinate.

  a Write down a differential equation to express this relationship.

  b Show that $y = \sqrt{x}$ is a solution of your differential equation.

  c Show that $y = a\sqrt{x}$ is a solution of your differential equation for any value of a.

3 The gradient at any point on the graph of $y = f(x)$ is proportional to the product of the two coordinates.

  a Write down a differential equation to express this relationship.

  b Show that $y = e^{\frac{1}{2}x^2}$ is a solution of your differential equation.

(M) **4** The rate of increase of the population of a country at any time is equal to 0.5% of the population size at that time.

  a If y is the population in millions and t is time in years, write down a differential equation to show this fact.

  b Show that a solution of the equation is $y = a e^{0.005t}$, where a is a constant.

  c The population now is 50 million. Estimate the population in 20 years' time.

(M) **5** A particle is moving away from point A.

  The speed of the particle is inversely proportional to the distance, x m, from A.

  When the distance is 5 m, the speed is $0.4\,\text{m s}^{-1}$.

  a Show that $x\dfrac{dx}{dt} = 2$.

  b Find the acceleration of the particle when the distance is 5 m.

(M) **6** A spherical balloon is being inflated by pumping in air at a constant rate.

  a The rate of increase of the radius of the sphere with time is inversely proportional to the square of the radius.

  Write down a differential equation to show this. Use r for the radius in cm and t for the time in seconds.

  b When the radius is 10 cm, the rate of increase of the radius is $0.5\,\text{cm s}^{-1}$.

  Show that the differential equation can be written as $r^2\dfrac{dr}{dt} = 50$.

  c Show that $r = a\sqrt[3]{t}$ is a solution to the differential equation, where $a = \sqrt[3]{150}$.

SUMMARY OF KEY POINTS

› The product rule: if $y = uv$ where u and v are functions of x, then $\dfrac{dy}{dx} = u\dfrac{dv}{dx} + v\dfrac{du}{dx}$

› The quotient rule: if $y = \dfrac{u}{v}$ then $\dfrac{dy}{dx} = \dfrac{v\dfrac{du}{dx} - u\dfrac{dv}{dx}}{v^2}$

› The derivatives of trigonometric functions:

› $\dfrac{d}{dx}(\tan x) = \sec^2 x$ › $\dfrac{d}{dx}(\arcsin x) = \dfrac{1}{\sqrt{1 - x^2}}$

› $\dfrac{d}{dx}(\cot x) = -\operatorname{cosec}^2 x$ › $\dfrac{d}{dx}(\arccos x) = -\dfrac{1}{\sqrt{1 - x^2}}$

› $\dfrac{d}{dx}(\sec x) = \sec x \tan x$ › $\dfrac{d}{dx}(\arctan x) = \dfrac{1}{1 + x^2}$

› $\dfrac{d}{dx}(\operatorname{cosec} x) = -\operatorname{cosec} x \cot x$

› If x and y are defined in terms of a parameter t, then $\dfrac{dy}{dx} = \dfrac{dy}{dt} \div \dfrac{dx}{dt}$

› Implicit equations in x and y can be differentiated to find $\dfrac{dy}{dx}$.

EXAM-STYLE QUESTIONS 7   Answers page 452

1 Differentiate $x^2 \cos x$ with respect to x.   **[4 marks]**

2 Differentiate $\dfrac{x^2 - 1}{x^2 + 1}$.   **[4 marks]**

3 Differentiate with respect to x:

a $\cot 2x$ **[2 marks]**

b $\cot^2 2x$ **[3 marks]**

(PF) 4 A particle is moving in such a way that its distance (x m) from a fixed point after t s is given by the formula $x = 0.3 \sin 2t - 0.4 \cos 2t$.

Show that x satisfies the differential equation $\dfrac{d^2x}{dt^2} + 4x = 0$.   **[5 marks]**

(CM) 5 A student is differentiating $y = 2 \sin x \cos x$. Here is the student's working:

$y = 2 \sin x \cos x$

Use the product rule to differentiate $\sin x \cos x$.

$u = \sin x$ and $v = \cos x$

$\dfrac{du}{dx} = \cos x$ and $\dfrac{dv}{dx} = -\sin x$

$$\frac{dy}{dx} = 2(\sin x \times (-\sin x) - \cos x \times \cos x)$$

$$= 2(-\sin^2 x - \cos^2 x)$$

$$= -2 \text{ because } \sin^2 x + \cos^2 x = 1$$

a The student has made a mistake. Correct the error. **[3 marks]**

b Explain an easier method for finding the derivative. **[3 marks]**

6 The graph shows the path of a point on the circumference of a rotating wheel as it moves along the ground.

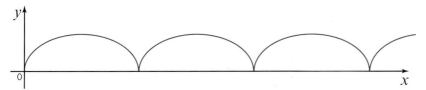

The parametric equations are $x = \theta - \sin \theta$ and $y = 1 - \cos \theta$.

a Find $\frac{dy}{dx}$ in terms of θ. **[4 marks]**

b Show that where the gradient of the curve is 0.5, $2\sin \theta + \cos \theta = 1$. **[2 marks]**

7 The equation of this graph is $y = e^{-x} \sin 10x$.

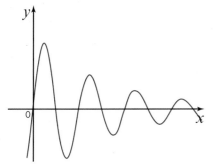

a Find $\frac{dy}{dx}$. **[4 marks]**

b Show that the x-coordinate of the first maximum point is 0.147 to 3 d.p. **[3 marks]**

c Find the x-coordinate of the first minimum point. **[2 marks]**

8 The speed of a car over a 10-second interval is given by the equation

$y = kt(10 - t)^2$, $0 \le t \le 10$, where t is the time in seconds, y is the speed in m s⁻¹ and k is a constant.

a When $t = 5$, the speed is 12.5 m s⁻¹. Find the value of k. **[2 marks]**

b Find the initial acceleration of the car. **[4 marks]**

c Find the maximum speed of the car. **[3 marks]**

CM 9 The equation of this curve is $y = \sin^3 x + \cos^3 x$ where x is in radians.

A student writes: 'The curve has a stationary point whenever x is a multiple of $\frac{\pi}{4}$.'

Show that this is not correct and suggest a suitable alternative. **[4 marks]**

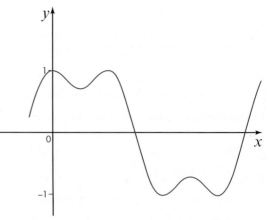

10 This is a logistic curve, a shape that often occurs in practical situations such as population growth and learning curves.

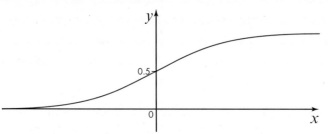

The equation of the curve is $y = \dfrac{e^x}{e^x + 1}$.

a Find $\dfrac{dy}{dx}$. **[4 marks]**

b Show that the equation satisfies the differential equation $\dfrac{dy}{dx} = y(1 - y)$. **[4 marks]**

PS 11 The equation of this ellipse is $x^2 - 2xy + 2y^2 = 9$.

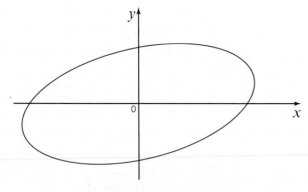

a Find the coordinates of the maximum point. **[7 marks]**

b Find the equation of the normal at the points where the graph crosses the x-axis. **[5 marks]**

12 The parametric equations of a curve are $x = \dfrac{2t}{1 + t^2}$ and $y = \dfrac{1 - t^2}{1 + t^2}$.

Show that $\dfrac{dy}{dx} = \dfrac{-x}{y}$. **[4 marks]**

13 The equation of a curve is $x^2 - xy + y^2 = 300$.

Find the coordinates of the stationary points on the curve. **[3 marks]**

8 INTEGRATION

In the 19th century Michael Faraday carried out a series of experiments that showed the connections between magnetism and electricity. In particular, he showed that an electric current has an associated magnetic field, and that a changing magnetic field will produce an electric current. These ideas were the basis of the invention of the dynamo and the electric motor.

James Clerk Maxwell formulated a set of equations that described these relationships. He was able to solve these equations and show that the solution was in the form of a moving wave – what we now call electromagnetic radiation. This was a completely unexpected result.

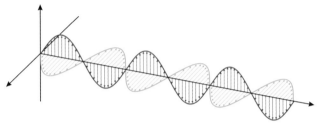

Maxwell was even able to calculate the speed of movement of these waves from his equations. Using only theoretical calculations, he predicted the speed of light. Albert Einstein called these equations the most important event in physics since the time of Isaac Newton.

Maxwell's equations are examples of differential equations. These are the best mathematical way to describe the connection between variables that change over time. By the end of this chapter you will have learnt how to solve simple differential equations.

LEARNING OBJECTIVES

You will learn how to:

› use your knowledge of differentiation to recognise integrals of various types

› use new techniques of integration, including integration by substitution and integration by parts

› integrate an increased range of trigonometric functions

› integrate algebraic fractions of different types

› solve simple differential equations.

TOPIC LINKS

In this chapter you will need to use the definitions of trigonometric functions from **Chapter 5 Trigonometry**. You should already be able to differentiate a range of functions and apply different techniques to solve differentiation problems (see **Chapter 6 Differentiation** and **Chapter 7 Further differentiation**). You encountered sigma notation for a sum in **Chapter 4 Sequences and series** and partial fractions in **Chapter 2 Algebra and functions 2: Partial fractions**.

PRIOR KNOWLEDGE

You should already know how to:

- ❭ differentiate polynomial, trigonometric and exponential functions and combinations of these
- ❭ manipulate trigonometric expressions using a range of standard identities, including double angle formulae and Pythagoras' theorem in different trigonometric forms
- ❭ use the fact that integration is the in reverse of differentiation to find the integrals of a range of simple trigonometric and exponential functions
- ❭ evaluate a definite integral to find the area under a curve.

You should be able to complete the following questions correctly:

1 Differentiate with respect to x:

   a $\sin(x^2 + 1)$ **b** $e^{-x}\cos 2x$ **c** $\ln(3x + 2)$

2 Find:

   a $\int \dfrac{4}{\sqrt{x}}\,dx$ **b** $\int e^{4x}\,dx$ **c** $\int \sec^2 x\,dx$

3 Find the area between the curve $y = (x - 1)(x + 4)$ and the x-axis.

4 Write $\dfrac{4}{x(x + 2)}$ as partial fractions.

8.1 Recognising integrals

Remember that integration is the inverse of differentiation.

You know how to differentiate functions such as x^n, e^{ax}, $\sin ax$ and $\cos ax$.

This means that you should be able to recognise the derivatives of these functions, and so integrate them.

You know that $\int x^n\,dx = \frac{1}{n+1}x^{n+1} + c$ where $n \neq -1$ and c is a constant.

What about $\int e^{ax}\,dx$, where a is a non-zero constant?

You know that $\dfrac{d}{dx}(e^{ax}) = ae^{ax}$ and it follows that

$$\frac{d}{dx}\left(\frac{1}{a}e^{ax}\right) = \frac{1}{a} \times ae^{ax} = e^{ax}$$

So $$\int e^{ax}\,dx = \frac{1}{a}e^{ax} + c$$

What about $\int \sin ax\,dx$?

You know that $\dfrac{d}{dx}(\cos ax) = -a\sin ax$

It follows that $\int \sin ax\,dx = -\dfrac{1}{a}\cos ax + c$

Similarly, $\int \cos ax\,dx = \dfrac{1}{a}\sin ax + c$

> **KEY INFORMATION**
>
> $\int x^n\,dx = \dfrac{1}{n+1}x^{n+1} + c,\, n \neq -1$
>
> $\int e^{ax}\,dx = \dfrac{1}{a}e^{ax} + c$
>
> $\int \sin ax\,dx = -\dfrac{1}{a}\cos ax + c$
>
> $\int \cos ax\,dx = \dfrac{1}{a}\sin ax + c$

For many integrals, including the examples above, you can guess the form of the integral and then find the value of a constant to multiply by to get the correct solution.

Example 1

Find: **a** $\int 2(4x+3)^2\mathrm{d}x$ **b** $\int 2\sin(4x+3)\mathrm{d}x$

Solution

a You could guess that the integral is a multiple of $(4x+3)^3$.

Differentiate that expression using the chain rule.

$$\frac{\mathrm{d}}{\mathrm{d}x}(4x+3)^3 = 3(4x+3)^2 \times 4$$

$$= 12(4x+3)^2$$

You want the coefficient to be 2 and not 12, so multiply by $\frac{2}{12}$.

$$\int 2(4x+3)^2\,\mathrm{d}x = \frac{2}{12}(4x+3)^3 = \frac{1}{6}(4x+3)^3 + c$$

> You can check this is correct by differentiating. Remember to always include a constant.

b In this case you might guess a multiple of $\cos(4x+3)$.

Differentiate using the chain rule.

$$\frac{\mathrm{d}}{\mathrm{d}x}\cos(2x+3) = -\sin(4x+3) \times 4$$

$$= -4\sin(4x+3)$$

You want the coefficient to be 2, so multiply by $\frac{2}{-4} = -\frac{1}{2}$.

$$\int 2\sin(4x+3)\,\mathrm{d}x = -\frac{1}{2}\cos(4x+3) + c$$

You know that $\int x^n\mathrm{d}x = \frac{1}{n+1}x^{n+1}+c$ where $n \neq -1$, but what if $n = -1$?

What is $\int \frac{1}{x}\mathrm{d}x$?

Suppose $\qquad \dfrac{\mathrm{d}y}{\mathrm{d}x} = \dfrac{1}{x}$

Then $\qquad \dfrac{\mathrm{d}x}{\mathrm{d}y} = x$

But you know that if $x = e^y$ then $\dfrac{\mathrm{d}x}{\mathrm{d}y} = e^y = y$.

> $\dfrac{\mathrm{d}y}{\mathrm{d}x} = \lim\limits_{\delta x \to 0}\dfrac{\delta y}{\delta x}$ and $\dfrac{\mathrm{d}x}{\mathrm{d}y} = \lim\limits_{\delta x \to 0}\dfrac{\delta x}{\delta y}$ so one is the reciprocal of the other.

So an integral is $\qquad x = e^y$.

Take logarithms.

$$\ln x = y$$

Therefore $\qquad \int \frac{1}{x}\mathrm{d}x = \ln x + c$

But $\ln x$ is only defined if $x > 0$, so the result so far is

$$\int \frac{1}{x}\mathrm{d}x = \ln x + c, \; x > 0 \qquad \textcircled{1}$$

However, if $x < 0$ then $\ln(-x)$ exists. Using the chain rule:

$$\frac{d}{dx}\ln(-x) = \frac{1}{-x} \times -1 = \frac{1}{x}$$

This means that $\int \frac{1}{x}\,dx = \ln(-x) + c,\ x < 0$ ②

Combining results ① and ②, you can write

$$\int \frac{1}{x}\,dx = \ln|x| + c,\ x \neq 0$$

> **Remember that $|x|$ is the modulus of x. It is the value of x ignoring any minus sign. For example,**
> $$|2.5| = |-2.5| = 2.5$$

Example 2

Find $\int \dfrac{1}{4x+1}\,dx$.

Solution

In this case you might guess a multiple of $\ln|4x+1|$.

Use the chain rule.

$$\frac{d}{dx}\ln|4x+1| = \frac{1}{4x+1} \times 4$$

$$= \frac{4}{4x+1}$$

Therefore $\int \dfrac{1}{4x+1}\,dx = \dfrac{1}{4}\ln|4x+1| + c$

> **KEY INFORMATION**
>
> $$\int \frac{1}{x}\,dx = \ln|x| + c, \qquad x \neq 0$$

Exercise 8.1A

Answers page 453

1 Find:

  a $\int \sqrt[3]{x}\,dx$ **b** $\int \sqrt[3]{4x+3}\,dx$ **c** $\int \dfrac{1}{\sqrt[3]{2x+5}}\,dx$

2 Find these integrals:

  a $\int 4e^{-2x}\,dx$ **b** $\int (e^{0.5x} - e^{-0.5x})\,dx$ **c** $\int (2x + 3e^{4x+5})\,dx$

3 Integrate:

  a $\int (\sin 2x - \cos 2x)\,dx$ **b** $\int 4\cos(0.1t + 1.3)\,dt$ **c** $\int 0.1\sin(2 - \theta)\,d\theta$

(PS) 4 **a** Find the value of $\int_0^\pi 2\cos 0.5x\,dx$.

    b Sketch a graph to show what area the definite integral in **part a** represents.

5 Find:

  a $\int (3x-2)^5\,dx$ **b** $\int \dfrac{1}{(5x+1)^2}\,dx$

  c $\int \sqrt{0.1x + 0.3}\,dx$

(PS) 6 This is a graph of $y = \dfrac{4}{x}$.

The region bounded by the curve, the x-axis and the lines $x = 1$ and $x = a$ has an area of 10.

Work out the value of a.

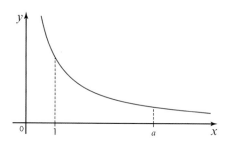

(M) Modelling   (PS) Problem solving   (PF) Proof   (CM) Communicating mathematically

7 Work out:

a $\int \dfrac{2}{x+1}\,dx$ **b** $\int \dfrac{3}{2x-1}\,dx$ **c** $\int \dfrac{x^2-6}{2x}\,dx$

8 This is a graph of $y = e^{0.5x} - 2$.

Calculate the shaded area.

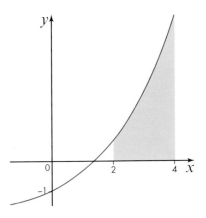

 9 **a** Differentiate $\sin(x^2)$.

     b Use your answer to **part a** to find $\int x\cos(x^2)\,dx$.

     c Work out $\int x\sin(x^2)\,dx$.

 10 Two students are asked to work out $\int \dfrac{1}{2x}\,dx$.

Alice writes:

$$\int \frac{1}{2x}\,dx = \frac{1}{2}\int \frac{1}{x}\,dx = \frac{1}{2}\ln x + c$$

Bhaskar writes:

Using the chain rule, $\dfrac{d}{dx}(\ln 2x) = \dfrac{1}{2x}\times 2 = \dfrac{1}{x}$

so $\dfrac{d}{dx}\left(\dfrac{1}{2}\ln 2x\right) = \dfrac{1}{2}\times\dfrac{1}{x} = \dfrac{1}{2x}$

Therefore $\int \dfrac{1}{2x}\,dx = \dfrac{1}{2}\ln 2x + c$

Who is correct? Give a reason for your answer.

8.2 Integration with trigonometric functions

You have already encountered the derivatives of a range of trigonometric functions.

The inverse of each of these functions will provide a corresponding integral.

For example, $\dfrac{d}{dx}(\tan x) = \sec^2 x$

It follows that $\int \sec^2 x\,dx = \tan x + c$

Trigonometric expressions can occur in situations where you might not expect them.

$$\frac{d}{dx}(\arcsin x) = \frac{1}{\sqrt{1-x^2}}$$

so $\int \dfrac{1}{\sqrt{1-x^2}}\,dx = \arcsin x + c$

also $\dfrac{d}{dx}(\arctan x) = \dfrac{1}{1+x^2}$

so $\int \dfrac{1}{1+x^2}\,dx = \arctan x + c$

> There is a list of derivatives of trigonometric functions in **Chapter 7 Further differentiation**.

> **KEY INFORMATION**
>
> $\int \dfrac{1}{\sqrt{1-x^2}}\,dx = \arcsin x + c$
>
> $\int \dfrac{1}{1+x^2}\,dx = \arctan x + c$

Example 3

Integrate with respect to x:

a $\operatorname{cosec} 0.5x \cot 0.5x$ **b** $\tan^2 x$.

Solution

a Recall that $\dfrac{d}{dx}(\operatorname{cosec} x) = -\operatorname{cosec} x \cot x$

This suggests trying $\operatorname{cosec} 0.5x$.

$$\frac{d}{dx}(\operatorname{cosec} 0.5x) = -\operatorname{cosec} 0.5x \cot 0.5x \times 0.5$$

$$= -0.5 \operatorname{cosec} 0.5x \cot 0.5x$$

So $\displaystyle\int \operatorname{cosec} 0.5x \cot 0.5x \, dx = -2 \operatorname{cosec} 0.5x + c$

b Use the identity $\tan^2 x + 1 = \sec^2 x$.

Therefore $\tan^2 x = \sec^2 x - 1$

So $\displaystyle\int \tan^2 x \, dx = \int (\sec^2 x - 1) \, dx$

$$= \tan x - x + c$$

In **Example 3b** you used a trigonometric identity to transform the integral into one which you could find. You can use a similar idea to find other integrals.

For example, you can calculate $\displaystyle\int \sin^2 x \, dx$ by first using the double angle formula

$$\cos 2x = 1 - 2\sin^2 x$$

Rearrange.

$$2\sin^2 x = 1 - \cos 2x$$

$$\sin^2 x = \frac{1}{2} - \frac{1}{2}\cos 2x$$

Integrate.

$$\int \sin^2 x \, dx = \int \left(\frac{1}{2} - \frac{1}{2}\cos 2x \right) dx$$

Therefore $\displaystyle\int \sin^2 x \, dx = \frac{1}{2}x - \frac{1}{4}\sin 2x + c$

In a similar way, you can find $\displaystyle\int \cos^2 x \, dx$ by using the formula

$$\cos 2x = 2\cos^2 x - 1$$

This gives the result $\displaystyle\int \cos^2 x \, dx = \frac{1}{2}x + \frac{1}{4}\sin 2x + c$

Stop and think

You know that $\sin^2 x + \cos^2 x = 1$.

Show that the formulae for $\displaystyle\int \sin^2 x \, dx$ and $\displaystyle\int \cos^2 x \, dx$ are consistent with this fact.

Exercise 8.2A   Answers page 453

1 Find:

   a $\int \sec^2 x \, dx$ **b** $\int \sec^2 (2x) \, dx$ **c** $\int \sec^2 (2x - 1) dx$

2 Find:

   a $\int \sec x \tan x \, dx$ **b** $\int \sec 4x \tan 4x \, dx$

3 Find:

   a $\int 3 \operatorname{cosec} x \cot x \, dx$ **b** $\int 3 \operatorname{cosec} 5x \cot 5x \, dx$

4 Find $\int \operatorname{cosec}^2 (2x - 0.4) \, dx$.

5 Find $\int \dfrac{\sin x}{1 - \sin^2 x} dx$.

(PF) 6 Show that $\int \cos^2 x \, dx = \dfrac{1}{2}x + \dfrac{1}{4}\sin 2x + c$.

7   **a** Differentiate $\arcsin 2x$ with respect to x.

   b Use the result of **part a** to find $\int \dfrac{1}{\sqrt{1 - 4x^2}} \, dx$.

8   **a** Differentiate $\ln |\cos x|$ with respect to x.

   b Use the result of **part a** to find $\int \tan x \, dx$.

   c Find $\int \cot x \, dx$.

9 Work out:

   a $\int \cos^2 2x \, dx$ **b** $\int \sin^2 0.5x \, dx$

10 Work out $\int \cot^2 x \, dx$.

8.3 Definite integrals

In **Book 1, Chapter 9 Integration** you learnt how to use a definite integral to find the area between a curve and the x-axis. You can extend this idea to find the area between two curves.

Example 4

This graph shows $y = \sin x$ and $y = \cos x$.

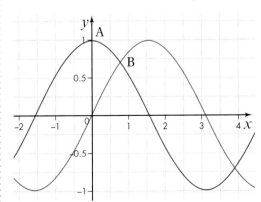

Calculate the area of the shape with vertices at the origin and points A and B.

Solution

The area required is the difference between the areas between the two curves and the x-axis.

First find the x-coordinate of B.

Where the curves cross, $\sin x = \cos x$

Divide by $\cos x$.

$$\tan x = 1$$

The smallest positive solution is $x = \dfrac{\pi}{4}$

so the limits for the integration are 0 and $\dfrac{\pi}{4}$.

The required area is $\int_0^{\frac{\pi}{4}} \cos x \, dx - \int_0^{\frac{\pi}{4}} \sin x \, dx$.

It is easier to treat this as a single integral rather than two separate ones.

$$\text{Area} = \int_0^{\frac{\pi}{4}} (\cos x - \sin x) \, dx$$

$$= \Big[\sin x + \cos x \Big]_0^{\frac{\pi}{4}}$$

$$= \Big[\sin \frac{\pi}{4} + \cos \frac{\pi}{4} \Big] - \Big[\sin 0 - \cos 0 \Big]$$

$$= \Big[\frac{1}{\sqrt{2}} + \frac{1}{\sqrt{2}} \Big] - [0 - 1]$$

$$= \frac{2}{\sqrt{2}} - 1$$

$$= \sqrt{2} - 1$$

> If you are dealing with definite integrals, you do not need to include a constant.

There is another way to think about integration.

Suppose you want to find the area under this curve, which is $\int_a^b y \, dx$.

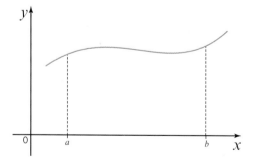

The area is approximately equal to the area of a large number of narrow rectangles, all with the same width, δx, and all with heights equal to the y-coordinate that corresponds to an x value in the rectangle.

Just a few of the rectangles are shown in the diagram on the right. The area of each rectangle is $y\,\delta x$, with a different value for y each time.

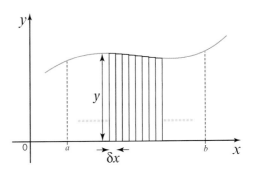

The sum of the areas of the rectangles is $\sum y\,\delta x$.

Now let $\delta x \to 0$. The sum of the areas of the rectangles will tend to the area under the curve.

Therefore $\int_a^b y\,\mathrm{d}x = \lim_{\delta x \to 0} \sum y\,\delta x$.

You can think of the definite integral as the limit of a sum as the number of terms increases.

This explains why an elongated S is used for an integral. In fact, $\sum$ is the Greek letter 'sigma', which is also S. Both characters stand for 'sum'.

PROOF

Many proofs involving differentiation and integration start with a small change, δx, and then look for a limit as $\delta x \to 0$.

Using an integral in this way, you can find more than just the area under a curve.

Here is a proof of the formula that the volume of a sphere is $\frac{4}{3}\pi r^3$.

The sphere in the diagram has radius r and has its centre at the origin.

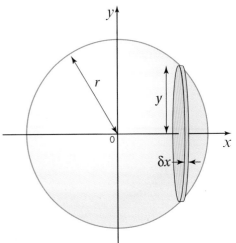

The volume of the sphere is approximately equal to the total volume of a large number of thin discs perpendicular to the x-axis with thickness δx and radius y.

The volume of one disc is

the area of the circular face $\times$ the thickness $= \pi y^2\,\delta x$

The total volume of all the discs is $\sum \pi y^2\,\delta x$.

Use the idea that an integral is the limit of this sum.

The volume of the sphere is $\lim_{\delta x \to 0} \sum \pi y^2 \delta x = \int_{-r}^{r} \pi y^2 \mathrm{d}x$.

To find the integral, you need to use the fact that $x^2 + y^2 = r^2$.

So $\qquad y^2 = r^2 - x^2$

This is the case because it is a circle of radius r.

and the volume $= \int_{-r}^{r} \pi(r^2 - x^2)\mathrm{d}x$

$$= \pi\left[r^2 x - \frac{1}{3}x^3 \right]_{-r}^{r}$$

$$= \pi\left[-r^3 - \frac{1}{3}r^3 \right] - \pi\left[r^3 + \frac{1}{3}r^3 \right]$$

$$= \frac{2}{3}\pi r^3 + \frac{2}{3}\pi r^3$$

$$= \frac{4}{3}\pi r^3 \text{ which is the required result.}$$

1 **a** Find the points where the curve $y = x^2$ and the straight line $y = 2 - x$ intersect.

b Find the area between the curve and the straight line.

2 This graph shows $y = \cos 0.5\pi x$ and $y = x^2 - 1$.

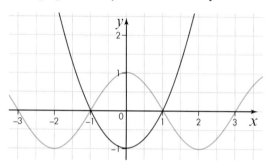

Find the area between the two curves.

PS **3** **a** Find the area between the curves $y = x^2$ and $y = x^{\frac{1}{2}}$.

b Find the area between the curves $y = x^n$ and $y = x^{\frac{1}{n}}$ where n is an integer $\geqslant 2$.

c What can you say about the area in **part b** when n is large?

4 The graph shows the lines $y = 2\sin 0.5x$ and $y = \dfrac{2x}{\pi}$.

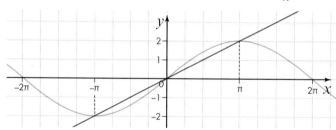

Find the total area between the two lines.

PS **5** This graph shows the lines $y = 4x - x^2$ and $y = x^2 - 4x + 6$.

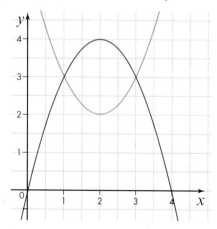

Show that the area between the curves is $2\frac{2}{3}$.

6 The graph shows $y = 2^{2-x}$ and $y = 2x$.

   a Show that the lines cross at $(1, 2)$.

   b Find the area of the region enclosed by the lines and the y-axis.

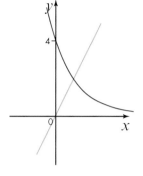

7 The graph shows $y = \sin x$ and $y = \sin 2x$.

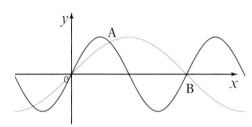

   a The curves cross at A. Show that the x-coordinate of A is $\dfrac{\pi}{3}$.

   b Find the area between the two curves between the origin and A.

   c The curves meet on the x-axis at B.

     Find the area between the curves between points A and B.

(PF) 8 The height of a cone is h and the radius of the base is r.

The vertex of the cone is at the origin and the x-axis passes through the centre of the base.

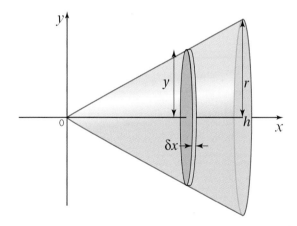

The cone is approximated by a large number of discs with thickness δx and radius y.

   a Find an expression for the volume of one disc.

   b Prove that a formula for the volume (V) of the cone is $V = \dfrac{1}{3}\pi r^2 h$.

8.4 Integration by substitution

Suppose you want to find $\int x\sqrt{x^2 + 1}\ \mathrm{d}x$.

It is not obvious how to do this because of the square root sign.
You can make it easier by making a substitution and changing the variable.

Remember the formula for the chain rule. This introduces another variable u and the fact that

$$\frac{dy}{dx} = \frac{dy}{du} \times \frac{du}{dx}$$

You can rearrange this as

$$\frac{dy}{du} = \frac{dy}{dx} \div \frac{du}{dx}$$

Now use the fact that $\frac{du}{dx}$ is the reciprocal of $\frac{dx}{du}$ and you can write

$$\frac{dy}{du} = \frac{dy}{dx} \times \frac{dx}{du} \qquad ①$$

Now if $y = \int x\sqrt{x^2 + 1}\, dx$ then $\frac{dy}{dx} = x\sqrt{x^2 + 1}$.

Put this into equation ①.

$$\frac{dy}{du} = x\sqrt{x^2 + 1} \times \frac{dx}{du}$$

and so $\quad y = \int x\sqrt{x^2 + 1} \times \frac{dx}{du}\, du \qquad ②$

This does not look very helpful. However, if you choose $u = x^2 + 1$

then $\frac{du}{dx} = 2x$ and so $\frac{dx}{du} = \frac{1}{2x}$

If you substitute these into equation ② you get

$$y = \int x\sqrt{u} \times \frac{1}{2x}\, du$$

and the xs cancel to give

$$y = \int \frac{1}{2}u^{\frac{1}{2}}\, du$$

Integrate.

$$y = \frac{1}{2} \times \frac{2}{3}u^{\frac{3}{2}} + c$$

Write u in terms of x.

$$y = \frac{1}{3}(x^2 + 1)^{\frac{3}{2}} + c$$

Stop and think Differentiate $y = \frac{1}{3}(x^2 + 1)^{\frac{3}{2}} + c$ using the chain rule to show that it is correct.

In practice, you can write this much more concisely.

Just write $\int x\sqrt{x^2 + 1}\, dx$ as $\int x\sqrt{x^2 + 1} \times \frac{dx}{du}\, du$ and then look for a suitable substitution for u that will make the integration simpler.

Knowing what to choose for u becomes easier with practice. Your choice may not always work. Be prepared to try again with something different. In this case, the fact that the xs cancel is important to produce a function of u that can be integrated.

If dx and du were numbers, you can see that the du terms would cancel. They are *not* numbers, but this is a helpful way to remember the formula.

This method is called **integration by substitution**. There are no rules about how to choose the substitution, which is one reason why integration is harder than differentiation.

Example 5

Find $\int x\sqrt{x+1}\ \mathrm{d}x$.

Solution

Write $\int x\sqrt{x+1}\ \mathrm{d}x = \int x\sqrt{x+1}\,\dfrac{\mathrm{d}x}{\mathrm{d}u}\ \mathrm{d}u$.

The square root sign makes this difficult, so try $u = x + 1$.

Then $\dfrac{\mathrm{d}u}{\mathrm{d}x} = 1$ and so also $\dfrac{\mathrm{d}x}{\mathrm{d}u} = 1$.

Substitute.

$$\int x\sqrt{x+1}\,\frac{\mathrm{d}x}{\mathrm{d}u}\ \mathrm{d}u = \int x\sqrt{u}\ \mathrm{d}u$$

Unfortunately, there is still an x which has not cancelled. You need to replace it with a function of u before you can integrate.

Luckily, you can do that because $u = x + 1$ and so $x = u - 1$. Substitute this.

$$\int x\sqrt{u}\ \mathrm{d}u = \int (u-1)\sqrt{u}\ \mathrm{d}u$$

$$= \int \left(u^{\frac{3}{2}} - u^{\frac{1}{2}} \right) \mathrm{d}u$$

$$= \frac{2}{5}u^{\frac{5}{2}} - \frac{2}{3}u^{\frac{3}{2}} + c$$

$$= \frac{2}{5}(x+1)^{\frac{5}{2}} - \frac{2}{3}(x+1)^{\frac{3}{2}} + c$$

Your answer must be in terms of x and not u.

Sometimes, as with $\int x\sqrt{x^2+1}\ \mathrm{d}x$, the substitution will eliminate any x terms immediately.

Sometimes, as with $\int x\sqrt{x+1}\ \mathrm{d}x$, you may need to do more work before the x terms are eliminated.

Example 6

Find $\int \dfrac{1}{\sqrt{9-x^2}}\ \mathrm{d}x$.

Solution

Make the substitution $x = 3\sin u$.

Then $\dfrac{\mathrm{d}x}{\mathrm{d}u} = 3\cos u$.

KEY INFORMATION

Integration by substitution:

$$\int \mathrm{f}(x)\ \mathrm{d}x = \int \mathrm{f}(x)\frac{\mathrm{d}x}{\mathrm{d}u}\,\mathrm{d}u$$

Notice that in this case, x is written in terms of u rather than the other way round.

Substitute.

$$\int \frac{1}{\sqrt{9-x^2}}\,dx = \int \frac{1}{\sqrt{9-9\sin^2 u}} \times 3\cos u\,du$$

$$= \int \frac{1}{\sqrt{9(1-\sin^2 u)}} \times 3\cos u\,du$$

$$= \int \frac{1}{3\cos u} \times 3\cos u\,du$$

$$= \int 1\,du$$

$$= u + c$$

In this case, all of the terms have cancelled.

Finally, as $x = 3\sin u$ then $u = \arcsin\frac{x}{3}$.

So $\int \frac{1}{\sqrt{9-x^2}}\,dx = \arcsin\frac{x}{3} + c$

The substitution is not an obvious one to make, but it works because it eliminates the square root in the denominator.

Stop and think
Can you generalise the result in **Example 6** to give a formula for $\int \frac{1}{\sqrt{a^2-x^2}}\,dx$ where $a > 0$?

Exercise 8.4A

Answers page 454

1 Find $\int x(x+2)^3\,dx$ using the substitution $u = x + 2$.

2 Find $\int \frac{x}{\sqrt{x^2+4}}\,dx$ using the substitution $u = x^2 + 4$.

3 Work out $\int x^2(2x^3+5)^4\,dx$.

4 Find $\int x\sin x^2\,dx$.

5 Work out:

a $\int \frac{1}{x+2}\,dx$ **b** $\int \frac{x}{x+2}\,dx$ **c** $\int \frac{x^2}{x+2}\,dx$

6 Work out:

a $\int x\sqrt{x^2+3}\,dx$ **b** $\int x\sqrt{x+3}\,dx$ **c** $\int x^2\sqrt{x+3}\,dx$

7 **a** Use the substitution $u = \cos x$ to find $\int \tan x\,dx$.   **b** Find $\int \cot x\,dx$.

8 Find $\int \frac{1}{\sqrt{100-x^2}}\,dx$.

PS **9** **a** Find $\int \frac{1}{4+x^2}\,dx$. Use the substitution $x = 2\tan u$.

b Generalise your result of **part a** to find $\int \frac{1}{a^2+x^2}\,dx$, $a > 0$.

8.5 Integration by parts

Look at this integration: $\int x \sin x \, dx$.

This is the product of two different functions. Although you can easily integrate each function separately, this does not help you to integrate the product.

> You cannot integrate each function separately – the integral is *not* $\frac{1}{2}x^2 \times -\cos x$.

The product rule for differentiation provides a useful tool for integrations of this type.

$$\frac{d}{dx}(uv) = u\frac{dv}{dx} + v\frac{du}{dx}$$

Write this in an integral form.

$$uv = \int u\frac{dv}{dx}\,dx + \int v\frac{du}{dx}\,dx$$

Rearrange.

$$\int u\frac{dv}{dx}\,dx = uv - \int v\frac{du}{dx}\,dx$$

This may not look very useful, but it can transform an integral into something more manageable.

Look again at $\int x \sin x \, dx$ and compare it with $\int u\frac{dv}{dx}\,dx$.

You are going to make two substitutions.

$$u = x \quad \text{and} \quad \frac{dv}{dx} = \sin x$$

Differentiate one and integrate the other.

$$\frac{du}{dx} = 1 \quad \text{and} \quad v = -\cos x$$

Now use the formula $\int u\frac{dv}{dx}\,dx = uv - \int v\frac{du}{dx}\,dx$.

$$\int x \sin x \, dx = x \times -\cos x - \int -\cos x \times 1 \, dx$$

$$= -x \cos x + \int \cos x \, dx$$

> **KEY INFORMATION**
>
> Integration by parts:
>
> $\int u\frac{dv}{dx}\,dx = uv - \int v\frac{du}{dx}\,dx$

> Don't worry about including a constant yet.

The integration is now straightforward.

$$\int x \sin x \, dx = -x \cos x + \sin x + c$$

> **Stop and think** Differentiate $-x\cos x + \sin x + c$ to show that it is correct.

This method is called **integration by parts**.

Use integration by parts with a product, equating one term to u and the other to $\frac{dv}{dx}$.

Choosing $u = x$ gave $\frac{du}{dx} = 1$, and that made the integration simpler.

> It is important to get these the right way round.

> **Stop and think** Show that choosing $u = \sin x$ and $\frac{dv}{dx} = x$ makes the integration more difficult.

Example 7

Find $\int e^{-x}\sin x\,dx$.

Solution

Make the substitutions $u = \sin x$ and $\dfrac{dv}{dx} = e^{-x}$

Then $\dfrac{du}{dx} = \cos x$ and $v = -e^{-x}$

$$\int u\frac{dv}{dx}\,dx = uv - \int v\frac{du}{dx}\,dx$$

So $\int e^{-x}\sin x\,dx = -e^{-x}\sin x + \int e^{-x}\cos x\,dx$ ①

Now you have to find $\int e^{-x}\cos x\,dx$.

Use integration by parts again, this time with $u = \cos x$ and $\dfrac{dv}{dx} = e^{-x}$.

This time $\dfrac{du}{dx} = -\sin x$ and $v = -e^{-x}$.

$$\int u\frac{dv}{dx}\,dx = uv - \int v\frac{du}{dx}\,dx$$

$$\int e^{-x}\cos x\,dx = -e^{-x}\cos x - \int e^{-x}\sin x\,dx$$

Now the integral is the one you started with. Substitute into equation ①.

$$\int e^{-x}\sin x\,dx = -e^{-x}\sin x - e^{-x}\cos x - \int e^{-x}\sin x\,dx$$

Rearrange.

$$2\int e^{-x}\sin x\,dx = -e^{-x}\sin x - e^{-x}\cos x$$

So $\int e^{-x}\sin x\,dx = -\dfrac{1}{2}e^{-x}(\sin x + \cos x) + c$

Stop and think Check the answer is correct by differentiation. Why did using integration by parts twice result in the initial integration appearing again?

One useful result that you can find by using integration by parts is $\int \ln x\,dx$.

You know that the derivative of $\ln x$ is $\dfrac{1}{x}$, but what about the integral?

At first glance, $\ln x$ does not look like a product, but you can think of it as $1 \times \ln x$.

Then if $u = \ln x$ and $\dfrac{dv}{dx} = 1$

$\dfrac{du}{dx} = \dfrac{1}{x}$ and $v = x$

$$\int u\frac{dv}{dx}\,dx = uv - \int v\frac{du}{dx}\,dx$$

$$\int \ln x\,dx = x\ln x - \int x \times \frac{1}{x}\,dx$$

$$= x\ln x - \int 1\,dx$$

$$= x\ln x - x + c$$

KEY INFORMATION

$\int \ln x\,dx = x\ln x - x + c$

This is an important result that you should remember, or remember how to derive it using the product rule.

Stop and think Check, by differentiation, that this result is correct.

You now have a number of tools for finding integrals. Knowing which to try is a matter of experience, but here are some guidelines:

❯ Is the function a type that you recognise as similar to a result you know? If so, try to guess the answer and differentiate to check or adjust your guess.

❯ Would a substitution help to simplify an 'awkward' term? If so, that might be worth a try.

❯ Is the **integrand** a product of two terms? If so, the product rule might help. Decide carefully which is u and which is $\dfrac{dv}{dx}$.

> **KEY INFORMATION**
> The integrand is the expression you are trying to integrate.

Exercise 8.5A

Answers page 454

1 Find:

   a $\int x \cos x \, dx$ **b** $\int (x+1) \cos x \, dx$

2 Find:

   a $\int x \sin 2x \, dx$ **b** $\int x \cos 4x \, dx$

3 Find:

   a $\int x \, e^x \, dx$ **b** $\int x \, e^{-x} \, dx$ **c** $\int x \, e^{0.5x} \, dx$

4 Find $\int x^2 e^x \, dx$.

5 Find $\int (2x^2 - 1) \, e^{-x} \, dx$.

6 Find $\int e^x \cos x \, dx$.

7 Find:

   a $\int \ln x \, dx$ **b** $\int \ln 2x \, dx$ **c** $\int \ln x^2 \, dx$ **d** $\int \ln (x+1) \, dx$

8 **a** Find $\int e^{-x} \sin 2x \, dx$.

   b Here is a graph of $y = e^{-x} \sin 2x$, $x \geqslant 0$.

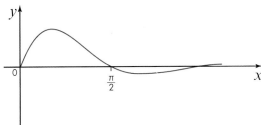

   Find the area bounded by the curve and the x-axis from 0 to $\dfrac{\pi}{2}$.

9 **a** Use a substitution to find $\int x\sqrt{x+1}\,dx$.

b Use integration by parts to find $\int x\sqrt{x+1}\,dx$.

c Show that your answers to **parts a** and **b** are the same.

8.6 Integrating algebraic fractions

$\dfrac{x}{x^2+1}$, $\dfrac{1}{x^2+1}$ and $\dfrac{1}{x^2-1}$ are all examples of algebraic fractions.

How can we integrate each one?

Method 1: $\int \dfrac{x}{x^2+1}\,dx$

Try the substitution $u = x^2 + 1$.

$$\frac{du}{dx} = 2x \text{ and so } \frac{dx}{du} = \frac{1}{2x}$$

$$\int \frac{x}{x^2+1}\,dx = \int \frac{x}{u}\frac{dx}{du}\,du$$

$$= \int \frac{x}{u} \times \frac{1}{2x}\,du$$

$$= \frac{1}{2}\int \frac{1}{u}\,du$$

$$= \frac{1}{2}\ln|u| + c$$

So $\int \dfrac{x}{x^2+1}\,dx = \dfrac{1}{2}\ln(x^2+1) + c$

> The modulus sign is not needed because $x^2 + 1$ is always positive.

This is an example of a more general result: $\int \dfrac{f'(x)}{f(x)}\,dx = \ln|f(x)| + c$

This is easy to check by differentiating and using the chain rule:

$$\frac{d}{dx}\ln|f(x)| = \frac{1}{f(x)} \times f'(x)$$

Method 2: $\int \dfrac{1}{x^2+1}\,dx$

This looks similar to the first integral and you might be tempted to try the same substitution.

However, $u = x^2 + 1$ leads to $\int \dfrac{1}{x^2+1}\,dx = \int \dfrac{1}{u} \times \dfrac{1}{2x}\,du$, and the x does not cancel.

> Looking at the integral, you might have noticed that the numerator was a multiple of the derivative of the denominator and been able to guess the form of the answer immediately. However, this is not essential. The right substitution will lead to the answer too.

Substituting $x = \sqrt{u-1}$ gives $\int \dfrac{1}{2u\sqrt{u-1}}\,du$, which is too complex,

so you need to try something else.

Think about the standard derivatives you have learnt. Recall that

$$\frac{d}{dx}(\arctan x) = \frac{1}{1+x^2}$$

This immediately gives the result $\int \dfrac{1}{x^2+1}\,dx = \arctan x + c$

Alternatively, if you did not think of this, a substitution would still find the answer.

> See **Chapter 7 Further differentiation** and Section 8.2.

$x = \tan u$ gives $\dfrac{\mathrm{d}x}{\mathrm{d}u} = \sec^2 u$

and $\displaystyle\int \dfrac{1}{x^2+1}\,\mathrm{d}x = \int \dfrac{1}{\tan^2 u + 1} \times \dfrac{\mathrm{d}x}{\mathrm{d}u}\,\mathrm{d}u$

$\displaystyle\qquad\qquad\qquad = \int \dfrac{1}{\sec^2 u} \times \sec^2 u\,\mathrm{d}u$

$\displaystyle\qquad\qquad\qquad = \int 1\,\mathrm{d}u$

$\qquad\qquad\qquad = u + c$

$\qquad\qquad\qquad = \arctan x + c$

Method 3: $\displaystyle\int \dfrac{1}{x^2-1}\,\mathrm{d}x$

> This is not an obvious substitution to try. Make sure you learn the standard results.

After the second integral, you might expect this to be a standard result too. Unfortunately you will not find it in your list.

However, in this case the denominator factorises and you can write the algebraic fraction in partial fractions, using the methods of **Chapter 2**.

Suppose $\dfrac{1}{(x-1)(x+1)} \equiv \dfrac{A}{x-1} + \dfrac{B}{x+1}$

Therefore $\dfrac{1}{(x-1)(x+1)} \equiv \dfrac{A(x+1) + B(x-1)}{(x-1)(x+1)}$

Equate the numerators.

$$1 \equiv A(x+1) + B(x-1)$$

Substituting $x = 1$ gives $1 = 2A$ so $A = \dfrac{1}{2}$.

Substituting $x = -1$ gives $1 = -2B$ so $B = -\dfrac{1}{2}$.

So you can change the integration using $\dfrac{1}{(x-1)(x+1)} = \dfrac{\frac{1}{2}}{x-1} - \dfrac{\frac{1}{2}}{x+1}$

$$\int \dfrac{1}{(x-1)(x+1)}\,\mathrm{d}x = \dfrac{1}{2}\int \dfrac{1}{x-1}\,\mathrm{d}x - \dfrac{1}{2}\int \dfrac{1}{x+1}\,\mathrm{d}x$$

$$= \dfrac{1}{2}\ln|x-1| - \dfrac{1}{2}\ln|x+1| + c$$

The result could be written differently by using the laws of logarithms.

$$\int \dfrac{1}{(x-1)(x+1)}\,\mathrm{d}x = \dfrac{1}{2}\ln\left|\dfrac{x-1}{x+1}\right| + c \text{ or } \ln\left|\dfrac{x-1}{x+1}\right|^{\frac{1}{2}} + c.$$

Example 8

Find $\displaystyle\int \dfrac{3x-6}{(x+2)(x-1)}\,\mathrm{d}x$.

Solution

Since the degree of the numerator is less than the degree of the denominator, you can write this in partial fraction form.

$$\dfrac{3x-6}{(x+2)(x-1)} \equiv \dfrac{A}{x+2} + \dfrac{B}{x-1} = \dfrac{A(x-1) + B(x+2)}{(x+2)(x-1)}$$

Equate the numerators.

$$3x - 6 = A(x-1) + B(x+2)$$

Let $x = -2$: $-12 = -3A$ so $A = 4$.

Let $x = 1$: $-3 = 3B$ so $B = -1$.

So $\qquad \dfrac{3x-6}{(x+2)(x-1)} \equiv \dfrac{4}{x+2} - \dfrac{1}{x-1}$

and $\displaystyle\int \dfrac{3x-6}{(x+2)(x-1)}\,dx = \int \dfrac{4}{x+2} - \dfrac{1}{x-1}\,dx$

$$= 4\ln|x+2| - \ln|x-1| + c$$

You could use the laws of logarithms to write this as

$$\int \dfrac{3x-6}{(x+2)(x-1)}\,dx = \ln|x+2|^4 - \ln|x-1| + c$$

$$= \ln\left|\dfrac{(x+2)^4}{x-1}\right| + c$$

or even as $\ln\dfrac{(x+2)^4}{|x+1|} + c$, since a fourth power is never negative.

Exercise 8.6A

Answers page 455

1 Find:

   a $\displaystyle\int \dfrac{1}{2x}\,dx$ **b** $\displaystyle\int \dfrac{1}{2x+5}\,dx$

2 Find:

   a $\displaystyle\int \dfrac{1}{(x+2)^2}\,dx$ **b** $\displaystyle\int \dfrac{x}{(x+2)^2}\,dx$

3 Find $\displaystyle\int \dfrac{x-3}{x+3}\,dx$.

4 Find $\displaystyle\int \dfrac{1+x}{1+x^2}\,dx$.

5 Find $\displaystyle\int \dfrac{x-2}{x^2-4x+1}\,dx$.

6 **a** Show that $\dfrac{x^2}{x+1} = x - \dfrac{x}{x+1}$.

   b Hence, or otherwise, find $\displaystyle\int \dfrac{x^2}{x+1}\,dx$.

7 Find:

   a $\displaystyle\int \dfrac{4}{4x^2+1}\,dx$ **b** $\displaystyle\int \dfrac{4}{4x^2-1}\,dx$

(PS) 8 **a** Find $\displaystyle\int \dfrac{x}{(x+1)(x+2)}\,dx$

   b Hence find $\displaystyle\int \dfrac{x-1}{x(x+1)}\,dx$.

8.7 Solving differential equations

You saw examples of differential equations in **Chapter 7 Further differentiation**. Here are three differential equations:

$$\frac{dy}{dx} = 10\,e^{-0.5x}$$

$$\frac{dy}{dx} = \sin^2 2y$$

$$\frac{dy}{dx} = xy^2$$

You can check whether a particular function is a solution of a differential equation by differentiating it, but finding a general solution requires integration.

Look at each of these differential equations in turn.

Method 1: $\dfrac{dy}{dx} = 10\,e^{-0.5x}$

In this case $\dfrac{dy}{dx}$ is a function of x so you can simply integrate it:

$$y = \int 10\,e^{-0.5x}\,dx$$

You could guess that the answer is a multiple of $e^{-0.5x}$.

$$\frac{d}{dx}\left(e^{-0.5x}\right) = -0.5\,e^{-0.5x}$$

So $\displaystyle\int 10\,e^{-0.5x}\,dx = \frac{10}{-0.5}\,e^{-0.5x} + c$

$$= -20\,e^{-0.5x} + c$$

Notice that there is a constant in the answer. Because solving a differential equation involves integration, there will always be a constant in the general solution.

Method 2: $\dfrac{dy}{dx} = \sin^2 2y$

In this case $\dfrac{dy}{dx}$ is a function of y and does not involve x.

You know that the reciprocal of $\dfrac{dy}{dx}$ is $\dfrac{dx}{dy}$ so you can write

$$\frac{dx}{dy} = \frac{1}{\sin^2 2y} = \operatorname{cosec}^2 2y$$

Now $\dfrac{dx}{dy}$ is a function of y and not x so you can integrate.

$$x = \int \operatorname{cosec}^2 2y\,dy$$

This looks like a standard result, $\dfrac{d}{d\theta}(\cot\theta) = -\operatorname{cosec}^2\theta$, so you might guess the answer is a multiple of $\cot 2y$.

$$\frac{d}{dy}(\cot 2y) = -\operatorname{cosec}^2 2y \times 2$$

So $x = -\dfrac{1}{2}\cot 2y + c$

Here the equation for y is given implicitly.

Method 3: $\dfrac{dy}{dx} = -xy^2$

Put the x and y terms on opposite sides of the equation by dividing by : $-y^2$

$$-\dfrac{1}{y^2}\dfrac{dy}{dx} = x$$

> The y^2 term needs to be on the same side of the equation as $\dfrac{dy}{dx}$

Integrate both sides with respect to x:

$$\int -\dfrac{1}{y^2}\dfrac{dy}{dx}\,dx = \int x\,dx$$

The integral on the left hand side can be changed to an integral with respect to y:

$$\int -\dfrac{1}{y^2}\,dy = \int x\,dx$$

Hence $\dfrac{1}{y} = \dfrac{1}{2}x^2 + c$

You could leave the answer in this implicit form or you could rearrange it as

> Don't forget the constant – but you only need one, not two.

$$y = \dfrac{1}{\frac{1}{2}x^2 + c} \text{ or } y = \dfrac{2}{x^2 + 2c}$$

You might also replace $2c$ with a different constant, say, and write $y = \dfrac{2}{x^2 + k}$.

Check by differentiation that this is the solution to the differential equation.

> **Stop and think** A student treats $\dfrac{dy}{dx}$ as if it is a fraction and rewrites the equation as
>
> $$-\dfrac{1}{y^2}\,dy = x\,dx$$
>
> The student then puts in integral signs: $\int -\dfrac{1}{y^2}\,dx = \int x\,dx$ and integrates both sides. Is this a valid method?

This method is called **separating the variables**.

You can use it if you can put the x and y terms on different sides of the equation.

This is not always possible. For example, you cannot separate the variables in the equation $\dfrac{dy}{dx} = x + y$.

However, given a possible solution you could check by differentiation whether it is correct.

> **Stop and think** Show that $y = ce^x - x$ is the general solution to the differential equation $\dfrac{dy}{dx} = x + y$.

Up to this point, the constant has tended to be a letter that is just added on at the end of the solution. This is not usually the case with differential equations, where the constant may occur anywhere in the solution.

The **general solution** for $\dfrac{dy}{dx} = x + y$ gives a whole family of curves, one for each value of the constant c.

Here are four curves in the family, with $c = 2, 1, 0$ and -1.

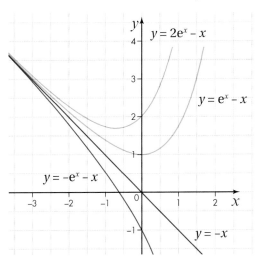

In a given context, there may be just one curve which is the required solution. This called a **particular solution**.

For example, the particular solution that passes through $(0, 1)$ is $y = e^x - x$.

A common type of differential equation occurs when the rate of change of a variable is proportional to the value of the variable. The equation will be of the form $\dfrac{dx}{dt} = kx$ where k is a constant.

Example 9

The number of fish in a lake is increasing annually. The population, x fish, in t years' time, is modelled by the equation $\dfrac{dx}{dt} = 0.2x$.

This year there are 300 fish.

a Work out a formula for the number of fish after t years.

b Describe any limitations in this solution.

Solution

a You can write the equation as $\dfrac{dt}{dx} = \dfrac{5}{x}$

Integrate: $t = \displaystyle\int \dfrac{5}{x}\,dx = 5\ln x + c$ where c is a constant.

$$\ln x = \dfrac{t - c}{5}$$

$$x = e^{0.2(t - c)}$$

> A modulus sign is not necessary because x must be positive.

You could leave the solution in this form, but is more usual to write it more neatly like this:

$$x = e^{0.2t - 0.2c} = e^{0.2t} \times e^{-0.2c}$$

You can write this as $x = k e^{0.2t}$, where you now have a different constant, k.

When $t = 0$, $x = 300$ so $300 = k \times 1$ and therefore $k = 300$.

The formula is $x = 300\,\mathrm{e}^{0.2t}$.

b This model implies that the population increases by more and more fish each year. This cannot continue for a long period because of limitations on space and food in the lake.

The solution will then no longer be valid.

The differential equation in **Example 9** could also have been solved by separating the variables:

$\dfrac{\mathrm{d}x}{\mathrm{d}t} = 0.2x$ becomes $\dfrac{1}{x}\,\mathrm{d}x = 0.2\,\mathrm{d}t$

Integrate: $\displaystyle\int \frac{1}{x}\,\mathrm{d}x = \int 0.2\,\mathrm{d}t$

Hence $\ln x = 0.2t + c$ or $x = \mathrm{e}^{0.2t + c}$

Therefore $x = \mathrm{e}^{0.2t} \times \mathrm{e}^{c} = k\,\mathrm{e}^{0.2t}$ as before.

> The constant 'c' here is different from the 'c' in **Example 9**.

Example 10

A stone is dropped from rest from the top of a high building.

The speed $v\,\mathrm{m\,s^{-1}}$ after $t\,\mathrm{s}$ is given by the differential equation $\dfrac{\mathrm{d}v}{\mathrm{d}t} = 10 - 0.5v$.

a Find an expression for v in terms of t.

b Show that the speed approaches a limiting value.

c Explain any limitations to this solution.

Solution

a You can write $\dfrac{\mathrm{d}t}{\mathrm{d}v} = \dfrac{1}{10 - 0.5v}$.

Then $t = \displaystyle\int \frac{1}{10 - 0.5v}\,\mathrm{d}v$.

It looks as if the solution is a multiple of $\ln|10 - 0.5v|$.

Use the chain rule.

$$\frac{\mathrm{d}}{\mathrm{d}v}\big(\ln|10 - 0.5v|\big) = \frac{1}{10 - 0.5v} \times -0.5$$

So $t = \displaystyle\int \frac{1}{10 - 0.5v}\,\mathrm{d}v = -2\ln|10 - 0.5v| + c$

Rearrange.

$$-2\ln|10 - 0.5v| = t - c$$

$$\ln|10 - 0.5v| = -\frac{1}{2}(t - c)$$

$$10 - 0.5v = \mathrm{e}^{-\frac{1}{2}(t - c)}$$

> Making the substitution $u = 10 - 0.5v$ would be another way to find the solution.

$$0.5v = 10 - e^{-\frac{1}{2}(t-c)}$$

Multiply by 2.

$$v = 20 - 2e^{-\frac{1}{2}(t-c)}$$

Write $2e^{-\frac{1}{2}(t-c)}$ as $2e^{-0.5t} \times e^{0.5c} = Ae^{-0.5t}$, where the constant is now $A = 2e^{0.5c}$.

> You can put the constant in any convenient form.

So $v = 20 - Ae^{-0.5t}$.

This is the general solution.

For the particular solution in this case, use the fact that the stone starts from rest, which means $v = 0$ when $t = 0$. Substitute these into the general solution.

$$0 = 20 - A \text{ so } A = 20$$

Therefore $v = 20 - 20e^{-0.5t}$.

> This is the particular solution required in this case.

b As t increases, $e^{-0.5t} \rightarrow 0$.

Therefore the speed approaches a limiting value of $20 - 0 = 20\,\text{m}\,\text{s}^{-1}$.

c The solution is only valid while the stone is in the air. If the building is not very high then the stone could hit the ground before it gets close to the limiting value of $20\,\text{m}\,\text{s}^{-1}$.

Exercise 8.7A

Answers page 455

1 Find the general solution of the equation $\dfrac{dy}{dx} = 2 - 0.3x^2$.

2 Solve the differential equation $\dfrac{dy}{dx} = 0.1y^2$.

3 Find the general solution of the differential equation $\dfrac{dy}{dx} = \sqrt{y}$.

4 Solve the differential equation $\dfrac{dy}{dx} = 3x^2y$.

5 **a** Solve the differential equation $\dfrac{dy}{dx} + \dfrac{y}{x} = 0$.

   b A curve passes through the point (6, 4) and $\dfrac{dy}{dx} + \dfrac{y}{x} = 0$.

    Find the equation of the curve.

 6 The equation of an ellipse satisfies the differential equation $\dfrac{dy}{dx} = -\dfrac{2x}{3y}$.

   a Show that the equation of the curve can be written as $x^2 + ny^2 = a$, where a is a constant and n is a number which you must find.

   b The point (5, 4) is on the curve.

    Find the coordinates of the points where the ellipse crosses the x-axis.

7 There are initially 1000 bacteria in a colony in a laboratory.

The number of bacteria, y, after t days is modelled by the differential equation $\dfrac{dy}{dt} = 1.5y$.

a Find a formula for the number of bacteria after t days.

b Find the number of bacteria after one week.

c Comment on any limitations on the validity of the solution.

8 For this curve, $\dfrac{dy}{dx} = -xy$ and it crosses the y-axis at $(0, 2)$.

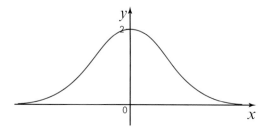

Find the equation of the curve in the form $y = f(x)$.

M **9** A stone is dropped from rest from the top of a high building.

The speed $v\,\mathrm{m\,s}^{-1}$ after $t\,\mathrm{s}$ is given by the differential equation $\dfrac{dv}{dt} = 10 - 0.5v$.

a Find an expression for v in terms of t, using the substitution $u = 10 - 0.5v$

b Show that the speed approaches a limiting value.

c Comment on any limitations to the solution.

M **10** A radioactive substance is decaying.

The mass, $x\,\mathrm{mg}$, left after t days satisfies the differential equation $\dfrac{dx}{dt} = -\lambda x$, where $\lambda > 0$.

a Show that a general solution can be written in the form $x = a\,e^{-\lambda t}$ and explain the interpretation of the constant a.

b The half-life of a radioactive element is the time for half of the substance to decay.

Show that the half-life of this element is $\dfrac{\ln 2}{\lambda}$.

SUMMARY OF KEY POINTS

- $\int \frac{1}{x} \, dx = \ln|x| + c, \quad x \neq 0$

- $\int \frac{1}{\sqrt{1-x^2}} \, dx = \arcsin x + c$

- $\int \frac{1}{1+x^2} \, dx = \arctan x + c$

Integration by substitution: $\int f(x) \, dx = \int f(x) \frac{dx}{du} \, du$

Integration by parts: $\int u \frac{dv}{dx} \, dx = uv - \int v \frac{du}{dx} \, dx$

EXAM-STYLE QUESTIONS 8

Answers page 455

1 Find $\int \frac{1}{5x+10} \, dx$. **[4 marks]**

2 **a** Find $\int \sec^2 0.5x \, dx$. **[4 marks]**

   b Use the result of **part a** to find $\int \tan^2 0.5x \, dx$. **[2 marks]**

3 The graph shows the region bounded by the lines $y = e^{0.5x}$, $y = e^{-0.5x}$ and $x = 3$.

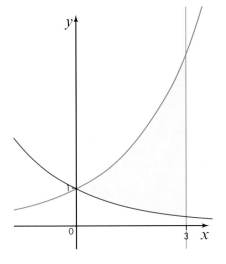

Find the area of the region. **[5 marks]**

4 Find $\int 2x\sqrt{2x+1} \, dx$. **[5 marks]**

5 **a** Find $\int x \sin 0.5x \, dx$. **[6 marks]**

   b The graph shows a graph of part of the curve $y = x \sin 0.5x$

   Calculate the area of the region between the curve and the x-axis. **[3 marks]**

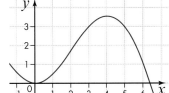

6 Work out:

a $\int \dfrac{1}{x^2+9}\,dx$ **[3 marks]**

b $\int \dfrac{x}{x^2+9}\,dx$ **[4 marks]**

c $\int \dfrac{x^2}{x^2+9}\,dx$ **[4 marks]**

(M) **7** A car is slowing down.

The speed, $v\,\mathrm{m\,s^{-1}}$ after $t\,$s satisfies the differential equation $\dfrac{dv}{dt}=-\dfrac{v^2}{100}$, $t \geqslant 0$.

The initial speed is $20\,\mathrm{m\,s^{-1}}$.

a Find an expression for the speed after $t\,$s. **[6 marks]**

b Explain why the solution may no longer be valid as t becomes large. **[1 mark]**

8 The graph shows the curve with the equation $y=\dfrac{100}{x^2-4}$.

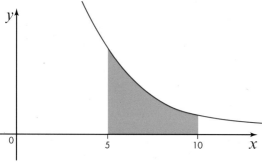

Calculate the area bounded by the curve, the x-axis and the lines $x=5$ and $x=10$. **[5 marks]**

(PS) **9** The normal to a curve at the point (x, y) has a gradient of $\dfrac{x}{2y}$.

a Explain why $\dfrac{dy}{dx}=-\dfrac{2y}{x}$. **[2 marks]**

b The points $(2, 5)$ and $(4, a)$ are on the curve. Find the value of a. **[8 marks]**

10 Find $\int \dfrac{x+7}{x^2-x-6}\,dx$. **[7 marks]**

11 Find $\int x^2 e^{-\frac{1}{2}x}\,dx$. **[6 marks]**

(M) **12** A car is moving away from traffic lights.

After t seconds the speed of the car is $v\,\mathrm{m\,s^{-1}}$ and the acceleration is proportional to the difference between the speed and $20\,\mathrm{m\,s^{-1}}$.

a Explain why $\dfrac{dv}{dt}=k(20-v)$ where k is a constant. **[1 mark]**

b Solve the differential equation $\dfrac{dv}{dt}=k(20-v)$ to give an equation for v in terms of t. **[5 marks]**

c What does the model imply about the speed as time increases? **[1 mark]**

9 NUMERICAL METHODS

A mobile phone needs software in order to work. One element of the software in a mobile phone manages the security of the phone. Whenever any new mobile phone goes on sale, it contains new software which will inevitably contain bugs. Some of these bugs might be in the security software and may, for example, allow someone to eavesdrop on your calls, or even to steal all your passwords. Over time, the mobile phone manufacturer will release software upgrades for your mobile phone and with each new release of software the number of bugs should decrease, so that eventually there are no bugs in the software. A number of iterations of new releases of software will be needed to get to the point where all the bugs have been eradicated.

Numerical methods start with an approximation to a solution and then use iteration (repeated processes) to refine the approximation to get one that is closer to the actual solution. In certain circumstances, though, iterations can take you further away from the solution rather than closer to it.

LEARNING OBJECTIVES

You will learn how to:

› locate roots of f(x) = 0 by considering changes of sign of f(x) in an interval of x on which f(x) is sufficiently well behaved

› solve equations approximately using simple iterative methods and to draw associated cobweb and staircase diagrams

› solve equations using the Newton–Raphson method and other recurrence relations of the form $x_{n+1} = \text{g}(x_n)$

› understand and use numerical integration of functions, including the use of the trapezium rule and estimating the approximate area under a curve and limits that it must lie between

› use numerical methods to solve problems in context.

TOPIC LINKS

The skills and techniques that you learn in this chapter will help you to approximate areas under curves, allowing the checking of solutions to problems in **Chapter 8 Integration**.

PRIOR KNOWLEDGE 9

You should already know how to:

- › simplify and manipulate algebraic expressions
- › work with coordinates in all four quadrants
- › identify and interpret roots (solutions) of equations
- › use and understand the graphs of functions
- › sketch curves defined by simple equations, including quadratics, cubics, quartics and reciprocals
- › interpret the algebraic solution of equations graphically.

You should be able to complete the following questions correctly:

1 Sketch the graph of $y = x^2 - 4x - 5$.

2 Sketch the graph of $y = x^3 - x^2 - 2x$.

3 Sketch the graph of $y = \dfrac{-3}{x+1}$, clearly indicating any asymptotes.

4 Sketch the graph of $y = -\dfrac{2}{(2x-1)^2}$, clearly indicating any asymptotes.

9.1 Finding roots

In **Book 1, Chapter 2 Algebra and functions 2: Equations and inequalities** you solved quadratic equations to find their roots (solutions). The method was to ensure that $f(x) = 0$ and then use either factorisation, the quadratic formula or completing the square to solve the equation and find the root(s).

For example, solve $f(x) = x^2 + x - 6$.

$$x^2 + x - 6 = 0$$

$$(x - 2)(x + 3) = 0$$

$$x = -3 \text{ or } x = 2$$

So $f(x) = x^2 + x - 6$ has solutions $x = -3$ or $x = 2$ when $f(x) = 0$ – that is, when the curve cuts the x-axis. If you hadn't been asked to solve $f(x) = x^2 + x - 6$ but instead to show that $f(x)$ has a root between -4 and -2, how could you have done this?

Here is a graph of this function.

What is significant about the value of $f(x)$ between -4 and the root (which we know to be at -3)?

$f(x) > 0$, because the curve is above the x-axis so all the y values are positive.

Similarly, what is significant about the value of $f(x)$ between the root (which we know to be at -3) and -2?

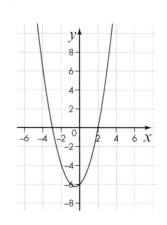

$f(x) < 0$, because the curve is below the x-axis so all the y values are negative.

If you can demonstrate that there is a sign change in $f(x)$ between -4 and $x = -2$ then you have shown that there is a root between these values.

$$f(x) = x^2 + x - 6$$

$$f(-4) = 16 - 4 - 6 = 6$$

$$f(-2) = 4 - 2 - 6 = -4$$

There is a sign change between $f(-4)$ and $f(-2)$, therefore there is a root of $f(x)$ between -4 and -2.

Watch out, though! There are some instances, which will be discussed in **Section 9.2**, that are exceptions to this rule.

Example 1

Show that $x^2 + x - 6 = 0$ has a root in the interval $1 < x < 3$.

Solution

$$f(x) = x^2 + x - 6$$

$$f(1) = 1 + 1 - 6 = -4$$

$$f(3) = 9 + 3 - 6 = 6$$

There is a sign change between $f(1)$ and $f(3)$, therefore there is a root of $f(x)$ in the interval $1 < x < 3$.

If there hadn't been a sign change when the interval limits were substituted into the function then you would have shown that $x^2 + x - 6 = 0$ did not have a root in the interval $1 < x < 3$ or that it had two roots in this interval.

Stop and think   Would you have identified a root if the interval given was $-4 < x < 3$? How does the interval relate to the identification of roots?

Example 2

Show that $\sin x = 0$ has a root in the interval $\dfrac{\pi}{2} < x < \dfrac{3\pi}{2}$.

Solution

$$f(x) = \sin x$$

$$f\left(\frac{\pi}{2}\right) = 1$$

$$f\left(\frac{3\pi}{2}\right) = -1$$

There is a sign change between $f\left(\dfrac{\pi}{2}\right)$ and $f\left(\dfrac{3\pi}{2}\right)$, therefore there is a root of $f(x)$ in the interval $\dfrac{\pi}{2} < x < \dfrac{3\pi}{2}$.

Note that the interval has been given in radians so you need to ensure that your calculator is in radian mode.

Remember that trigonometric functions have an infinite number of roots so be wary that any intervals you use are narrow enough to identify a sign change.

Example 3

a   On the same pair of axes, sketch the graphs of $y = x^3$ and $y = x^2 + 2x$, where $-3 \leqslant x \leqslant 3$. Use your graphs to show that the equation $x^3 = x^2 + 2x$ has three roots.

b Show that one root of the equation $x^3 = x^2 + 2x$ lies in the interval $1.8 < x < 2.1$.

Solution

a Using your knowledge from **Book 1, Chapter 3 Algebra and functions 3: Sketching curves**, you need to sketch the graphs of $y = x^3$ and $y = x^2 + 2x$ on the same pair of axes.

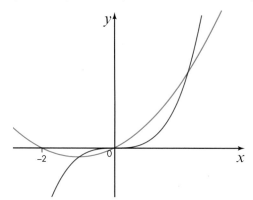

You need to highlight the three points of intersection on the graph as these are the roots of the equation $x^3 = x^2 + 2x$.

b Rearrange the equation into the form $f(x) = 0$.

$$x^3 - x^2 - 2x = 0$$

$$f(x) = x^3 - x^2 - 2x$$

Substitute $x = 1.8$ (the lower bound of the given interval) into $f(x)$.

$$f(1.8) = 5.832 - 3.24 - 3.6 = -1.008$$

Substitute $x = 2.1$ (the upper bound of the given interval) into $f(x)$.

$$f(2.1) = 9.261 - 4.41 - 4.2 = 0.651$$

There is a sign change between $f(1.8)$ and $f(2.1)$, therefore there is a root of $x^3 = x^2 + 2x$ in the interval $1.8 < x < 2.1$.

> **TECHNOLOGY**
>
> Using a graphic calculator or graphing software package, plot the graph of $y = x^3 - x^2 - 2x$. How many times does the curve cross the x-axis? Is there a root in the interval $1.8 < x < 2.1$?

Exercise 9.1A

Answers page 456

 1 Show that each of the following equations has a root in the given interval.

    a $3x^2 + 4x - 11 = 0$ $1.2 < x < 1.4$

    b $x^3 + 6x^2 + 11x + 6 = 0$ $-1.4 < x < -0.8$

    c $8x^4 + 2x^3 - 53x^2 + 37x - 6 = 0$ $1 < x < 3$

(PF) 2 Given that $f(x) = 1 - e^x - \ln x$, show that the equation $f(x) = 0$ has a root in the interval $0.2 < x < 0.7$.

(PF) 3 Show that $f(x) = \dfrac{\sin x}{e^x}$ does not have a root in the interval $-0.8 < x < -0.7$ when $f(x) = 0$.

4 **a** On the same pair of axes, sketch the graphs of $y = \dfrac{1}{x^2}$ and $y = e^x$ where $-4 \leqslant x \leqslant 4$.

Use your graphs to show that the equation $\dfrac{1}{x^2} = e^x$ has one root.

b Show that the root of the equation $\dfrac{1}{x^2} - e^x = 0$ lies in the interval $0.6 < x < 0.8$.

 5 Does the equation $\sin^2 x - \cos^2 x = 0$, where $0 < x < 2\pi$, have a root near to $x = \dfrac{\pi}{4}$?

You must show and explain all your mathematical reasoning.

 6 **a** On the same pair of axes, sketch the graphs of $y = 7 + \dfrac{5}{x}$ and $y = \dfrac{13}{x^2}$, where $-6 \leqslant x \leqslant 6$.

Using your graph, write down the number of roots for the equation $7 + \dfrac{5}{x} = \dfrac{13}{x^2}$.

b Show that a root of the equation $7 + \dfrac{5}{x} = \dfrac{13}{x^2}$ lies in the interval $0.95 < x < 1.1$.

c Explain how solving the equation $7x^2 + 5x - 13 = 0$ gives the solutions to $7 + \dfrac{5}{x} = \dfrac{13}{x^2}$.

Hence find the roots of $7 + \dfrac{5}{x} = \dfrac{13}{x^2}$ to 3 significant figures.

9.2 How change-of-sign methods can fail

In **Section 9.1** you saw that $x^2 + x - 6 = 0$ has two roots: one in the interval $-4 < x < -2$ and the other in the interval $1 < x < 3$. Would these roots have been identified if the interval had instead been $-4 < x < 3$?

$$f(x) = x^2 + x - 6$$

$$f(-4) = 16 - 4 - 6 = 6$$

$$f(3) = 9 + 3 - 6 = 6$$

There is no sign change between f(−4) and f(3) but this does *not* mean that there are no roots of f(x) between −4 and 3. There isn't a sign change because there are *two* roots. At the first root the sign changes from positive to negative and at the second it changes from negative to positive. The interval used is too wide and needs to be narrowed.

> **KEY INFORMATION**
>
> If the number of roots for f(x) = 0 in the given interval is even, then a sign change will not be detected.

Example 4

This is a sketch of $f(x) = 5 - 7x - 3x^2$. By evaluating intervals, show that f(x) = 0 has two roots in the interval $-4 < x < 2$.

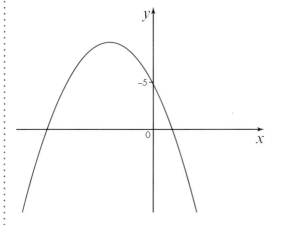

Solution

$$f(x) = 5 - 7x - 3x^2$$

$$f(-3) = 5 + 21 - 27 = -1$$

$$f(-2) = 5 + 14 - 12 = 7$$

There is a sign change between f(−3) and f(−2), therefore there is a root of f(x) in the interval $-3 < x < -2$.

$$f(0) = 5$$

$$f(1) = 5 - 7 - 3 = -5$$

There is a sign change between f(0) and f(1), therefore there is a root of f(x) in the interval $0 < x < 1$.

There are cases when f(x) changes sign but there is not a root. Consider $f(x) = \dfrac{1}{x}$ in the interval $-1 < x < 1$.

$$f(-1) = \frac{1}{-1} = -1$$

$$f(1) = \frac{1}{1} = 1$$

There is a sign change between f(−1) and f(1) but there isn't a root.

Here is the graph of $f(x) = \dfrac{1}{x}$:

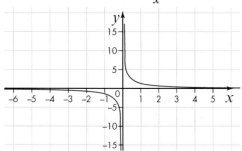

In the interval $-1 < x < 1$ there is a sign change but there isn't a root. Instead, there is a **discontinuity** at $x = 0$. The line $x = 0$ is an **asymptote**.

> **KEY INFORMATION**
>
> If there is a sign change, check that a root exists by sketching the function, and determine that the sign change does not represent a discontinuity.

Example 5

Show that f(x) = tan x does not have a root in the interval $\dfrac{\pi}{4} < x < \dfrac{3\pi}{4}$.

Solution

$$f(x) = \tan x$$

$$f\left(\frac{\pi}{4}\right) = 1$$

$$f\left(\frac{3\pi}{4}\right) = -1$$

There is a sign change between $f\left(\dfrac{\pi}{4}\right)$ and $f\left(\dfrac{3\pi}{4}\right)$.

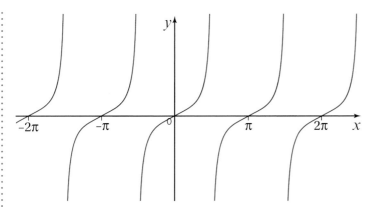

There is a discontinuity at $x = \dfrac{\pi}{2}$, so although there is a sign change between $f\left(\dfrac{\pi}{4}\right)$ and $f\left(\dfrac{3\pi}{4}\right)$ there isn't a root.

Exercise 9.2A

Answers page 457

1 This is a sketch of $f(x) = 6x^2 - x - 2$. By evaluating intervals, show that $f(x) = 0$ has two roots in the interval $-4 < x < 2$.

(PF) 2 By evaluating intervals, show that the function $f(x) = \dfrac{1}{x-3}$ has a discontinuity in the interval $2.4 < x < 3.3$.

(CM) 3 Explain why evaluation of roots cannot be used to show that $f(x) = (x \pm a)^2$ has a root in the interval $a - 1 < x < a + 1$.

(PF) 4 The function $f(x) = \dfrac{1}{x} + 2$ has a root and an asymptote in the interval $-1 < x < 1$. Sketch the function on a pair of axes.

    a Use a sub-interval to show a sign change for the root.

    b Use a sub-interval to show a sign change for the asymptote.

(PS) 5 For $f(x) = \tan x$, where, $\dfrac{\pi}{4} < x < \dfrac{3\pi}{2}$, choose suitable intervals to demonstrate the sign change of the function for each root and each asymptote.

(PF) 6 This is a sketch of $f(x) = \dfrac{1}{(3x+2)(2x-1)}$. Find intervals, to 1 decimal place, for each of the discontinuities and show that the sign of $f(x)$ changes within the intervals.

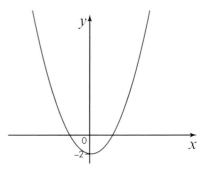

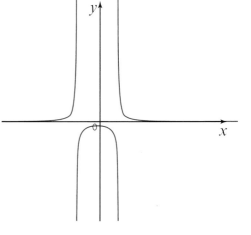

9.3 Iterative methods

An **iteration** is the repetition of a process or action. You can use iteration to find a sequence of approximations, each one getting closer to a root of $f(x) = 0$. An approximation in the sequence is found using a specified approach and then the result of this approximation is used in the next iteration until the required level of accuracy is found. You will learn about a general iteration method and the Newton–Raphson method.

To solve an equation in the form $f(x) = 0$ by an iterative method, you first need to rearrange $f(x) = 0$ into a form where $x = g(x)$. You then use the iterative formula $x_{n+1} = g(x_n)$ for the iterations. There may be more than one rearrangement of $f(x) = 0$.

$x_{n+1} = g(x_n)$ converges to a root a provided that the initial approximation x_0 is close to a and that $-1 < g'(a) < 1$, where $g'(a)$ is the value of the first derivative of $g(x_n)$ when $x_n = a$.

> **KEY INFORMATION**
>
> To solve an equation in the form $f(x) = 0$ by an iterative method, rearrange $f(x) = 0$ into a form where $x = g(x)$. Subsequently use the formula $x_{n+1} = g(x_n)$ for the iterations.

Example 6

a Show that one root of the equation $x^2 = 7x + 3$ lies in the interval $7 < x < 8$.

b Show that $x^2 = 7x + 3$ can be written in the form $x = 7 + \dfrac{3}{x}$.

c Find the root of $x^2 = 7x + 3$ in the interval $7 < x < 8$, to 3 decimal places.

d Show graphically the first two iterations for this formula.

Solution

a Rearrange the equation so that $f(x) = 0$.

$x^2 - 7x - 3 = 0$

$f(7) = 49 - 49 - 3 = -3$

$f(8) = 64 - 56 - 3 = 5$

There is a sign change between $f(7)$ and $f(8)$, therefore there is a root of $x^2 = 7x + 3$ in the interval $7 < x < 8$.

b $x^2 = 7x + 3$

Divide both sides of the equation by x.

$x = 7 + \dfrac{3}{x}$

c Specify the iteration formula.

As $x = 7 + \dfrac{3}{x}$, the iteration formula will be

$x_{n+1} = 7 + \dfrac{3}{x_n}$

Let $x_0 = 7$ (the lower bound of the interval containing the root).

$x_1 = 7 + \dfrac{3}{x_0}$

$= 7 + \dfrac{3}{7} = 7.428\,571\,429$

$x_2 = 7 + \dfrac{3}{x_1}$

$= 7 + \dfrac{3}{7.428\,571\,429} = 7.403\,846\,154$

> **KEY INFORMATION**
>
> If each iteration gets closer to the root, the sequence is **convergent**.

You need to be able to use the memory function on your calculator to store and recall the value from each iteration.

Each iteration gets closer to the root.

$x_3 = 7.405\,194\,805$

$x_4 = 7.405\,121\,01$

The last two iterations gave the same answer to 3 decimal places, so you can now stop.

So a root of $x^2 = 7x + 3$ in the interval $7 < x < 8$ is 7.405, to 3 decimal places.

d To show the iterations for the formula graphically, you first need to draw the graphs of $y = x$ and $y = 7 + \dfrac{3}{x}$ on the same pair of axes. The graphs intersect when $y = 7 + \dfrac{3}{x}$, which is the iterative formula you have used.

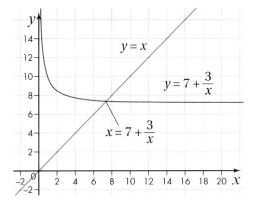

If you substitute $x = 7$ into $y = 7 + \dfrac{3}{x}$, then $y = 7.428\,571\,429$. On the graph this is the same as moving vertically from $x = 7$ to $y = 7.428\,571\,429$, shown by line ① on the graph below.

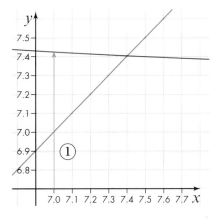

Now let $x = 7.428\,571\,429$; this is a horizontal movement to $y = x$ (line ② below).

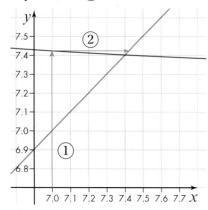

Substitute $x = 7.428\,571\,429$ into $y = 7 + \dfrac{3}{x}$; then

$y = 7.403\,846\,154$. On the graph this is the same as moving vertically from $y = 7.428\,571\,429$ to $y = 7.403\,846\,154$ (line ③).

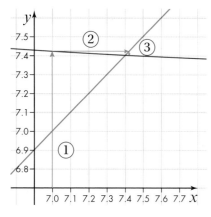

Now let $x = 7.403\,846\,154$, which is a horizontal movement to $y = x$ (line ④).

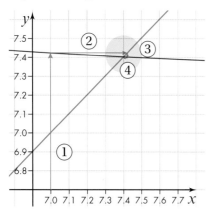

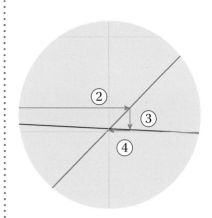

You have produced a **cobweb diagram**. A more general form is shown below.

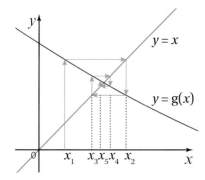

In **Example 6b** the function was rearranged into a specific iterative formula. Are there other ways in which this function could have been rearranged? Generally, can there be more than one iteration formula for f(x) = 0?

Example 7

a Show that $x^3 - 5x^2 - 4x + 7 = 0$ can be written in the form $x = a + \dfrac{b}{x} + \dfrac{c}{x^2}$. State the values of the constants a, b and c.

b Hence, or otherwise, use a suitable iteration formula to find an approximate solution of the equation, to 2 decimal places, with $x_0 = 1$.

c Show graphically the first two iterations for this formula.

Solution

a $x^3 - 5x^2 - 4x + 7 = 0$

Divide all terms in the equation by x^2.

$$x - 5 - \frac{4}{x} + \frac{7}{x^2} = 0$$

Rearrange the equation to make x the subject.

$$x = 5 + \frac{4}{x} - \frac{7}{x^2}$$

State the values of a, b and c.

$$a = 5, b = 4, c = -7$$

b Specify the iteration formula.

As $x = 5 + \dfrac{4}{x} - \dfrac{7}{x^2}$,

the iteration formula will be

$$x_{n+1} = 5 + \frac{4}{x_n} - \frac{7}{(x_n)^2}$$

$$x_0 = 1$$

$$x_1 = 5 + \frac{4}{1} - \frac{7}{1} = 2$$

$$x_2 = 5.25$$

$$x_4 = 5.495\,486$$

$$x_5 = 5.496\,085$$

So a root of $x^3 - 5x^2 - 4x + 7 = 0$ with $x_0 = 1$ is 5.50, to 2 decimal places.

c To show the iterations for the formula graphically, you first need to draw the graphs of $y = x$ and $y = 5 + \dfrac{4}{x} - \dfrac{7}{x^2}$ on the same pair of axes.

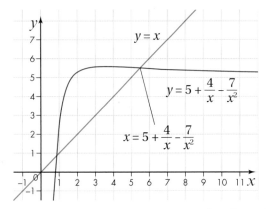

The graphs intersect when $x = 5 + \dfrac{4}{x} - \dfrac{7}{x^2}$, which is the iterative formula you have used.

If you substitute $x = 1$ into $y = 5 + \dfrac{4}{x} - \dfrac{7}{x^2}$ then $y = 2$. On the graph this is the same as moving vertically from $x = 1$ to $y = 2$, shown by line ① on the graph below.

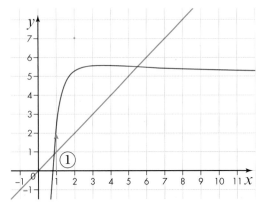

Now let $x = 2$; this is a horizontal movement to $y = x$ (line ②
below)

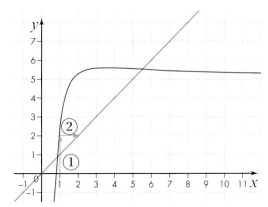

Substitute $x = 2$ into $y = 5 + \dfrac{4}{x} - \dfrac{7}{x^2}$; then $y = 5.25$. On the graph
this is the same as moving vertically from $y = 2$ to
$y = 5.25$ (line ③).

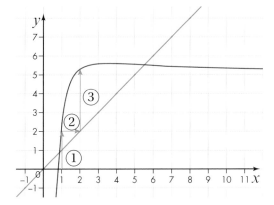

Now let $x = 5.25$; this is a horizontal movement to
$y = x$ (line ④).

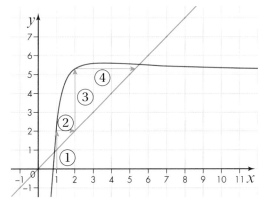

You have produced a **staircase diagram**. A more general form is shown below.

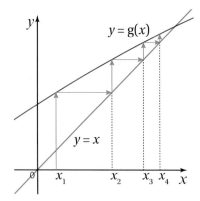

Stop and think

What is the relationship between g(x), the iteration formula and the type of graphical representation? Is it possible to have a combined diagram? If so, under what circumstances?

Exercise 9.3A

Answers page 458

(PF) 1 Show that $x^2 = 11 - 5x$ can be written in each of the following forms:

   a $x = \sqrt{11 - 5x}$ **b** $x = \dfrac{11}{x} - 5$ **c** $x = \dfrac{x^2 - 11}{-5}$

2 The equation $x^3 - 6 = x^2$ has a root in the interval $2 < x < 3$. Use the iteration formula

   $x_{n+1} = \sqrt{\dfrac{6}{x_n - 1}}$ starting with $x_0 = 2$ to find x_3 to 3 decimal places.

(PF) 3   **a** Show that one root of the equation $x^2 + 4x - 7 = 0$ lies in the interval $-6 < x < -5$.

   b Show that $x^2 + 4x - 7 = 0$ can be written in the form $x = a + \dfrac{b}{x}$, where a and b are constants to be found.

   c Find the root of $x^2 + 4x - 7 = 0$ in the interval $-6 < x < -5$, to 2 decimal places.

   d Show graphically the first two iterations for this formula.

(PS) 4   **a** Show that the equation $2x^3 - 5x + 1 = 0$ has a root in the interval $-2 < x < -1$.

(PF)   **b** Using a suitable iteration formula, find the root of $2x^3 - 5x + 1 = 0$ in the interval $-2 < x < -1$, to 3 decimal places.

   c Show graphically the first two iterations for this formula.

(PS) 5   **a** Find the root of $x^5 + 3x^2 - 7 = 0$ near $x = 1$, to 2 decimal places.

   b Show graphically the first two iterations for this formula.

(PS) 6 Use a suitable general iteration formula to find the two positive roots of the equation $x^4 - 4x^2 + x = 0$, to 2 decimal places.

9.4 The Newton–Raphson method

The Newton–Raphson method was described and published, albeit independently, by mathematicians Isaac Newton and Joseph Raphson in the 17th century. It is a method to find the root of an equation in a given interval using the tangent to the curve. The method uses a **recurrence relation**, where the next term is defined by a function of the previous term.

You need to consider the point at which the tangent meets the curve. At this point the gradient is given by:

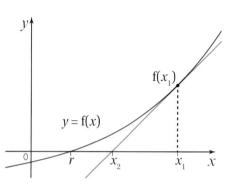

$$m = \frac{\text{change in } y}{\text{change in } x}$$

$$= \frac{f(x_2) - f(x_1)}{x_2 - x_1}$$

The gradient is given by the first derivative so m can be replaced by $f'(x_1)$.

$$f'(x_1) = \frac{f(x_1)}{x_1 - x_2}$$

If you rearrange this equation and change it into a general formula, you have:

$$x_{n+1} = x_n - \frac{f(x_n)}{f'(x_n)}$$

This is the Newton–Raphson formula.

Example 8

Use the Newton–Raphson method to find the root of $x^2 - 7x - 3 = 0$ in the interval $7 < x < 8$, to 3 decimal places.

Solution

Let $\quad f(x) = x^2 - 7x - 3$

so $\quad f'(x) = 2x - 7$

Let $x_0 = 7$ (the lower bound of the interval containing the root).

So $\quad x_1 = x_0 - \dfrac{f(x_0)}{f'(x_0)}$

$$= 7 - \frac{7^2 - (7)(7) - 3}{(2)(7) - 7}$$

$$= 7.428\,571\,429$$

So $\quad x_2 = x_1 - \dfrac{f(x_1)}{f'(x_1)}$

$$= 7.428\,571\,429 - \frac{7.428\,571\,429^2 - (7)(7.428\,571\,429) - 3}{(2)(7.428\,571\,429) - 7}$$

$$= 7.405\,194\,805$$

$x_3 = 7.405\,124\,839$

> **KEY INFORMATION**
>
> The Newton–Raphson formula is $x_{n+1} = x_n - \dfrac{f(x_n)}{f'(x_n)}$.
>
> You will be given this formula in exams.

> **KEY INFORMATION**
>
> The Newton–Raphson formula is very efficient if $|x_{n+1} - x_n|$ is very small.

You now have two iterations that have given the same answer to 3 decimal places, so you can stop.

So a root of $x^2 = 7x + 3$ in the interval $7 < x < 8$ is 7.405, to 3 decimal places.

If you compare this method with the general iterative method used in **Example 6** above, the Newton–Raphson method results in one fewer iterations in this case.

Example 9

Use the Newton–Raphson method to find an approximate solution of $f(x) = x^3 - 5x^2 - 4x + 7$ near to 5, correct to 2 decimal places.

Solution

Let $f(x) = x^3 - 5x^2 - 4x + 7$

so $f'(x) = 3x^2 - 10x - 4$

Let $x_0 = 5$.

So $x_1 = x_0 - \dfrac{f(x_0)}{f'(x_0)}$

$= 5 - \dfrac{5^3 - (5)(5^2) - (4)(5) + 7}{(3)(5^2) - (10)(5) - 4}$

$= 5.619\,048$

$x_2 = 5.501\,198$

$x_3 = 5.496\,067$

You now have two iterations that have given the same answer to 2 decimal places, so you can stop.

So an approximate solution of $x^3 - 5x^2 - 4x + 7 = 0$ near to 5 is 5.50, to 2 decimal places.

Exercise 9.4A

Answers page 459

1 **a** Given that $f(x) = x^2 + 5x - 11$, find the first derivative $f'(x)$ of $f(x)$.

  b Use the Newton–Raphson method to find the root of $f(x) = 0$ in the interval $1 < x < 2$, correct to 1 decimal place.

2 Use the Newton–Raphson process three times to find the root of the equation $x^5 + 3x^2 - 7 = 0$ near $x = 1$. Give the answer correct to 2 decimal places.

3 The equation $x^3 - x^2 - 6 = 0$ has a root in the interval $2.2 < x < 2.4$. Use the Newton–Raphson method starting with $x_0 = 2.2$ to find the root, correct to 3 decimal places.

4 Apply the Newton–Raphson method to find the negative root of the equation $2x^3 - 5x + 1 = 0$, correct to 2 decimal places.

 5 The equation $x^2 + 4x - 7 = 0$ has both a negative and positive root. Use the Newton–Raphson method to find both roots, correct to 2 decimal places.

 6 a Use a suitable general iteration formula to find the negative root of the equation $x^4 - 4x^2 + x = 0$, correct to 2 decimal places.

b Use the Newton–Raphson method to find the negative root of the equation $x^4 - 4x^2 + x = 0$, correct to 2 decimal places.

c Compare the efficiency and accuracy of each method in this case and state a preference, giving clear justification for your choice.

9.5 How iterative methods can vary

As you have seen, equations can be rearranged in a number of different ways to generate an iteration formula. Sometimes different iteration formulae for $f(x) = 0$ may result in different roots of the equation from the same value of x_0.

> **KEY INFORMATION**
>
> Different rearrangements of $f(x) = 0$, giving different iteration formulae, may result in different roots.

Example 10

a $x^2 - 7x - 3 = 0$ can be written in the form $x = 7 + \dfrac{3}{x}$.

Show that it can also be written in the form $x = \dfrac{x^2 - 3}{7}$.

b Show that the iteration formulae $x_{n+1} = 7 + \dfrac{3}{x_n}$ and $x_{n+1} = \dfrac{(x_n)^2 - 3}{7}$ give different roots of the equation to 1 decimal place when $x_0 = 3$.

Solution

a $x^2 - 7x - 3 = 0$

Add $7x$ to both sides of the equation.

$$x^2 - 3 = 7x$$

Divide both sides of the equation by 7.

$$x = \frac{x^2 - 3}{7}$$

b $x_{n+1} = 7 + \dfrac{3}{x_n}$

$x_0 = 3$

$x_1 = 8$

$x_2 = 7.375$

$x_3 = 7.406\,779\,661$

$x_4 = 7.405\,034\,325$

This formula converges to a root of 7.4, to 1 decimal place.

$$x_{n+1} = \frac{(x_n)^2 - 3}{7}$$

$x_0 = 3$

$x_1 = 0.857\,142\,857\,1$

$x_2 = -0.323\,615\,160\,3$

$x_3 = -0.413\,610\,461\,1$

$x_4 = -0.404\,132\,340\,9$

This formula converges to a root of -0.4, to 1 decimal place.

Stop and think   Is there any relationship between the iteration formula chosen and the root to which it converges?

Sometimes, although a value for x_0 is chosen that is close to a root, repeated iterations may not converge to a root. Instead, they may diverge. This is true both for general iterative methods and for the Newton–Raphson process.

KEY INFORMATION

A suitable iteration formula and a value of x_0 close to the root do not always result in approximations that converge to the root – instead, they may diverge.

Example 11

The equation $e^x - x - 2 = 0$ has a root in the interval $1 < x < 2$. By using the iteration formula $x_{n+1} = e^{x_n} - 2$ with $x_0 = 2$, find the root to 2 decimal places.

Solution

$x_{n+1} = e^{x_n} - 2$

$x_0 = 2$

$x_1 = 5.389\,056\,099$

$x_2 = 216.996\,576\,9$

$x_3 = 1.739\,465\,916 \times 10^{94}$

This is a divergent sequence, so it is not possible to find the root to 2 decimal places starting with $x_0 = 2$.

Example 12

a   Sketch the curve of $y = \dfrac{1}{x^2} - \dfrac{1}{24}$.

b   The curve has a root between 4 and 18. Use the Newton–Raphson process three times to find the root of the equation $\frac{1}{x^2} - \frac{1}{24} = 0$ near $x = 11$. Give the answer correct to 2 decimal places.

Solution

a   Use your knowledge from **Book 1, Chapter 3 Algebra and functions 3: Sketching curves** to sketch the graph.

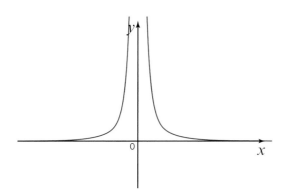

b Let $f(x) = \dfrac{1}{x^2} - \dfrac{1}{24}$

So $f'(x) = -\dfrac{2}{x^3}$

Let $x_0 = 11$.

So $x_1 = x_0 - \dfrac{f(x_0)}{f'(x_0)}$

$x_1 = -11.229\,166\,67$

$x_2 = 12.654\,846$

$x_3 = -23.238\,812\,76$

The Newton–Raphson method fails in this case, because it is not converging towards a root.

> **KEY INFORMATION**
>
> The Newton–Raphson process fails if $f'(a)$ is close to zero.

In general, the Newton–Raphson method fails when $f'(a)$ is close to zero.

Exercise 9.5A

Answers page 459

1 $x^2 = 11 - 5x$ can be written in each of the following forms:

a $x = \sqrt{11 - 5x}$ **b** $x = \dfrac{11}{x} - 5$ **c** $x = \dfrac{x^2 - 11}{5}$

The equation $x^2 + 5x - 11 = 0$ has a root in the interval $1 < x < 2$. Using each of the above forms as the basis for a general iteration formula, and with $x_0 = 1$, determine whether each formula results in this root or an alternative root.

2 The equation $x^3 - 6 = x^2$ has a root in the interval $2 < x < 3$. Use the iteration formula

$x_{n+1} = \sqrt{\dfrac{6}{x_n - 1}}$ starting with $x_0 = 5$ four times. Does this combination of iteration formula and starting value result in a convergent or divergent sequence of results?

PF 3 Show that two different forms of the equation $x^2 - 4x - 7 = 0$ with $x_0 = -5$ result in convergence to two different roots.

PS 4 The equation $2x^3 - 5x + 1 = 0$ has a root in the interval $-2 < x < -1$. Using the Newton–Raphson method, find a value of x_0 such that a divergent sequence of results is produced.

5 The equation $x^5 + 3x^2 - 7 = 0$ has a root near $x = 1.5$. Is there a general iteration formula that produces a divergent sequence of results when $x_0 = 1.5$?

 6 The equation $x^4 - 4x^2 + x = 0$ has a root in the interval $1.8 < x < 1.9$. Using the Newton–Raphson method, find the first integer value of $x_0 > 1.9$ such that a divergent sequence of results is produced.

9.6 Numerical integration

In **Chapter 8 Integration** you integrated functions between two limits and consequently found the area under the curve. However, some functions cannot be integrated because they cannot be written as **elementary functions** that can subsequently be integrated. For example, $\sin(x^2)$ cannot be integrated. Consequently, if you cannot integrate the function then you need to be able to approximate the area under the curve. You can do this using a number of different methods, including the trapezium rule.

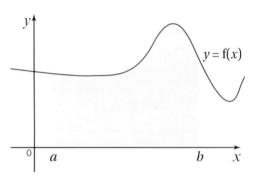

The graph of $f(x)$ is shown in the first diagram on the right. If $f(x)$ cannot be integrated, how can you find the area of the curve between the limits a and b?

If the area is divided up into n strips of equal width, then you now have the second diagram. The width of each strip will be $h = \dfrac{b-a}{n}$.

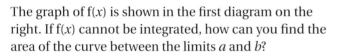

If a line is drawn on each strip joining the top left-hand corner to the top right-hand corner each strip becomes a trapezium.

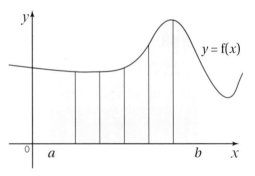

If you find the area of each trapezium and then add them together you will have an approximation for the area under the curve between a and b. Notice that there are some sections of the area under the curve that are not included in some of the trapezia and some extra area above the curve that are included in others. Where there are numerous areas under the curve that are not included in trapezia, this will result in an underestimation. Conversely, where there are numerous extra areas above the curve that are included in trapezia, this will result in an overestimation.

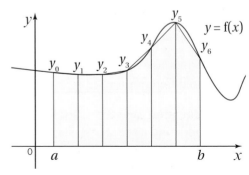

190

Working from left to right, the area of the first trapezium is $\frac{1}{2}h(y_0 + y_1)$, the area of the second trapezium is $\frac{1}{2}h(y_1 + y_2)$, and so on. So an approximation of the area under the curve is given by:

$$\int_a^b y\,dx \approx \frac{1}{2}h(y_0 + y_1) + \frac{1}{2}h(y_1 + y_2) + \cdots + \frac{1}{2}h(y_{n-1} + y_n)$$

Notice that each y value is shared by two trapezia, with the exception of y_0 and y_n. Consequently, the formula can be simplified as follows:

$$\int_a^b y\,dx \approx \frac{1}{2}h\left[y_0 + y_n + 2(y_1 + y_2 + \cdots + y_{n-1})\right]$$

This is known as the trapezium rule.

> The wiggly equals sign is used to indicate approximation.

KEY INFORMATION

The trapezium rule is:

$$\int_a^b y\,dx \approx$$

$$\frac{1}{2}h\left[y_0 + y_n + 2(y_1 + y_2 + \cdots + y_{n-1})\right]$$

where $h = \dfrac{b - a}{n}$.

Stop and think   What is the relationship between the number of strips and the accuracy of the approximation?

Example 13

Find an approximation for the area under the curve $f(x) = \sqrt{\operatorname{cosec} x}$ between $x = \dfrac{\pi}{6}$ and $x = \dfrac{5\pi}{6}$ using the trapezium rule with:

a 4 strips   **b** 8 strips.

Solution

a $h = \dfrac{\dfrac{5\pi}{6} - \dfrac{\pi}{6}}{4}$

$= \dfrac{\pi}{6}$

You need to work out the y values at the boundaries of each strip.

x	$\dfrac{\pi}{6}$	$\dfrac{\pi}{3}$	$\dfrac{\pi}{2}$	$\dfrac{2\pi}{3}$	$\dfrac{5\pi}{6}$
y	1.41	1.07	1	1.07	1.41

State the trapezium rule formula.

$$\int_a^b y\,dx \approx \frac{1}{2}h\left[y_0 + y_n + 2(y_1 + y_2 + \cdots + y_{n-1})\right]$$

Substitute the values for h and y_m into the trapezium rule formula.

$$\int_a^b y\,dx \approx \left(\frac{1}{2}\right)\left(\frac{\pi}{6}\right)\left[1.41 + 1.41 + 2(1.07 + 1 + 1.07)\right]$$

Calculate the approximation.

$$\int_a^b y\,dx \approx \frac{91}{120}\pi \text{ or } 2.38 \text{ (to 2 decimal places)}$$

b $h = \dfrac{\dfrac{5\pi}{6} - \dfrac{\pi}{6}}{8}$

$= \dfrac{\pi}{12}$

You need to work out the y values at the boundaries of each strip.

x	$\dfrac{\pi}{6}$	$\dfrac{\pi}{4}$	$\dfrac{\pi}{3}$	$\dfrac{5\pi}{12}$	$\dfrac{\pi}{2}$	$\dfrac{7\pi}{12}$	$\dfrac{2\pi}{3}$	$\dfrac{3\pi}{4}$	$\dfrac{5\pi}{6}$
y	1.41	1.19	1.07	1.02	1	1.02	1.07	1.19	1.41

State the trapezium rule formula.

$$\int_a^b y \, dx \approx \frac{1}{2}h\left[y_0 + y_n + 2\left(y_1 + y_2 + \cdots + y_{n-1}\right) \right]$$

Substitute the value for h and y_m into the trapezium rule formula.

$$\int_a^b y \, dx \approx \left(\frac{1}{2}\right)\left(\frac{\pi}{12}\right)[1.41 + 1.41$$
$$+ 2(1.19 + 1.07 + 1.02 + 1 + 1.02 + 1.07 + 1.19)]$$

Calculate the approximation.

$$\int_a^b y \, dx \approx \frac{299}{400}\pi \text{ or } 2.35 \text{ (to 2 decimal places)}$$

Exercise 9.6A

Answers page 459

1 By copying and completing the following table, use the trapezium rule to estimate $\int_1^3 \dfrac{1}{x} \, dx$.

x	1	$\dfrac{3}{2}$	2	$\dfrac{5}{2}$	3
y	1		$\dfrac{1}{2}$		

CM 2 **a** Use integration to find the exact value of $\int_0^2 (x^4 + 7)\,dx$.

b By copying and completing the following table, use the trapezium rule to estimate $\int_0^2 (x^4 + 7)\,dx$.

x	0	$\dfrac{1}{2}$	1	$\dfrac{3}{2}$	2
y	7		8		

c Compare the answers to **parts a** and **b** and comment on your results.

3 **a** Use the trapezium rule with six strips to estimate the value of $\int_1^3 \frac{1}{(1+3x)^2}$.

b Is your answer to **part a** an overestimate or an underestimate? Justify your answer.

 4 **a** Use integration to find the exact value of $\int_0^{\frac{\pi}{2}} \sin x \, dx$.

b Use the trapezium rule with four strips to estimate the value of $\int_0^{\frac{\pi}{2}} \sin x \, dx$.

c Use the trapezium rule with eight strips to estimate the value of $\int_0^{\frac{\pi}{2}} \sin x \, dx$.

d Compare the answers to **parts a, b** and **c** and comment on your results.

 5 Find the percentage error when the trapezium rule is used to evaluate $\int_{-2}^{-1} (e^x - 1)^2 \, dx$ using four strips.

 6 Find the number of strips required to ensure that the percentage error is less than 2% when the trapezium rule is used to evaluate $\int_0^{\frac{\pi}{2}} \cos x \, dx$.

9.7 Using numerical methods to solve problems

You need to be able to solve problems in context using numerical methods.

Example 14

Use the Newton–Raphson process to approximate $\sqrt{2}$ to 3 decimal places.

Solution

Let $x = \sqrt{2}$.

Rearrange.
$$x^2 - 2 = 0$$

Let $f(x) = x^2 - 2$

so $f'(x) = 2x$

Let $x_0 = 1$.

So $\quad x_1 = x_0 - \dfrac{f(x_0)}{f'(x_0)}$

$\qquad = 1.5$

$\quad x_2 = 1.416\,667$

$\quad x_3 = 1.414\,216$

$\quad x_4 = 1.414\,214$

x_3 and x_4 are the same to 3 decimal places so $\sqrt{2} = 1.414$, to 3 decimal places.

Example 15

A cliff is subject to coastal erosion and is in danger of collapse. The diagram shows the profile of the cliff before (in blue) and after a number of years of erosion has taken place. The profile cliff at the end of the period is shown by the red curve given by $f(x) = \ln\dfrac{2x+1}{2} + 6$.

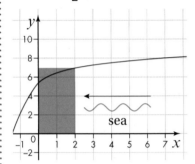

Using the trapezium rule with six strips, work out an estimate of the area, in m², of the profile of the cliff that has been lost to erosion.

Is this an over estimate or an underestimate of the actual amount lost?

Solution

$h = \dfrac{2-0}{6} = \dfrac{1}{3}$

You need to work out the y values at the boundaries of each strip.

x	0	$\frac{1}{3}$	$\frac{2}{3}$	1	$\frac{4}{3}$	$\frac{5}{3}$	2
y	5.31	5.82	6.15	6.41	6.61	6.77	6.92

State the trapezium rule formula.

$$\int_a^b y\,\mathrm{d}x \approx \frac{1}{2}h\left[y_0 + y_n + 2(y_1 + y_2 + \cdots + y_{n-1})\right]$$

Substitute the values for h and y_m into the trapezium rule formula.

$$\int_0^2 y\,\mathrm{d}x \approx \left(\frac{1}{2}\right)\left(\frac{1}{3}\right)\Big[5.31 + 6.92$$
$$+2\left(5.82 + 6.15 + 6.41 + 6.61 + 6.77\right)\Big]$$

Calculate the approximation.

$$\int_0^2 y\,\mathrm{d}x \approx \frac{101}{8} \text{ or } 12.625\,\text{m}^2 \text{ (to 3 decimal places)}$$

This will be an underestimate, as the areas of all the trapezia underneath the curve.

Exercise 9.7A Answers page 460

 1 Use the Newton–Raphson method to approximate $\sqrt{11}$ to 3 decimal places.

2 **a** Use the trapezium rule with four strips to approximate the area under the curve $f(x) = x^3 - 3x - 4$ between $x = 3$ and 4.

    b Use integration to work out the exact value of $\int_3^4 (x^3 - 3x - 4)\,dx$.

    c Work out the percentage error of the estimate.

3 Using the Newton–Raphson method, show that if $f(x) = x^2 - a$ then $x_{n+1} = \dfrac{1}{2}\left(x_n + \dfrac{a}{x_n} \right)$.

 4 The area of the triangle shown is 24 units^2.

Use a general iterative method to find the lengths of the two labelled sides of the triangle, correct to 2 decimal places.

5 A pharmaceutical manufacturer produces a drug and sells it at a price of £2.20 per gram.

The cost (in pence per gram) of producing is $f(m) = m^{\frac{1}{5}} + m + 150$.

Use a general iterative method to find the number of grams at which sales exceed production costs.

 6 You have been asked to write a computer program to find reciprocals.

Use the Newton–Raphson method to approximate the reciprocal of 1.9 to 2 decimal places.

SUMMARY OF KEY POINTS

❯ An iteration is a repeated process or action.

❯ Generally, if there is an interval in which the sign changes, then the interval contains a root of $f(x) = 0$.

❯ If the number of roots for $f(x) = 0$ in the given interval is even, then a sign change will not be detected.

❯ If there is a sign change, check that a root exists (and not a discontinuity) by sketching the function.

❯ To solve an equation in the form $f(x) = 0$ by an iterative method, rearrange $f(x) = 0$ into a form where $x = g(x)$. Then use the formula $x_{n+1} = g(x_n)$ for the iterations.

❯ Different rearrangements of $f(x) = 0$, giving different iteration formulae, may result in different roots.

❯ $x_{n+1} = g(x_n)$ converges to a root a provided that the initial approximation x_0 is close to a and $-1 < g'(a) < 1$, where $g'(a)$ is the value of the first derivative of $g(x_n)$ when $x_n = a$.

❯ The Newton–Raphson formula is $x_{n+1} = x_n - \dfrac{f(x_n)}{f'(x_n)}$.

> ❯ The Newton–Raphson process fails if $f'(a)$ is close to zero.

> ❯ The Newton–Raphson is very good if $|x_{n+1} - x_n|$ is very small.

❯ If each iteration gets closer to the root, the sequence is convergent.

❯ A suitable iteration formula and a value of x_0 close to the root do not always result in approximations that converge to the root; instead they may diverge.

❯ The trapezium rule is:

$$\int_a^b y \, dx \approx \frac{1}{2}h\left[y_0 + y_n + 2(y_1 + y_2 + \cdots + y_{n-1})\right] \text{ where } h = \frac{b-a}{n}.$$

❯ Where there are numerous areas under the curve that are not included in trapezia, this will result in an underestimation. Conversely, where there are numerous extra areas above the curve that are included in trapezia this will result in an overestimation.

EXAM-STYLE QUESTIONS 9 Answers page 460

(CM) 1   **a**   Given that $f(x) = x + \sin(x) - 3$, where x is in radians, find $f(2.1)$ and $f(2.3)$. Explain why $f(x)$ has a root in the interval $2.1 < x < 2.3$. **[3 marks]**

      b   Using the iteration formula $x_{n+1} = 3 - \sin(x_n)$ and $x_0 = 2.1$, find x_3 to 2 decimal places. **[3 marks]**

2   **a**   Given that $f(x) = x^2 + 6x - 13$, find the first derivative $f'(x)$ of $f(x)$. **[1 mark]**

      b   Use the Newton–Raphson method to find the root of $f(x) = 0$ in the interval $1 < x < 2$, correct to 1 decimal place. **[3 marks]**

(CM) 3   **a**   Use integration to find the exact value of $\int_{-1}^{1}(2x^3 + 3)\,dx$. **[2 marks]**

(PS)   **b**   Use the trapezium rule with four strips to estimate the value of $\int_{-1}^{1}(2x^3 + 3)\,dx$. **[2 marks]**

c Use the trapezium rule with eight strips to estimate the value of $\int_{-1}^{1} (2x^3 + 3)\,dx$. **[3 marks]**

d Compare the answers to **parts a** and **b** and comment on your results. **[2 marks]**

(PS) 4 Use the Newton–Raphson method to approximate $\sqrt{7}$ to 4 decimal places.

(PF) 5 **a** Show that one root of the equation $x^5 + 2x - 7 = 0$ lies in the interval $1 < x < 2$. **[2 marks]**

b Show that $x^5 + 2x - 7 = 0$ can be written in the form $x_{n+1} = (a - bx_n)^{\frac{1}{5}}$, where a and b are constants to be found. **[3 marks]**

c Find the root of $x^5 + 2x - 7 = 0$ in the interval $1 < x < 2$ to 2 decimal places. **[3 marks]**

d Show graphically the first two iterations for this formula. **[2 marks]**

(PS) 6 The graph shows the intersection of the curves $y = \cos x$ and $y = x$, where x is in radians. Using the Newton–Raphson method, find the coordinates of the point of intersection of the two curves to 2 decimal places.

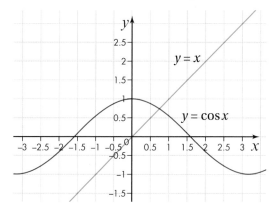

[8 marks]

(PS) 7 Find the percentage error when the trapezium rule is used to evaluate $\int_{-1}^{1} \sqrt{x+1}\,dx$ using four strips. **[7 marks]**

(PS) 8 The solutions to a quadratic equation are $x = 1 \pm \sqrt{8}$. Use the Newton–Raphson method to approximate the value of each root to 2 decimal places. **[8 marks]**

(PS) 9 **a** Use a suitable general iteration formula to find the root of the equation $x^4 - x^3 - 6 = 0$ in the interval $1 < x < 2$ to 3 decimal places. **[4 marks]**

b Show graphically the first two iterations for this formula. **[2 marks]**

(PS) 10 The volume of a sphere is 23.83 units^3 to 2 decimal places. Use the Newton–Raphson method to find the radius of the sphere to 2 decimal places. **[8 marks]**

(PS) 11 Find the number of strips required to ensure that the percentage error is less than 3% when the trapezium rule is used to evaluate $\int_{1}^{2} \frac{1}{x^4}\,dx$. **[9 marks]**

(PS) 12 The purchase price of a car is £14 500.00. Thirteen years later, the value of the car is £1503.14, to the nearest penny. Assuming a constant rate of depreciation, use the Newton–Raphson method to find the rate of depreciation to the nearest whole per cent. **[7 marks]**

10 THREE-DIMENSIONAL VECTORS

From the earliest attempt at describing the hunting of a beast through the medium of cave painting, humans have sought to represent the world and Universe in which they live in two dimensions. Cartographers have mapped the planet and astronomers have charted the stars. However, two-dimensional representations have obvious limitations. The Mercator map was created in 1596 to aid navigation, and is now the standard map of the world displayed in atlases. It shows countries with the correct shapes but distorted by the use of projection so that it makes countries nearer the poles look larger than they are. For example, Greenland appears to be of a similar size to Africa, whereas it is actually 14 times smaller. Similarly, charts of the skies show stars that are literally astronomical distances apart right next to each other. There are difficulties representing three dimensions as two. Two dimensions are fine for a local map or a sat nav, where the curvature of the Earth is less significant – but two dimensions do not give a true reflection over larger scales.

Architects need to visualise their plans in three dimensions. Pilots flying aeroplanes or meteorologists forecasting the weather need to understand the world as it really is. Computer-generated images displayed via virtual reality headsets can place users within a simulated environment. What underpins this mapping of reality? Three-dimensional vectors.

LEARNING OBJECTIVES

You will learn how to:

> write a vector in **i j k** notation or as a column vector in three dimensions

> calculate the magnitude of a vector in three dimensions

> find the position vector of a point

> find the midpoint between two position vectors in three dimensions

> identify types of triangles and quadrilaterals in three dimensions.

TOPIC LINKS

This chapter extends the material from **Book 1, Chapter 10 Vectors** on two-dimensional vectors. This chapter moves from **i j** notation to **i j k** notation by introducing a third axis.

PRIOR KNOWLEDGE

You should already know how to:

> write a vector in **i j** notation or as a column vector

> calculate the magnitude and direction of a vector in two dimensions

> identify a unit vector and scale it up as appropriate

> find the position vector of a point

> find the midpoint between two position vectors

> identify whether two vectors are parallel

> identify types of triangles and quadrilaterals in two dimensions

> find the area of a triangle or quadrilateral.

You should be able to complete the following questions correctly:

1 The vector **a** is given by $(143\mathbf{i} - 24\mathbf{j})$.

Find:

a the magnitude of **a**

b the unit vector in the direction of **a**.

2 The position vectors **b** and **c** are given by $\mathbf{b} = \begin{bmatrix} -3 \\ 17 \end{bmatrix}$ and $\mathbf{c} = \begin{bmatrix} 7 \\ -7 \end{bmatrix}$.

a Find the position vector of the midpoint between **b** and **c**.

b Find the distance between the position vectors **b** and **c**.

3 **a** Prove that the quadrilateral ABCD with A(31, 81), B(86, 33), C(38, −22) and D(−17, 26) is a square.

b Find the area of the square.

10.1 Vectors in three dimensions

Using a third axis, a **vector** can be extended to three dimensions. Instead of just **i** and **j**, you need a third direction, **k**, so this is called **i j k notation**. Similarly, a three-dimensional **column vector** will have three numbers instead of two.

To find the **magnitude** of a vector in three dimensions, you can still use Pythagoras' theorem but you must square and add all three lengths.

> **KEY INFORMATION**
>
> Pythagoras' theorem can be used in three dimensions. Use $a^2 + b^2 + c^2 = d^2$ to find the magnitude of a vector.

Proof of Pythagoras' theorem in three dimensions

The diagram shows the cuboid ABCDEFGH.

The base ABCD is a rectangle, so the triangle ABC is right-angled.

Using Pythagoras' theorem, $AB^2 + BC^2 = AC^2$.

Because the base ABCD is horizontal and the edge CG is vertical, ACG is also a right-angled triangle.

Hence, you also have $AC^2 + CG^2 = AG^2$.

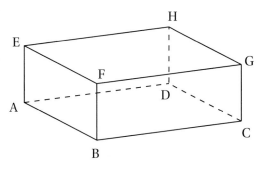

By substituting for AC^2, you get AB^2 + BC^2 + CG^2 = AG^2, which confirms that Pythagoras' theorem can be applied in three dimensions.

PROOF

This is an example of proof by deduction where, by applying logical steps using known concepts in two dimensions, you can derive a new formula in three dimensions.

Example 1

Find the vector in the direction $(-2\mathbf{i} + \mathbf{j} + 2\mathbf{k})$ which has a magnitude of 21.

Solution

Use Pythagoras' theorem to calculate the magnitude of $(-2\mathbf{i} + \mathbf{j} + 2\mathbf{k})$.

$$\text{Magnitude} = \sqrt{(-2)^2 + 1^2 + 2^2} = \sqrt{4 + 1 + 4} = \sqrt{9} = 3$$

Divide $(-2\mathbf{i} + \mathbf{j} + 2\mathbf{k})$ by 3 to obtain the unit vector.

$$\frac{1}{3}(-2\mathbf{i} + \mathbf{j} + 2\mathbf{k})$$

Next multiply the unit vector by 21.

$$\frac{1}{3}(-2\mathbf{i} + \mathbf{j} + 2\mathbf{k}) \times 21 = 7(-2\mathbf{i} + \mathbf{j} + 2\mathbf{k})$$

$$= -14\mathbf{i} + 7\mathbf{j} + 14\mathbf{k}$$

KEY INFORMATION

A **unit vector** is a vector with a magnitude of 1.

Midpoint

To find the midpoint between two **position vectors** in three dimensions, find the mean of each of the $\mathbf{i}$, $\mathbf{j}$ and $\mathbf{k}$ coefficients individually. For example, the midpoint between the position vectors $(2\mathbf{i} - 9\mathbf{j} + \mathbf{k})$ and $(12\mathbf{i} + \mathbf{j} + 4\mathbf{k})$ is given by:

$$\frac{1}{2}[(2\mathbf{i} - 9\mathbf{j} + \mathbf{k}) + (12\mathbf{i} + \mathbf{j} + 4\mathbf{k})] = \frac{1}{2}(14\mathbf{i} - 8\mathbf{j} + 5\mathbf{k})$$

$$= 7\mathbf{i} - 4\mathbf{j} + \frac{5}{2}\mathbf{k}$$

Example 2

Triangle XYZ has vertices X(29, 7, 2), Y(9, −4, 20) and Z(1, 2, −4).

a Show that XYZ is isosceles.

b Find the area of the triangle.

Solution

a Write the points X, Y and Z as position vectors.

$$\mathbf{x} = \begin{bmatrix} 29 \\ 7 \\ 2 \end{bmatrix}, \mathbf{y} = \begin{bmatrix} 9 \\ -4 \\ 20 \end{bmatrix} \text{ and } \mathbf{z} = \begin{bmatrix} 1 \\ 2 \\ -4 \end{bmatrix}$$

$$\overrightarrow{XY} = \mathbf{y} - \mathbf{x} = \begin{bmatrix} 9 \\ -4 \\ 20 \end{bmatrix} - \begin{bmatrix} 29 \\ 7 \\ 2 \end{bmatrix} = \begin{bmatrix} -20 \\ -11 \\ 18 \end{bmatrix}$$

$$\text{Magnitude} = \sqrt{(-20)^2 + (-11)^2 + 18^2} = \sqrt{845}$$

To show that XYZ is an isosceles triangle, it is necessary to show that two sides of the triangle have the same length and the other side has a different length.

Alternatively, XY could be found using the formula
$d^2 = (x_1 - x_2)^2 + (y_1 - y_2)^2 + (z_1 - z_2)^2$

For example,
$d^2 = (29 - 9)^2 + (7 - (-4))^2 + (2 - 20)^2$
$= 20^2 + 11^2 + (-18)^2$
$= 845$, from which $d = \sqrt{845}$.

$$\overrightarrow{XZ} = \mathbf{z} - \mathbf{x} = \begin{bmatrix} 1 \\ 2 \\ -4 \end{bmatrix} - \begin{bmatrix} 29 \\ 7 \\ 2 \end{bmatrix} = \begin{bmatrix} -28 \\ -5 \\ -6 \end{bmatrix}$$

Magnitude $= \sqrt{(-28)^2 + (-5)^2 + (-6)^2} = \sqrt{845}$

$$\overrightarrow{YZ} = \mathbf{z} - \mathbf{y} = \begin{bmatrix} 1 \\ 2 \\ -4 \end{bmatrix} - \begin{bmatrix} 9 \\ -4 \\ 20 \end{bmatrix} = \begin{bmatrix} -8 \\ 6 \\ -24 \end{bmatrix}$$

Magnitude $= \sqrt{(-8)^2 + 6^2 + (-24)^2} = \sqrt{676} = 26$

Since XY = XZ ≠ YZ, exactly two sides are the same and triangle XYZ is isosceles.

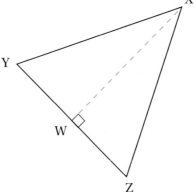

b To find the area of an isosceles triangle, you can use the formula $A = \frac{1}{2}bh$. The base b is the length of the third side (in this case, YZ) and the height h is the distance between the midpoint of YZ and the vertex X, as shown in the diagram.

Let the midpoint of YZ be W.

$$\mathbf{W} = \frac{1}{2}\left(\begin{bmatrix} 9 \\ -4 \\ 20 \end{bmatrix} + \begin{bmatrix} 1 \\ 2 \\ -4 \end{bmatrix} \right) = \frac{1}{2} \begin{bmatrix} 10 \\ -2 \\ 16 \end{bmatrix} = \begin{bmatrix} 5 \\ -1 \\ 8 \end{bmatrix}$$

$$\overrightarrow{WX} = \begin{bmatrix} 29 \\ 7 \\ 2 \end{bmatrix} - \begin{bmatrix} 5 \\ -1 \\ 8 \end{bmatrix} = \begin{bmatrix} 24 \\ 8 \\ -6 \end{bmatrix}$$

Magnitude $= \sqrt{24^2 + 8^2 + (-6)^2} = \sqrt{676} = 26$

Area of triangle XYZ $= \frac{1}{2} \times 26 \times 26 = 338s$

Stop and think   Is it possible for an equilateral triangle to have three vertices with integer coordinates?

Exercise 10.1A

1 a Find the magnitude of each vector.

i $12\mathbf{i} - 8\mathbf{j} - 9\mathbf{k}$

ii $46\mathbf{i} - 46\mathbf{j} + 23\mathbf{k}$

iii $\begin{bmatrix} 5 \\ -4 \\ 3 \end{bmatrix}$

iv $\begin{bmatrix} -7 \\ -7 \\ -7 \end{bmatrix}$

b Find the unit vector in the direction:

i $12\mathbf{i} + 21\mathbf{j} - 16\mathbf{k}$

ii $-72\mathbf{i} + 33\mathbf{j} + 56\mathbf{k}$

iii $\begin{bmatrix} 11 \\ 13 \\ 8 \end{bmatrix}$

iv $\begin{bmatrix} 5 \\ -4 \\ -2 \end{bmatrix}$

(PS) (CM) 2 A hotel is set out as a coordinate grid in a cube measuring 12 units along each edge. The manager has insisted that no bedroom is permitted to be further than 9 units from a toilet block. The lifts along one vertical edge are modelled as the origin for each floor. Room 168 is on Floor 7 and is modelled as $(2\mathbf{i} + 9\mathbf{j})$. The two closest toilet blocks to Room 168 are on Floor 5 at $(7\mathbf{i} + \mathbf{j})$ and Floor 10 at $(8\mathbf{i} + 3\mathbf{j})$.

a Is the manager's rule satisfied for Room 168? Explain your answer.

b State an assumption that you made in answering this question.

3 For each pair of position vectors **a** and **b**, find:

i the midpoint

ii the distance between them.

a $\mathbf{a} = (-3\mathbf{i} + 4\mathbf{j} + 12\mathbf{k})$ and $\mathbf{b} = (7\mathbf{i} + 39\mathbf{j} - 2\mathbf{k})$

b $\mathbf{a} = (19\mathbf{i} - 3\mathbf{j} - 4\mathbf{k})$ and $\mathbf{b} = (7\mathbf{i} + 3\mathbf{j})$

c $\mathbf{a} = \begin{bmatrix} 20 \\ 23 \\ -15 \end{bmatrix}$ and $\mathbf{b} = \begin{bmatrix} -8 \\ 100 \\ 29 \end{bmatrix}$

(CM) 4 Identify which of these four pairs of vectors are parallel, justifying your choice.

a $\begin{bmatrix} 15 \\ 10 \\ -20 \end{bmatrix}$ and $\begin{bmatrix} 18 \\ 12 \\ 24 \end{bmatrix}$

b $\begin{bmatrix} 27 \\ -42 \\ -36 \end{bmatrix}$ and $\begin{bmatrix} 9 \\ -16 \\ -12 \end{bmatrix}$

c $\begin{bmatrix} -6 \\ -4 \\ 10 \end{bmatrix}$ and $\begin{bmatrix} -3 \\ -1 \\ 13 \end{bmatrix}$

d $\begin{bmatrix} 12 \\ -8 \\ 32 \end{bmatrix}$ and $\begin{bmatrix} 21 \\ -14 \\ 56 \end{bmatrix}$

5 Find a vector which has a magnitude of 45 in direction $(4\mathbf{i} - 8\mathbf{j} + \mathbf{k})$.

(PS) (PF) 6 The point A(7, −3, 3) lies on the surface of the sphere S_1, which has its centre at the origin.

a Does point B(2, −8, −1) lie inside, outside or on the surface of S_1?

(M) Modelling   (PS) Problem solving   (PF) Proof   (CM) Communicating mathematically

Points J(–4, 6, –1) and K(–2, 2, 17) are such that JK is a diameter of the sphere S_2.

Air vents are to be placed at two points on the sphere.

b Verify that vents could be placed at the points L(2, 10, 3) and M(4, –2, 7).

(PS) **7** The vector $\begin{bmatrix} a \\ b \\ c \end{bmatrix}$ has a magnitude of 13.

    a If a, b and c are non-zero integers, how many possible sets of values are there for a, b and c?

    b How would the answer differ if zero was permitted?

(PS) **8** A, B and C are points on a straight line such that C is three times as far from B as A is from B.

A has the coordinates (–5, 1, –3) and C has the coordinates (19, –15, 5).

Find all possible coordinates for B.

(PS)
(PF) **9** Triangle FGH has vertices F(–7, 10, 1), G(9, 32, 5) and H(3, 20, 29).

    a Show that FG = GH.

M is the midpoint of FH.

    b Find the coordinates of M.

    c Show that the area of FGH is given by $\sqrt{125\,460}$.

10.2 Vectors and shapes

By applying the concepts of the magnitude of a vector and **parallel** vectors to three dimensions, you can solve problems involving both two-dimensional and three-dimensional shapes. Note that a two-dimensional shape can still be described using vectors with three dimensions. All vertices of a two-dimensional shape lie in the same plane – but not necessarily the x–y plane.

The properties of shapes are often described in terms of their angles. For example, a square has four right angles. In two dimensions, it is straightforward to identify a right angle by comparing the gradients of the line segments, but this is not the case in three dimensions. The most common method for finding an angle or checking if two vectors are perpendicular is the scalar product. The scalar product is covered in **Further Mathematics** but is not included in this book.

You should be aware of the properties of triangles and quadrilaterals so that you can identify them from their coordinates or position vectors. The properties listed below only refer to the lengths of line segments and whether line segments are parallel to each other. No knowledge of their angles is required.

A triangle is:

> equilateral if all sides are the same length

> isosceles if two sides are the same length and the third side is a different length

> scalene if all sides are different lengths

> right-angled if its sides satisfy Pythagoras' theorem.

KEY INFORMATION

You need to know the properties of triangles and quadrilaterals.

A quadrilateral is:

- a square if all sides are the same length and both diagonals are the same length

- a rectangle if both diagonals are the same length and it has two pairs of equal length sides

- a rhombus if all sides are the same length but the diagonals are different lengths

- a parallelogram if the diagonals are different lengths and it has two pairs of equal length sides

- a trapezium if one pair of sides are parallel and of different lengths

- a kite if it has two pairs of equal sides which are adjacent to each other.

> **Stop and think**
>
> A quadrilateral is a parallelogram if it has two parallel sides of the same length. Why is this condition not sufficient?

Example 3

The points L, M and N have position vectors $\mathbf{l} = (4\mathbf{i} - 10\mathbf{j} + 3\mathbf{k})$, $\mathbf{m} = (-3\mathbf{i} - 6\mathbf{j} + \mathbf{k})$ and $\mathbf{n} = (5\mathbf{i} + 3\mathbf{j} - 9\mathbf{k})$ respectively.

a Prove that LMN is a right-angled triangle.

b Find the area of the triangle LMN, giving the answer in the form $a\sqrt{345}$.

Solution

a $\vec{LM} = \mathbf{m} - \mathbf{l} = (-3\mathbf{i} - 6\mathbf{j} + \mathbf{k}) - (4\mathbf{i} - 10\mathbf{j} + 3\mathbf{k}) = (-7\mathbf{i} + 4\mathbf{j} - 2\mathbf{k})$

Magnitude $= \sqrt{(-7)^2 + 4^2 + (-2)^2} = \sqrt{69}$

$\vec{LN} = \mathbf{n} - \mathbf{l} = (5\mathbf{i} + 3\mathbf{j} - 9\mathbf{k}) - (4\mathbf{i} - 10\mathbf{j} + 3\mathbf{k}) = (\mathbf{i} + 13\mathbf{j} - 12\mathbf{k})$

Magnitude $= \sqrt{1^2 + 13^2 + (-12)^2} = \sqrt{314}$

$\vec{MN} = \mathbf{n} - \mathbf{m} = (5\mathbf{i} + 3\mathbf{j} - 9\mathbf{k}) - (-3\mathbf{i} - 6\mathbf{j} + \mathbf{k}) = (8\mathbf{i} + 9\mathbf{j} - 10\mathbf{k})$

Magnitude $= \sqrt{8^2 + 9^2 + (-10)^2} = \sqrt{245}$

Since LM^2 + MN^2 = 69 + 245 = 314 = LN^2, Pythagoras' theorem is satisfied and LMN is a right-angled triangle.

> To show that LMN is a right-angled triangle, it is sufficient to demonstrate that the three sides satisfy Pythagoras' theorem.

b To find the area of a right-angled triangle, use the formula $A = \frac{1}{2}bh$, where the base and height are the two shorter sides of the triangle.

$A = \frac{1}{2}bh$

$= \frac{1}{2} \times \sqrt{69} \times \sqrt{245}$

$= \frac{1}{2} \times \sqrt{69} \times 7\sqrt{5}$

$= \frac{7}{2} \times \sqrt{345}$

Distance between two points

The formula introduced in **Book 1** for the distance between two points, $d^2 = (x_1 - x_2)^2 + (y_1 - y_2)^2$, can now be extended to three dimensions:

$$d^2 = (x_1 - x_2)^2 + (y_1 - y_2)^2 + (z_1 - z_2)^2$$

Example 4

A quadrilateral STUV has vertices S(1, 9, −17), T(−7, 1, −3), U(−5, 17, 5) and V(3, 25, −9).

What type of quadrilateral is STUV?

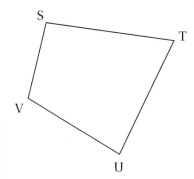

Solution

Start by finding the lengths of the sides.

$$ST^2 = (1 - (-7))^2 + (9 - 1)^2 + (-17 - (-3))^2$$
$$= 8^2 + 8^2 + (-14)^2 = 324$$
$$ST = \sqrt{324} = 18$$

$$VU^2 = (3 - (-5))^2 + (25 - 17)^2 + (-9 - 5)^2$$
$$= 8^2 + 8^2 + (-14)^2 = 324$$
$$VU = \sqrt{324} = 18$$

$$TU^2 = (-7 - (-5))^2 + (1 - 17)^2 + (-3 - 5)^2$$
$$= (-2)^2 + (-16)^2 + (-8)^2 = 324$$
$$TU = \sqrt{324} = 18$$

$$SV^2 = (1 - 3)^2 + (9 - 25)^2 + (-17 - (-9))^2$$
$$= (-2)^2 + (-16)^2 + (-8)^2 = 324$$
$$SV = \sqrt{324} = 18$$

Since all four sides are the same length, the quadrilateral is either a square or a rhombus.

Find the lengths of the diagonals.

$$SU^2 = (1 - (-5))^2 + (9 - 17)^2 + (-17 - 5)^2$$
$$= 6^2 + (-8)^2 + (-22)^2 = 584$$
$$SU = \sqrt{584}$$

$$TV^2 = (-7 - 3)^2 + (1 - 25)^2 + (-3 - (-9))^2$$

$$= (-10)^2 + (-24)^2 + 6^2 = 712$$

$$TV = \sqrt{712}$$

Since all four sides are the same length but the diagonals are different lengths, the quadrilateral is a rhombus.

Stop and think   How could you find the fourth vertex of a parallelogram given the other three vertices?

Exercise 10.2A

Answers page 462

(PF) 1 D(2, –8, 5), E(–11, 3, –7) and F(–6, –1, 2) are three vertices of a triangle.

Prove that the triangle DEF is isosceles.

(PF) 2 The points T, U and V have position vectors $(10\mathbf{i} + 8\mathbf{j} - 3\mathbf{k})$, $(13\mathbf{i} + 12\mathbf{j} - 14\mathbf{k})$ and $(3\mathbf{i} + 27\mathbf{j} + 2\mathbf{k})$ respectively.

Prove that the triangle TUV is:

a scalene   **b** right-angled.

(PF) 3 **a** Show that the quadrilateral EFGH with vertices E(12, 27, 16), F(26, 20, 2), G(12, 6, –5) and H(–2, 13, 9) is a square.

b Show that the quadrilateral JKLM with vertices J(6, –1, 3), K(5, 11, –2), L(3, 9, –7) and M(4, –3, –2) is a parallelogram.

(PS) (M) 4 An exhibition is set up so that four of the exhibits are located at the vertices of a square. On a scale model, with the centre of the exhibition at the origin, three of the exhibits are located at (9, –7, 18), (–6, –1, 28) and (–12, 9, 13).

Find the coordinates of the fourth exhibit.

(PF) 5 A quadrilateral has vertices with position vectors $\mathbf{a} = (6\mathbf{i} - 11\mathbf{j} + 4\mathbf{k})$, $\mathbf{b} = (3\mathbf{i} - \mathbf{j} - \mathbf{k})$, $\mathbf{c} = (10\mathbf{i} + \mathbf{j} + 8\mathbf{k})$ and $\mathbf{d} = (23\mathbf{i} - 17\mathbf{j} + 27\mathbf{k})$.

a Show that ABCD is a kite.

b Hence find the position vector of the point where the diagonals intersect.

(PS) 6 Determine what type of quadrilateral WXYZ is, with W(–44, 34, 80), X(19, 16, 66), Y(1, 2, 3) and Z(–62, 20, 17).

7 P(7, 5, 9), Q(5, 6, 7) and R(6, 8, 9) are three vertices of a triangle. Angle PQR = θ.

Given that $\theta = \arccos \frac{4}{9}$, find the exact area of triangle PQR.

8 A cube has vertices A(5, 3, 1), B(–6, 25, 23), C(16, 47, 12), D(27, p, –10), E(–17, 14, –21), F(–28, 36, 1), G(–6, 58, –10) and H(5, 36, q).

a Find the values of p and q.

b Find the position vector of the centre of the cube.

c Find the volume of the cube.

SUMMARY OF KEY POINTS

> Vectors can be represented in three dimensions using **i j k** notation or as a column vector.

> The magnitude of a vector in three dimensions can be found using Pythagoras' theorem, either with the formula $a^2 + b^2 + c^2 = d^2$ or, for coordinates of points, $d^2 = (x_1 - x_2)^2 + (y_1 - y_2)^2 + (z_1 - z_2)^2$.

> A unit vector is a vector with a magnitude of 1.

> The midpoint between two position vectors is the mean of each of the **i**, **j** and **k** coefficients.

> A triangle is:

> > equilateral if all sides are the same length

> > isosceles if two sides are the same length and the third side is a different length

> > scalene if all sides are different lengths

> > right-angled if its sides satisfy Pythagoras' theorem.

> A quadrilateral is:

> > a square if all sides are the same length and both diagonals are the same length

> > a rectangle if both diagonals are the same length and it has two pairs of equal length sides

> > a rhombus if all sides are the same length but the diagonals are different lengths

> > a parallelogram if the diagonals are different lengths and it has two pairs of equal length sides

> > a trapezium if one pair of sides are parallel and of different lengths

> > a kite if it has two pairs of equal sides which are adjacent to each other.

EXAM-STYLE QUESTIONS 10

Answers page 463

(M)(CM) **1** A mountaineer is modelling a mountain on a computer. The foot of the mountain, where the mountaineer plans to begin her journey, and the peak of the mountain, have position vectors $\begin{bmatrix} 314 \\ -215 \\ 1 \end{bmatrix}$ m and $\begin{bmatrix} 981 \\ 981 \\ 1509 \end{bmatrix}$ m respectively.

    a Find the distance between the foot and the peak of the mountain. **[2 marks]**

    b Explain why the actual journey is likely to be longer than this distance. **[2 marks]**

(PF) **2** The points P, Q and R are (59, -19, 13), (43, 36, -26) and (19, -4, -11) respectively.

    a Show that the triangle PQR is isosceles. **[3 marks]**

    b Show that the triangle PQR is right-angled. **[3 marks]**

(PF) **3** The point A(-12, 7, 5) lies on the surface of a sphere, S, which has centre X(2, -10, 8).

    Show that the midpoint between B(23, -6, -7) and C(-5, 8, -13) also lies on the surface of S. **[5 marks]**

(M) **4** A helicopter travels in the direction $(10\mathbf{i} - 15\mathbf{j} + 6\mathbf{k})$ at $38\,\mathrm{m\,s^{-1}}$.

    Find the velocity of the helicopter. **[3 marks]**

CM **5** Jayanie has been asked to show that the quadrilateral with vertices A(−12, 35, −97), B(36, 101, 79), C(69, 13, 103) and D(21, −53, −73) is a rectangle. Her solution is given as follows:

For AB, $d^2 = (-12 - 36)^2 + (35 - 101)^2 + (-97 - 79)^2 = (-48)^2 + (-66)^2 + (-176)^2 = 1942$

For CD, $d^2 = (69 - 21)^2 + (13 - (-53))^2 + (103 - (-73))^2 = (48)^2 + (66)^2 + (176)^2 = 1942$

For AD, $d^2 = (-12 - 21)^2 + (35 - (-53))^2 + (-97 - (-73))^2 = (-33)^2 + (88)^2 + (-24)^2 = 972$

For BC, $d^2 = (36 - 69)^2 + (101 - 13)^2 + (79 - 103)^2 = (-33)^2 + (88)^2 + (-24)^2 = 972$

Since $AB^2 = CD^2$ and $AD^2 = BC^2$, the quadrilateral is a rectangle.

  a Explain why Jayanie's solution is incomplete. **b** Complete the solution. **[5 marks]**

PF **CM** **6** Verify that the points L(50, −75, 61), M(113, 69, 173), N(−31, 181, 110) and O(−94, 37, −2) are the vertices of a square with an area of 37 249 square units. **[7 marks]**

CM **PF** **7** Explain why the quadrilateral WXYZ with vertices W(5, 33, −2), X(10, 26, −10), Y(1, 15, −6) and Z(−4, 22, 2) is a rectangle. **[6 marks]**

CM **PF** **8** P, Q, R and S have position vectors $\mathbf{p} = -3\mathbf{i} + 11\mathbf{j} - 9\mathbf{k}$, $\mathbf{q} = 12\mathbf{i} + 21\mathbf{j} - 34\mathbf{k}$, $\mathbf{r} = 13\mathbf{i} - 14\mathbf{k}$ and $\mathbf{s} = 4\mathbf{i} - 6\mathbf{j} + \mathbf{k}$.

  a Prove that PQ is parallel to SR. **b** State the ratio PQ : SR.

  c State what type of quadrilateral PQRS is, giving a reason for your answer. **[5 marks]**

CM **9** F, G and H are points such that $4\overrightarrow{FG} = 3\overrightarrow{GH}$.

G has the position vector $(17\mathbf{i} + a\mathbf{j} - 2\mathbf{k})$, where $a > 10$.
H has the position vector $(9\mathbf{i} + 13\mathbf{j} + 22\mathbf{k})$.

The magnitude of GH is 28.

  a Explain how you know that F, G and H are collinear.

  b Find the position vector of F. **[5 marks]**

PF **PS** **10** The position vectors of the points S, T and U are given by $\mathbf{s} = \begin{bmatrix} -2 \\ 9 \\ 10 \end{bmatrix}$, $\mathbf{t} = \begin{bmatrix} -3 \\ 14 \\ 17 \end{bmatrix}$ and $\mathbf{u} = \begin{bmatrix} 3 \\ 18 \\ 15 \end{bmatrix}$.

  a Find the vector $\overrightarrow{ST}$.

  b Prove that ST and TU are perpendicular.

  c Find the position vector of V such that STUV is a rectangle.

  d Find the coordinates of the centre of the rectangle, X.

  e Show that the area of the triangle XST is given by $k\sqrt{42}$, where k is a constant to be found. **[11 marks]**

PS **PF** **11** A quadrilateral EFGH has vertices E(9, −2, −1), F(13, 6, −9), G(5, 20, −8) and H(−3, 4, 8).

  a Show that EF is parallel to HG.

  b Show that EH is equal in length to FG.

  c Find the area of the quadrilateral. **[9 marks]**

11 PROOF

What is a contradiction in mathematics? It is an idea which is always untrue (false). As a real-life example, suppose you want to prove that you have did pay for an ice cream from an ice-cream van.

To prove by contradiction that you did pay for your ice cream, start by assuming that you did not pay for your ice cream. However, if you did not pay for your ice cream, then the ice-cream vendor would not have given the ice cream to you. As you are eating the ice cream, you must have paid for it.

You will see other examples of proof by contradiction in this chapter.

LEARNING OBJECTIVES

You will learn how to:

› understand and use the structure of mathematical proof, proceeding from given assumptions through a series of logical steps to a conclusion

› use proof by contradiction.

TOPIC LINKS

Proof by deduction – the drawing of a conclusion by using the general rules of mathematics – is the most commonly used form of proof throughout this course. You will find many examples of proof by deduction in individual chapters – for example, the proof in **Chapter 10 Three-dimensional vectors** that Pythagoras' theorem can be applied in three dimensions. There are more examples of proof by deduction in the Exam-style questions at the end of this chapter, but this topic is not revisited specifically.

PRIOR KNOWLEDGE

You should already know how to:

› argue mathematically to show that algebraic expressions are equivalent, and use algebra to support and construct arguments and proofs

› apply angle facts, triangle congruence, similarity and properties of quadrilaterals to conjecture and derive results about angles and sides, including Pythagoras' theorem and the fact that the base angles of an isosceles triangle are equal, and use known results to obtain simple proofs

› use vectors to construct geometric arguments and proofs.

You should be able to complete the following questions correctly:

1 Show that the sum of any three consecutive odd numbers is always a multiple of 3.

2 Show that the sum of two consecutive integers is always an odd number.

3 Show that the difference between the squares of any two consecutive positive integers is equal to the sum of the two integers.

4 Prove that the sum of the squares of any two consecutive even numbers is always a multiple of 4.

5 Prove that $(2n - 1)^2 - (2n + 1)^2$ is a multiple of 8.

11.1 Proof by contradiction

The steps for **proof by contradiction** are:

❯ Assume that the opposite of what you are trying to prove is true (that is, assume that what you are trying to prove is false).

❯ Follow logical steps in the proof. If the assumption at the start of the proof results in a contradiction, then what you are trying to prove cannot be false. Consequently, whatever you are trying to prove must be true.

Example 1

Prove by contradiction that if p^2 is even, then p is even.

Solution

Start by assuming that p is odd.

So $p = 2n + 1$, where n is a positive whole number.

Then $p^2 = (2n + 1)^2$

$$= 4n^2 + 4n + 1$$

$$= 2(2n^2 + 2n) + 1$$

$2(2n^2 + 2n)$ is even, so $2(2n^2 + 2n) + 1$ must be odd, and therefore p^2 is odd.

This is a contradiction, so p must be even.

> This is a contradiction of what you have been asked to prove.

Example 2

Prove by contradiction that $\sqrt{2}$ is irrational.

Solution

As you are using proof by contradiction, you need to assume that $\sqrt{2}$ is rational.

Therefore, you can write $\sqrt{2}$ as $\frac{p}{q}$, where p and q are integers with no common factors other than 1.

If $\sqrt{2} = \frac{p}{q}$, then $2 = \frac{p^2}{q^2}$

Rearrange.

$$2q^2 = p^2$$

Hence p^2 is even, and so p is even.

You can write $p = 2n$, so $p^2 = 4n^2$.

You have already shown that $2q^2 = p^2$, so now rewrite this.

$2q^2 = 4n^2$

$q^2 = 2n^2$

So q is even.

If p and q are both even, then they have a common factor of 2, which contradicts the assumption that they don't have a common factor other than 1.

So $\sqrt{2}$ is *not* rational – consequently, it must be irrational.

> A rational number is one that can be written as a fraction in **irreducible form** (where the numerator and the denominator have no common factors other than 1).

> You proved this in **Example 1**.

Example 3

Prove by contradiction that there are infinitely many prime numbers.

Solution

Start by assuming the converse, that there are not infinitely many prime numbers. Since there is at least one prime number (for example, 2), this implies that there must be a largest prime number. Let this largest prime number be called P.

Consider the product of all the prime numbers from 2 to P. Let this integer be called X.

Define Y as $X + 1$.

Firstly, note that Y is larger than P.

Secondly, note that since X is a product of all the primes from 2 to P, and Y is 1 more than X, then Y is not a multiple of any of the primes from 2 to P.

Now, Y is either prime or it is not prime. Look at each case in turn.

If Y is prime, then you have found a prime number larger than P, which immediately contradicts the assumption that P is the largest prime.

If Y is not prime, then it must have at least one prime factor. Any prime factor of Y cannot be one of the primes from 2 to

P because Y is not a multiple of any of the primes from 2 to P. Hence any prime factor of Y must be another prime number, but if P is the largest prime number, then there are not any other prime numbers which could be a prime factor of Y. Hence this is also a contradiction.

Since Y cannot be prime and it cannot be not prime, this means that the initial assumption, that there is a largest prime number, must be false. Hence it must be true that there are infinitely many prime numbers.

Example 4

Prove by contradiction that the hypotenuse of a right-angled triangle is smaller in length than the sum of the other two sides.

Solution

Assume the converse – that the hypotenuse, c, is the same length as or longer than the sum of the other two sides, a and b.

$$c \geq a + b$$

Since a, b and c are positive lengths, squaring both sides will not change the inequality.

$$c^2 \geq (a + b)^2$$

Expand the brackets.

$$c^2 \geq a^2 + 2ab + b^2$$

By Pythagoras' theorem, it is already known that $c^2 = a^2 + b^2$.

Substitute.

$$a^2 + b^2 \geq a^2 + 2ab + b^2$$

Subtract $a^2 + b^2$ from both sides.

$$0 \geq 2ab$$

Since a and b are positive lengths, $2ab$ cannot be zero or negative, so there is a contradiction. This means that the initial assumption, that that the hypotenuse is the same length as or longer than the other two sides, must be false.

Hence it must be true that the hypotenuse of a right-angled triangle is smaller in length than the sum of the other two sides.

Example 5

Prove by contradiction that there is no smallest positive rational number.

Solution

Assume the converse – that there is a smallest positive rational number. Let this number be called R.

Let the number S be R divided by 10. So $S = \frac{1}{10}R$, and S is smaller than R.

If R is a positive rational number, then so is S, but S is a smaller positive rational number than R. This is a contradiction, since R is the smallest positive rational number. This means that the initial assumption, that there is a smallest positive rational number, must be false.

Hence it must be true that there is no smallest positive rational number.

Example 6

Prove by contradiction that a triangle has at most one obtuse angle.

Solution

Let the angles in a triangle be A, B and C.

Assume the converse – that a triangle has two or three obtuse angles. Let A and B both be obtuse angles.

Since A and B are obtuse, $A > 90°$ and $B > 90°$.

Hence $A + B > 90° + 90°$, so $A + B > 180°$.

But the angles in a triangle sum to 180°, so $A + B + C = 180°$. For $A + B$ to be greater than 180°, C must be negative. But an angle in a triangle cannot be negative, so there is a contradiction. This means that the initial assumption, that a triangle has two or three obtuse angles, must be false.

Hence it must be true that a triangle has at most one obtuse angle.

Exercise 11.1A

Answers page 465

(PF) **1** Prove by contradiction that if p^2 is odd, then p is odd.

(PF) **2** Prove by contradiction that $\sqrt{3}$ is irrational.

(PF) (CM) **3** Prove by contradiction that $\sqrt{9}$ is rational. Comment on the effectiveness of this method of proof in this case.

(PF) **4** Prove by contradiction that every non-zero real number has a unique reciprocal.

(PF) **5** Prove by contradiction that $(x - a)^2 \geqslant 0$, $\{a, x \in \mathbb{R}\}$.

(PF) **6** Prove by contradiction that $x^3 < 0$, $\{x \in \mathbb{R}\}$ when $x < 0$.

SUMMARY OF KEY POINTS

› Proof by deduction involves drawing a conclusion by using the general rules of mathematics.

› Proof by contradiction involves assuming that the opposite of what you are trying to prove is true. By following logical steps in the proof, this will lead to a contradiction. As the assumption at the start of the proof results in a contradiction, then what you are trying to prove cannot be false. Consequently, whatever you are trying to prove must be true.

EXAM-STYLE QUESTIONS 11   **Answers page 466**

(PF) 1 Five people, Alf, Bukunmi, Carlos, Dupika and Elsa, are to be seated around a circular table for a dinner party. Alf is to sit next to Bukunmi. Carlos must not sit next to Dupika. Prove by contradiction that Elsa cannot sit next to Alf.   **[4 marks]**

(PF) 2 **(CM)**
   a Using proof by deduction, show that if n^2 is even, then n is even.

   b Using proof by contradiction, show that if n^2 is even, then n is even.

   c Compare and comment on the two methods of proof used in **parts a** and **b**.   **[5 marks]**

(PF) 3
   a Deduce Pythagoras' theorem using the following diagram of a square containing another square.

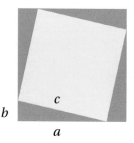

   b Write down any assumptions you made in **part a**.   **[3 marks]**

(PF) 4 Prove by contradiction that a quadrilateral has at most one reflex angle.   **[4 marks]**

(PF) 5 Complete the following proof by contradiction that $\sqrt[3]{7}$ is irrational.

   Assume that $\sqrt[3]{7}$ is rational.

   Therefore $\sqrt[3]{7} = \dfrac{p}{q}$, where p and q are integers with no common factors other than 1.

   If $\sqrt[3]{7} = \dfrac{p}{q}$, then $7 = \ldots$ **[4 marks]**

(PF) 6 A right-angled triangle has sides a, b and c, where a, b and c are integers. Prove by contradiction that at least one of the shorter sides has an even length.   **[4 marks]**

(PF) 7 Prove by contradiction that if x and y are positive real numbers, then $x + y \geqslant 2\sqrt{xy}$.   **[4 marks]**

(PF) 8 Show that if n is even, then $n^2 + 3n + 2$ is even.   **[4 marks]**

PF **9** The functions f and g are defined as $f(x) = 2x + 5$ and $g(x) = x^2 + 3$. Prove that $gf(x)$ is a multiple of 4 for all positive integer values of x. **[3 marks]**

PF **10** Prove by contradiction that if x and y are positive real numbers, then $x^2 + 4y^2 \geqslant 4xy$. **[4 marks]**

PF **11** Complete the following proof that if the three sides of a triangle, p, q and r, satisfy $p^2 + q^2 = r^2$ and all the angles are acute, then the triangle is right-angled.

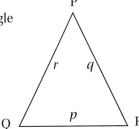

> Assume that the triangle is not right-angled.
>
> Draw in a point S so that S forms a right-angled triangle with two of the other points.

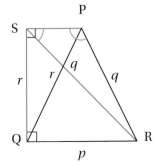

Using Pythagoras $(SR)^2 = r^2 + p^2 = q^2$ and so $SR = q$. ... **[6 marks]**

PF **12** Prove by contradiction that there exist no integers x and y for which $15x + 20y = 1$. **[4 marks]**

PF **13** Prove by contradiction that there do not exist integers x and y such that $3x - 9y = 4$. **[4 marks]**

PF **14** Prove that the equation of a circle with centre (a, b) and point (c, d) on its circumference can be written as $(y - d)(b - d) + (x - c)(a - c) = 0$. **[4 marks]**

PF **15** Prove that the curve $y = \dfrac{x}{x^2 + A}$ has turning points when $x = \pm\sqrt{A}$. **[6 marks]**

PF **16** Prove by contradiction that the sum of any two odd square numbers cannot be a square number. **[6 marks]**

PF **17** Eric has been asked to find the volume of a cuboid. He has been told that the faces of the cuboid have areas of 72 cm², 96 cm² and 108 cm². 'That's easy,' says Eric. '$72 \times 96 \times 108 = 746\,496$ and the square root of $746\,496$ is 864. The volume is 864 cm³.'

Prove that Eric's method will work for any cuboid. **[4 marks]**

PF **18** The diagonal of a square has a length of d m.

Prove that the area of the square is given by $A = \frac{1}{2}d^2$.

PF **19** By considering the line segment between the points $(0, 0)$ and (h, r), where $h > 0$ and $r > 0$, show by integration that the volume of a cone is given by $V = \frac{1}{3}\pi r^2 h$. **[6 marks]**

PF **20** Prove by contradiction that $\sin x + \cos x \geqslant 1$ for $0 \leqslant x \leqslant \frac{\pi}{2}$. **[6 marks]**

PF **21** Prove by contradiction that if a, b and c are odd integers, then there are no integer solutions for the equation $ax^2 + bx + c = 0$. **[6 marks]**

12 PROBABILITY

To find the probability of dying of a heart attack in the next 10 years you would need to make some conditions to stop this being a very open-opened task. You would need to specify whether it applies to both genders, all ethnicities, all abilities and all ages, or whether you were focusing on just one set group of people, such as adults in one particular country.

It's important to be clear about the conditions of your experiment. In this chapter you will learn how to formally define statements about probability. For example, you would expect that the probability of dying of a heart attack in the next 10 years would be much less for teenagers than for people at retirement age.

In order to truly quantify research, many studies focus on particular groups of people. This results in conditional probabilities with fairly restricted conditions. Unfortunately, more often than not this restriction is not emphasised enough when conclusions are reported. For example, a company trying to bring out a new medicine for people with a heart condition might be restricted to working with women between the ages of 65 and 70 who have previously had a heart attack. Based on the outcome of the research, a GP may decide to prescribe the medication to a man who is 70 years old and has no previous record of heart attacks. By doing this, the GP would be extrapolating – the research has involved only women between 65 and 70 who have already had a heart attack, which is quite different from a man of 70 without any previous history of heart conditions.

LEARNING OBJECTIVES

You will learn how to:

> use formal notation and formulae
> use conditional probability, including the use of tree diagrams, Venn diagrams and two-way tables
> use the conditional probability formula
> model situations involving probability, including critiquing assumptions made.

TOPIC LINKS

You will need the skills and techniques that you have learnt in **Book 1** to manipulate algebraic expressions and calculate probabilities and binomial probabilities. You will use Venn diagrams and tree diagrams with formal notation to calculate probabilities, and this will help in **Chapter 14 Statistical hypothesis testing**.

PRIOR KNOWLEDGE

You should already know:

> how to interpret a Venn diagram and a tree diagram
> how to use the complement of an event, where $P(A') = 1 - P(A)$
> that if A and B are mutually exclusive (do not overlap) then $P(A \text{ or } B) = P(A) + P(B)$.

You should be able to complete the following questions correctly:

1 Use the Venn diagram to calculate the following probabilities:

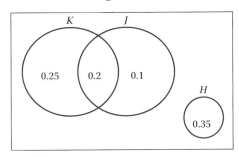

a P(*K* and *J*)   **b** P(*K* and *H*)   **c** P(*H′*)

d P(*K′* and *J′* and *H′*)   **e** P(*K* or *H*)

2 A bag contains 10 counters; 6 are blue and 4 are yellow. One counter is selected at random, then replaced. A second counter is then selected.

a Draw a tree diagram to represent this information.

b Find the probability that both counters are yellow.

c Find the probability that one counter is blue and the other is yellow.

12.1 Set notation to describe events and outcomes

From **Book 1, Chapter 13 Probability and statistical distributions,** you should be familiar with representing **probability** problems as Venn diagrams. At A-level you need to understand and use **set notation**.

Using set notation, P(*A* and *B*) is written as P(*A* ∩ *B*).

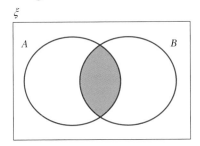

> You read this as '*A* intersection *B*.'

P(*A* or *B*) is written as P(*A* ∪ *B*).

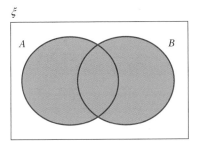

> You read this as '*A* union *B*.'

P(not A) is written as P(A').

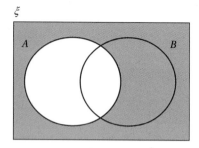

P(not A and not B) is written as P($A' \cap B'$).

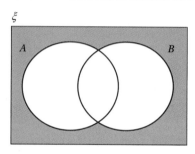

Mutually exclusive, independent and dependent events

Mutually exclusive events are events that cannot both occur at once. For example, a coin toss can have the outcome heads or tails, but not both, or a die can land on a 6 or a 1, but not on both at once.

Independent events are two events that do not affect each other. This means that the probability of one event occurring, such as rolling a 6 on a fair die, would not affect the probability of the other event occurring, such as flipping a head on a coin.

Dependent events are events that will influence each other, such as picking a pair of socks out of your drawer on a dark morning. If you have a pair of red socks, a pair of blue socks and a pair of white socks in a drawer, the probability of picking a white sock would be $\frac{2}{6}$. If you pick out one white sock, the probability that the next sock is also white is now only $\frac{1}{5}$. The two probabilities are different, showing that the two events are dependent. The probability of the second event depends on what happens in the first event.

When dealing with more than one event, there are certain rules that you must follow when studying probability of these events. These rules depend greatly on whether the events you are looking at are independent or dependent on each other. You will encounter more of these in the next section on conditional probability.

A formula you can use for two events which links the probabilities of the union and the intersection is the **addition rule**.

In probability we refer to the addition operator (+) as 'or'. Thus, when you want to define some event such that the event can be A or B, to find the probability of that event:

KEY INFORMATION

If A and B are two events then:

› $A \cap B$ represents the event 'both A and B occur'.

› $A \cup B$ represents the event 'either A or B (or both) occur'.

$P(A \text{ or } B) = P(A \cup B)$

$P(A \cup B)$ can be worked out using the addition rule:

$P(A \cup B) = P(A) + P(B) - P(A \cap B)$

The reason for subtracting $P(A \cap B)$, the probability of both A and B occurring, is so that you don't include events twice in your addition if they fall within $A \cap B$. This can be shown using Venn diagrams:

> **KEY INFORMATION**
> For any two events A and B,
> $P(A \cup B) = P(A) + P(B) - P(A \cap B)$.

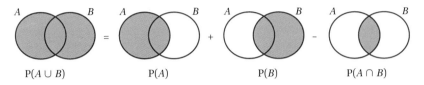

For example, if a card is picked at random from a pack of 52 cards, the probability that an ace or a spade is picked is represented by $P(\text{ace or spade})$.

$P(\text{ace} \cup \text{spade}) = P(\text{ace}) + P(\text{spade}) - P(\text{ace} \cap \text{spade})$

$$= \frac{4}{52} + \frac{13}{52} - \frac{1}{52}$$

$$= \frac{16}{52} = \frac{4}{13}$$

Example 1

A and B are two events such that $P(A) = 0.65$, $P(B) = 0.75$ and $P(A \cup B) = 0.85$.

Find:

a $P(A \cap B)$ **b** $P(A')$

c $P(A' \cup B)$ **d** $P(A' \cap B)$

Solution

a $P(A \cap B) = P(A) + P(B) - P(A \cup B)$

$\qquad\qquad\quad = 0.65 + 0.75 - 0.85$

$\qquad\qquad\quad = 0.55$

b $P(A') = 1 - P(A)$

$\qquad\quad = 1 - 0.65$

$\qquad\quad = 0.35$

> You can rearrange the addition rule to find the intersection:
>
> $P(A \cap B) = P(A) + P(B) - P(A \cup B)$

Draw a Venn diagram, now that you know all the probabilities.

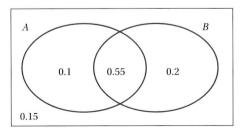

c $P(A' \cup B) = 0.2 + 0.15 + 0.55$

$= 0.9$

d $P(A' \cap B) = P(A') + P(B) - P(A' \cup B)$

$= 0.35 + 0.75 - 0.9$

$= 0.2$

> This means 'not A' together with B, so you add up the regions outside of A and add the remaining region in B.

> You can identify this region on your Venn diagram as the area outside of A but inside of B.

If A and B are mutually exclusive, the intersection of A and B is empty, so $P(A \cap B) = 0$.

You can substitute $P(A \cap B) = 0$ into

$P(A \cup B) = P(A) + P(B) - P(A \cap B)$ to give:

$P(A \cup B) = P(A) + P(B)$

KEY INFORMATION

The addition rule for mutually exclusive events is $P(A \cup B) = P(A) + P(B)$.

Example 2

After conducting a questionnaire, event Y was being in the 11–16 age category and event Z was being in the 21–25 category. Y and Z are mutually exclusive. $P(Y) = 0.3$ and $P(Z) = 0.5$. Find:

a $P(Y \cup Z)$ **b** $P(Y \cap Z')$ **c** $P(Y' \cap Z')$

Solution

a

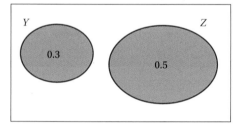

$P(Y \cup Z) = P(Y) + P(Z)$

$= 0.3 + 0.5 = 0.8$

b $P(Y \cap Z') = P(Y)$

$= 0.3$

c $P(Y' \cap Z') = 1 - P(Y \cup Z)$

$= 1 - 0.8$

$= 0.2$

Example 3

The probability of choosing your favourite flavour of crisps is called event P. The probability of choosing your favourite flavour of squash is called event Q. Events P and Q are independent. Event P has a probability of $\frac{1}{4}$. Event Q has the probability $\frac{1}{3}$. Find:

a $P(P \cap Q)$ **b** $P(P \cap Q')$ **c** $P(P' \cap Q')$

Solution

a $P(P \cap Q) = P(P) \times P(Q)$

$$= \frac{1}{4} \times \frac{1}{3} = \frac{1}{12}$$

b $P(P \cap Q') = P(P) - P(P \cap Q)$

$$= \frac{1}{4} - \frac{1}{12}$$

$$= \frac{2}{12} = \frac{1}{6}$$

c $P(P' \cap Q') = P(P) + P(Q) - P(P \cap Q)$

$$= \frac{1}{4} + \frac{1}{3} - \frac{1}{12}$$

$$= \frac{6}{12} = \frac{1}{2}$$

Using the large data set 12.1

  a Using the large data set, choose two categories that are not mutually exclusive. Explain any dependencies.

  b By calculating appropriate probabilities, display the data from your categories in a Venn diagram.

Exercise 12.1A Answers page 469

1 Two events, A and B, are mutually exclusive, and $P(A) = \frac{1}{5}$ and $P(B) = \frac{1}{4}$. Find:

  a $P(A \cap B)$ **b** $P(A \cap B')$ **c** $P(A' \cap B')$

2 E and F are two events such that $P(E) = 0.2$, $P(F) = 0.4$ and $P(E' \cap F) = 0.4$. Find:

  a $P(E \cup F)$ **b** $P(E' \cap F')$

  c The relationship between E and F.

3 A card is selected at random from a pack of 52 playing cards. D is the event 'the card selected is a diamond' and Q is the event 'the card selected is a queen'.

  Find the following:

  a $P(Q)$ **b** $P(D)$ **c** $P(D \cup Q)$

  d $P(D \cap Q)$ **e** $P(Q')$ **f** $P(D \cap Q')$

4 In the UK 40% of households have a juicer and 50% have a laptop. 25% of households have both. Calculate the probability that a household chosen at random has either a juicer or a laptop but not both.

 5 Two fair dice are rolled and the results are recorded. Show that the event 'the sum of the scores on the dice is 4' and 'both dice land on the same number' are not mutually exclusive.

6 C and D are two events such that $P(C) = 0.5$, $P(D) = 0.2$ and $P(C \cap D) = 0.1$. Find:

   a $P(C \cup D)$ **b** $P(D')$ **c** $P(C \cap D')$ **d** $P(C \cup D')$

7 K and S are two events such that $P(K) = P(S) = 3P(K \cap S)$ and $P(K \cup S) = 0.75$. Find:

   a $P(K \cap S)$ **b** $P(K)$ **c** $P(S')$ **d** $P(K' \cap S')$ **e** $P(K \cap S')$

12.2 Conditional probability

You have already defined dependent and independent events and seen how the probability of one event relates to the probability of the other event. With those concepts in mind, you can now look at **conditional probability**. Conditional probability deals with dependent events by looking at the probability of an event, given that another event occurs first.

For example, you would express the phrase 'the probability of picking a white sock and another white sock' as 'the probability of picking a white sock, given that the first sock picked was white'.

In formal notation for a conditional probability, a vertical bar stands for 'given'. The sock example would be written as:

     P(pick a white sock | first sock is white)

Generally, the probability of event A happening, given that event B has already happened, is written $P(A|B)$.

Event B must have happened in order for event A to happen. The only possible outcomes which correspond to (A, given B has happened) must be the intersection, $A \cap B$. This is shown in the Venn diagrams below.

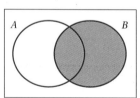
You know that event B has already happened.

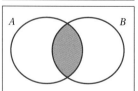
You want the probability of event A, knowing that B has happened. This is the same as $P(A \cap B)$.

Therefore you can work out the probability of A given B using the formula:

$$P(A|B) = \frac{P(A \cap B)}{P(B)}$$

Stop and think   What would the probability formula look like for B, given A?

If the formula for conditional probability is rearranged you get an important probability law called the **product rule**. For events A and B:

$$P(A \cap B) = P(A) \times P(B|A)$$

You can also use $P(B \cap A) = P(B) \times P(A|B)$ when A and B are swapped.

If A and B are independent, the probability of A is the same whether or not B happens.

Therefore $P(A|B) = P(A)$.

Using conditional probability, this gives you

$$P(A) = \frac{P(A \cap B)}{P(B)}$$

Rearranging this gives

$$P(A \cap B) = P(A) \times P(B)$$

> The formula is usually stated in this way, rather than as a formula for conditional probability.

> **KEY INFORMATION**
>
> The product rule for independent events is $P(A \cap B) = P(A) \times P(B)$.

Example 4

J and M are two events such that $P(J) = 0.6$, $P(M) = 0.2$ and $P(M|J) = 0.3$.

Find:

a $P(J|M)$ **b** $P(J' \cap M')$ **c** $P(M' \cap J)$

Solution

Use the product rule to find the missing values for the Venn diagram.

$$P(J \cap M) = P(M|J) \times P(J)$$
$$= 0.3 \times 0.6$$
$$= 0.18$$

Now draw a Venn diagram.

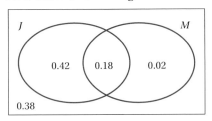

---

a $P(J|M) = \dfrac{P(J \cap M)}{P(M)}$

  $= \dfrac{0.18}{0.2}$

  $= 0.9$

b $P(J' \cap M') = 0.38$

c $P(M' \cap J) = 0.42$

> The intersection of not J and not M.

> The intersection of not M with J.

Example 5

100 students were interviewed to find out if they were learning to drive. Here are the results:

	Male	Female	Total
Learning to drive	19	48	67
Not learning to drive	24	9	33
Total	43	57	100

M stands for 'male' and L stands for 'learning to drive'.

Find the following probabilities:

a $P(M)$ **b** $P(L)$ **c** $P(M \cap L)$

d $P(L|M')$ **e** $P(L'|M)$

Solution

a $P(M) = \dfrac{43}{100}$ as there are 43 males and 100 students in total.

b $P(L) = \dfrac{67}{100}$ as there are 67 learning to drive and 100 students in total.

c $P(M \cap L) = \dfrac{19}{100}$ as 19 students are male and learning to drive out of the 100 students.

d $P(L|M') = \dfrac{48}{57}$ as 48 students are female and learning to drive out of the 57 females.

e $P(L'|M) = \dfrac{24}{43}$ as 24 males are not learning out of the 43 males.

> Remember that the probability of complement of an event A is $P(A') = 1 - P(A)$.

Example 6

You can get sunburn in July if the UV index is high. Getting sunburn is called event J. A high UV index is called event K. The probability of event J is $P(J) = 0.76$ and the probability of event K is $P(K) = 0.35$.

a Find $P(J|K)$.   **b** Find $P(K|J)$.

Solution

a Using the product rule:

$$P(J|K) = \frac{P(J \cap K)}{P(K)}$$

Substitute in the values.

$$P(J|K) = \frac{0.25}{0.35} = 0.714$$

This means that there is a 0.714 chance of getting sunburnt, given that it's a high UV index day.

b Using the product rule:

$$P(K|J) = \frac{P(K \cap J)}{P(J)}$$

Substitute in the values.

$$P(K|J) = \frac{0.25}{0.76} = 0.329$$

This means that there is a 0.329 chance of it being a high UV index day, given that you have been sunburnt.

Example 7

To qualify for a competition you have to complete two tasks. The probability of completing the first and then the second task is 0.53, and the probability of completing the first task is 0.71. What is the probability of completing the second task, given that you complete the first?

Solution

$$P(\text{task 2}|\text{task 1}) = \frac{P(\text{task 1 and task 2})}{P(\text{task 1})} = \frac{0.53}{0.71} = 0.746$$

Using the large data set 12.2

Using the large data set, choose two categories that are not mutually exclusive:

a Calculate the probability of one of the categories, given the probability of the second.

b Can you assume the probability in **part a** is the same in other areas? State your reasons carefully.

Exercise 12.2A

Answers page 469

1 A card is drawn at random from a pack of 52 playing cards. Given that the card is a spade, find the probability that the card is a jack.

(PS) 2 In May 2015 the probability of moderate winds in Leuchars was $\frac{14}{31}$ and the probability of moderate winds and a humidity of more than 95% was $\frac{6}{31}$. Find the probability of a humidity of more than 95%, given moderate winds.

3 Two coins are flipped and the results are recorded. Given that the coin lands on tails, find the probability of:

a two tails   **b** a head and a tail.

4 M and P are two events such that $P(M) = 0.6$, $P(P) = 0.5$ and $P(M \cap P) = 0.4$. Find:

a $P(M \cup P)$ **b** $P(P|M)$

c $P(M|P)$ **d** $P(M|P')$

5 Let E and F be events such that $P(E) = \frac{1}{4}$, $P(F) = \frac{1}{2}$ and $P(E \cup F) = \frac{3}{5}$. Find:

a $P(E|F)$ **b** $P(E' \cap F)$ **c** $P(E' \cap F')$

Tree diagrams for independent events

If A and B are independent, then $P(B)$ is the same whether A happens or not, and vice versa. This means $P(B|A) = P(B|A') = P(B)$.

> See **Book 1, Chapter 13 Probability and statistical distributions** for more on tree diagrams.

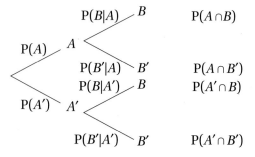

Using conditional probability, this means: $P(B|A) = P(B) = \dfrac{P(A \cap B)}{P(A)}$

For independent events, the product law becomes

$$P(A \cap B) = P(A)P(B|A) = P(A)P(B)$$

Tree diagrams for dependent events

When you create a tree diagram for dependent events, the probabilities on the second set of branches depend on the outcome from the first set of branches. This means they are conditional, so you need to work carefully to label the diagram. A generic tree diagram for two events would look like this:

You can use this tree diagram to confirm the formula for conditional probability. As you have to multiply along the branches, you get $P(A) \times P(B|A) = P(A \cap B)$.

Example 8

For events F and G, $P(F) = 0.3$, $P(G|F) = 0.42$ and $P(G'|F') = 0.61$.

a Draw a tree diagram representing the events and the corresponding probabilities.

b Find $P(G)$.

c Find $P(F'|G')$.

Solution

a You will need two sets of branches: one for event F and another for event G. As you already know $P(F)$, put event F on the first set of branches.

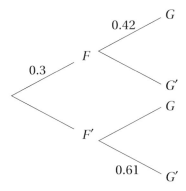

Now calculate the missing probabilities.

$$P(F') = 1 - P(F) = 1 - 0.3 = 0.7$$

$$P(G'|F) = 1 - P(G|F) = 1 - 0.42 = 0.58$$

$$P(G|F') = 1 - P(G'|F') = 1 - 0.61 = 0.39$$

Write the probabilities onto the branches on the tree diagram.

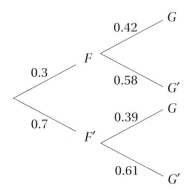

b $P(G) = 0.3 \times 0.42 + 0.7 \times 0.39 = 0.126 + 0.273 = 0.399$

c $P(F'|G') = \dfrac{P(F' \cap G')}{P(G')}$

$P(G') = 1 - 0.399 = 0.601$

$P(F' \cap G') = 0.7 \times 0.61 = 0.427$

So $P(F'|G') = \dfrac{0.427}{0.601} = 0.7105$

Example 9

Nola either cycles or walks to college. The probability that she walks is 0.61. If she cycles, the probability she will be late is 0.3. If she walks, the probability she will be late is 0.21.

a Draw a tree diagram to represent the events and their corresponding probabilities.

b Find the probability that Nola has walked, given that she is late.

Solution

a

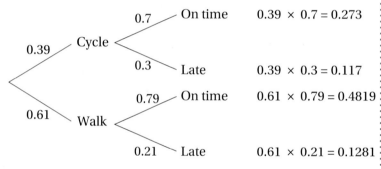

0.39 — Cycle
- 0.7 — On time — $0.39 \times 0.7 = 0.273$
- 0.3 — Late — $0.39 \times 0.3 = 0.117$

0.61 — Walk
- 0.79 — On time — $0.61 \times 0.79 = 0.4819$
- 0.21 — Late — $0.61 \times 0.21 = 0.1281$

b Probability of being late $= (0.39 \times 0.3) + (0.61 \times 0.21) = 0.2451$

Probability of walking and being late $= 0.61 \times 0.21 = 0.1281$

So the probability of walking, given that Nola is late, is found using the conditional probability formula:

$$P(W|L) = \frac{P(W \cap L)}{P(L)}$$

$$= \frac{0.1281}{0.2451} = 0.5226$$

Exercise 12.2B

Answers page 469

 1 A bag contains four yellow beads and five blue beads. A bead is selected at random and its colour is recorded. It is not replaced in the bag. A second bead is taken and the colour is noted.

Find the probability that:

a the second bead is blue, given that the first is yellow

b the second bead is yellow, given that the first is blue

c both beads are yellow

d one yellow and one blue bead are chosen.

(PS) 2 A new business is setting up to make artisan chocolates.

Machine J makes 25% of the chocolates.
Machine K makes 45% of the chocolates.
Machine L makes the remaining chocolates.

It is known that 2% of the chocolates made by machine J are misshapen, 3% of the chocolates made by machine K are misshapen and 5% of the chocolates made by machine L are misshapen.

a Draw a tree diagram to illustrate all the possible outcomes and associated probabilities.

A chocolate is selected at random.

b Calculate the probability that the chocolate is made by machine J and is not misshapen.

c Calculate the probability that the chocolate is misshapen.

d Given that the chocolate is misshapen, find the probability it was not made by machine K.

3 A local supermarket sells peppers. Bag X contains 12 red and yellow peppers in equal numbers. Another bag, Y, contains 4 peppers, of which 3 are red and 1 is yellow.

A pepper is taken out at random from bag X and placed in bag Y. This process is repeated. Finally a pepper is taken from the 6 peppers now inside bag Y, and its colour is noted.

Event A – when the two peppers taken from bag X are the same colour.
Event B – when the pepper drawn from bag Y is red.

a Complete a tree diagram to show all the probabilities.

b Find P(B).

c Calculate P($A \cap B$).

d Hence find P($A \cup B$).

e Find the probability that all three peppers are red, given that they are the same colour.

4 A game consists of spinning a spinner and then flipping a coin. The spinner has two colours: red and blue. The probability of spinning red is $\frac{2}{9}$ and the probability of spinning blue is $\frac{7}{9}$.

When the spinner lands on red, a biased coin with a probability of $\frac{3}{5}$ of landing on heads is flipped.

When the spinner lands on blue, a biased coin with a probability of $\frac{1}{5}$ of landing on heads is flipped.

a Complete a tree diagram.

Mirunalini plays the game.

b Find the probability she obtains tails.

c Given that Hilary played the game and obtained a head when she flipped the coin, find the probability that Hilary's spinner landed on blue.

Mirunalini and Hilary play the game again.

d Find the probability that the spinner lands on the same colour for both girls.

12.3 Modelling real-life problems with probability

A **model** allows you to simplify a real event or situation and then use the results to make predictions. Using a model, you can carry out experiments without the difficulty, expense or risk involved in setting up a real-life test.

Using a model means that certain assumptions have to be made, which means that not all the features of the real event are included. As the model is a simplified version of a real event, it may only work in certain situations as not all variables can be controlled.

Example 10

When flipping a coin, state the probability that you would flip a tail. State any assumptions you make.

Consider the reliability of your probability by critically analysing your assumptions. State two more realistic assumptions you could make and the effect that these assumptions would have on the probability.

Solution

The probability would be 0.5 as a coin only has two sides: heads or tails. This would be under the assumption that both sides of the coin are equally likely to appear and the coin is tossed fairly.

One side of the coin could, in theory, weigh more than the other. If the heads side weighed slightly more than the tails side, then spinning the coin would mean it was more likely to land with the heavy side side downwards – so the probability of a tail would increase. This means the coin should always be flipped using the same method to ensure a fair result.

Example 11

Jen is a 24-year-old woman, living and born in England. In 2012, Bulgaria recorded 3000 heart attacks in women under 25, and there were 1 500 000 women under 25 living in Bulgaria at that time. What is the probability that Jen will have a heart attack?

Critically analyse your answer, stating any other assumptions you would make and the effect they would have on the probability.

Solution

Assume that there is an equal probability of any woman under 25 having a heart attack, and that there are equal probabilities for women in both Bulgaria and England of suffering from heart attacks.

Therefore the probability would be $\dfrac{3000}{1\,500\,000} = 0.002$.

The reliability of the probability should be questioned, as it is not only age that affects the chance of someone having a heart attack. For example, you could consider whether:

> someone living in England is less likely to have a heart attack than someone living in Bulgaria

> Jen's weight should be accounted for, as her chances of a heart attack will be higher if she is overweight

> Jen may have a family history of heart attacks, which would increase the likelihood of her having a heart attack.

Example 12

Liam and Samir are playing a game that involves choosing one of three dice, coloured orange, black and white. Then each player rolls his die and the one showing the higher number is the winner.

Each die has a different number on each face:

Orange	5, 7, 8, 9, 10, 18
Black	2, 3, 4, 15, 16, 17
White	1, 6, 11, 12, 13, 14

The total on each die is 57, so the game would appear fair. Do you agree?

Solution

In order to model this situation, add in the assumption that the dice are selected randomly to play against each other. The individual dice are unbiased, but each die will lose against one die and win against one die.

To determine if this game is fair, the coloured dice can be played against each other and the winning die can be recorded in a **sample space** diagram. Then you will see whether any of the dice has an advantage.

The winning dice are shown by the initials in the sample space diagrams.

First match: Orange vs. black

Probability of orange die winning

$= \frac{21}{36} = \frac{7}{12}$

Probability of black die winning

$= \frac{15}{36} = \frac{5}{12}$

	Orange die					
Black die	5	7	8	9	10	18
2	O	O	O	O	O	O
3	O	O	O	O	O	O
4	O	O	O	O	O	O
15	B	B	B	B	B	O
16	B	B	B	B	B	O
17	B	B	B	B	B	O

The first match results show that the orange die has a higher probability of winning against the black die.

See **Book 1, Chapter 14 Statistical sampling and hypothesis testing** for more on sample space diagrams.

Second match: Orange vs. white

Probability of orange die winning

$= \frac{5}{12}$

Probability of white die winning $= \frac{7}{12}$

	Orange die					
White die	5	7	8	9	10	18
1	O	O	O	O	O	O
6	W	O	O	O	O	O
11	W	W	W	W	W	O
12	W	W	W	W	W	O
13	W	W	W	W	W	O
14	W	W	W	W	W	O

The second match results show that the white die has a higher probability of winning against the orange die.

Third match: White vs. black

Probability of white die winning $= \frac{5}{12}$

Probability of black die winning $= \frac{7}{12}$

	White die					
Black die	1	6	11	12	13	14
1	B	W	W	W	W	W
6	B	W	W	W	W	W
11	B	W	W	W	W	W
12	B	B	B	B	B	B
13	B	B	B	B	B	B
14	B	B	B	B	B	B

The third match results show that the black die has a higher probability of winning against the white die.

The three matches show that the dice are not evenly matched. In each match, one of the dice has a higher probability of winning, so the game is biased.

To make the game fair, Liam and Samir could select their die randomly, so that they don't know which die they are choosing and therefore which one has the higher probability of winning each match.

Exercise 12.3A

Answers page 470

1 To win a game you need to roll a 7 on two dice.

    a What is the probability of winning in one throw?

    b State any assumptions you make.

(PS) 2 Arthur is an 18-year-old man, living and born in England. In 2016, unemployment in Australia was at 13.19% for 16–24 year olds, and there were 1.3 million men aged 16–24 living in Australia at that time.

What is the probability that Arthur will get a job?

Critically analyse your answer, stating any other assumptions you would make and the effect they would have on the probability.

3 Two players each choose one of three spinners. Each player spins his or her chosen spinner and the one showing the higher number is the winner.

The faces of the three spinners are labelled as follows:

Spinner 1	10, 10, 10, 20, 20, 30
Spinner 2	5, 10, 15, 20, 50
Spinner 3	1, 9, 19, 21, 50

The total on each spinner is 100, so the game would appear fair.

Do you agree? Explain your answer.

SUMMARY OF KEY POINTS

Notation:

- $P(A \cup B)$ means the probability that A or B or both occur, or A union B.
- $P(A \cap B)$ means the probability that both A and B occur, or A intersection B.
- $P(A')$ means the probability that A does not occur.
- $P(A|B)$ means the probability that A occurs, given that B definitely occurs.

Formulae that are always true:

- Addition rule: $P(A \cup B) = P(A) + P(B) - P(A \cap B)$
- Product rule: $P(A|B) = \dfrac{P(A \cap B)}{P(B)}$
- $P(A') = 1 - P(A)$

Formulae that are true only when A and B are independent:

- $P(A \cap B) = P(A) \times P(B)$
- $P(A) = P(A|B) = P(A|B')$

Formulae that are true only when A and B are mutually exclusive:

- $P(A \cap B) = 0$
- $P(A \cup B) = P(A) + P(B)$

EXAM-STYLE QUESTIONS 12

Answers page 470

 1   There are two ways to watch a film in a hotel: digital download or DVD. The table shows the number of films of each type that a hotel has, in each of four subject categories.

	Crime	Romantic comedy	Science fiction	Documentary	Total
DVD	8	16	18	18	60
Digitial download	16	40	14	30	100
Total	24	56	32	48	160

A film is selected at random. Calculate the probability that the film is:

a a digital download [1 mark]

b not a romantic comedy [1 mark]

c crime or science fiction [2 marks]

d a digital download, given that it is a documentary. [3 marks]

PS 2 Three students, Hamed, Melbin and Zineb, travel independently to college from their homes using one of three methods: walking, cycling or catching the bus.

	Walk	Cycle	Bus
Hamed	0.65	0.1	0.25
Melbin	0.4	0.45	0.15
Zineb	0.25	0.55	0.2

a Calculate the probability that, on any given day

i all three walk [1 mark]

ii only Zineb catches the bus [1 mark]

iii at least of them two cycle. [2 marks]

b Tahlia, a friend of Zineb, never travels to college by bus. The probability that Tahlia walks is 0.9 when Zineb walks to college. The probability that Tahlia cycles is 0.7 when Zineb cycles to college.

Calculate the probability that, on any given college day, Zineb and Tahlia travel to college by

i the same method [2 marks]

ii different methods. [3 marks]

PS 3 Kath visits the supermarket every Saturday morning to do her weekly shopping. Items she may buy are sausages, eggs and mushrooms.

The probability, P(S), that she buys sausages on any given Saturday is 0.85.
The probability, P(E), that she buys eggs on any given Saturday is 0.60.
The probability, P(M), that she buys milk on any given Saturday is 0.55.

Calculate the probability that, on any given Saturday, Kath buys:

a none of those items [3 marks]

b exactly two or three items. [4 marks]

4 In a group of 32 students, 24 take art and 16 take music. One student takes neither art nor music. The Venn diagram below shows the events art and music. The values p, q, r and s represent numbers of students.

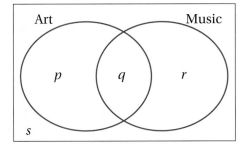

a i Write down the value of s. [1 marks]

ii Find the value of q. [2 marks]

iii Write down the value of p and the value of r. [3 marks]

b i A student is selected at random. Given that the student takes music, write down the probability that the student takes art. **[2 marks]**

ii Hence, show that taking music and taking art are not independent events. **[2 marks]**

c Two students are selected at random, one after the other. Find the probability that the first student takes only music and the second student takes only art. **[4 marks]**

CM 5 In a class of 100 boys, 47 boys cycle and 62 boys swim. Each boy must do at least one sport from cycling and swimming.

a Find the number of boys who do both sports. **[1 mark]**

b One boy is selected at random.

i Find the probability that he does only one sport from cycling and swimming. **[2 marks]**

ii Given that the boy selected does only one sport, find the probability that he cycles. **[2 marks]**

Let C be the event that a boy cycles and S be the event that a boy swims.

c Explain why C and S are not mutually exclusive. **[2 marks]**

d Show that C and S are not independent. **[3 marks]**

PS 6 A health club has a number of facilities, which include a gym and a sauna. Nick and Jo visit the health club together every Tuesday evening.

On any visit, Nick uses either the gym or the sauna or both, but no other facilities. The probability that he uses the gym, $P(G)$, is 0.7. The probability that he uses the sauna, $P(S)$, is 0.55. The probability that he uses both the gym and the sauna is 0.25.

a Calculate the probability that, on a particular visit:

i Nick does not use the gym **[1 mark]**

ii he uses the gym but not the sauna **[2 marks]**

iii he uses either the gym or the sauna but not both. **[2 marks]**

b Assuming that Nick's decision on which facility to use is independent from visit to visit, calculate the probability that, during a month in which there are exactly four Tuesdays, he does not use the gym. **[3 marks]**

c The probability that Jo uses the gym when Nick uses the gym is 0.6, but it is only 0.1 when he does not use the gym.

Calculate the probability that, on a particular visit, Jo uses the gym. **[3 marks]**

d On any visit to the gym, Jo uses exactly one of the club's facilities.

The probability that she uses the sauna is 0.35.

Calculate the probability that, on a particular visit, she uses a facility other than the gym or the sauna. **[3 marks]**

7 Dave and his neighbour Ange work at the same place.

On any day Dave travels to work, he uses one of three options: his car only, a tram only or both his car and a tram. The probability that he uses his car, either on its own or with a tram, is 0.8. The probability that he uses both his car and a tram is 0.36.

a Calculate the probability that, on any particular day when Dave travels to work, he:

  i does not use his car   **[1 mark]**

  ii uses his car only   **[2 marks]**

  iii uses a tram.   **[2 marks]**

b On any day, the probability that Ange travels to work with Dave is 0.56 when Dave uses his car only, 0.66 when Dave uses both his car and a tram, and 0.25 when he uses a tram only.

  i Calculate the probability that, on any particular day when Dave travels to work, Ange travels with him.   **[3 marks]**

  ii Assuming that the choices are independent from day to day, calculate, to 3 decimal places, the probability that, during any particular 5-day week when Dave travels to work every day, Ange never travels with him.   **[3 marks]**

8 A game is played in which teams have to find two flags that are hidden on a beach. There are 9 square flags and 3 hexagonal flags. Once a team find a flag it is not replaced. They then look for a second flag.

a Draw a tree diagram to represent this information.   **[3 marks]**

Find the probability that:

b the second flag is hexagonal   **[2 marks]**

c both flags found are square, given that the second flag is square.   **[2 marks]**

9 For the events G and H, $P(G \cap H') = 0.33$, $P(G' \cap H) = 0.21$ and $P(G \cup H) = 0.75$.

a Draw a Venn diagram for the events G and H.   **[3 marks]**

b Find $P(G)$ and $P(H)$.   **[3 marks]**

c Find $P(G \mid H')$.   **[2 marks]**

d Are events G and H independent?   **[3 marks]**

13 STATISTICAL DISTRIBUTIONS

Data such as the marks of students in an exam will have a mean and a standard deviation showing the spread of the data. The majority of students taking the exam will get within two standard deviations of the mean.

You can use these statistics to predict other information about the data and the real-life situation. If you are told that the bottom 5% of students will fail the course, you could work out the lowest mark that a student can have and still be awarded a passing grade, or the probability of a student's mark being in the top 10%.

In this chapter you will learn how to use the normal distribution to calculate missing information such as in the example above and to find the probability of certain data items occurring.

LEARNING OBJECTIVES

You will learn how to:

› understand and use the normal distribution

› find probabilities involving the normal distribution by using a calculator

› analyse the shape, symmetry and points of inflection of the normal distribution

› link the normal distribution to the binomial distribution

› apply continuity corrections

› select an appropriate probability distribution for a particular context.

TOPIC LINKS

You will need the skills and techniques that you learnt in **Book 1, Chapter 12 Data presentation and interpretation** to interpret histograms and calculate the mean and standard deviation. The content from **Book 1, Chapter 13 Probability and statistical distributions** will help you calculate probabilities. From that same chapter, the ability to use binomial expansions will help you calculate exact and cumulative probabilities.

PRIOR KNOWLEDGE

You should already know how to:

› interpret the mean and standard deviation and draw histograms

› infer properties of populations or distributions from a sample, while knowing the limitations of sampling

› apply statistics to describe a population.

You should be able to complete the following questions correctly:

1   The numbers of calls received by the emergency services in 20 consecutive minutes were as follows: 6, 1, 3, 4, 0, 4, 2, 3, 9, 4, 4, 3, 5, 0, 4, 6, 6, 4, 2, 5

   Calculate: **a** the mean   **b** the standard deviation.

   In the next 5 minutes the following numbers of calls were made: 0, 0, 2, 5, 1

   c Calculate the new mean and standard deviation.

13.1 The normal distribution

If you measured the heights of all the people at a concert on any particular night and plotted the results, the graph would look something like this **histogram**:

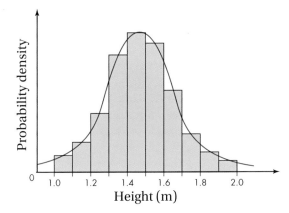

By plotting the **continuous variable** of height, you can represent all the possible probabilities. A randomly chosen person being of a particular height would be a continuous distribution. Continuous distributions are distributions for continuous variables and are more commonly known as probability densities. A probability density shaped like a bell like this is called the normal distribution.

> Normal distributions deal with continuous variables.

Each end of the bell curve extends **asymptotically**. The y-axis in the **normal distribution** represents the probability density. For example, in the diagram above, the area under the curve represents 1. From this you can see the probability of a person being about 1.2 m is about half the probability of a person being about 1.4 m.

> 'Asymptotically' means that the curve never touches the axis.

You do not need to understand the concept of probability density in detail as it is beyond the scope of the course. However, you should learn the following features of a curve that describes a continuous distribution like the normal distribution.

> The area under the curve equals 1.

   Using the example of heights at a concert, you would conclude that the probability of having a height is 1!

> The probability of any particular value of X is 0.

In the concert example, the probability that a particular person will have height of exactly 1.3 m is essentially zero.

> You can make the probability as close to zero as you like, by making the height measurement more and more precise.

> The area under the curve and bounded between two given points on the x-axis is the probability that a number chosen at random will fall between the two points.

In the concert example, suppose that the probability of measuring between 1.3 m and 1.4 m is $\frac{1}{10}$. Then the continuous distribution for possible measurements would have a shape that places 10% of the area below the curve in the region bounded by 1.3 m and 1.4 m on the x-axis.

Stop and think   What would the histograms look like if the heights of males and females had been shown separately?

> Normal distributions are defined by two parameters: the mean (μ) and the **standard deviation** (σ).

  > 68% of the area of a normal distribution lies within 1 standard deviation of the mean.

  > Approximately 95% of the area of a normal distribution lies within 2 standard deviations of the mean.

  > 99.75% of the data lies within 3 standard deviations of the mean.

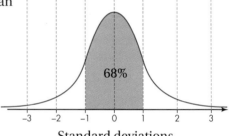

Standard deviations

> The mean, μ, and the standard deviation, σ, determine the shape of the normal curve.

  > The mean gives the central location of the data, which is the line of symmetry.

  > The smaller the standard deviation, the less spread out are the data. The larger the standard deviation, the more spread out are the data.

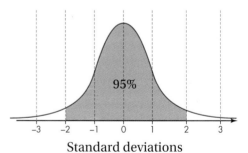

Standard deviations

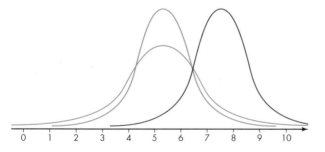

In the diagram above, the red and green curves have the same shape and width. This means that the standard deviations are the same. The green and blue curves have the same mean. The blue curve is wider, so it has the larger standard deviation, so its data values are more spread out.

A **point of inflection** is where the gradient of the curve stops increasing and starts decreasing (or vice versa).

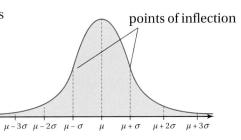

points of inflection

The inflection points are the between the mean, μ, and the standard deviation, σ. This means that the inflection points are between $\mu - \sigma$ and $\mu + \sigma$.

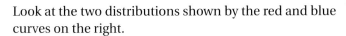

Look at the two distributions shown by the red and blue curves on the right.

Both distributions in the diagram have the same mean. The red curve has a smaller standard deviation, meaning its distribution is less spread out on either side of the mean. The blue curve has a larger standard deviation, meaning its distribution is more spread out on either side of the mean.

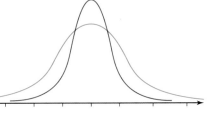

Example 1

The marks in an examination are normally distributed.

a What percentage of the marks would be above the mean mark?

b What percentage of the marks would be between -1 and $+1$ standard deviations of the mean?

c What percentage of the marks would be between the mean and $+2$ standard deviations of the mean?

d What percentage of the marks would be between -2 standard deviations and $+1$ standard deviation of the mean?

Solution

a As the normal distribution is symmetrical either side of the mean, you would expect 50% of the marks to be above the mean.

b You would expect that 68% of the marks would be between -1 and $+1$ standard deviations either side of the mean.

c 95% of the data is between -2 and $+2$ standard deviations either side of the mean, so 47.5% of the marks would be between the mean and $+2$ standard deviations.

d Between -2 and $+2$ standard deviations there is 95% of the data and between -1 and $+1$ standard deviations there is 68% of the data.

So from -2 standard deviations to the mean is 47.5% of the data and from the mean to $+1$ standard deviation is 34% of the data. Combining these gives you 81.5% of the data.

Exercise 13.1A **Answers page 471**

1 Correct the following list of properties of a normal distribution.

  a The distribution is symmetrical about the standard deviation.

  b The mode, median and mean are all different.

  c The total area above the curve is 1.

  d The distribution is defined by three parameters: the mean, the median and the standard deviation.

(CM) 2 Comment on the features of the two distributions in the diagram.

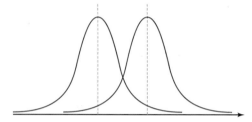

(PS) 3 The heights of Year 13 students are normally distributed.

  a What percentage of the Year 13 students would be above the mean height?

  b What fraction of the Year 13 students would have heights between the mean and +1 standard deviation?

  c What percentage of the Year 13 students would have heights between the mean and +2 standard deviations?

  d What fraction of the Year 13 students would have heights between −3 standard deviations and +2 standard deviations of the mean?

(CM) 4 Describe each of the curves in this diagram, using the terms 'mean' and 'standard deviation'.

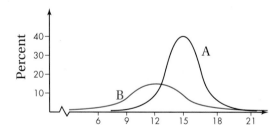

13.2 Using the normal distribution

The diagram on the right shows a normal distribution with a mean of 24 and a standard deviation of 1. The shaded area between 23 and 25 contains 68% of the distribution – so the probability of getting a value between 23 and 25 would be 0.68. Similarly, 95% of the data lies between 22 and 26, and anything below 21 or above 27 would be classed as an outlier.

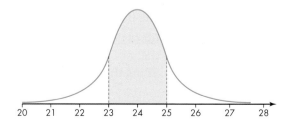

The area under a normal curve shows probabilities between two values. The probability of the variable taking a value between two limits is the area under the curve between those limits. Not all probabilities will be as straightforward as 1 or 2 standard deviations, so you will need to use a formula.

If a variable X has a normal distribution, then you can write this using mathematical notation:

$$X \sim N(\mu, \sigma^2)$$

where

› X is the random variable (this can be any letter)

› ~ is shorthand for 'is distributed'

› N tells you that it has a normal distribution

› μ is the mean

› σ is the standard deviation

› σ^2 is the square of the standard deviation and is called the variance.

For example, if $X \sim N(25, 1^2)$, this tells you that:

› the random variable X has a normal distribution

› with a mean $\mu = 25$

› and a standard deviation $\sigma = 1$.

This means that 68% of the values lie between 24 (= 25 − 1) and 26 (= 25 + 1).

Don't forget that the second parameter is the variance. The standard deviation is the square root of this number.

Example 2

The heights of males and females in a town were recorded and had results with normal distributions as follows:

Males: $X \sim N(178, 7^2)$

Females: $Y \sim N(167, 64)$

Write down mean, variance and standard deviation for each gender.

Solution

Using $X \sim N(\mu, \sigma^2)$, you know the first parameter is the mean and the second is the variance. With that in mind:

Males: $X \sim N(178, 7^2)$ has a mean of 178 cm, a variance of 49 (or 7^2) and a standard deviation of 7.

Females: $Y \sim N(167, 64)$ has a mean of 167 cm, a variance of 64 (or 8^2) and a standard deviation of 8.

Exercise 13.2A

Answers page 471

1   During a kickboxing tournament, two different fighters had scores with normal distributions as follows:

Fighter A: mean = 3, variance = 4
Fighter B: mean = 2, standard deviation = 3

Use the correct mathematical notation to describe the distribution of each of the fighters' scores.

2   A school took part in a Maths Challenge over two different years. The scores had normal distributions as follows:

Year 1: $\mu = 0$, $\sigma^2 = 5$
Year 2: $\mu = -1$, $\sigma = 2$

Use the correct mathematical notation to describe the distribution of the scores for each year.

CM   3   A mathematics student was asked to write down the mathematical notation for the scores on an app, which have a normal distribution with mean $\mu = 7$ and standard deviation $\sigma = 11$.

The student wrote $N \sim X(11, 49)$.

Write down everything that is wrong with the student's notation, then write down the correct notation.

4   Two mathematics students sat a series of tests and had results with normal distributions as follows:

Student A: $X \sim N(5, 7^2)$
Student B: $Y \sim N(6, 81)$

a   Write down the mean, variance and standard deviation of each student's scores.

b   Sketch this as a graph.

The standard normal distribution

It is not necessarily the case that normal distributions have the same means and standard deviations. A normal distribution with a mean of 0 and a standard deviation of 1 is called a **standard normal distribution**.

The standard normal distribution can be used to work out probabilities for variables with a normal distribution. The standard normal distribution is defined as follows:

$$Z \sim N(0, 1)$$

This tells you that:

❯ the standard normal random variable is denoted by Z

❯ Z has a normal distribution

❯ with mean $\mu = 0$

❯ and standard deviation $\sigma = 1$.

The diagram shows the standard normal distribution.

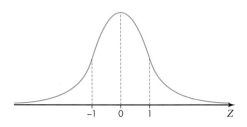

Any normal distribution can be made to fit the standard normal distribution. The process is called **standardising**. Since the distribution has a mean of 0 and a standard deviation of 1, the value of Z is equal to the number of standard deviations below (or above) the mean. This is often referred to as the z-score. For example, a z-score of -2.5 represents a value 2.5 standard deviations below the mean.

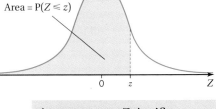

Area $= P(Z \leqslant z)$

On the right is a graph of the standard normal distribution. It is symmetrical about the mean, 0. The shaded area under the curve shows the probability that Z takes a value that is less than or equal to z. To calculate the probability of this, you would need to find the area.

The z-scores can all be calculated on your calculator. The calculator shows values of the normal cumulative distribution function, $\Phi(z)$, which correspond to the area under the curve to the left of Z for different values of z.

> An uppercase Z signifies the random variable which is being standardised. A lowercase z signifies the z-score.

There are three different cases to consider:

> Z is less than a value a ($Z < a$)

> Z is greater than a value a ($Z > a$)

> Z is between values a and b ($a < Z < b$).

If you are given that $Z \sim N(0, 1)$, you can say that the probability that Z is less than a value a (where $a > 0$) is shown by

$$P(Z < a)$$

where P is the probability.

KEY INFORMATION

$\Phi(z) = P(Z \leqslant z)$

You know that the mean of the standard normal distribution is 0 so the diagram will look like the left-hand diagram below.

If you are given that $Z \sim N(0, 1)$, you can say that the probability that Z is greater than a value a (where $a > 0$) is shown by

$$P(Z > a)$$

where P is the probability.

TECHNOLOGY

Your calculator should have a 'distribution' mode, with various settings for the normal distribution. Make sure that you know how to find values of the cumulative distribution function.

You know that the mean of the standard normal distribution is 0 so the diagram will look like the right-hand diagram below.

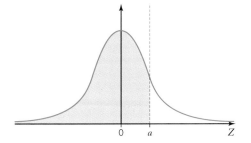

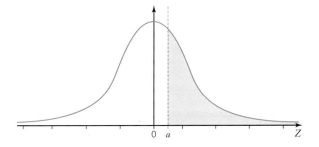

If you are given that $Z \sim N(0, 1)$, you can say that the probability that Z is between the value a and b (where a, $b > 0$ and $b > a$) is shown by

$$P(a < Z < b)$$

where P is the probability.

You know that the mean of the standard normal distribution is 0 so the diagram will look like the one on the right.

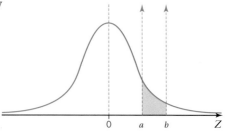

Exercise 13.2B   Answers page 471

1 Given that $Z \sim N(0, 1)$, draw a clear sketch diagram for:

  a $P(Z > a)$ where $a < 0$

  b $P(Z < a)$ where $a < 0$

  c $P(a < Z < b)$ where $a < 0$, $b > 0$

  d $P(a < Z < b)$ where a, $b < 0$ and $b > a$

  e $P(0 < Z < a)$ where $a > 0$.

2 A mathematics student was asked to write down the mathematical notation for the standard normal distribution.

The student wrote $N \sim Z(1, 0^2)$.

List everything that is wrong with what the student wrote and then write down the correct notation.

Using a calculator for the normal distribution

Cumulative probabilities associated with the z-score can all be calculated using your calculator.

If you are given that $Z \sim N(0, 1)$ and asked to find, for example, $P(Z < 0.43)$, begin by drawing a clear diagram.

The expression for the probability is:

$$P(Z < 0.43) = \Phi(0.43)$$

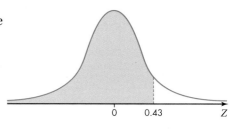

where Φ represents the area to the left of the value in brackets – that is, any given value of Z.

> Φ is the capital Greek letter F called phi and pronounced 'fi'.

On your calculator, set the mode to 'normal cumulative probability', as you want to know the probability up to and including 0.43. Set the lower bound as a low negative number (for example –999, as the curve is an asymptote) and the upper bound as 0.43. The standard deviation should be 1 as this is a standard normal distribution.

So $P(Z < 0.43) = 0.666\,402$.

This means that the probability of Z having a value less than 0.43 is 0.666 402.

How do you find $P(Z > 0.43)$? The calculator only gives the probabilities to the left of a value, i.e. less than the value.

However, the total area under the curve is 1, so you can calculate the difference.

$$P(Z > 0.43) = 1 - \Phi(0.43)$$

$$= 1 - 0.666\,402$$

$$= 0.333\,598$$

So the probability of Z having a value greater than 0.43 is 0.333 598. Note that $P(Z < 0.43) + P(Z > 0.43) = 1$.

How do you find $P(Z < -0.43)$?

You know that the normal distribution is symmetrical about the mean, so:

$$P(Z < -0.43) = P(Z > 0.43)$$

$$= 1 - P(Z < 0.43)$$

$$= 1 - \Phi(0.43)$$

$$= 1 - 0.666\,402$$

$$= 0.333\,598$$

This means the probability of Z having a value less than -0.43 is 0.333 598. Note that this is the same as $P(Z > 0.43)$.

How do you find $P(0.1 < Z < 0.31)$?

To find the probability (area) between 0.31 and 0.1 you need to find the probability less than 0.31 and subtract the probability less than 0.1.

$$P(0.1 < Z < 0.31) = P(Z < 0.31) - P(Z < 0.1)$$

$$= \Phi(0.31) - \Phi(0.1)$$

$$= 0.621\,72 - 0.539\,83$$

$$= 0.081\,89$$

Finally, what is the probability that Z will have a value greater than the mean but less than 1.5?

To find the probability (area) between the mean (0) and 1.5, you need to find the probability that Z will have a value less than 1.5 and subtract the probability that Z will have a value less than 0.

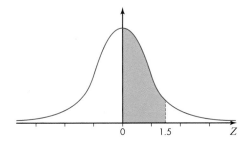

TECHNOLOGY

This time you can select the lower bound as 0.43 and the upper bound as a large number (for example 999, as the curve is an asymptote).

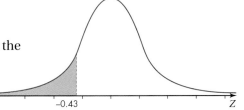

TECHNOLOGY

Select the lower bound to a small number, the upper bound to -0.43 and the standard deviation to 1.

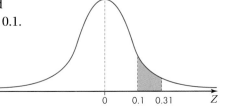

It is important to show the stages in your working and not rely solely on writing down the answer from your calculator.

$$P(0 < Z < 1.5) = P(Z < 1.5) - P(Z < 0)$$

The probability that Z is less than 0 is 0.5 since 0 is the mean and half of a normal distribution lies below the mean and half above it.

$$P(0 < Z < 1.5) = P(Z < 1.5) - P(Z < 0)$$
$$= \Phi(1.5) - 0.5$$
$$= 0.933\,19 - 0.5$$
$$= 0.433\,19$$

> **TECHNOLOGY**
>
> On your calculator, set the lower bound to 0, the upper bound to 1.5 and the standard deviation to 1.

Example 3

Find the following probabilities using your calculator:

a $P(Z \leqslant 0.34)$ **b** $P(Z < 0.2)$ **c** $P(Z > 0.6)$

Solution

a Select a small lower bound, an upper bound of 0.34 and a standard deviation of 1.
$$P(Z \leqslant 0.34) = 0.633\,07$$

b Select a small lower bound, an upper bound of 0.2 and a standard deviation of 1.
$$P(Z < 0.2) = 0.579\,26$$

c Select a lower bound of 0.6, a large upper bound and a standard deviation of 1.
$$P(Z > 0.6) = 0.274\,25$$

Exercise 13.2C

Answers page 472

For this exercise, use the correct notation and draw a diagram in each case.

1. Find the probability that $Z \sim N(0, 1)$ will be less than 1.4.

2. Find the probability that $Z \sim N(0, 1)$ will be less than 0.87.

3. Find the probability that $Z \sim N(0, 1)$ will be greater than 1.06.

4. Find the probability that $Z \sim N(0, 1)$ will be less than −3.

5. Find the probability that $Z \sim N(0, 1)$ will be greater than −1.32.

6. Find the probability that $Z \sim N(0, 1)$ will be between 1.1 and 2.1.

7. Find the probability that $Z \sim N(0, 1)$ will be between −1.32 and 1.21.

8. Find the probability that $Z \sim N(0, 1)$ will be between −2.65 and −1.43.

13.3 Non-standardised variables

You know that the area under a normally distributed curve (that is, with mean 0 and standard deviation 1) is equal to 1. To find the probability that a random variable x is in any interval within the curve, you need to calculate the area of that interval. To find the area of any interval under any normal curve, convert to a z-score. A z-score finds the value of a **non-standarised variable**.

Use the following formula to find a z-score:

$$z = \frac{\text{value} - \text{mean}}{\text{standard deviation}}$$

$$= \frac{x - \mu}{\sigma}$$

The horizontal scale of the curve of the standard normal distribution corresponds to z-scores.

If the values of the random variable are converted into z-scores, the result will be the standard normal distribution. Following this conversion, the area that falls in the interval under the non-standard normal curve is the same as the area under the standard normal curve within the corresponding boundaries.

Example 4

Given the random variable $X \sim N(15, 4^2)$, find:

a $P(X < 25)$ **b** $P(X > 25)$ **c** $P(10 < X < 25)$

Solution

a Draw a clear diagram:

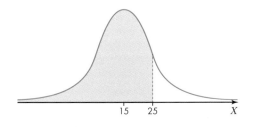

$$P(X < 25) = P\left(Z < \frac{25 - 15}{4}\right)$$

$$= P(Z < 2.5)$$

$$= \Phi(2.5)$$

$$= 0.993\,79$$

So, the probability of X having a value less than 25 is 0.993 79.

b Draw a clear diagram:

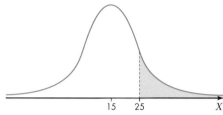

$$P(X > 25) = 1 - P(X < 25)$$

$$= 1 - P\left(Z < \frac{25 - 15}{4}\right)$$

$$= 1 - P(Z < 2.5)$$

$$= 1 - \Phi(2.5)$$

$$= 1 - 0.993\,79$$

$$= 0.006\,21$$

So, the probability of X having a value greater than 25 is 0.006 21.

c Draw a clear diagram:

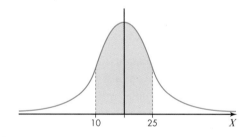

$$P(10 < X < 25) = P\left(\frac{10 - 15}{4} < Z < \frac{25 - 15}{4}\right)$$

$$= P(-1.25 < Z < 2.5)$$

$$= P(Z < 2.5) - P(Z < -1.25)$$

$$= \Phi(2.5) - \Phi(-1.25)$$

$$= 0.993\,79 - 0.105\,65$$

$$= 0.888\,14$$

So, the probability of X having a value between 10 and 25 is 0.888 14.

TECHNOLOGY

Set the lower bound to 2.5, the upper bound to a large number and the standard deviation to 1.

TECHNOLOGY

Set the lower bound to −1.25, the upper bound to 2.5 and the standard deviation to 1.

> **Using the large data set 13.1**
>
> Using your large data set, choose a category of interest. Assuming that your category follows a normal distribution, calculate the mean and standard deviation.
>
> Find:
>
> **a** the value of the item of data which is $P(Z > 2.5)$
>
> **b** the value of the item of data which is $P(Z < -1.2)$
>
> **c** the probability of $P(-1.2 < Z < 2.5)$.

Exercise 13.3A

Answers page 472

For this exercise, present your working as shown in the examples above and use the correct notation.

1 Find the area of the shaded region under each standard normal curve.

a

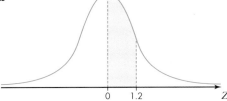

b

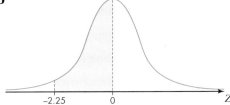

c

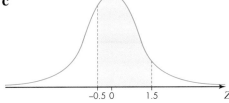

d

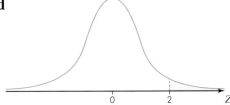

2 Given that $X \sim N(30, 100)$, find the following probabilities:

a $P(X < 35)$ **b** $P(X > 38.6)$ **c** $P(X > 20)$

d $P(35 < X < 40)$ **e** $P(15 < X < 32)$ **f** $P(17 < X < 19)$

(PS) 3 The distribution of heights of 20-year-old men is modelled using the normal distribution with a mean of 177 cm and a standard deviation of 7 cm. Find the probability that the height of a randomly selected 20-year-old man is:

a under 172 cm   **b** over 180 cm   **c** between 172 cm and 180 cm.

(PS) 4 When Tim makes a hot chocolate with instant powder, he uses a spoon to add it into a mug. The weight of powder in grams may be modelled by the normal distribution with a mean of 6 g and a standard deviation of 1 g. Tim has found that if he uses more than 8 g Jenni says it's too strong and if he uses less than 4.5 g she says it's too weak.

Find the probability that he makes the hot chocolate:

a too weak   **b** too strong   **c** just right.

5 A particular breed of hens produce eggs with a mean mass of 60 g and a standard deviation of 4 g, and the mass is found to be normally distributed.

The eggs are classified as small, medium, large or extra large depending on their weight, as follows:

Classification	Weight
small	less than 55 g
medium	between 55 g and 65 g
large	between 65 g and 70 g
extra large	greater than 70 g

   a Find the proportion of eggs that are classified as small.

   b Find the proportion of eggs that are classified as medium.

   c Find the proportion of eggs that are classified as large or extra large.

   d If the classification for medium eggs is changed to 'between 55 g and 68 g', by how much does the proportion of eggs classified as medium increase?

6 The police regularly monitor the speeds of cars on a section of motorway and it is found that the speeds are normally distributed. The mean speed is 68.5 mph with a standard deviation of 5 mph.

   a Find the proportion of motorists that break the speed limit (70 mph).

   b A motorist believes that people aren't fined unless they are 10% over the speed limit. What proportion of motorists would be fined on this basis?

   c A motorist is given a fixed penalty fine for driving at 85 mph. What percentage of motorists exceed this speed?

Finding values of variables from known probabilities
Sometimes you may know the probability of an event and then need to work backwards to find the corresponding value of X.

TECHNOLOGY

Your calculator now needs to be in 'inverse normal distribution' mode.

Example 5

A random variable X is normally distributed with a mean of 9 and a standard deviation of 0.27. Find the value of x so that $P(X < x) = 0.751\,75$.

Solution

First, write down the distribution in the question as a standardised normal distribution.

$$P\left(Z < \frac{x-9}{0.27}\right) = 0.75175$$

On your calculator, set the area (the probability up to and including the value you are trying to find) as 0.751 75. The standard deviation is 0.27 and the mean is 9.

$x = 9.1836$

Using the large data set 13.2

Using your large data set, choose a category of interest. Assume that the data in your chosen category is normally distributed.

a Find the mean.

b Find the standard deviation.

c Find the value of x so that $P(X < x) = 0.850$.

d How would the value in **part c** be different if the location of your category had been elsewhere? Explain your reasons and what assumptions you have made.

Exercise 13.3B

Answers page 472

1 The lifetime of an electrical component may be modelled by a normal distribution with mean 150 hours and variance 52 hour².

   a Find the probability that a component lasts:

     i   more than 150 hours

     ii less than 100 hours

     iii between 100 and 150 hours.

   b How long would you expect:

     i   15% of the components to last

     ii 30% of the components to last

     iii 90% of the components to last?

2 A factory produces bags of sugar. The actual weights may be modelled by a normal distribution with mean 498.7 g and standard deviation 7.3 g. Bags are rejected if they are 15% underweight or 90% overweight.

   What weights are these?

3 Assume that scores in a test are normally distributed with a mean of 19 and a standard deviation of 2.4. Let X be the distribution of test scores. Find the values of a and b such that:

   a $P(X > a) = 0.432$ **b** $P(X < b) = 0.205$

4 Given that $X \sim N(30, 100)$, find the value of x such that:

   a $P(X < x) = 0.99$ **b** $P(X < x) = 0.97982$ **c** $P(X > x) = 0.19489$

   d $P(X < x) = 0.75$ **e** $P(X) > 0.99$ **f** $P(X > x) = 0.05$

(PS) 5 A manufacturer produces a new fluorescent light bulb and claims that it has an average lifetime of 4000 hours with a standard deviation of 375 hours.

   Find the lifetime such that 90% of light bulbs will last for less than this duration.

(PS) 6 A double bed is the right length for 92.9% of men.

   The heights of men in the UK are normally distributed with mean height 1.753 m and standard deviation 0.1 m.

   What is the greatest height a man can be to fit into a double bed?

(PS) 7 The police regularly monitor the speeds of cars on a section of motorway and it is found that the speeds are normally distributed. The mean speed is 68.5 mph with a standard deviation of 5 mph.

   Find the range within which 80% of the speeds lie.

8 An examination has a mean mark of 80 and a standard deviation of 12. The marks can be assumed to be normally distributed.

    a What is the lowest mark needed, to be in the top 25% of the students taking this examination?

    b Between which two marks will the middle 95% of students lie?

    c 200 students take this examination. Calculate the number of students likely to score 90 or more.

Determining the mean or the standard deviation

There are instances where a random variable is known to be normally distributed but the value of the mean or the standard deviation (or both) is not known. When you have information about the probabilities, it is possible to determine the unknown parameters.

Example 6

A farm which produces potatoes models the weights (in g) of a particular variety of potato by a random variable, $J \sim N(\mu, 25)$. The processing company observes that 2% of the potatoes weigh more than 350.7 g. Determine the mean, μ.

Solution

Draw a sketch of the given information:

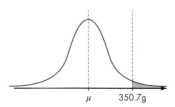

You need to find the area of the shaded region. The question states that 2% of the potatoes have a weight of more than 350.7 g, therefore the area to the right is 2%.

This gives $P(Z > 2.05375) = 0.02$

The standardised value of 350.7 satisfies the equation

$$\frac{350.7 - \mu}{5} = 2.05375$$

Multiply by 5 to give $350.7 - \mu = 10.2685$

Solving this for μ gives $\mu = 340.4315 \approx 340\,\text{g}$.

TECHNOLOGY

On your calculator, set the area to 0.98 (remembering that the calculator is set to read from the left of the graph); as it is modelled using the normal distribution, the mean will be 0 and the standard deviation will be 1.

Example 7

A new process is introduced at the farm. This changes the variance so that the weight of the potatoes is now modelled by the random variable $M \sim N(340, \sigma^2)$. The processing company reveals that 90% of the potatoes have a weight above 322.3 g. Determine the variance of the random variable.

Solution

Draw a sketch:

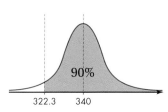

You now need to find the area of the shaded region. The question states that 90% of the potatoes have a weight of more than 322.3 g, therefore the area to the right is 90%.

This gives $P(Z > -1.2816) = 0.90$

The standardised value of 322.3 satisfies the equation

$$\frac{322.3 - 340}{\sigma} = -1.2816$$

Multiply by σ to give $322.3 - 340 = -1.2816\sigma$

$$-17.7 = -1.2816\sigma$$

Solving this for σ gives $\sigma = 13.81$

So the variance, $\sigma^2 = 190.72 \, \text{g}^2$.

TECHNOLOGY

On your calculator, set the area to 0.1 (remembering that the calculator is set to read from the left of the graph); as it is modelled using the normal distribution, the mean will be 0 and the standard deviation will be 1.

Example 8

The weight of potatoes put into bags by the processing plant is modelled by a normal distribution with a mean of μ g and a standard deviation of σ g.

The processing company observes that in the bags 83% contain less than 3548 g and 8.5% contain less than 3431 g. Determine μ and σ.

Solution

As there are two pieces of information, two sketches are required.

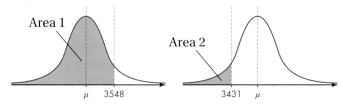

You need to find the area of the shaded regions.

Area 1: The question states that 83% of the potatoes have a weight of less than 3548 g, therefore the area to the left is 83%.

This gives $P(Z < 0.954) = 0.83$

The standardised value of 3548 satisfies the equation

$$\frac{3548 - \mu}{\sigma} = 0.954$$

Area 2: The question states that 8.5% of the potatoes have a weight of less than 3431 g, therefore the area to the left is 8.5%.

This gives $P(Z < -1.372) = 0.085$

The standardised value of 3431 satisfies the equation

$$\frac{3431 - \mu}{\sigma} = -1.372$$

You can now rearrange both of these equations to set up a pair of simultaneous equations in μ and σ.

$$3548 - \mu = 0.954\sigma$$

$$3431 - \mu = -1.372\sigma$$

Subtract the equations to eliminate μ.

$$117 = 2.326\sigma$$

$$\sigma = 50.3$$

Therefore, substituting back into one of the two derived equations:

$$3548 - \mu = 0.954\sigma$$

$$3548 - \mu = 47.99$$

$$\mu = 3500$$

Using the large data set 13.3

Using your large data set, choose a category of interest. Assume your category is modelled by a normal distribution with a mean of μ and a standard deviation of σ.

a   Calculate the figure which represents the top 5% of the data.

b   Calculate the figure which represents 85% of the data.

c   Determine μ and σ based on the answers to **parts a** and **b**.

Exercise 13.3C

Answers page 472

(PS) 1   Assume that scores in a test are normally distributed with a mean of 43 and a standard deviation of σ. Let X be the distribution of test scores.

Find σ if the probability of getting a score above 48 is 0.2.

(PS) 2   The length of a bike frame may be modelled by a normal distribution with a standard deviation of 13 mm. 11% of the frames are longer than 47 cm.

Calculate the mean length of a frame.

(PS) 3   The volume of the contents of cans of fizzy drink may be modelled by a normal distribution. 18% of the cans contain a volume of more than 325.42 ml and 72% contain a volume of more than 332.91 ml. Find the mean and standard deviation of the volumes.

(PS) **4** The quartiles of a normal distribution are known to be 9.92 and 12.24.

Find the mean and standard deviation of the distribution.

(PS) **5** Engineers at a company make certain components for aeroplanes. The company believes that the time taken to make this component may be modelled by the normal distribution with a mean of 90 minutes and a standard deviation of 3 minutes.

Assuming the company's beliefs to be true, find the probability that the time taken to make one of these components, selected at random, was:

a over 95 minutes **b** under 87 minutes **c** between 87 and 95 minutes.

Stan believes that the company is allowing too long for the job and times himself manufacturing the component. He finds that only 10% of the components take him over 85 minutes to make, and that 20% take him less than 65 minutes.

d Estimate Stan's mean and standard deviation.

6 A powerlifter knows from experience that she can deadlift at least 140 kg once in every 5 attempts. She also knows that she can deadlift at least 130 kg on 8 out of 10 attempts.

Find the mean and standard deviation of the weights the powerlifter can lift.

13.4 Using the normal distribution to approximate the binomial distribution

In **Book 1, Chapter 13 Probability and statistical distributions** you learnt about the **binomial distribution**. You are now going to use binomials with the normal distribution.

A fair coin has the same probability of landing on heads as it does on tails. If you want to know the probability of getting exactly 9 heads out of 15 flips, you could use the binomial distribution.

> The mean is $\mu = np = 15 \times 0.5 = 7.5$.

> The variance is $\sigma^2 = np(1 - p) = 15 \times 0.5 \times 0.5 = 3.75$.

> The standard deviation is $\sigma = \sqrt{3.75} = 1.9364$.

To find the exact probability of flipping 9 heads out of 15, you would use $X \sim B(n, p)$ on your calculator, or

$$^{15}C_9 \times 0.5^9 \times 0.5^6 = 0.15274$$

So P(exactly 9 heads) = 0.15274

Remembering to find the z-score,

$$z = \frac{\text{value} - \text{mean}}{\text{standard deviation}}$$

$$= \frac{x - \mu}{\sigma},$$

the z-score for exactly 9 heads is $\dfrac{9 - 3.75}{1.9364} = 2.712$ standard deviations above the mean of the distribution.

You already know that the probability of any one specific value is 0, as the normal distribution is a continuous distribution. However, the binomial distribution is a **discrete probability distribution**, so you need to use something called a **continuity correction**.

To find P(exactly 9 heads) you need you to consider any value from 8.5 to 9.5. Once you are considering a range, you will have an interval, so you can work out the area under a normal curve from 8.5 to 9.5.

The area of the interval is 0.151 94, which is the approximation of the binomial probability. For these parameters, the approximation is very accurate.

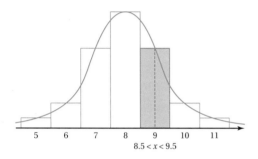

$8.5 < x < 9.5$

You may use the normal distribution as an approximation for the binomial B(n, p) (where n is the number of trials, each having probability p of success) when:

> n is large and

> p is close to 0.5.

This ensures that the distribution is reasonably symmetrical and not skewed at either end.

The parameters for the normal distribution are then:

> mean $= \mu = np$

> variance $= \sigma^2 = np(1 - p)$

> standard deviation $= \sigma = \sqrt{np(1 - p)}$.

If $np \geqslant 5$ and $n(1 - p) \geqslant 5$ then the binomial random variable x is approximately normally distributed, with mean $\mu = np$ and standard deviation $\sigma = \sqrt{np(1 - p)}$.

TECHNOLOGY
Using the 'normal cumulative probability' mode on your calculator, set the lower bound to 8.5 and the upper bound to 9.5; the mean is 7.5 and the standard deviation 1.9364.

Example 9

In the following two surveys, respondents answer 'yes' or 'no'. In each case, decide whether you can use the normal distribution to approximate x, the number of people who answer 'yes'. If you can, find the mean and standard deviation. If you cannot, explain why.

a 34 per cent of people in Lincoln say that they are likely to make a New Year's resolution. You randomly select 15 people in Lincoln and ask each of them if they are likely to make a New Year's resolution.

b 6 per cent of people in Lincoln who made a New Year's resolution resolved to exercise more. You randomly select 65 people in Lincoln who made a resolution and ask each of them if they have resolved to exercise more.

Solution

a In this binomial experiment, $n = 15$, $p = 0.34$, and $(1 - p) = 0.66$.

So, $np = 15 \times 0.34 = 5.1$ and $n(1 - p) = 15 \times 0.66 = 9.9$.

Because np and $n(1 - p)$ are greater than 5, you can use the normal distribution with $\mu = 5.1$ and $\sigma = \sqrt{np(1 - p)} = 1.83$.

b In this binomial experiment, $n = 65$, $p = 0.06$, and $(1 - p) = 0.94$.

So, $np = 65 \times 0.06 = 3.9$ and $n(1 - p) = 65 \times 0.94 = 61.1$.

You should notice here that $np < 5$, which means you cannot use the normal distribution to approximate the distribution of x.

Example 10

Use a continuity correction to convert each of the following binomial intervals to a normal distribution interval.

a The probability of getting between 170 and 290 successes.

b The probability of at least 58 successes.

c The probability of getting fewer than 23 successes.

Solution

a The continuous normal distribution is $169.5 < x < 290.5$.

b The continuous normal distribution is $x > 57.5$.

c The continuous normal distribution is $x < 22.5$.

Example 11

You are told that 38 per cent of people in the UK admit that they look at other people's mobile phones. You randomly select 200 people in the UK and ask each of them if they look at other people's mobile phones. If the statement is true, what is the probability that at least 70 will answer 'yes'?

Solution

Because $np = 200 \times 0.38 = 76$ and $n(1 - p) = 200 \times 0.62 = 124$, the binomial variable x is approximately normally distributed, with

$$\mu = np = 76$$

$$\sigma = \sqrt{200 \times 0.38 \times 0.62} = 6.86$$

Using a continuity correction, you can rewrite the discrete probability $P(x > 70)$ as the continuous probability function $P(x > 69.5)$. The graph shows a normal curve with $\mu = 76$ and $\sigma = 6.86$ and a shaded area to the right of 69.5.

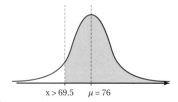

x > 69.5 $\mu = 76$

The z-score that corresponds to 69.5 is $z = \dfrac{69.5 - 76}{6.86} = -0.9475$.
So, the probability that at least 70 will answer 'yes' is

$$P(x > 69.5) = P(z > -0.9475)$$

$$= 1 - P(z < -0.9475)$$

$$= 1 - 0.171\,69 = 0.8283$$

TECHNOLOGY

This can also be done on your calculator. Set the lower bound to –0.9475, the upper bound to a large number, the mean to 0 and the standard deviation to 1.

Your calculator may also let you change the mean and standard deviation, in which case you can check your calculation in one step using the lower bound as 69.5, the upper bound as a large number, the mean as 76 and the standard deviation as 6.86.

Example 12

A survey reports that 95% of teenagers eat a burger each week. You randomly select 200 teenagers and ask each of them whether they have eaten a burger this week.

What is the probability that exactly 194 will have eaten a burger?

Solution

Because $np = 200 \times 0.95 = 190$ and $n(1 - p) = 200 \times 0.05 = 10$, the binomial variable x is approximately normally distributed, with

$$\mu = np = 190$$

$$\sigma = \sqrt{200 \times 0.95 \times 0.05} = 3.0822$$

Using a continuity correction, you can rewrite the discrete probability $P(x = 194)$ as the continuous probability function $P(193.5 < x < 194.5)$.

The z-score that corresponds to 193.5 is $z = \dfrac{193.5 - 190}{3.0822} = 1.135\,55$.

The z-score that corresponds to 194.5 is $z = \dfrac{194.5 - 190}{3.0822}$
$= 1.459\,996$.

TECHNOLOGY

This can also be done on your calculator. Set the lower bound to 1.135 55, the upper bound to 1.459 996, the mean to 0 and the standard deviation to 1.

So, the probability that exactly 194 teenagers will say they have eaten a burger is

$$P(193.5 < x < 194.5) = P(1.135\,55 < z < 1.459\,996)$$

$$= P(x < 1.459\,996) - P(z < 1.135\,55)$$

$$= 0.9279 - 0.871\,93$$

$$= 0.0560$$

There is a probability of about 0.06 that exactly 194 of the teenagers will say they have eaten a burger this week.

Exercise 13.4A

Answers page 473

 1 You decide to use the normal distribution to approximate the binomial distribution. You want to know the probability of getting exactly 16 tails out of 20 coin flips.

    a Calculate the mean number of tails in 20 coin flips.

    b Calculate the variance and standard deviation.

    c Using the normal distribution, calculate the probability of 16 tails out of 20 coin flips.

2 IQ scores are given as an integer value, X. Tests are designed so that the distribution of scores has a mean of 95 and a standard deviation of 11.

    Determine the following probabilities:

    a $P(X < 100)$ **b** $P(85 < X < 105)$

3 The discrete random variable X has a binomial distribution with $n = 200$ and $p = 0.32$.

    Determine, by using a suitable approximation, the following probabilities:

    a $P(X < 150)$ **b** $P(X > 30)$

    c $P(X > 145)$ **d** $P(75 < X < 150)$

 4 **a** State the conditions under which a binomial probability model can be well approximated by a normal model.

    T is a random variable with the distribution B(15, 0.4).

    b Rahman uses the binomial distribution to calculate the probability that $T < 5$ and gives his answer to 4 significant figures. What answer does he get?

    c Cynthia uses a normal distribution to calculate an approximation for the probability that $T < 5$ and gives her answer to 4 significant figures. What answer does she get?

    d Assuming that Cynthia has worked everything out correctly, calculate the percentage error in her approximation.

5 Advertisements for a popular brand of moisturiser claim that 72% of people using it see noticeable effects after just two weeks.

Assuming that the manufacturer's claim is correct for the population using the moisturiser, calculate the probability that at least 13 of a random sample of 15 people using the moisturiser can see noticeable effects:

a using the binomial distribution

b using the normal approximation to the binomial.

c Comment on the agreement or disagreement between your two values.

d Would the agreement be better or worse if the proportion had been 85% instead of 72%?

6 A multiple-choice theory test consists of 35 questions, for each of which the candidate is required to tick one of five possible answers. Exactly one answer to each question is correct. A correct answer gains one mark and an incorrect answer gains no marks.

a One candidate guesses every answer without looking at the question first. What is the probability that they get a particular answer correct?

Calculate the mean and variance of the number of questions they answer correctly.

The panel of examiners want to ensure that no more than 1% of guessers pass the examination in this way.

b Use the normal approximation to the binomial, working to 3 decimal places, to establish the pass mark that meets this requirement.

SUMMARY OF KEY POINTS

> A random variable X that is normally distributed, with mean μ and variance σ^2, is notated $X \sim N(\mu, \sigma^2)$.

> The standard normal distribution is defined as $Z \sim N(0, 1)$, as the mean is 0 and the standard deviation is 1.

> To standardise the variable, use $z = \dfrac{x - \mu}{\sigma}$.

> The normal distribution may be used to approximate discrete distributions but continuity corrections are required.

> The binomial distribution $B(n, p)$ may be approximated by $N(np, np(p - 1))$ provided that n is large and p is close to 0.5.

EXAM-STYLE QUESTIONS 13

Answers page 473

1 The mass of sugar in a 1 kg bag may be assumed to have a normal distribution with mean 1005 g and standard deviation 2 g. Find the probability that:

  a a 1 kg bag will contain less than 1000 g of sugar   **[2 marks]**

  b a 1 kg bag will contain more than 1007 g of sugar   **[2 marks]**

  c a 1 kg bag will contain between 1000 g and 1007 g of sugar.   **[3 marks]**

2 The heights of the UK adult female population are normally distributed with mean 164.5 cm and standard deviation 8.75 cm.

  a Find the probability that a randomly chosen adult female is taller than 160 cm.   **[2 marks]**

Nancy is a student in Year 7. She is at the 45th percentile for her height.

  b Assuming that Nancy remains at the 45th percentile, estimate her height as an adult.   **[2 marks]**

3 The weights of people using a lift are normally distributed with mean 72 kg and standard deviation 10 kg. The lift has a maximum allowed load of 320 kg. If 4 people from this population are in the lift, determine the probability that the maximum load is exceeded.   **[3 marks]**

(PS) 4 A university laboratory is lit by a large number of fluorescent tube light bulbs with lifetimes modelled by a normal distribution with a mean 1000 hours and standard deviation 110 hours. The bulbs remain on continuously.

  a What proportion of bulbs have lifetimes that exceed 900 hours?   **[1 mark]**

  b What proportion of bulbs have lifetimes that exceed 1200 hours?   **[1 mark]**

  c Given that a bulb has already lasted 900 hours, what is the probability it will last another 100 hours?   **[3 marks]**

The university replace the bulbs periodically after a fixed interval.

  d To the nearest day, how long should this interval be if, on average, 1% of the bulbs are to burn out between successive replacement times?   **[3 marks]**

PS 5 A factory produces blades for wind turbines. The lengths of these blades are normally distributed, with 33% of them measuring 212.6 ft or more and 12% of them measuring 211.8 ft or less.

Write down simultaneous equations for the mean and standard deviation of the distribution, and solve to find the values. Hence estimate the proportion of blades that measure 212 ft or more.

The blades are acceptable if they measure between 211.8 ft and 212.8 ft. What percentage of blades are rejected as being outside the acceptable range? **[6 marks]**

PS 6 A machine is used to fill oil into gearboxes with a nominal volume of 3.54 litres. Suppose the machine delivers a quantity of oil which is normally distributed with a mean of 3.58 litres and a standard deviation of 0.13 litres.

a Find the probability that a randomly selected gearbox contains less than the nominal volume. **[2 marks]**

It is required that no more than 4% of gearboxes contain less than the nominal volume.

b Find the lowest value of μ which will comply when $\sigma = 0.13$. **[2 marks]**

c Find the greatest value of σ which will comply when $\mu = 3.58$ litres. **[3 marks]**

7 A preserve manufacturer produces a pack containing 8 assorted pots of different flavours. The actual weight of each pot may be taken to have an independent normal distribution with mean 416 g and standard deviation σ g. Find the value of σ such that 99% of packs weigh more than 400 g. **[3 marks]**

8 The weight, X g, of beans put in a tin by machine K is normally distributed with a mean of 300 g and a standard deviation of 11 g.

A tin is selected at random.

a Find the probability that this tin contains more than 308 g. **[2 marks]**

The weight of beans stated on the tin is w g.

b Find w such that $P(W < w) = 0.02$. **[4 marks]**

The weight, F g, of beans put into a cardboard carton by machine J is trialled, and is normally distributed with a mean μ g and a standard deviation σ g.

c Given that $P(F < 290) = 0.95$ and $P(F < 295) = 0.97$, find the values of μ and σ. **[6 marks]**

9 A restaurant monitors how long diners spend in the restaurant. The manager suggests that the length of a visit can be modelled by a normal distribution with mean 100 minutes. Only 15% of diners stay for more than 115 minutes.

a Find the standard deviation of the normal distribution. **[4 marks]**

b Find the probability that a visit lasts less than 40 minutes. **[3 marks]**

The restaurant increases its opening times so shift workers can also dine out. It introduces a closing time of 11:00 p.m. Alison arrives at the restaurant at 10:00 p.m.

c Explain whether or not this normal distribution is still a suitable model for the length of her visit. **[2 marks]**

10 A computer game takes a mean of 120 hours to complete, with a standard deviation of 13 hours. The completion time may be assumed to be normally distributed.

Find the time, t hours, such that 1 gamer in 10 takes longer than t to complete the game. **[6 marks]**

14 STATISTICAL HYPOTHESIS TESTING

According to a leading car manufacturer's website, the average mileage for the latest model is 49.4 miles per gallon. If you wanted to test the claim that the mean miles per gallon for all the latest models is 49.4, you would perform a hypothesis test at a given significance level.

After randomly sampling 100 cars from the population of the manufactured cars you would first need to calculate the mean mileage of your sample. It would be very unlikely for your sample mean to be equal to the mean of the population. Your sample may well yield a mean with a higher average mileage per gallon, or even a much lower mileage per gallon – either way, it would not equal the population mean exactly. If you were to repeat this with a different sample of 100 cars from the same population, the mean would not equal the mean of the first sample or that of the population.

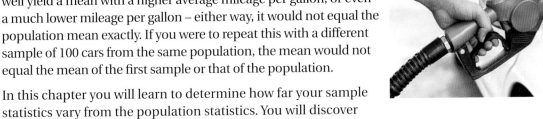

In this chapter you will learn to determine how far your sample statistics vary from the population statistics. You will discover that these are based on the sampling distributions.

LEARNING OBJECTIVES

You will learn how to:

› interpret a given correlation coefficient using a given p-value or critical value
› conduct a hypothesis test with a null hypothesis involving the population correlation coefficient
› conduct a statistical hypothesis test for the mean of a normal distribution
› find the gradient and intercept using non-linear regression.

TOPIC LINKS

You will need the skills and techniques from **Book 1, Chapter 12 Data presentation and interpretation** to draw scatter diagrams and interpret your findings. In **Book 1, Chapter 14 Statistical sampling and hypothesis testing** you learnt how to set up a hypothesis test and use given significance levels. You will also need to use your knowledge of normal distributions from **Chapter 13 Statistical distributions** to carry out a hypothesis test using the normal distribution.

PRIOR KNOWLEDGE

You should already know how to:

› draw and interpret a scatter diagram
› carry out a hypothesis test
› find probabilities from the normal distribution

> find the mean and standard deviation from a normal distribution

You should be able to complete the following questions correctly:

1

Engine size (litres)	1.4	1.6	1.8	2.0	2.5
Miles per gallon	44.6	42.7	40.1	37.2	31.5

    a Draw a scatter diagram for this data.

    b Add a line of best fit.

    c What type of correlation is there?

    d Describe the relationship between engine size and miles per gallon.

14.1 Correlation coefficients

In **Book 1, Chapter 12 Data presentation and interpretation** you saw that when you have two variables in your sample (**bivariate data**), you can use a scatter diagram to look for a connection or **correlation** between the variables. Now you are going to learn how this can be quantified.

The method of quantifying correlation is by means of a **correlation coefficient**. There are a number of different correlation coefficients, but you are going to use Pearson's **product moment correlation coefficient (PMCC)**, notated as *r*. The PMCC gives a numerical measure of the strength of the linear association between two variables.

> **KEY INFORMATION**
>
> A correlation coefficient is a number used to assess the strength of a correlation.

Calculations of correlation coefficients use the differences between data points on a scatter diagram and the mean point $(\bar{x}, \bar{y})$. If you select a plotted point with coordinates (x_i, y_i) and then measure the distances to the two means then

> the horizontal distance will be given by $x_i - \bar{x}$

> the vertical distance will be given by $y_i - \bar{y}$

where $\bar{x}$ represents the mean of the *x* values $\bar{y}$ represents the mean of the *y* values.

If you break a scatter diagram into four **quadrants**, with each quadrant formed around the mean point, $(\bar{x}, \bar{y})$, for the data points in the first quadrant (top right):

> all the *x* values are greater than $\bar{x}$

> all the *y* values are greater than $\bar{y}$.

Consequently, $x_i - \bar{x}$ will always be positive in the first quadrant and $y_i - \bar{y}$ will also be positive in the first quadrant.

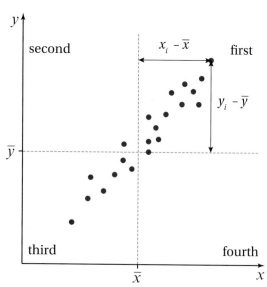

The following table summarises the results for all four quadrants:

Quadrant	$x_i - \bar{x}$	$y_i - \bar{y}$
First	+	+
Second	−	+
Third	−	−
Fourth	+	−

If you now multiply $x_i - \bar{x}$ by $y_i - \bar{y}$, you get the following results:

Quadrant	$(x_i - \bar{x})(y_i - \bar{y})$
First	$+ \times + = +$
Second	$- \times + = -$
Third	$- \times - = +$
Fourth	$+ \times - = -$

Data with a positive correlation has most points in the first and third quadrants, where $(x_i - \bar{x})(y_i - \bar{y})$ is positive.

Note that $(x_i - \bar{x})(y_i - \bar{y})$ is positive if and only if x_i and y_i lie on the same side of their respective means. The correlation coefficient is positive if x_i and y_i both tend to be greater than, or both tend to be less than, their respective means.

Stop and think What would the four quadrants look like if the correlation was negative?

Interpreting the PMCC

The PMCC (r) tells you the strength of a correlation between two variables. It ranges from −1 to 1 and can have any value in this range.

› If $r = -1$ then this is a perfect negative linear correlation. For negative r, the closer the value is to −1, the stronger the correlation; the closer it is to 0, the weaker the correlation.

› If $r = 0$ then there is no linear correlation. However, another type of correlation might exist.

› If $r = 1$ then this is a perfect positive linear correlation. For positive r, the closer the value is to 1, the stronger the correlation; the closer it is to 0, the weaker the correlation.

Using the large data set 14.1

Using the large data set, choose two variables you believe to be related.

a Use this data to draw a scatter diagram and add on to it a line of best fit.

b Interpret your scatter diagram – what does it tell you about the correlation between the variables?

Exercise 14.1A

Answers page 473

1 Two variables have a positive linear correlation.

What is the range of values for the product moment correlation coefficient for such a relationship?

2 Write an accurate mathematical description of the correlation for each of the following values of the product moment correlation coefficient:

   a 0.37   **b** −0.96

   c −1.00   **d** 0.07

3 Which of the following values of the product moment correlation coefficient (PMCC) shows the strongest correlation and which shows the weakest?

   0.70   −0.05

   0.5   −0.85

4 For each of the sets of data below:

   i draw a scatter diagram and comment on whether there appears to be any linear correlation

   ii compare the given product moment correlation coefficient with your comment based on the scatter diagram.

   a The Maths and Further Maths exam scores of 9 students, given as percentages. PMCC = 0.392 485

Maths	56	77	34	67	88	91	56	59	68
Further Maths	46	71	47	61	35	87	63	33	57

   b The results of 7 students in the for the AS-level mock exam. PMCC = 0.519 396

Pure Paper 1 result	96	84	75	93	85	90	72
Applied Paper 2 result	79	93	73	85	82	94	74

   c The annual salaries of workers in a steel fabrication factory and their corresponding average number of hours worked. PMCC = 0.779 31

Salary (£1000s)	6	11	13	22	27	33	48	63
Hours worked per week	18	24	20	30	36	37	42	34

5 Draw a scatter graph and then comment on the type and degree of the correlation.

Candidate	A	B	C	D	E	F	G
Maths Paper 1 result	96	84	75	93	85	90	72
Art Paper 1 result	25	32	50	78	12	45	48

Testing a hypothesis about a PMCC

You have just seen that if the value of the correlation coefficient, r, is close to $+1$ or -1 you can be satisfied that there is a linear correlation, and that if r is close to 0 there is probably little or no correlation.

If r has a value such as 0.5 you need to understand that this value only relates to the members of a sample from the whole population of the bivariate distribution. There will be a level of correlation within the whole population, denoted by ρ.

ρ is the Greek letter 'rho'.

This diagram shows a population with a sample of 10 points highlighted. The correlation, r, is calculated using these 10 points. This sample correlation coefficient can be used to estimate the population correlation coefficient. By using r, you can carry out a **hypothesis test** on the value of ρ, the population correlation coefficient.

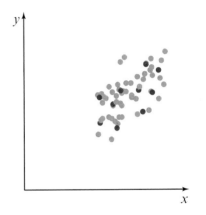

If you assume there is no correlation between the variables within the population, the **null hypothesis** will look like this:

$$H_0: \rho = 0$$

You already know from **Book 1, Chapter 14 Statistical sampling and hypothesis testing** that alternative hypotheses can either be **1-tail** or **2-tail**. This then generates three possible **alternative hypotheses**:

$H_1: \rho > 0$, where the two variables in the population have a positive correlation.

$H_1: \rho < 0$, where the two variables in the population have a negative correlation.

$H_1: \rho \neq 0$, where the two variables in the population have a correlation which may be either positive or negative.

KEY INFORMATION

r measures the correlation within the sample and ρ measures the correlation within the entire population.

By using a table of critical values, you can see whether the value of r is significant or not. For your hypothesis test you will need to know the size of your sample, the significance level and what your alternative hypothesis is. You will also need to remember that if your alternative hypothesis is just 'there is a correlation', that this is a 2-tail test. In this case you will need to remember to *halve* the significance as you will be sharing it between the upper and lower tails. The critical region is the area of the sampling distribution that will lead to the rejection of the null hypothesis.

Critical values

The value of r is the sample correlation coefficient; you want to know how this fits with the population. If you assume there is no correlation between the variables you get the null hypothesis, $H_0: \rho = 0$. Using a **significance level** and your sample size, you can see from the critical value whether the value acquired is significant or not and therefore whether to accept or reject H_0.

You need to know how to interpret the critical values that you will be given to work with.

Example 1

Find the critical regions for the following alternative hypotheses:

a $H_1: \rho > 0$ There is a positive correlation between the variables (1-tail test).

Sample size, $n = 7$

Significance level 1%

b $H_1: \rho < 0$ There is a negative correlation between the variables (1-tail test).

Sample size, $n = 10$

Significance level 5%

c $H_1: \rho \neq 0$ There is correlation between the variables (2-tail test).

Sample size, $n = 14$

Significance level 10%

Solution

a Using the table, identify $n = 7$ and significance level 1%:

Product Moment Correlation Coefficient					Sample size
0.10	0.05	Level 0.025	0.01	0.005	
0.8000	0.9000	0.9500	0.9800	0.9900	4
0.6870	0.8054	0.8783	0.9343	0.9587	5
0.6084	0.7293	0.8114	0.8822	0.9172	6
0.5509	0.6694	0.7545	0.8329	0.8745	7
0.5067	0.6215	0.7067	0.7887	0.8343	8
0.4716	0.5822	0.6664	0.7498	0.7977	9
0.4428	0.5494	0.6319	0.7155	0.7646	10

Remember that you are looking for a positive correlation.

So the critical region for a sample size of 7 at a 1% significance level is 0.8329.

b Using the table, identify $n = 10$ and significance level 5%:

Product Moment Correlation Coefficient					Sample size
0.10	0.05	Level 0.025	0.01	0.005	
0.8000	0.9000	0.9500	0.9800	0.9900	4
0.6870	0.8054	0.8783	0.9343	0.9587	5
0.6084	0.7293	0.8114	0.8822	0.9172	6
0.5509	0.6694	0.7545	0.8329	0.8745	7
0.5067	0.6215	0.7067	0.7887	0.8343	8
0.4716	0.5822	0.6664	0.7498	0.7977	9
0.4428	0.5494	0.6319	0.7155	0.7646	10
0.4187	0.5214	0.6021	0.6851	0.7348	11

Remember that you are looking for a negative correlation.

So the critical region for a sample size 10 and significance level 5% is –0.5494.

c Using the table, identify $n = 14$ and significance level 10%. Remember this is a 2-tail test so the 10% is split between both tails, i.e. 5% at each end:

Product Moment Correlation Coefficient					Sample size
0.10	0.05	Level 0.025	0.01	0.005	
0.8000	0.9000	0.9500	0.9800	0.9900	4
0.6870	0.8054	0.8783	0.9343	0.9587	5
0.6084	0.7293	0.8114	0.8822	0.9172	6
0.5509	0.6694	0.7545	0.8329	0.8745	7
0.5067	0.6215	0.7067	0.7887	0.8343	8
0.4716	0.5822	0.6664	0.7498	0.7977	9
0.4428	0.5494	0.6319	0.7155	0.7646	10
0.4187	0.5214	0.6021	0.6851	0.7348	11
0.3981	0.4973	0.5760	0.6581	0.7079	12
0.3802	0.4762	0.5529	0.6339	0.6835	13
0.3646	0.4575	0.5324	0.6120	0.6614	14
0.3507	0.4409	0.5140	0.5923	0.6411	15

Remember that you are looking for a correlation.

So the critical region for a sample sized 14 and significance level 10% shared between both tails (giving 5%) is $x < -0.4575$ and $x > 0.4575$.

As you saw in **Book 1, Chapter 14 Statistical sampling and hypothesis testing**, to carry out a hypothesis test first you must write down the null and alternative hypotheses. Include the definition of ρ and write down the significance level.

Look up the critical value from tables, stating the value of n, the significance level and whether the test is 1-tail or 2-tail, and then compare the value of r with the critical value. If r is less than the critical value, then accept H_0. If it is greater than the critical value, then reject H_0.

When you compare the value of r with the critical value, make sure to include this in your written answer. Also, write your conclusion in words in terms of the original problem.

Example 2

A gymnastics coach believes that people can balance better on one foot rather than the other. The table shows the results of an experiment that is conducted with a random sample of the gym members to see if there is a connection between how long a client can balance on their right foot, then on their left foot, whilst blindfolded.

Gym member	A	B	C	D	E	F	G	H
Right foot (seconds)	55	47	31	67	78	32	46	17
Left foot (seconds)	42	43	54	78	59	67	22	46

The PMCC is calculated and is found to be $r = 0.252\,46$. Conduct a hypothesis test at the 5% level to see if the coach's claim is true.

Solution

$H_0: \rho = 0$ There is no connection between the times balancing on the right and the left foot.

$H_1: \rho > 0$ There is a positive correlation between the times balancing on the right and the left foot.

$n = 8$, 1-tail test and significance level 5%

> Look up the critical value from tables, stating the value of n, the significance level and whether the test is 1-tail or 2-tail.

Product Moment Correlation Coefficient					Sample size
0.10	0.05	Level 0.025	0.01	0.005	
0.8000	0.9000	0.9500	0.9800	0.9900	4
0.6870	0.8054	0.8783	0.9343	0.9587	5
0.6084	0.7293	0.8114	0.8822	0.9172	6
0.5509	0.6694	0.7545	0.8329	0.8745	7
0.5067	0.6215	0.7067	0.7887	0.8343	8
0.4716	0.5822	0.6664	0.7498	0.7977	9

The critical value is 0.6215.

Since $0.252\,46 < 0.6215$, which is the critical value at the 5% significance level, the null hypothesis is accepted.

Accept H_0: there is not sufficient evidence to suggest that there is positive correlation between time spent balancing on the right foot compared with the left foot.

Example 3

A researcher wishes to find out if there is any connection between the length of time teenagers spend gaming and their reaction times. She collects data on 100 fourteen-year-olds and calculates a PMCC value of $r = 0.2311$.

Carry out a hypothesis test at the 5% level to determine whether there is any connection between time spent gaming and reaction times.

Solution

$H_0: \rho = 0$ There is no correlation between time spent gaming and reaction times.

$H_1: \rho \neq 0$ There is a correlation between time spent gaming and reaction times.

Significance level 5%

The critical value for a 2-tail test at the 5% significance level for $n = 100$ is 0.1966. Since $0.2311 > 0.1966$, reject H_0. There is evidence to suggest that there is a correlation between time spent gaming and reaction times.

Using the large data set 14.2	A statistician believes there is a relationship between two of the variables in the large data set.

a Select two variables from your large data set. Determine whether there is a connection between these variables.

b Would you expect this connection to be the same in different time periods or different locations? State your reasons carefully.

Exercise 14.1B

Answers page 474

 1 A teacher wants to see if there is a connection between students' marks in a chemistry practical and their marks in the written papers for a chemistry examination. The results from a random selection of 20 students are as follows:

Practical mark (max 20)	17	14	13	19	12	16	17	16	13	6	4	20	13	16	10	12	18	19	6	11
Written paper mark (max 100)	78	62	58	63	73	41	75	57	72	29	25	77	53	34	70	68	91	94	24	51

a State suitable null and alternative hypotheses.

b The product moment correlation coefficient is calculated to be $r = 0.695$. Carry out the hypothesis test at the 5% significance level.

c Would your conclusion be the same at the 10% significance level? Explain your answer.

2 A teacher notices that the taller a Year 13 student is, the heavier they are. She picks 15 Year 13 students at random and measures them. The results are as follows:

Height (cm)	151	142	178	198	174	168	147	196	147	198	174	166	145	185	167
Weight (kg)	54	49	80	88	67	51	53	104	56	79	70	68	41	82	71

a State the null and alternative hypotheses.

b The correlation between the variables is $r = 0.896\,209$. Carry out the hypothesis test at the 10% level. What conclusion do you reach?

3 A newspaper article suggests that social media is having a negative impact upon students' exam results. In order to investigate the claim, a random sample of ten Year 13 students is taken. They are asked their scores in their last maths test and how long they spend using social media in a normal week. Here are the results:

Score in test (max 100)	50	65	78	99	11	16	48	36	74	88
Hours on social media	4	2.5	2	1.5	2	5.5	3.5	4	2.5	2

a Represent this information on a scatter diagram.

b State appropriate null and alternative hypotheses for the test. Justify the alternative hypothesis you have given.

c The correlation coefficient is calculated to be $r = -0.64721$. Carry out the hypothesis test at the 2.5% level of significance. State clearly the conclusion reached.

d Is there an outlier? By omitting a possible outlier, how does this affect your results?

14.2 Hypothesis tests for the mean of a normal distribution

You saw in **Chapter 13 Statistical distributions** that the normal distribution can be used to approximate the binomial distribution in certain cases.

If $X \sim B(n, p)$ and n is large and p is close to 0.5 then X can be approximated by $Y \sim N(np, np(p - 1))$.

Remember that, as the binomial distribution is discrete but the normal distribution is continuous, you need to apply a continuity correction.

Binomial distribution	Normal distribution
$P(X = b)$	$P(b - 0.5 < Y < b + 0.5)$
$P(X \leqslant b)$	$P(Y < b + 0.5)$
$P(X < b)$	$P(Y < b - 0.5)$
$P(X \geqslant b)$	$P(Y > b - 0.5)$
$P(X > b)$	$P(Y > b + 0.5)$

When you are doing a hypothesis test it is assumed that the population has a normal distribution with mean μ and standard deviation σ. This is written as $N(\mu, \sigma^2)$.

If a sample is taken from a normally distributed population, the population mean would be μ and its standard deviation $\frac{\sigma}{\sqrt{n}}$.

This is called the sampling distribution of the means and is written $N\left(\mu, \frac{\sigma^2}{n}\right)$.

> **KEY INFORMATION**
>
> A population with a normal distribution is denoted $N(\mu, \sigma^2)$.
>
> The sampling distribution of the means is denoted $N\left(\mu, \frac{\sigma^2}{n}\right)$.

For instance, if you spin a spinner labelled 1 to 4, with each section being equal in size so it is fair, then there are four possible samples you could obtain: 1, 2, 3 or 4, from a population of 4.

Each of these samples is equally likely to occur, as it is a fair spinner. In each case, the sample mean is just the value on the section.

> Remember that the standard normal distribution is denoted $N(0, 1)$, where the mean is 0, the standard deviation is 1 and the variance is 1.

So the probability distribution of the sample means for a sample of size 1 is:

$\bar{x}$	1	2	3	4
$P(\bar{X} = \bar{x})$	0.25	0.25	0.25	0.25

Note the uniform distribution of the means.

Using your calculator, you can find the mean and the variance of the population:

Mean $= 2.5$

Variance $= \dfrac{5}{4}$

If you spin two spinners of the same kind and find the mean of the numbers on both spinners, then there are 16 possible samples you could obtain. Some of these repeat – for example, a 3 and a 4 is considered the same as a 4 and a 3.

4	2.5	3	3.5	4
3	2	2.5	3	3.5
2	1.5	2	2.5	3
1	1	1.5	2	2.5
+	1	2	3	4

So the probability distribution of the sample means for a sample of size 2 is:

$\bar{Y}$	1	1.5	2	2.5	3	3.5	4
$P(\bar{Y} = \bar{y})$	$\dfrac{1}{16}$	$\dfrac{2}{16}$	$\dfrac{3}{16}$	$\dfrac{4}{16}$	$\dfrac{3}{16}$	$\dfrac{2}{16}$	$\dfrac{1}{16}$

Using your calculator you can find the mean and the variance of the population:

Mean $= 2.5$

Variance $= \dfrac{5}{16}$

Note the distribution of the means is now starting to become symmetrical around the population mean.

If you spin three spinners of the same kind, then there are 64 possible samples you could obtain. If you make a complete list of all the possible samples and calculate the sample mean for each, you can find the probability distribution of the sample means in the same way as for samples of size 2.

The samples of size 2 and 3 have a peak in the centre corresponding to the mean value of 2.5. The larger the sample size, the more the distribution starts to look like a bell-shaped curve because the larger the sample size becomes, the smaller the standard deviation.

As you have just seen, all three probability distributions have mean 2.5, which is the same as the population mean, μ. This gives rise to the fact that in a given population with mean μ and standard deviation σ, the sampling distribution of the mean has a mean of μ and a standard deviation of $\dfrac{\sigma^2}{n}$.

Example 4

The speed limit outside Paul's school is 30 mph. Paul thinks the speeds of cars outside his school are increasing. He records the speeds of 10 cars at various times throughout the day. The results in mph are:

32, 36, 24, 41, 36, 33, 44, 45, 29, 32

Research from the highways team shows that the speeds on this road are normally distributed with a standard deviation of 3.5 mph.

Use this data set to test, at the 1% significance level, Paul's belief that speeds are increasing.

Solution

Use the normal distribution of the sample means to decide which values of the test statistic are sufficiently extreme as to suggest that the alternative hypothesis, not the null, is true.

Null hypothesis, H_0: $p = 30$ The speed on the road is normal.

Alternative hypothesis, H_1: $p > 30$ The speed has increased.

1-tail test at the 1% significance level.

The distribution of sample means, $\overline{X}$, is $N(\mu, \dfrac{\sigma^2}{n})$.

According to the null hypothesis, $\mu = 30$, and it is known that $\sigma = 3.5$ and $n = 10$.

So this distribution is $N(30, \dfrac{3.5^2}{10})$.

The sample mean is 35.2 mph.

The probability of the mean, $\overline{X}$, of a randomly chosen sample being greater than the sample mean is given by:

$$P(\overline{X} \geqslant 35.2) = 1 - \Phi\left(\frac{35.2 - 30}{\left(\frac{3.5}{\sqrt{10}}\right)}\right)$$

$$= 1 - \Phi(4.6982)$$

$$= 1 - 0.999\,998\,69$$

$$= 0.000\,001\,31$$

Since $0.000\,001\,31 < 1\%$, the required significance level, the null hypothesis is rejected in favour of the alternative hypothesis.

TECHNOLOGY

You will need to have your calculator in Normal CD mode as you are finding the probability of everything that is at least 35.2 mph.

Known and estimated standard deviation

In the previous example you were told the value of the standard deviation of the population, σ. It is only possible to use this method of hypothesis testing when this information is known.

However, if the sample size is large enough, the sample standard deviation may be used as an estimate of the population standard deviation.

Example 5

GCSE mathematics examinations have had a mean score of 65 for several years. A politician puts forward a theory that the examinations are not rigorous enough and have now become too easy to pass and should therefore change. He believes there has recently been an increase in the mean score.

A random sample of 250 students is taken from the population of students who sat the same GCSE mathematics examination.

Statistics from the raw scores, x, that are collected are as follows:

$$n = 250, \Sigma x = 17\,425, \Sigma x^2 = 1\,262\,934$$

Carry out a suitable hypothesis test on the politician's claim at the 1% significance level.

State your assumptions carefully.

Solution

H_0: The population mean is unchanged, i.e. $\mu = 65$.

H_1: The population mean has increased, i.e. $\mu > 65$.

Calculate the mean and the standard deviation:

$$\bar{x} = \frac{\Sigma x}{n}$$
$$= \frac{17\,425}{250} = 69.7$$

$$s = \sqrt{\frac{\Sigma x^2 - n\bar{x}^2}{n}}$$

$$= \sqrt{\frac{\Sigma 1\,262\,934 - 250 \times 69.7^2}{250}} = 13.9$$

Now standardise the z-value corresponding to $\bar{x} = 69.7$ calculated using $\mu = 65$ and by approximating σ by $s = 13.9$.

$$z = \frac{\bar{x} - \mu}{\frac{\sigma}{\sqrt{n}}}$$

$$= \frac{69.7 - 65}{\frac{13.9}{\sqrt{250}}} = 5.3$$

The critical region value is $z = 2.236$ when the significance level is 1%.

You now need to make a comparison of your score against the critical value.

TECHNOLOGY

On your calculator, select inverse normal and set area to 99%, mean to 0 and standard deviation to 1, to find the z-score.

5.3 > 2.236, therefore the null hypothesis is rejected. The evidence suggests that the politician's claim is true – the examinations are not rigorous enough.

You have made the assumption that the test scores are normally distributed.

Exercise 14.2A

Answers page 475

1   A machine is designed to make pins with a mean mass of 2 g and a standard deviation of 0.05 g. Assume the distribution of the masses of the pins is a normal distribution.

   a   A pin chosen at random has a mass greater than 2.01 g. Find the probability of this outcome.

   During an experiment, a student weighs a random sample of 30 pins and finds their total mass to be 59.4 g.

   b   Conduct a hypothesis test at the 5% significance level of whether this provides evidence of a decrease in the mean mass of the pins.

2   A certain breed of guinea pig has a mass which is normally distributed with a mean of 750 g and a standard deviation of 11.2 g. A breeder is not convinced her guinea pigs are of this breed. In order to test this, she collects a sample of eight guinea pigs and weighs them, with the following results:

       752, 738, 756, 771, 729, 756, 739, 749

   a   Write down the breeder's null and alternative hypotheses.

   b   State whether a 1-tail or 2-tail test is needed, and why.

   c   Carry out the test at the 10% significance level and write down your conclusion.

3   A machine at a local DIY store puts paint into cans. The store owner electronically controls the mean volume of the cans. The standard deviation is fixed at 0.11 litres. The mean volume is suspected to be lower than it should be. If a random sample of 42 cans has a total volume of 323.82 litres, is there evidence to support the suspicion if the mean volume is set to 8 litres?

4   The time taken for a bus to go from Newark to Lincoln is normally distributed with a mean time of 28 minutes at rush hour. A new road network has been created which it is hoped will speed up the journey during rush hour traffic.

   A group of students monitor the traffic flow and claim that the journey is now taking longer than 28 minutes. The students take 250 observations at rush hour and find the mean is 30.3 minutes, with a sample standard deviation of 7 minutes.

   Carry out an investigation on the students' claim, stating the null and alternative hypotheses, and carry out the test at the 5% level of significance, stating your conclusion carefully.

5   A local theatre group needs to buy some new stage lighting. The supplier of Brite light bulbs claims that the mean life of their light bulbs is 288 hours. The theatre group buys 150 of their bulbs and find the mean to be only 254.5 hours, with a sample standard deviation of 11 hours. Is there evidence at a 5% level that the mean is lower than 288 hours?

(PS) **6** In order for a local shop to sell samphire to a quality restaurant, it must be of a certain length. The requirements are that the length should be normally distributed with mean 16 cm and standard deviation 0.8 cm. The shop manager finds a new area to collect from and wants to know if the samphire would be suitable to sell to the restaurant. She collects and measures 80 samples and finds that the total of their lengths is 1256 cm.

Carry out a test at the 5% level. What should the shop manager conclude?

7 A student measures the mass of individual jelly sweets. She finds that they are normally distributed with mean 11.4 g and standard deviation 0.65 g. The jelly sweets are in packets of 18.

a What is the probability that the mass of a sweet lies between 10 g and 12 g?

b Calculate the probability that the total contents of a packet have a mass between 202 g and 208 g.

(PS) **8** The masses of adult male silverback gorillas in the wild are known to be normally distributed with mean 230 kg and standard deviation 21.3 kg. A sample of 14 males are weighed and the mean mass is found to be 212 kg. Assuming that the standard deviation is unchanged, test, at the 5% significance level, whether the mean mass of adult male silverback gorillas has decreased.

14.3 Non-linear regression

So far all of the bivariate data you have dealt with has fitted nicely into a linear model. Not all data will display a linear relationship, but that is not to say they do not have a relationship. A way to look at relationships between two variables which are not linear is to model them using logarithms.

Exponential growth is where the amount being added is in proportion to the amount already present – the bigger the amount, the bigger the growth. For example, if something is growing exponentially, the relationship between x and y on a graph is a curve. To find the relationship between x and y, you need to use non-linear regression.

Modelling curves of the form $y = ax^n$

Examples of situations involving exponential growth include the number of micro-organisms in a colony at a particular time, a virus when no immunisation is available or positive feedback in electrical amplification. When carrying out experiments, you will often plot a graph of one variable against another. With exponential growth, the graph will not be a straight line so the relationship will be difficult to see.

In order to overcome this, you can use logarithms to change the relationship into one that is linear. If the relationship is of the form $y = ax^n$, plotting $\log y$ against $\log x$ will give you a straight line graph.

Using knowledge of $y = mx + c$, n represents the gradient of the graph. To find a, you take the intercept of the graph and then find its inverse logarithm.

$$y = ax^n$$
$$\Rightarrow \quad \log y = \log ax^n$$
$$= \log a + \log x^n$$
$$= \log a + n \log x$$
$$= n \log x + \log a$$

This is now of the form $y = mx + c$, so plotting $\log y$ against $\log x$ gives a straight line with gradient n and intercept $\log a$.

See **Book 1, Chapter 7 Exponentials and logarithms** for more about the laws of logarithms.

Example 6

The relationship between two variables in an experiment, x and y, is believed to be of the form $y = ax^n$, where a and n are constants. The values of x and y are recorded as follows:

X	2	4	6	8	10	12	14	16
Y	0	1.5	2.6	3.8	5.1	6.5	8.4	11.7

Verify that the model $y = ax^n$ is appropriate, and find the approximate values of the constants a and n.

Solution

$$y = ax^n$$

Take logarithms.

$$\log y = \log ax^n$$
$$= \log a + \log x^n$$
$$= \log a + n \log x$$

This is the equation of a straight line graph with gradient n and intercept $\log a$.

Tabulate the values of $\log y$ and $\log x$, and plot them as a graph.

x	log x	y	log y
2	0.301	0	0
4	0.602	1.5	0.176091259
6	0.78	2.6	0.414973348
8	0.9	3.8	0.579783597
10	1	5.1	0.707570176
12	1.08	6.5	0.812913357
14	1.15	8.4	0.924279286
16	1.2	11.7	1.068185862

As you can see, plotting $\log x$ against $\log y$ creates an approximate straight line graph, so the relationship $y = ax^n$ is an appropriate model.

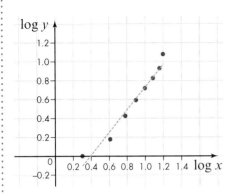

$$\text{Gradient} = n = \frac{0.5}{0.4} = 1.25$$

$$\text{Intercept} = \log a = -0.46 \quad \Rightarrow \quad a = 10^{-0.46} \approx 0.35$$

The relationship is approximately $y = 0.35x^{1.25}$.

If the relationship is exponential, of the form $y = kb^x$, plotting $\log y$ against x gives a straight line graph.

$$y = kb^x$$

$$\Rightarrow \quad \log y = \log kb^x$$

$$= \log k + \log b^x$$

$$= \log k + x \log b$$

$$= (\log b)x + \log k$$

Example 7

The relationship between two variables in an experiment, c and d, is believed to be of the form $d = ab^c$, where a and b are constants. The following values of c and d are recorded:

c	1.5	2.0	2.5	3.0	3.5	4.0
d	12	19	30	46	74	116

Verify that the model $d = ab^c$ is appropriate, and estimate the values of a and b.

Solution

$$d = ab^c$$

Take logarithms.

$$\log d = \log ab^c$$

$$= \log a + \log b^c$$

$$= \log a + c \log b$$

This is the equation of a straight line with gradient $\log b$ and intercept $\log a$. Tabulate the values of $\log d$, and plot them against c.

c	1.5	2.0	2.5	3.0	3.5	4.0
d	12	19	30	46	74	116
$\log d$	1.08	1.28	1.48	1.66	1.87	2.06

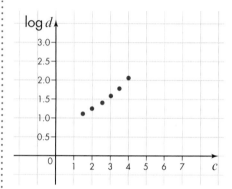

The graph is an approximate straight line, so the relationship $d = ab^c$ is an appropriate model.

Gradient $= \log b = \dfrac{1.08}{2.7} \quad \Rightarrow \quad b = 10^{0.4} \approx 2.512$

Intercept $= \log a = 0.5 \quad \Rightarrow \quad a = 10^{0.5} \approx 3.16$

The relationship is approximately $d = 3.16(2.512)^c$.

Exercise 14.3A

Answers page 475

1 Two variables, g and h, are related by the formula $g = ah^c$, where a and c are constants.

  a Show that this relationship can be written as $\log g = \log a + c \log h$.

  b Explain why the model can be tested by plotting $\log g$ against $\log h$.

  Values of g and h are recorded in an experiment:

h	6	11	16	21	23	34
g	11	12	15	17	19	23

  c Plot the graph of $\log h$ against $\log g$ and use your graph to estimate the values of a and c.

2 Two variables, s and k, are related by the formula $k = mn^s$, where m and n are constants.

a Show that this relationship can be written as $\log k = \log m + s \log n$.

b Explain why the model can be tested by plotting $\log k$ against s.

In an experiment, the following values of s and k are obtained:

s	0.25	0.5	0.75	1.0	1.25	1.5	1.75
k	4.3	4.1	3.9	3.5	3.1	2.9	2.4

c Plot the graph of $\log k$ against s and use your graph to estimate the values of m and n.

3 The relationship between two variables, x and y, is believed to be of the form $y = kx^n$, where k and n are constants.

In an experiment, the following values of x and y are recorded:

x	1	2	3	4	5	6	7
y	1	7	17	29	45	65	91

a Plot the graph of $\log y$ against $\log x$ and explain why this tells you that the model $y = kx^n$ is appropriate.

b Use your graph to estimate the values of k and n.

c Estimate the value of y when $x = 10$.

(PS) 4 A sofa manufacturer has invested in new springs which have been super heat treated. The manager investigates the breaking strength of the metal after different durations of treatment and the following data is obtained (in suitable units):

Treatment, x	0.113	0.151	0.215	0.255	0.326
Breaking strain, y	0.079	0.156	0.531	0.973	1.879

a Sketch a graph of y against x, and suggest a form for a relationship connecting them.

b Write your rule in terms of logarithms, and hence plot a suitable graph and use it to find an approximate form of the relationship.

SUMMARY OF KEY POINTS

❭ The product moment correlation coefficient (PMCC), r, tells you the strength of a correlation. It ranges from -1 to 1 and can have any value in this range.

❭ r measures the correlation within the sample and ρ measures the correlation within the entire population.

❭ To set up a hypothesis test to compare the sample to the population:

  ❭ H_0: $\rho = 0$ There is no correlation between the two variables.

❭ The alternative hypothesis takes one of the following three forms:

  ❭ H_1: $\rho \neq 0$ There is a correlation between the variables (2-tail test) which may be either positive or negative.

  ❭ H_1: $\rho > 0$ There is a positive correlation between the variables (1-tail test).

  ❭ H_1: $\rho < 0$ There is a negative correlation between the variables (1-tail test).

  ❭ If $X \sim \mathrm{B}(n, p)$ and n is large and p is not too close to either 0 or 1 then X can be approximated by $Y \sim \mathrm{N}(np, npq)$, where $q = 1 - p$.

❭ The distribution of the means of a sample from a normally distributed population would give a mean of μ and standard deviation $\dfrac{\sigma}{\sqrt{n}}$ and is denoted by $\mathrm{N}\left(\mu, \dfrac{\sigma^2}{n}\right)$.

❭ If a relationship is of the form $y = ax^n$, then plotting $\log y$ against $\log x$ gives a straight line graph.

  ❭ $y = ax^n \;\Rightarrow\; \log y = n \log x + \log a$

❭ If a relationship is of the exponential form $y = kb^x$, then plotting $\log y$ against x gives a straight line graph.

  ❭ $y = kb^x \;\Rightarrow\; \log y = (\log b)x + \log k$

EXAM-STYLE QUESTIONS 14

Answers page 476

 1 A travel company tracks the number of UK residents flying to Europe and the value of money in pounds that they exchange to euros. The table shows data on the number of travellers, t, and the amount of money they exchange, e, for eight consecutive months.

Number of travellers, t (thousands)	2124	2234	3412	3654	2986	2765	2341	2094
Amount of money exchanged, e (£ millions)	1214	1352	1545	1601	1439	1504	1328	1162

  a Draw a scatter diagram and describe the correlation and what this means in terms of the variables.

  [4 marks]

  b The correlation coefficient is calculated and is $r = 0.924975$. Explain if this agrees with your answer to **part a**.

  [4 marks]

2 Below are the blood pressures, P (in mm Hg), and the ages, q (in years), of eight hospital patients.

q	56	53	68	42	83	74	59	68
P	182	107	132	77	96	109	87	105

   a The correlation coefficient is calculated at $r = 0.036$. What relationship does this suggest the data shows?   **[2 marks]**

   b Draw a scatter diagram and interpret your findings.   **[2 marks]**

3 An experiment is carried out in which the relationship between two variables, J and h, is believed to be of the form $J = ab^h$, where a and b are constants.

A student plots a graph and finds that this gives a straight line with gradient m and intercept c.

   a What graph did the student plot?   **[1 mark]**

   b Describe how to find the value of a.   **[2 marks]**

   c Describe how to find the value of b.   **[2 marks]**

4 An engineer wants to see if there is a connection between the speed of a car and its fuel consumption. The results from a random selection of 10 speeds of the same car are as follows.

Speed (mph)	75	30	52	98	26	88	106	49	65	76
Fuel consumption (mpg)	36	57	64	25	48	35	22	49	44	39

   a State suitable null and alternative hypotheses.   **[2 marks]**

The product moment correlation coefficient was calculated to be −0.859.

   b Carry out the hypothesis test at the 5% significance level.   **[4 marks]**

   c Would your conclusion be the same at the 10% significance level? Explain your answer.   **[2 marks]**

5 A teacher wished to see if there is a positive relationship between a teenager's KS2 maths score and their score in their GCSE maths exam. A sample of 15 students from a school is taken and their scores are shown in the table below.

KS2 score (max 120)	110	101	108	95	99	115	118	100	101	109	114	92	111	107	94
GCSE score (max 240)	204	173	193	128	183	211	219	106	124	182	204	54	201	151	91

   a Represent this information on a scatter diagram.   **[2 marks]**

   b State appropriate null and alternative hypotheses for the test. Justify the alternative hypothesis you have given.   **[2 marks]**

The product moment correlation coefficient is calculated to be $r = 0.878$.

   c Carry out the hypothesis test at the 5% level of significance. State clearly the conclusion reached.   **[4 marks]**

6 A new coffee machine makes coffee at 82°C. The data shows the temperature of a cup of coffee as it cools.

Time, x (minutes)	0	5	11	15	18	25	30	34	38	45	50
Temperature, y (°C)	81.9	75.9	65.1	60.9	57.0	50.8	46.8	45.1	42.8	39.0	38.1

a Plot the graph of log y against log x and explain why this tells you that the model $y = kx^n$ is appropriate. **[3 marks]**

b Use your graph to estimate the values of k and n. **[3 marks]**

c How long does it take the coffee to cool to a temperature of 35°C? **[3 marks]**

(PS) **7** The results of an examination, in which there were 3519 candidates, are modelled by a normal distribution with mean 62 and standard deviation 15.

a If the pass mark is 50, what is the approximate number of candidates (to the nearest whole number) that you would expect to pass? Use a continuity correction. **[2 marks]**

b What is the approximate number of candidates that you would expect to gain marks between 57 and 67 inclusive? **[4 marks]**

c After the introduction of a new computer package the results for 100 students had a new mean of 66. Is there evidence to suggest the computer package has improved the results at a 5% level of significance?

8 The product moment correlation coefficient for 10 pairs of bivariate data was calculated to be 0.7192 in a 2-tail test. Which of the following statements is true?

A The null hypothesis is rejected at the 10% level but accepted at the 5% level. **[1 mark]**

B The null hypothesis is rejected at the 5% level but accepted at the 2% level. **[1 mark]**

C The null hypothesis is rejected at the 2% level but accepted at the 1% level. **[1 mark]**

D The null hypothesis is rejected at the 1% level. **[1 mark]**

(PS) **9** There is a theory that students who are good at mathematics are also good at physics. In order to investigate this theory a random sample of 10 students was chosen. Their marks in the two subjects are recorded below:

Student	A	B	C	D	E	F	G	H	I	J
Mathematics (x)	47	15	59	95	47	48	49	71	80	23
Physics (y)	46	32	50	89	45	55	39	76	81	35

a What is the null hypothesis? **[1 mark]**

b What is the alternative hypothesis? **[1 mark]**

The product moment correlation coefficient is calculated and is found to be $r = 0.936695$.

c What is the conclusion if the 5% significance level is used? **[4 marks]**

10 The time taken for a cyclist to go from Woodall Spa to Boston is normally distributed with a mean time of 68 minutes. A new cycle path is introduced, which it is hoped will speed up the journey. It is claimed that the journey is more than 7 minutes shorter than before.

From 100 observations taken from cyclists using the new path, the mean was found to be 65.3 minutes, with a sample standard deviation of 3 minutes. Investigate the cyclists' claim that the journey is faster than advertised, state suitable null and alternative hypotheses for the test, and carry out the test at the 5% level of significance, stating your conclusion carefully. **[10 marks]**

11 The masses of adult male cats are known to be normally distributed with mean of 3.4 kg and a standard deviation of 0.8 kg. A sample of size 75 is taken and the mean is found to be 3.55 kg. Assuming that the standard deviation is unchanged, test, at the 1% significance level, whether the mean mass of adult male cats has increased. **[10 marks]**

15 KINEMATICS

The film *Bend it like Beckham* was named after the way in which David Beckham could spin a football so that it found its way into the back of the net. Many sports include the possibility of a ball being launched into the air with the intention of it returning to the ground at a certain place or distance. Scoring a six in cricket, for example, requires the batsman to strike the ball in such a way that it passes the boundary before its first bounce. Netball and basketball require the ability to launch the ball so that after it reaches its maximum height it passes through the net as it is travelling downwards. In golf, the object is to strike the ball so that it gets closer to the hole, without landing in a bunker or a lake.

In none of these examples is the ball travelling vertically upwards with no horizontal motion. They are all examples of projectiles, which are modelled as particles travelling with both a horizontal and a vertical component to their motion.

Projectiles are not limited to sport. A pirate using a cannon to fire a cannonball needs to aim it so that it hits the other ship. Too far or too near and a cannonball is wasted. If it were fired vertically upward, the cannonball would come back to the pirate's vessel. In order to launch a cannonball successfully and with the minimum of wastage, the pirate needs to control the angle and velocity of launch.

LEARNING OBJECTIVES

You will learn how to:

› model motion using vectors in two dimensions

› apply the equations of constant acceleration to problems involving vectors

› apply the equations of constant acceleration to problems in which the displacement, velocity and acceleration take place in two dimensions and are functions of time

› apply the equations of constant acceleration to problems involving projectile motion.

TOPIC LINKS

This chapter is the first of three chapters on Mechanics and follows directly on from **Book 1, Chapter 15 Kinematics** – but whereas the motion in **Book 1** all took place in one dimension (either horizontal or vertical), the motion in this chapter takes place in two dimensions.

PRIOR KNOWLEDGE

You should already know how to:

› apply the equations of constant acceleration (SUVAT) from kinematics to horizontal and vertical motion

› convert between $m\,s^{-1}$ and $km\,h^{-1}$

> apply the formula $F = ma$ relating force, mass and acceleration

> use **i j** and column notation to represent vectors

> use Pythagoras' theorem and trigonometry to find the magnitude and direction of a vector

> differentiate and integrate polynomials, using techniques such as the product rule and integration by parts.

You should be able to complete the following questions correctly:

1 A particle travels 32 m in 4 s with a constant acceleration of $1.2\,\text{m}\,\text{s}^{-2}$.

   a Find the initial velocity of the particle.

   b Find the final velocity of the particle.

2 Convert $25\,\text{m}\,\text{s}^{-1}$ into $\text{km}\,\text{h}^{-1}$.

3 **a** Find the single force which is being applied to a particle of mass 2.5 kg that is accelerating at $\begin{bmatrix} 24 \\ -10 \end{bmatrix}\text{m}\,\text{s}^{-2}$.

   b Find the magnitude and direction of the force.

4 **a** Find $\dfrac{dv}{dt}$ given that $v = 5t^3 + t\sqrt{t}$.

   b Find $\int (3t - 2)^4\, dt$.

15.1 Equations of constant acceleration

In **Book 1, Chapter 15 Kinematics** you applied the equations of constant acceleration (SUVAT) to problems involving horizontal or vertical motion. The equations relate the vector quantities of **displacement** (s), initial **velocity** (u), final velocity (v) and **acceleration** (a), and the scalar quantity time (t).

The equations are:

$$v = u + at$$

$$v^2 = u^2 + 2as$$

$$s = \left(\frac{u + v}{2}\right)t$$

$$s = ut + \frac{1}{2}at^2$$

$$s = vt - \frac{1}{2}at^2$$

You were also shown how differentiation and integration could be used to relate displacement, velocity and acceleration when they were given as functions of time (t):

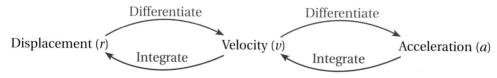

Displacement (r)   Differentiate → Integrate ←   Velocity (v)   Differentiate → Integrate ←   Acceleration (a)

Example 1

A cyclist accelerates uniformly from $V\,\text{m s}^{-1}$ to $3V\,\text{m s}^{-1}$ in $5\,\text{s}$, moving $30\,\text{m}$ in that time.

Find the cyclist's acceleration.

Solution

You are given $u = V\,\text{m s}^{-1}$, $v = 3V\,\text{m s}^{-1}$, $t = 5\,\text{s}$ and $s = 30\,\text{m}$.

Substitute the values into $s = \left(\dfrac{u+v}{2}\right)t$.

$$s = \left(\frac{u+v}{2}\right)t$$

$$30 = \left(\frac{V+3V}{2}\right) \times 5$$

$$6 = \left(\frac{4V}{2}\right)$$

$$6 = 2V$$

$$V = 3$$

Hence the initial velocity is $3\,\text{m s}^{-1}$ and the final velocity is $9\,\text{m s}^{-1}$.

Substitute the values into $v = u + at$.

$$9 = 3 + a \times 5$$

$$6 = 5a$$

$$a = 1.2\,\text{m s}^{-2}$$

The cyclist's acceleration was $1.2\,\text{m s}^{-2}$.

Example 2

A ball is thrown vertically upwards and moves freely under gravity, coming to instantaneous rest after $\dfrac{15}{7}\,\text{s}$.

a Find the greatest height achieved by the ball above its point of projection.

b Find the speed at which the ball was thrown.

Solution

a You are given $t = \dfrac{15}{7}\,\text{s}$.

Since the ball moves freely under gravity, $a = -9.8\,\text{m s}^{-2}$.

Since the ball comes to instantaneous rest, $v = 0\,\text{m s}^{-1}$.

Substitute the values into $s = vt - \dfrac{1}{2}at^2$.

$$s = vt - \frac{1}{2}at^2$$

$$s = 0 \times \frac{15}{7} - \frac{1}{2} \times (-9.8) \times \left(\frac{15}{7}\right)^2 = 22.5\,\text{m}$$

The ball's greatest height was $22.5\,\text{m}$.

> a is negative because gravity acts vertically downwards towards the centre of the Earth.

b Let the initial velocity of the ball be $U \, \text{m} \, \text{s}^{-1}$.

Substitute the values into $v = u + at$.

$$0 = U - 9.8 \times \frac{15}{7}$$

$$U = 9.8 \times \frac{15}{7} = 21 \, \text{m} \, \text{s}^{-1}.$$

The ball was thrown at $21 \, \text{m} \, \text{s}^{-1}$.

Exercise 15.1A

Answers page 477

(M) 1 A particle accelerates in a straight line from $2.4 \, \text{m} \, \text{s}^{-1}$ to $9.6 \, \text{m} \, \text{s}^{-1}$ at a constant rate of $1.2 \, \text{m} \, \text{s}^{-2}$.

    a Find the distance travelled by the particle.

    b Find the time taken.

(PS) (PF) 2 A lorry travels between two traffic lights $0.5 \, \text{km}$ apart, stopping at both. After the first set of traffic lights, the lorry driver accelerates at $1.5 \, \text{m} \, \text{s}^{-2}$ to a speed of $15 \, \text{m} \, \text{s}^{-1}$. He maintains this speed for $20 \, \text{s}$, then applies the brakes at a constant deceleration until he reaches the second set of traffic lights. The speed limit along the whole road is 30 miles per hour. Assume that 1 mile is about $1.6 \, \text{km}$.

    a Find the deceleration of the lorry.

    b Show that the lorry driver broke the speed limit.

(M) 3 A marble is dropped from the roof of a house.
It takes the marble $1.37 \, \text{s}$ to reach the ground.

Find the height from which the marble was dropped.

(M) (PF) 4 A toy rocket is launched vertically into the air at $39.2 \, \text{m} \, \text{s}^{-1}$ and moves freely under gravity.

    a Show that $2 \, \text{s}$ after it is launched, the rocket is travelling at $19.6 \, \text{m} \, \text{s}^{-1}$.

    b Find the maximum height reached by the rocket.

    c Find the time taken for the rocket to return to its starting position.

(M) 5 A ball is thrown vertically upwards at $13 \, \text{m} \, \text{s}^{-1}$ from $20 \, \text{m}$ above the ground.
The ball moves freely under gravity.

    a Find the times at which the ball is $25 \, \text{m}$ above the ground.

    b Find the time taken for the ball to hit the ground.

(M) (PF) 6 The displacement of a particle after t seconds is given by $r = t^2(15 - \sqrt{t}) \, \text{m}$.

    a Find, by differentiation, expressions for the velocity and acceleration in terms of t.

    b Find the velocity after $9 \, \text{s}$.

    c Show that the maximum velocity of the particle is $640 \, \text{m} \, \text{s}^{-1}$.

(M) Modelling   **(PS)** Problem solving   **(PF)** Proof   **(CM)** Communicating mathematically

7 The acceleration of a particle travelling in a straight line is given by $a = -6t \, \text{m s}^{-2}$.

Initially the particle is travelling at a velocity of $9 \, \text{m s}^{-1}$ with a displacement of $2 \, \text{m}$ from the origin O.

a Use integration to find expressions for the velocity and displacement of the particle in terms of t.

b Show that the particle is at instantaneous rest after $\sqrt{3}$ s.

c Show that the displacement of the particle after $2 \, \text{s}$ is $12 \, \text{m}$.

15.2 Velocity vectors

Before considering situations where objects are accelerating, this section will look at particles which are travelling at a steady speed.

Recall from **Book 1, Chapter 10 Vectors** that the **position vector** of a point is its position relative to the origin, and that to find the relative displacement of A from A($\overrightarrow{AB}$), you subtract the position vector of A from the position vector of B.

> **KEY INFORMATION**
>
> A velocity vector shows the direction of travel per unit time.

Whereas a position vector tells you where something is, like the coordinates of a point, a **velocity vector** tells you the direction it is travelling every second or hour (or other unit of time).

For example, when $\mathbf{i}$ is defined as one unit east and $\mathbf{j}$ as one unit north, the vector $(-4\mathbf{i} + 7\mathbf{j}) \, \text{km h}^{-1}$ means that an object is travelling from one position to another $4 \, \text{km}$ west and $7 \, \text{km}$ north every hour.

Consider a boat which starts at a position of $(10\mathbf{i} + 2\mathbf{j}) \, \text{km}$ and that has a velocity of $(-4\mathbf{i} + 7\mathbf{j}) \, \text{km h}^{-1}$.

After one hour the boat will have travelled $4 \, \text{km}$ west and $7 \, \text{km}$ north. Its position vector will be

$$(10\mathbf{i} + 2\mathbf{j}) + (-4\mathbf{i} + 7\mathbf{j}) = (6\mathbf{i} + 9\mathbf{j}) \, \text{km}$$

After two hours the boat will have travelled $8 \, \text{km}$ west and $14 \, \text{km}$ north. Its position vector will be

$$(10\mathbf{i} + 2\mathbf{j}) + 2(-4\mathbf{i} + 7\mathbf{j}) = (2\mathbf{i} + 16\mathbf{j}) \, \text{km}$$

After t hours the boat will have travelled $4t \, \text{km}$ west and $7t \, \text{km}$ north. Its position vector will be

$$(10\mathbf{i} + 2\mathbf{j}) + t(-4\mathbf{i} + 7\mathbf{j}) = [(10 - 4t)\mathbf{i} + (2 + 7t)\mathbf{j}] \, \text{km}$$

This leads to a general expression for the position of an object:

> Position vector ($\mathbf{r}$) = initial position vector
> $+ \, t \times$ velocity vector

> **KEY INFORMATION**
>
> You need to learn this formula.

When two particles are in the same position

Consider two particles travelling at a steady velocity. If the particles are to meet (or collide), then they must, at some point, be in the same position at the same time.

To find this position and time, put the coefficients of **i** equal to each other and solve the equation to find the value of t. If you repeat this for the coefficients of **j**, then if the particles are to meet you will obtain the same value of t. If you get a different value of t, then the particles will not be in the same position at the same time.

Example 3

The position vectors of particles E and F at time t s are given by $\mathbf{r}_E = [(-11 + 5t)\mathbf{i} + (23 - 3t)\mathbf{j}]$ m and $\mathbf{r}_F = [(13 + t)\mathbf{i} + (-7 + 2t)\mathbf{j}]$ m.

Find the time and position at which the particles meet.

Solution

Put the coefficients of **i** equal to each other.

i: $\quad -11 + 5t = 13 + t$

$\qquad 4t = 24$

$\qquad t = 6\,\text{s}$

Put the coefficients of **j** equal to each other.

j: $\quad 23 - 3t = -7 + 2t$

$\qquad 30 = 5t$

$\qquad t = 6\,\text{s}$

Since you get the same value of t from both **i** and **j**, the particles meet when $t = 6\,\text{s}$.

To find the position vector, substitute $t = 6$.

$\qquad \mathbf{r}_E = [(-11 + 5 \times 6)\mathbf{i} + (23 - 3 \times 6)\mathbf{j}]$ m

$\qquad\quad = [(-11 + 30)\mathbf{i} + (23 - 18)\mathbf{j}]$ m

$\qquad\quad = (19\mathbf{i} + 5\mathbf{j})$ m

The same result will be achieved when using $\mathbf{r}_F$.

$\qquad \mathbf{r}_F = [(13 + 6)\mathbf{i} + (-7 + 2 \times 6)\mathbf{j}]$

$\qquad\quad = (19\mathbf{i} + 5\mathbf{j})$ m

Bearings

You can determine when one particle is north (or another compass point) of another particle by considering the relative displacement (in terms of t), as shown in **Example 4**.

The **bearings** you need to consider are listed in the table.

north (000°)/south (180°)	coefficient of $\mathbf{i}$ equals zero
west (270°)/east (090°)	coefficient of $\mathbf{j}$ equals zero
northeast (045°)/southwest (225°)	$\mathbf{i}$ and $\mathbf{j}$ coefficients are the same
northwest (315°)/southeast (135°)	$\mathbf{i}$ and $\mathbf{j}$ coefficients are equal but have opposite signs

KEY INFORMATION

You need to know the rules for the different bearings.

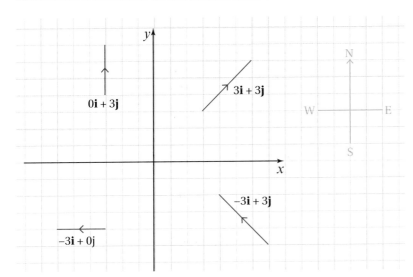

Example 4

The position vectors of two remote-controlled boats G and H after t s are given by $\mathbf{r}_G = [(10 + 3t)\mathbf{i} + (15 - t)\mathbf{j}]$ m and $\mathbf{r}_H = [(18 + t)\mathbf{i} + (-6 + 2t)\mathbf{j}]$ m.

Find the time at which G is:

a north of H　　　　　　**b** east of H

c northeast of H　　　　**d** southeast of H.

Solution

The relative displacement is given by $\mathbf{r}_G - \mathbf{r}_H$.

$$\mathbf{r}_G - \mathbf{r}_H = [(10 + 3t)\mathbf{i} + (15 - t)\mathbf{j}] - [(18 + t)\mathbf{i} + (-6 + 2t)\mathbf{j}]$$
$$= [(-8 + 2t)\mathbf{i} + (21 - 3t)\mathbf{j}] \text{ m}$$

a　For north, the coefficient of $\mathbf{i}$ equals zero.

$\mathbf{i}$:　$-8 + 2t = 0$

　　　$2t = 8$

　　　$t = 4$ s

b　For east, the coefficient of $\mathbf{j}$ equals zero.

$\mathbf{j}$:　$21 - 3t = 0$

　　　$21 = 3t$

　　　$t = 7$ s

c For northeast, the **i** and **j** coefficients are the same.

   i: $-8 + 2t$

   j: $21 - 3t$

$$-8 + 2t = 21 - 3t$$
$$5t = 29$$
$$t = 5.8\,\text{s}$$

d For southeast, the **i** and **j** coefficients are equal but have opposite signs.

   i: $-8 + 2t$

   j: $21 - 3t$

$$-8 + 2t = -(21 - 3t)$$
$$-8 + 2t = -21 + 3t$$
$$t = 13\,\text{s}$$

Distance between two points

If you are given the distance between two particles, you can apply Pythagoras' theorem to the relative displacement. So for a relative displacement of $[(-8 + 2t)\mathbf{i} + (21 - 3t)\mathbf{j}]$ m, $d^2 = (-8 + 2t)^2 + (21 - 3t)^2$.

Example 5

Yacht J is initially at $\begin{bmatrix} -5 \\ 6 \end{bmatrix}$ km and yacht K is initially at $\begin{bmatrix} 40 \\ p \end{bmatrix}$ km.

The two yachts set sail simultaneously at 9:20 a.m.

J has velocity $\begin{bmatrix} 4 \\ 10 \end{bmatrix}$ km h^{-1} and K has velocity $\begin{bmatrix} -6 \\ 2 \end{bmatrix}$ km h^{-1}.

Continuing on these courses, the yachts would have collided, but at 12:50 p.m., yacht K changes its direction and moves instead with a velocity of $\begin{bmatrix} -5 \\ -1 \end{bmatrix}$ km h^{-1}.

Find the distance between the yachts at the time when they would otherwise have collided.

Solution

The position of yacht J is given by $\mathbf{r}_J = \left[\begin{bmatrix} -5 \\ 6 \end{bmatrix} + t \begin{bmatrix} 4 \\ 10 \end{bmatrix} \right]$ km.

This can be rewritten as $\mathbf{r}_J = \begin{bmatrix} -5 + 4t \\ 6 + 10t \end{bmatrix}$ km.

The position of yacht K is given by $\mathbf{r}_K = \left[\begin{bmatrix} 40 \\ p \end{bmatrix} + t \begin{bmatrix} -6 \\ 2 \end{bmatrix} \right]$ km.

This can be rewritten as $\mathbf{r}_K = \begin{bmatrix} 40 - 6t \\ p + 2t \end{bmatrix}$ km.

You need to find the time at which the yachts would have collided. At this time, the yachts would have been in the same place, so the coefficients of **i** would have been the same, as would the coefficients of **j**.

Comparing the coefficients of **i**:

$$-5 + 4t = 40 - 6t$$

$$10t = 45$$

$$t = 4.5\,\text{h}$$

The yachts would have collided at 1:50 p.m.

Comparing the coefficients of **j**:

$$6 + 10t = p + 2t$$

Substitute $t = 4.5$.

$$6 + 10 \times 4.5 = p + 2 \times 4.5$$

$$6 + 45 = p + 9$$

$$51 = p + 9$$

$$p = 42$$

Find the position vector of yacht J at the time the yachts would have collided by substituting $t = 4.5$.

$$\mathbf{r}_J = \begin{bmatrix} -5 + 4t \\ 6 + 10t \end{bmatrix} = \begin{bmatrix} -5 + 4 \times 4.5 \\ 6 + 10 \times 4.5 \end{bmatrix} = \begin{bmatrix} -5 + 18 \\ 6 + 45 \end{bmatrix} = \begin{bmatrix} 13 \\ 51 \end{bmatrix}\,\text{km}$$

At 12:50 p.m. yacht K will have been travelling for 3.5 hours (i.e. $t = 3.5$).

$$\mathbf{r}_K = \begin{bmatrix} 40 - 6t \\ 42 + 2t \end{bmatrix} = \begin{bmatrix} 40 - 6 \times 3.5 \\ 42 + 2 \times 3.5 \end{bmatrix} = \begin{bmatrix} 40 - 21 \\ 42 + 7 \end{bmatrix} = \begin{bmatrix} 19 \\ 49 \end{bmatrix}$$

From this position yacht K will be travel for one hour at $\begin{bmatrix} -5 \\ -1 \end{bmatrix}$ km h^{-1} before the time when the yachts would have collided.

$$\mathbf{r}_K = \begin{bmatrix} 19 \\ 49 \end{bmatrix} + \begin{bmatrix} -5 \\ -1 \end{bmatrix} = \begin{bmatrix} 14 \\ 48 \end{bmatrix}\,\text{km}$$

The relative displacement of J from K is

$$\mathbf{r}_J - \mathbf{r}_K = \begin{bmatrix} 13 \\ 51 \end{bmatrix} - \begin{bmatrix} 14 \\ 48 \end{bmatrix} = \begin{bmatrix} -1 \\ 3 \end{bmatrix}\,\text{km}$$

The distance between the yachts can be found using Pythagoras' theorem.

$$d = \sqrt{(-1)^2 + 3^2}$$

$$= \sqrt{10}$$

$$= 3.16\,\text{km}$$

KEY INFORMATION

You need to know how to use differentiation or completing the square to find the minimum distance between two objects.

If you need to find the closest the particles get to each other, you can differentiate d^2 or complete the square to find the minimum distance, as shown in **Example 6**.

Example 6

Particle A has an initial position of $(12\mathbf{i} - 3\mathbf{j})$ m and a constant velocity of $(-4\mathbf{i} + 4\mathbf{j})$ m s^{-1}.

Particle B has an initial position of $(6\mathbf{i} - 7\mathbf{j})$ m and a constant velocity of $(-3\mathbf{i} + 5\mathbf{j})$ m s^{-1}.

a Find the time at which the particles are 10 metres apart.

b Find the time at which the particles are closest.

Solution

$\mathbf{r}_A = (12 - 4t)\mathbf{i} + (-3 + 4t)\mathbf{j}$

$\mathbf{r}_B = (6 - 3t)\mathbf{i} + (-7 + 5t)\mathbf{j}$

Relative displacement $= \mathbf{r}_A - \mathbf{r}_B = (6 - t)\mathbf{i} + (4 - t)\mathbf{j}$

$$d^2 = (6 - t)^2 + (4 - t)^2$$
$$= 36 - 12t + t^2 + 16 - 8t + t^2$$
$$= 2t^2 - 20t + 52$$

a $2t^2 - 20t + 52 = 100$

$2t^2 - 20t - 48 = 0$

$t^2 - 10t - 24 = 0$

$(t - 12)(t + 2) = 0$

$t = 12\,\text{s}$

> Why cannot t equal -2?

b $\dfrac{\mathrm{d}d^2}{\mathrm{d}t} = 4t - 20 = 0 \quad \Rightarrow \quad t = 5\,\text{s}$

> To differentiate d^2 with respect to t, write $\dfrac{\mathrm{d}d^2}{\mathrm{d}t}$.

Stop and think   Show how you could use completing the square in **Example 6** to find the same result.

Exercise 15.2A

Answers page 478

1 A boat is moored at $(-7\mathbf{i} + 9\mathbf{j})$ km at 5:00 a.m.

It sets off on its journey at a constant velocity of $(3\mathbf{i} - 4\mathbf{j})$ km h^{-1}.

a Find the displacement of the boat between 5:00 a.m. and 8:00 a.m.

b Find the distance travelled between 5:00 a.m. and 8:00 a.m.

c Write an expression for the boat's position t hours after 5:00 a.m.

d Determine the position of the boat at 2:00 p.m.

(M) 2 A particle has an initial position vector of $\begin{bmatrix} 7 \\ 5 \end{bmatrix}$ m. The particle moves with a constant velocity

of $\begin{bmatrix} a \\ b \end{bmatrix}$ m s^{-1}. After 3 seconds it has a position vector of $\begin{bmatrix} 10 \\ -1 \end{bmatrix}$ m.

Find the values of a and b.

(M) 3 At midnight, a helicopter is located at $\begin{bmatrix} 14 \\ -59 \end{bmatrix}$ km and is flying at a steady speed.

Three minutes later, the helicopter is located at $\begin{bmatrix} 10 \\ -57 \end{bmatrix}$ km.

a Find the velocity of the helicopter, giving the answer in km h^{-1}.

b Find the position of the helicopter at 1:30 a.m.

(M) 4 Particles A and B are initially at the position vectors $(8\mathbf{i} + \mathbf{j})$ m and $(12\mathbf{i} - 11\mathbf{j})$ m respectively.

Particles A and B start moving with constant velocities $(3\mathbf{i} - \mathbf{j})$ m s^{-1} and $(\mathbf{i} + 5\mathbf{j})$ m s^{-1} respectively.

a Write down expressions for the position vectors of A and B after t seconds.
   Give your answers in the form $a\mathbf{i} + b\mathbf{j}$, where a and b are in terms of t.

b Find the time at which A and B are in the same position and state the position vector of this point.

c Find the distance between A and B after 5 seconds.

(M) 5 Cyclists G and H set off simultaneously at 7:30 a.m.
(PF)   Cyclist G sets off from $\begin{bmatrix} 13 \\ -5 \end{bmatrix}$ km at a constant velocity of $\begin{bmatrix} -1 \\ 6 \end{bmatrix}$ km h^{-1}.

Cyclist H sets off from $\begin{bmatrix} 2 \\ 17 \end{bmatrix}$ km at a constant velocity of $\begin{bmatrix} 3 \\ -2 \end{bmatrix}$ km h^{-1}.

a Show that the cyclists are 15.7 km apart at 8:30 a.m.

b Find the distance between the cyclists at 11:30 a.m.

c Find the time when they meet.

(M) 6 Ships C and D set sail at 2:00 p.m.
(CM)
Ship C has initial position $(-8\mathbf{i} + 4\mathbf{j})$ km and ship D has initial position $(-5\mathbf{i} - 3\mathbf{j})$ km.
Ship C leaves with velocity $(3\mathbf{i} + \mathbf{j})$ km h^{-1} and ship D with $(\mathbf{i} + 2\mathbf{j})$ km h^{-1}.

a Find the relative displacement of ship C from ship D at time t.

b Find the time when:
   i   C is due north of D
   ii C is on a bearing of 045° from D
   iii C is due east of D.

c Show that C will never be southeast of D.

(M) 7 Boat A has a constant velocity of $(4\mathbf{i} + \mathbf{j})$ km h^{-1}. Boat B has a constant velocity of $(3\mathbf{i} + 2\mathbf{j})$ km h^{-1}.

a Which boat is faster?
   Boat A sets sail from $(-5\mathbf{i} + 8\mathbf{j})$ km at 2:50 p.m. and simultaneously boat B sets sail from $(-3\mathbf{i} - 4\mathbf{j})$ km.

b Find an expression for d^2, where d is the distance between the boats t hours after they set sail.

c Find the time at which the boats are closest and the distance between them.

8 A port P is situated at $(-18\mathbf{i} + 23\mathbf{j})$ km. A boat X departs from P at 6:25 p.m., travelling at $30\,\text{km h}^{-1}$ in the direction of $(4\mathbf{i} - 3\mathbf{j})$ km. Meanwhile, a boat Y, moored at a port 30 km east and 24 km south of P, sets sail at 6:45 p.m. at $(-20\mathbf{i} + 18\mathbf{j})\,\text{km h}^{-1}$. The captain of boat X notices that X and Y are on a collision course and takes evasive action so that at 6:55 p.m. his boat alters its velocity to $(15\mathbf{i} - 20\mathbf{j})\,\text{km h}^{-1}$.

a Had X not changed direction, where would the boats have collided and at what time?

b Find the distance between the boats at 7:25 p.m.

15.3 Equations of constant acceleration using vectors

Since displacement, velocity and acceleration are vector quantities, the SUVAT equations can also be used with vectors in **i j** notation (or column vectors).

Four of the equations can be rewritten using vector notation as:

$$\mathbf{v} = \mathbf{u} + \mathbf{a}t$$

$$\mathbf{s} = \left(\frac{\mathbf{u} + \mathbf{v}}{2}\right)t$$

$$\mathbf{s} = \mathbf{u}t + \frac{1}{2}\mathbf{a}t^2$$

$$\mathbf{s} = \mathbf{v}t - \frac{1}{2}\mathbf{a}t^2$$

Stop and think Why cannot the SUVAT equation $v^2 = u^2 + 2as$ be used with vectors?

Example 7

A particle accelerates at a constant $\begin{bmatrix} 11 \\ -13 \end{bmatrix}\,\text{m s}^{-2}$ for 4 s.

Given that its final velocity is $\begin{bmatrix} 43.5 \\ -49 \end{bmatrix}\,\text{m s}^{-1}$, find:

a the speed of the particle after 1 s

b its distance from its initial position after 2 s.

Solution

a You are told that $\mathbf{a} = \begin{bmatrix} 11 \\ -13 \end{bmatrix}\,\text{m s}^{-2}$, $\mathbf{v} = \begin{bmatrix} 43.5 \\ -49 \end{bmatrix}\,\text{m s}^{-1}$ and $t = 4\,\text{s}$.

Use $\mathbf{v} = \mathbf{u} + \mathbf{a}t$ to find $\mathbf{u}$.

Substitute into $\mathbf{u} = \mathbf{v} - \mathbf{a}t$.

$$\mathbf{u} = \begin{bmatrix} 43.5 \\ -49 \end{bmatrix} - \begin{bmatrix} 11 \\ -13 \end{bmatrix} \times 4$$

$$= \begin{bmatrix} 43.5 \\ -49 \end{bmatrix} - \begin{bmatrix} 44 \\ -52 \end{bmatrix} = \begin{bmatrix} -0.5 \\ 3 \end{bmatrix}\,\text{m s}^{-1}$$

Now use $\mathbf{v} = \mathbf{u} + \mathbf{a}t$ again to find $\mathbf{v}$ after 1 s.

$$\mathbf{v} = \begin{bmatrix} -0.5 \\ 3 \end{bmatrix} + \begin{bmatrix} 11 \\ -13 \end{bmatrix} \times 1 = \begin{bmatrix} 10.5 \\ -10 \end{bmatrix}$$

Use Pythagoras' theorem.

$$\text{Speed} = \sqrt{10.5^2 + (-10)^2} = 14.5\,\text{m}\,\text{s}^{-1}$$

The speed is the magnitude of the velocity.

b Find the displacement of the particle from its initial position by substituting into $\mathbf{s} = \mathbf{u}t + \frac{1}{2}\mathbf{a}t^2$.

$$\mathbf{s} = \begin{bmatrix} -0.5 \\ 3 \end{bmatrix} \times 2 + \frac{1}{2} \times \begin{bmatrix} 11 \\ -13 \end{bmatrix} \times 2^2$$

$$= \begin{bmatrix} -1 \\ 6 \end{bmatrix} + 2\begin{bmatrix} 11 \\ -13 \end{bmatrix} = \begin{bmatrix} -1 \\ 6 \end{bmatrix} + \begin{bmatrix} 22 \\ -26 \end{bmatrix} = \begin{bmatrix} 21 \\ -20 \end{bmatrix}$$

Apply Pythagoras' theorem again.

$$\text{Distance} = \sqrt{21^2 + (-20)^2} = 29\,\text{m}$$

The distance is the magnitude of the displacement.

Example 8

A particle is observed to have an initial velocity of $(13\mathbf{i} - 3\mathbf{j})\,\text{m}\,\text{s}^{-1}$ and an acceleration of $(\mathbf{i} + 4\mathbf{j})\,\text{m}\,\text{s}^{-2}$.

Find the time at which the particle is travelling parallel to the vector $(3\mathbf{i} + \mathbf{j})$.

Solution

Find an expression for $\mathbf{v}$ in terms of t by substituting into $\mathbf{v} = \mathbf{u} + \mathbf{a}t$.

$$\mathbf{v} = (13\mathbf{i} - 3\mathbf{j}) + (\mathbf{i} + 4\mathbf{j})t = (13 + t)\mathbf{i} + (-3 + 4t)\mathbf{j}$$

Since this must be parallel to $(3\mathbf{i} + \mathbf{j})$, you can write

$$(13 + t)\mathbf{i} + (-3 + 4t)\mathbf{j} = k(3\mathbf{i} + \mathbf{j})$$

Equating the coefficients of $\mathbf{i}$: $13 + t = 3k$

Equating the coefficients of $\mathbf{j}$: $-3 + 4t = k$

Hence $13 + t = 3(-3 + 4t)$

$$13 + t = -9 + 12t$$

$$11t = 22$$

$$t = 2$$

Two vectors are parallel if they share a common factor. Hence the particle will be travelling parallel to the vector $(3\mathbf{i} + \mathbf{j})$ when its velocity is a multiple of $(3\mathbf{i} + \mathbf{j})$.

Therefore the particle is travelling parallel to the vector $(3\mathbf{i} + \mathbf{j})$ after 2 s.

Answers page 478

(M) **1** A car has an initial velocity of $(-2\mathbf{i} + 9\mathbf{j})\,\mathrm{m\,s^{-1}}$ and an acceleration of $(3\mathbf{i} - 2\mathbf{j})\,\mathrm{m\,s^{-2}}$.

    a **i** Find the velocity of the car after 6 s.

       ii Find the speed of the car after 6 s.

    b **i** Find the displacement of the car from its initial position after 6 s.

       ii Find the distance of the car from its initial position after 6 s.

(M) **2** A car has an initial velocity of $(2\mathbf{i} - 7\mathbf{j})\,\mathrm{m\,s^{-1}}$ at point A and a final velocity of $(-12\mathbf{i} + 11\mathbf{j})\,\mathrm{m\,s^{-1}}$ at point B. The car travels with a constant acceleration.

    a Find the average speed of the car.

    Given that the car reaches its final velocity 5 s after passing point A:

    b find the final displacement of the car

    c find the acceleration of the car.

(M) (PF) **3** A particle of mass 2000 g experiences a force of $\begin{bmatrix} -10 \\ 12 \end{bmatrix}$ N.

    The initial position vector of the particle is $\begin{bmatrix} 10 \\ 1 \end{bmatrix}$ m and the initial velocity of the particle is $\begin{bmatrix} 7 \\ 3 \end{bmatrix}\,\mathrm{m\,s^{-1}}$.

    Show that after 4 s, the particle has a position vector of $\begin{bmatrix} -2 \\ 61 \end{bmatrix}$ m.

(M) (PS) **4** Two seconds after setting off with a constant acceleration of $(2\mathbf{i} + \mathbf{j})\,\mathrm{m\,s^{-2}}$, a motorcycle has a velocity of $(8\mathbf{i} - \mathbf{j})\,\mathrm{m\,s^{-1}}$.

    a Find the initial speed of the motorcycle.

    b Find the displacement of the motorcycle
       i between the fourth and eighth seconds
       ii during the tenth second.

(M) (PF) **5** A particle has an initial velocity of $(-\mathbf{i} + 7\mathbf{j})\,\mathrm{m\,s^{-1}}$.

    Given that its final displacement after 3 s is $(15\mathbf{i} - 6\mathbf{j})$ m, show that:

    a its acceleration at this time is $2\sqrt{13}\,\mathrm{m\,s^{-2}}$

    b its speed at this time is $11\sqrt{2}\,\mathrm{m\,s^{-1}}$.

(PS) (CM) **6** Stephen was asked to show that a cyclist with an initial velocity of $(-0.1\mathbf{i} + 0.3\mathbf{j})\,\mathrm{m\,s^{-1}}$ and a constant acceleration of $(0.5\mathbf{i} + 0.2\mathbf{j})\,\mathrm{m\,s^{-2}}$ is travelling parallel to the vector $(2\mathbf{i} + \mathbf{j})$ after T seconds, stating the value of T.

    This was his solution:

$$-0.1\mathbf{i} + t \times 0.5\mathbf{i} = (0.5T - 0.1)\mathbf{i}$$
$$0.3\mathbf{j} + t \times 0.2\mathbf{j} = (0.2T + 0.3)\mathbf{j}$$
$$2(0.5T - 0.1) = 0.2T + 0.3$$
$$T - 0.2 = 0.2T + 0.3$$
$$0.8T = 0.5$$
$$T = 0.5 \div 0.8 = 0.625\,\mathrm{s}$$

    Therefore the cyclist is travelling parallel to $(2\mathbf{i} + \mathbf{j})$ after 0.625 s.

    Assess Stephen's solution. Explain how it could be improved.

7 A particle has an initial velocity of $\begin{bmatrix} 6 \\ 2 \end{bmatrix}$ m s^{-1} and an acceleration of $\begin{bmatrix} -4 \\ 1 \end{bmatrix}$ m s^{-2}.

a Show that after $\frac{8}{3}$ s the particle is moving parallel to $\begin{bmatrix} -1 \\ 1 \end{bmatrix}$.

b How many seconds after the particle's journey commences is it moving parallel to

   i $\begin{bmatrix} -2 \\ 1 \end{bmatrix}$ **ii** $\begin{bmatrix} -3 \\ 1 \end{bmatrix}$?

c Prove that the particle will never be moving parallel to $\begin{bmatrix} -4 \\ 1 \end{bmatrix}$.

 8 A remote-controlled boat has a displacement from the origin of $(4\mathbf{i} - 4\mathbf{j})$ m after 2 s and $(6\mathbf{i} + 6\mathbf{j})$ after 6 s.

Find the displacement of the remote-controlled boat after 10 s.

15.4 Vectors with calculus

In **Book 1, Chapter 15 Kinematics** you were shown how to relate displacement, velocity and acceleration when they were written as functions of time, using calculus. The same concept can be applied when the displacement, velocity and acceleration are written as vectors, by differentiating or integrating the **i** and **j** terms individually.

Example 9

After t s, a particle has a position vector given by

$$\mathbf{r} = [(t^3 - 2t + 4)\mathbf{i} + (\tfrac{1}{2}t^3 + 5 - t^2)\mathbf{j}]\text{ m}$$

a Find the time at which the particle is travelling parallel to the vector $(\mathbf{i} + \mathbf{j})$.

b Find the magnitude of the acceleration when $t = 4$ s.

Solution

a For the particle to travel parallel to the vector $(\mathbf{i} + \mathbf{j})$, its velocity will be a multiple of $(\mathbf{i} + \mathbf{j})$.

Find an expression for **v** by differentiating the **i** and **j** terms of **r** individually.

$$\mathbf{r} = \left[(t^3 - 2t + 4)\mathbf{i} + \left(\frac{1}{2}t^3 + 5 - t^2\right)\mathbf{j} \right]\text{m}$$

$$\mathbf{v} = \frac{d\mathbf{r}}{dt} = \left[(3t^2 - 2)\mathbf{i} + \left(\frac{3}{2}t^2 - 2t\right)\mathbf{j} \right]\text{ms}^{-1}$$

Since **v** must be a multiple of $(\mathbf{i} + \mathbf{j})$, it can be written as $k(\mathbf{i} + \mathbf{j})$.

Hence $(3t^2 - 2)\mathbf{i} + \left(\frac{3}{2}t^2 - 2t\right)\mathbf{j} = k(\mathbf{i} + \mathbf{j})$

Equate the **i** and **j** coefficients.

   i: $3t^2 - 2 = k$

   j: $\frac{3}{2}t^2 - 2t = k$

Therefore $3t^2 - 2 = \frac{3}{2}t^2 - 2t$

$$\frac{3}{2}t^2 + 2t - 2 = 0$$

$$3t^2 + 4t - 4 = 0$$

$$(3t - 2)(t + 2) = 0$$

$$t = \frac{2}{3} \text{ or } -2$$

Since t is a time and must be positive, $t = \frac{2}{3}$ s.

b To find an expression for the acceleration, differentiate **v** with respect to time. Differentiate the **i** and **j** terms of **v** individually.

$$\mathbf{v} = \left[(3t^2 - 2)\mathbf{i} + \left(\frac{3}{2}t^2 - 2t\right)\mathbf{j}\right] \text{ms}^{-1}$$

$$\mathbf{a} = \frac{d\mathbf{v}}{dt} = [6t\mathbf{i} + (3t - 2)\mathbf{j}] \text{ m s}^{-2}$$

When $t = 4$, $\mathbf{a} = (6 \times 4)\mathbf{i} + (3 \times 4 - 2)\mathbf{j}$

$$= (24\mathbf{i} + 10\mathbf{j}) \text{ m s}^{-2}$$

The magnitude of the acceleration can be found using Pythagoras' theorem.

$$\text{Magnitude} = \sqrt{24^2 + 10^2} = 26 \text{ m s}^{-2}$$

Exercise 15.4A

Answers page 479

(M) 1 At time t s, a particle P has position vector **r** m relative to the origin, where

$$\mathbf{r} = [3(4t - 1)\mathbf{i} + (17t - 2t^2 + 7)\mathbf{j}] \text{ m}$$

Which of the following is the speed of P after 3 s?

A 13 m s^{-1} **B** 15 m s^{-1}

C 17 m s^{-1} **D** 32 m s^{-1}

(M) 2 At time t s, a particle has the position vector $\mathbf{r} = \begin{bmatrix} 7t - t^2 \\ 2t^2 - t - 3 \end{bmatrix}$ m relative to a fixed origin O.

(PF)

    a Find the velocity of the particle when $t = 2.5$.

    b Show that the acceleration is constant, stating its magnitude in the form $a\sqrt{5}$ m s^{-2}.

(PF) 3 A rocket has a position vector after t s of $\mathbf{r} = [(t^3 - 12t^2 + pt + 2)\mathbf{i} + (21t^2 + qt - 2t^3 + 1)\mathbf{j}] \text{ m}$, where **(M)** p and q are constants. After 2 s, the rocket has a velocity of $(-15\mathbf{i} - 12\mathbf{j}) \text{ m s}^{-1}$.

    a Show that $p = 21$.

    b Find the value of q.

    c Find the time at which the rocket is moving parallel to
       i the vector **i**   **ii** the vector **j**

(M) 4 A ball of mass 4 kg is moving under the action of a single force of **F** N. The ball has a position vector after t s given by $\mathbf{r} = [24t\sqrt{t}\,\mathbf{i} + (\frac{1}{2}t^3 - 10t^2 - 6)\mathbf{j}] \text{ m}$. Find the force after 9 s.

PF **5** A particle has a position vector given by $\mathbf{r} = \dfrac{1}{6}\begin{bmatrix} 9t^2 - 12t \\ 2t^3 \end{bmatrix}$ m relative to a fixed origin O at time t s, where $0 \leqslant t \leqslant 4$.

  a Show that the acceleration of the particle after 4 s is $\begin{bmatrix} 3 \\ 8 \end{bmatrix}$ m s^{-2}.

  The particle then remains at this constant acceleration for $4 < t \leqslant 10$.

  b Find the velocity of the particle when $t = 10$.

In **Book 1** you met the concept of using integration to find the displacement given the velocity as a function of time, or the displacement or velocity given the acceleration as a function of time. When you integrate a function you need to include a constant of integration. However, when the function is written in two dimensions in **i j** notation (or as column vector), **i** and **j** will each need their own constant of integration. It is usual to use C and D in this situation.

Also, when finding the displacement from the acceleration you need to integrate twice, so you need two constants of integration when finding the velocity and then two additional constants of integration when finding the displacement. In **Example 10**, these are called C_1 and D_1.

Example 10

A particle accelerates at $[(12t^2 - 6)\mathbf{i} + 5t\mathbf{j}]$ m s^{-2}, where t is the time in s. Initially, the particle is located at the point with position vector $2\mathbf{i}$ m, and the velocity of the particle is $(20\mathbf{i} - 2\mathbf{j})$ m s^{-1} when $t = 2$.

Find the distance of the particle from its starting point after 3 s.

Solution

$\mathbf{a} = [(12t^2 - 6)\mathbf{i} + 5t\mathbf{j}]$ m s^{-2}

$$\mathbf{v} = \int \mathbf{a}\, dt = \left[(4t^3 - 6t + C)\mathbf{i} + \left(\frac{5}{2}t^2 + D \right)\mathbf{j} \right] \text{m s}^{-1}$$

When $t = 2$, $\mathbf{v} = 20\mathbf{i} - 2\mathbf{j}$.

$$20\mathbf{i} - 2\mathbf{j} = (4(2)^3 - 6(2) + C)\mathbf{i} + \left(\frac{5}{2}(2)^2 + D \right)\mathbf{j}$$

$$= (20 + C)\mathbf{i} + (10 + D)\mathbf{j}$$

Equate the **i** and **j** coefficients.

  i: $20 = 20 + C$, so $C = 0$

  j: $-2 = 10 + D$, so $D = -12$

Therefore $\mathbf{v} = \left[(4t^3 - 6t)\mathbf{i} + \left(\frac{5}{2}t^2 - 12 \right)\mathbf{j} \right] \text{m s}^{-1}$.

Integrate again to find the displacement. Let the constants this time be C_1 and D_1.

$$\mathbf{v} = \left[(4t^3 - 6t)\mathbf{i} + \left(\frac{5}{2}t^2 - 12 \right)\mathbf{j} \right] \text{ms}^{-1}$$

$$\mathbf{r} = \int \mathbf{v} \, dt = \left[(t^4 - 3t^2 + C_1)\mathbf{i} + \left(\frac{5}{6}t^3 - 12t + D_1 \right)\mathbf{j} \right] \text{m}$$

When $t = 0$, $\mathbf{r} = 2\mathbf{i}$.

$$2\mathbf{i} = ((0)^4 - 3(0)^2 + C_1)\mathbf{i} + \left(\frac{5}{6}(0)^3 - 12(0) + D_1 \right)\mathbf{j}$$

$$= (C_1)\mathbf{i} + (D_1)\mathbf{j}$$

Equate the $\mathbf{i}$ and $\mathbf{j}$ coefficients.

$\mathbf{i}$: $\quad 2 = C_1$

$\mathbf{j}$: $\quad 0 = D_1$

Therefore $\quad \mathbf{r} = \left[(t^4 - 3t^2 + 2)\mathbf{i} + \left(\frac{5}{6}t^3 - 12t \right)\mathbf{j} \right] \text{m}$.

When $t = 3$, $\mathbf{r} = \left[((3)^4 - 3(3)^2 + 2)\mathbf{i} + \left(\frac{5}{6}(3)^3 - 12(3) \right)\mathbf{j} \right] \text{m}$.

$$= \left(56\mathbf{i} - \frac{27}{2}\mathbf{j} \right) \text{m}$$

This is the displacement of the particle from its starting point.

The distance is the magnitude of the displacement and can be found using Pythagoras' theorem.

$$\text{Distance} = \sqrt{56^2 + \left(-\frac{27}{2} \right)^2} = 57.6 \, \text{m}$$

TECHNOLOGY

Although you are not permitted to use a calculator to perform symbolic integration, you are allowed to use an integration facility to check numerical answers such as $\left(56\mathbf{i} - \frac{27}{2}\mathbf{j} \right)$.

Exercise 15.4B

Answers page 479

(M) (PF) **1** At time t s, a particle of mass 2.5 kg is subjected to a single force $\mathbf{F}$, given by $\mathbf{F} = (20t\mathbf{i} - 15\mathbf{j})$ N. The initial velocity of the particle is $(-6\mathbf{i} + 34\mathbf{j})$ ms^{-1}.

Show that the particle is travelling at 34 ms^{-1} when $t = 3$.

(M) (PF) **2** Software is being used to model the motion of a racing car in a computer simulation.

The initial position of the car on the grid is $\begin{bmatrix} 4 \\ -7 \end{bmatrix}$ m.

The velocity of the car is modelled using the expression $\mathbf{v} = \begin{bmatrix} t^2 - 2 \\ 14 - 3t \end{bmatrix}$ ms^{-1}.

Show that the model indicates that the car will be about 30.4 m from the origin after 4 s.

(M) **3** A motion detector is tracking the movements of an insect after t s. When $t = 0$, the insect is detected at the point with position vector $(-0.2\mathbf{i} + 0.5\mathbf{j})$ m. Its velocity satisfies the expression $\mathbf{v} = \frac{1}{10}[(5 - 6t^2)\mathbf{i} + (16 + \frac{4}{t+1})\mathbf{j}]$ ms^{-1}.

Find the exact position of the insect when $t = 1$.

(M) **4** The velocity of a particle at time t s is $\mathbf{v} = \left(8\mathbf{i} + t\sqrt{t^2 + 9}\mathbf{j} \right)$ ms^{-1}.

a Find the magnitude of the acceleration of the particle after 4 s.

The particle is initially at the point with position vector $3\mathbf{i}$ m.

b Find an expression for the displacement of the particle from the origin.

(M) **5** A particle P, with an initial velocity of $\begin{bmatrix} -6 \\ 4 \end{bmatrix}\,\mathrm{m\,s^{-1}}$, is accelerating at $\begin{bmatrix} 2t-1 \\ 2t-4 \end{bmatrix}\,\mathrm{m\,s^{-2}}$ at time $t\,\mathrm{s}$.

    a Find the times at which P is moving parallel to the vector $\begin{bmatrix} 3 \\ 2 \end{bmatrix}$.

    b Find the velocity of P at each of these times.

(M) (CM) **6** Chloe has been given this problem to solve:

The acceleration of an object X moving in a horizontal plane is given by $\mathbf{a} = (4t\mathbf{i} + 8\mathbf{j})\,\mathrm{m\,s^{-2}}$, where t is the time in seconds. Initially, X is at the fixed origin O with a velocity of $-\mathbf{i}\,\mathrm{m\,s^{-1}}$. Find the distance of X from the origin when $t = 3$.

Chloe's workings are as follows:

$$\mathbf{a} = (4t\mathbf{i} + 8\mathbf{j})$$
$$\mathbf{v} = \int \mathbf{a}\,\mathrm{d}t = \int (4t\mathbf{i} + 8\mathbf{j})\,\mathrm{d}t = 2t^2\mathbf{i} + 8t\mathbf{j}$$
$$\mathbf{r} = \int \mathbf{v}\,\mathrm{d}t = \int (2t^2\mathbf{i} + 8t\mathbf{j})\,\mathrm{d}t = \tfrac{2}{3}t^3\mathbf{i} + 4t^2\mathbf{j}$$

When $t = 3$, $\mathbf{r} = \tfrac{2}{3} \times 3^3\mathbf{i} + 4 \times 2^2\mathbf{j} = (18\mathbf{i} + 16\mathbf{j})\,\mathrm{m}$

Distance $= 18 + 16 = 34\,\mathrm{m}$

    a Identify any mistakes in Chloe's workings.

    b Provide the correct solution.

(M) (PS) **7** At $t\,\mathrm{s}$, a molecule has an acceleration of $[(p_1 t + p_2)\mathbf{i} + (q_1 t + q_2)\mathbf{j}]\,\mathrm{m\,s^{-2}}$.

Initially the molecule has position vector $(-3\mathbf{i} + 2\mathbf{j})\,\mathrm{m}$ and velocity $(4\mathbf{i} - 4\mathbf{j})\,\mathrm{m\,s^{-1}}$.

When $t = 2$, the velocity of the molecule is $12\mathbf{j}\,\mathrm{m\,s^{-1}}$.

When $t = 4$, the velocity of the molecule is $(8\mathbf{i} + 24\mathbf{j})\,\mathrm{m\,s^{-1}}$.

Find an expression for the position vector of the molecule in terms of t.

(M) (PF) **8** A particle P, of mass $500\,\mathrm{g}$, experiences a single force **F**.

$t\,\mathrm{s}$ after **F** is applied, the velocity of P is given by $[(5 + pt)\mathbf{i} + (qt^2 - 5)\mathbf{j}]\,\mathrm{m\,s^{-1}}$.

After $6\,\mathrm{s}$, $\mathbf{F} = (\mathbf{i} + 18\mathbf{j})\,\mathrm{N}$.

    a Find the values of p and q.

Initially, P has position vector $4\mathbf{j}\,\mathrm{m}$ and particle Q has position vector $(8\mathbf{i} + 12\mathbf{j})\,\mathrm{m}$.

Particle Q moves with a constant velocity of $(3\mathbf{i} - 5\mathbf{j})\,\mathrm{m\,s^{-1}}$.

    b Show that P and Q collide.

    c Find the position of P when they collide.

15.5 Projectiles

In **Book 1, Chapter 15 Kinematics** (and in the previous sections above), all motion was either horizontal or vertical. Vertical motion was affected by gravity but horizontal motion was not.

Consider throwing a ball in the air for your friend to catch. Your natural instinct is to throw the ball towards your friend (which is a horizontal direction) but with an upwards motion since you know that gravity is going to make the ball come back down again. If you applied no upwards motion to the ball then, when it arrived at your friend, it would be too low for them to catch.

When a particle is launched into the air such that its subsequent motion is neither horizontal nor vertical, it is known as a **projectile**. Usually this means that the particle is launched at an angle as shown in the diagram (but it could also be a person running off the side of a swimming pool, for example).

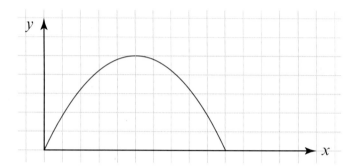

This graph is actually a parabola (a quadratic graph). This is because horizontal motion is dependent only upon t whereas vertical motion is dependent upon t^2.

Although the motion is neither horizontal nor vertical, the motion can be considered as horizontal and vertical components, and the SUVAT equations can be applied in each of these directions.

If the initial velocity is $U\,\text{m}\,\text{s}^{-1}$ at an angle of θ, then the initial horizontal velocity is given by $U\cos\theta\,\text{m}\,\text{s}^{-1}$ and the initial vertical velocity is given by $U\sin\theta\,\text{m}\,\text{s}^{-1}$.

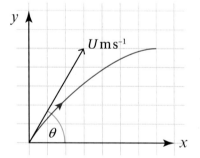
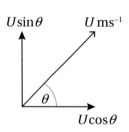

Horizontally, there is no acceleration (and as a result the horizontal velocity will be constant). Vertically, the acceleration is given by g.

At any instant, the direction that the particle is travelling in can be found as an angle from the horizontal and vertical components of the particle's velocity, by using trigonometry as shown in **Example 11**.

> **KEY INFORMATION**
> A projectile is an object propelled through the air such that its subsequent motion takes place in two dimensions rather than one.

> **KEY INFORMATION**
> Consider the horizontal and vertical components of the motion separately.

Example 11

A batsman hits a baseball at $53.9\,\text{m}\,\text{s}^{-1}$ at an angle of $60°$ above the horizontal. It is assumed that the baseball is hit from ground level and moves freely under gravity. Find:

a the horizontal distance travelled by the baseball after 6 s

b the vertical displacement of the baseball after 6 s

c the speed of the baseball after 6 s

d the direction the baseball is travelling after 6 s

e the times at which the baseball is 36.75 m above the ground, giving the answers in the form $a\sqrt{3}$ s.

Solution

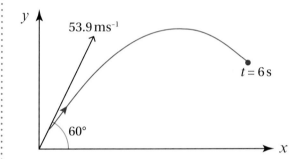

a To find the horizontal distance, start by considering the motion horizontally. You have $t = 6\,\text{s}$ and $a = 0\,\text{m}\,\text{s}^{-2}$ (since there is no acceleration horizontally) and you can work out the horizontal component of the initial velocity. Hence you have values for t, a and u and you can use $s = ut + \frac{1}{2}at^2$ to find the horizontal displacement (which will also be the horizontal distance in this situation).

Horizontally: $u = 53.9\cos 60°\,\text{m}\,\text{s}^{-1}$

$a = 0\,\text{m}\,\text{s}^{-2}$

$s = x\,\text{m}$

$t = 6\,\text{s}$

Substitute into $s = ut + \dfrac{1}{2}at^2$.

$x = 53.9 \times 6 \times \cos 60° + 0 = 162\,\text{m}$

b To find the vertical displacement, consider the motion vertically. Again, you have $t = 6\,\text{s}$ and you can work out the vertical component of the initial velocity, but this time $a = -9.8\,\text{m}\,\text{s}^{-2}$ (since gravity is acting vertically downwards). Hence you can use $s = ut + \frac{1}{2}at^2$ to find the vertical displacement. Note that this will not be the same as the vertical distance travelled since the baseball stops instantaneously at its maximum height and returns before 6 s have elapsed.

Vertically: $\quad u = 53.9 \sin 60° \, \text{m s}^{-1}$

$\qquad\qquad a = -9.8 \, \text{m s}^{-2}$

$\qquad\qquad s = y \, \text{m}$

$\qquad\qquad t = 6 \, \text{s}$

Substitute into $s = ut + \frac{1}{2}at^2$.

$$y = 53.9 \times 6 \times \sin 60° + \frac{1}{2} \times (-9.8) \times 6^2 = 104 \, \text{m}$$

c Recall that speed is the magnitude of the velocity. The speed of the baseball at any time during its flight can be found by applying Pythagoras' theorem to the horizontal and vertical components of the baseball's velocity.

Horizontally: $\quad u = 53.9 \cos 60° \, \text{m s}^{-1}$

$\qquad\qquad a = 0 \, \text{m s}^{-2}$

$\qquad\qquad t = 6 \, \text{s}$

Substitute into $v = u + at$.

$$v = 53.9 \times \cos 60° + 0 = 26.95 \, \text{m s}^{-1}$$

Hence the horizontal component of the velocity is $26.95 \, \text{m s}^{-1}$.

Vertically: $\quad u = 53.9 \sin 60° \, \text{m s}^{-1}$

$\qquad\qquad a = -9.8 \, \text{m s}^{-2}$

$\qquad\qquad t = 6 \, \text{s}$

Substitute into $v = u + at$.

$$v = 53.9 \times \sin 60° - 9.8 \times 6 = -12.12 \, \text{m s}^{-1}$$

Hence the vertical component of the velocity is $-12.12 \, \text{m s}^{-1}$.

The speed is the magnitude of the velocity and is found by applying Pythagoras' theorem.

$$\text{Speed} = \sqrt{26.95^2 + (-12.12)^2} = 29.6 \, \text{m s}^{-1}$$

d Similarly to the speed, the direction can be found using the horizontal and vertical components of the baseball's velocity, but this time by applying trigonometry. In the diagram below, $12.12 \, \text{m s}^{-1}$ is the opposite and $27.95 \, \text{m s}^{-1}$ is the adjacent, so you use the tangent ratio.

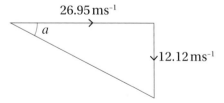

The angle will be $\tan^{-1}\left(\dfrac{12.12}{26.95}\right) = 24.2°$ below the horizontal.

e This question is similar to **part b** except that now you are given the vertical displacement and need to find the two possible values of t. There are two values because the motion is parabolic.

Vertically: $u = 53.9 \sin 60° = \dfrac{53.9\sqrt{3}}{2}\ \text{m s}^{-1}$

$\qquad\qquad a = -9.8\ \text{m s}^{-2}$

$\qquad\qquad s = 36.75\ \text{m}$

Substitute into $s = ut + \dfrac{1}{2}at^2$.

$$36.75 = \dfrac{53.9\sqrt{3}}{2}\,t + \dfrac{1}{2} \times (-9.8) \times t^2$$

$$73.5 = 53.9\sqrt{3}\,t - 9.8t^2$$

$$9.8t^2 - 53.9\sqrt{3}\,t + 73.5 = 0$$

Divide by 4.9.

$$2t^2 - 11\sqrt{3}\,t + 15 = 0$$

Factorise:

$$(2t - \sqrt{3})(t - 5\sqrt{3}) = 0$$

$$t = \dfrac{\sqrt{3}}{2}\ \text{s or } 5\sqrt{3}\ \text{s}$$

Example 12

A particle is launched at $(10\mathbf{i} + 10.5\mathbf{j})\ \text{m s}^{-1}$ from a height of 45 m above the ground.

Find the horizontal distance travelled by the particle before it lands on the ground.

Solution

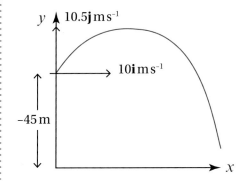

i j notation was first introduced in **Book 1, Chapter 10 Vectors** and the third dimension for **k** was added in this book, in **Chapter 10 Three-dimensional vectors**. When the velocity is given in **i j** notation, it has already been written as horizontal and vertical components. So horizontally $u = 10\,\text{m s}^{-1}$ and vertically $u = 10.5\,\text{m s}^{-1}$.

Vertically: $u = 10.5\,\text{m s}^{-1}$

$\qquad\quad a = -9.8\,\text{m s}^{-2}$

$\qquad\quad s = -45\,\text{m}$

Substitute into $s = ut + \dfrac{1}{2}at^2$.

$\qquad -45 = 10.5t + \dfrac{1}{2} \times (-9.8) \times t^2$

$\qquad -45 = 10.5t - 4.9t^2$

$\quad 4.9t^2 - 10.5t - 45 = 0$

$\quad 49t^2 - 105t - 450 = 0$

Solve by factorisation or by using the quadratic formula.

$\qquad (7t - 30)(7t + 15) = 0$

$\qquad\qquad t = \dfrac{30}{7}$

The particle lands on the ground after $\dfrac{30}{7}$ s.

Horizontally: $u = 10\,\text{m s}^{-1}$

$\qquad\qquad a = 0\,\text{m s}^{-2}$

$\qquad\qquad t = \dfrac{30}{7}\,\text{s}$

Substitute into $s = ut + \dfrac{1}{2}at^2$.

$s = 10 \times \dfrac{30}{7} + 0 = \dfrac{300}{7}\,\text{m} = 42.9\,\text{m}$

> The displacement is negative because upwards has been defined as positive and the particle finishes its journey lower than where it started.

Exercise 15.5A

Answers page 480

 1 A baseball is hit from ground level at a velocity of $30\,\text{m s}^{-1}$ and an angle of $53°$ above the horizontal.

The baseball is modelled as a particle moving freely under gravity.

   a Find the horizontal distance travelled by the baseball after 3 s.

   b Find the vertical displacement of the baseball after 3 s.

   c Find the speed of the baseball after 3 s.

 2 Find the horizontal distances travelled at the two times when the baseball in **question 1** is 20 m above the ground.

(M) (PF) **3** A rocket is launched from the ground at a velocity of $24.5\,\mathrm{m\,s^{-1}}$ and at an angle of α above the horizontal. The rocket is modelled as a particle moving freely under gravity. The rocket lands after 4 s, at a point $x\,\mathrm{m}$ from where it is launched.

    a Show that $\sin\alpha = \dfrac{4}{5}$.

    b Find the value of x.

(M) **4** A particle is launched from a point 49 m above the ground at a velocity of $29.4\,\mathrm{m\,s^{-1}}$ and at an angle of $30°$ above the horizontal.

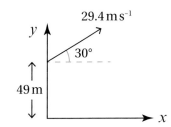

    a Find the time at which the particle hits the ground.

    b Find the horizontal distance travelled before the particle hits the ground.

    c Find the speed of the particle when it hits the ground.

    d Find the direction of the particle when it hits the ground.

(PS) (PF) **5** A particle is launched from a point 50 m above the ground at a velocity of $20\,\mathrm{m\,s^{-1}}$ and at an angle of $10°$ below the horizontal. Show that the horizontal distance travelled by the particle before it strikes the ground is about 56 m.

(M) (PS) **6** A stone is thrown horizontally from the top of a cliff at $U\,\mathrm{m\,s^{-1}}$. The stone lands in the sea after 3 seconds. The stone is modelled as a particle moving freely under gravity.

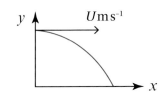

    a Find the height of the cliff.

When the stone lands in the sea it is 73.5 m from the top of the cliff.

    b Find the value of U.

(M) (PF) **7** The unit vectors $\mathbf{i}$ and $\mathbf{j}$ lie in a vertical plane, $\mathbf{i}$ being horizontal and $\mathbf{j}$ being vertical. A ball is launched at $t = 0\,\mathrm{s}$ from a point 35 m above the ground at a velocity of $(5\mathbf{i} + 12\mathbf{j})\,\mathrm{m\,s^{-1}}$ and moves freely under gravity.

    a Find the initial speed of the ball.

    b Find the speed of the ball when $t = 1.5\,\mathrm{s}$.

    c Show that the ball is travelling at $\arctan 0.54$ to the horizontal when $t = 1.5\,\mathrm{s}$.

    d Find the values of t for which the speed of the ball is $5\sqrt{2}\,\mathrm{m\,s^{-1}}$.

(M) (PS) **8** The unit vectors $\mathbf{i}$ and $\mathbf{j}$ lie in a vertical plane, $\mathbf{i}$ being horizontal and $\mathbf{j}$ being vertical. A package is launched from the point $28.5\mathbf{j}\,\mathrm{m}$ at an angle of α above the horizontal. 5 seconds later the package is at $(225\mathbf{i} + 26\mathbf{j})\,\mathrm{m}$. The package is modelled as a particle moving freely under gravity.

    a Find the value of $\tan\alpha$.

    b Find the initial speed of the package.

Specific formulae for projectiles

Starting from the SUVAT formulae, it is possible to derive specific formulae for calculating the time of flight, the range, the greatest height and the equation of the path for y in terms of x.

KEY INFORMATION

You should know how to derive each of the formulae.

Time of flight for a particle returning to the same height

Consider a particle launched with an initial velocity of $U \, \text{m s}^{-1}$ and at an angle of θ above the horizontal.

Vertically, $u = U \sin\theta \, \text{m s}^{-1}$ and $a = -g \, \text{m s}^{-2}$.

Let the particle return to its original height after T s. Hence $s = 0 \, \text{m}$ when $t = T$ s.

Substitute into $s = ut + \frac{1}{2}at^2$.

$$0 = UT\sin\theta - \frac{1}{2}gT^2$$

From which

$$0 = T(U\sin\theta - \frac{1}{2}gT)$$

Hence $U\sin\theta - \frac{1}{2}gT = 0$

Rearranging, $T = \dfrac{2U \sin\theta}{g}$

The time taken for a particle to return to the same height is given by

$$T = \frac{2U\sin\theta}{g} \, \text{s}$$

Range

Consider the same particle returning to its original height.

Horizontally, $u = U\cos\theta \, \text{m s}^{-1}$ and $a = 0 \, \text{m s}^{-2}$.

Let the horizontal distance travelled by the particle be given by x m.

Substitute into $s = ut + \frac{1}{2}at^2$.

$$x = Ut\cos\theta + 0, \text{ so } T = \frac{x}{U\cos\theta}$$

Hence $\dfrac{x}{U\cos\theta} = \dfrac{2U\sin\theta}{g}$

From which $x = \dfrac{2U^2\sin\theta\cos\theta}{g} = \dfrac{U^2\sin 2\theta}{g}$

The range of a particle returning to the same height is given by

$$x = \frac{U^2 \sin 2\theta}{g} \, \text{m}$$

Greatest height

Consider the same particle launched with an initial velocity of $U \, \text{m s}^{-1}$ and at an angle of θ above the horizontal.

Vertically, $u = U\sin\theta \, \text{m s}^{-1}$ and $a = -g \, \text{m s}^{-2}$.

The particle will reach its greatest height when its velocity is instantaneously zero. Let this happen when $t = T$ s.

Substitute into $v = u + at$.

$$0 = U\sin\theta - gt$$

Rearranging, $T = \dfrac{U\sin\theta}{g}$, which is half the time taken to return to its original height, as would be expected from its parabolic trajectory.

PROOF

The following proofs are examples of proof by deduction. Given certain initial conditions and applying one or more equations of constant acceleration, you are able to find a new formula for a specific situation.

Let the greatest height be H m.

Substitute $s = H$ m, $u = U\sin\theta$ m s^{-1}, $a = -g$ m s^{-2} and $t = \dfrac{U\sin\theta}{g}$ s into $s = ut + \frac{1}{2}at^2$.

$$H = U\sin\theta\left(\frac{U\sin\theta}{g}\right) - \frac{1}{2}g\left(\frac{U\sin\theta}{g}\right)^2$$

$$= \frac{U^2\sin^2\theta}{g} - \frac{U^2\sin^2\theta}{2g} = \frac{U^2\sin^2\theta}{2g}$$

The greatest height of a particle is given by

$$H = \frac{U^2\sin^2\theta}{2g}\text{ m}$$

Equation of the path of a projectile

Consider the same particle launched with an initial velocity of U m s^{-1} and at an angle of θ above the horizontal.

Assume that at time t s, the particle has a horizontal displacement of x m and a vertical displacement of y m. Recall that at any time T, t is given by $\dfrac{x}{U\cos\theta}$.

Vertically, $u = U\sin\theta$ m s^{-1} and $a = -g$ m s^{-2}.

Substitute $s = y$ m, $u = U\sin\theta$ m s^{-1}, $a = -g$ m s^{-2} and $t = \dfrac{x}{U\cos\theta}$ s into $s = ut + \frac{1}{2}at^2$.

$$y = U\sin\theta\left(\frac{x}{U\cos\theta}\right) - \frac{1}{2}g\left(\frac{x}{U\cos\theta}\right)^2$$

$$= x\tan\theta - \frac{gx^2}{2U^2\cos^2\theta}$$

$$= x\tan\theta - \frac{gx^2\sec^2\theta}{2U^2}$$

$$= x\tan\theta - \frac{gx^2(1+\tan^2\theta)}{2U^2}$$

The vertical displacement, y m, of the particle when the horizontal displacement is x m is given by

$$y = x\tan\theta - \frac{gx^2(1+\tan^2\theta)}{2U^2}\text{ m}.$$

TECHNOLOGY

You can use graphical software to model projectile motion. For a given initial velocity, U m s^{-1}, and a given angle, the formula

$$y = x\tan\theta - \frac{gx^2(1+\tan^2\theta)}{2U^2}$$

will plot the parabola of y in terms of x. From this graph, you can trace the range and the maximum height as well as the position of a particle at any instant of its motion.

Stop and think   Show how the formulae confirm what you would expect to happen if a particle is launched from the ground horizontally or vertically.

Example 13

Two stones are launched from the same place at ground level and land at the same point on the ground. The first stone is launched at an angle of $40°$ to the horizontal at a speed of 28 m s^{-1}. The second stone is launched at an angle of $55°$ to the horizontal.

Find the greatest height attained by the second stone.

Solution

There are two possible approaches to this question. You could start from the very beginning, using the SUVAT equations as in **Exercise 15.5A**, or you could use the formulae for range and greatest height, as derived above.

First consider how to apply SUVAT.

Begin with the first stone.

Vertically: $u = 28 \sin 40° \, \text{m s}^{-1}$

$\qquad\qquad a = -9.8 \, \text{m s}^{-2}$

$\qquad\qquad s = 0 \, \text{m}$

Substitute into $s = ut + \dfrac{1}{2}at^2$.

$$0 = 28t \sin 40° + \frac{1}{2} \times (-9.8) \times t^2$$

$$0 = 28t \sin 40° - 4.9t^2$$

$$4.9t^2 - 28t \sin 40° = 0$$

$$t(4.9t - 28 \sin 40°) = 0$$

Hence $\qquad\qquad t = 0 \text{ or } \dfrac{28 \sin 40°}{4.9}$

Horizontally: $u = 28 \cos 40° \, \text{m s}^{-1}$

$\qquad\qquad a = 0 \, \text{m s}^{-2}$

$\qquad\qquad t = \dfrac{28 \sin 40°}{4.9} \, \text{s}$

$\qquad\qquad s = x \, \text{m}$

Substitute into $s = ut + \dfrac{1}{2}at^2$.

$$x = 28 \cos 40° \times \frac{28 \sin 40°}{4.9}$$

$$= \frac{28^2 \sin 40° \cos 40°}{4.9} \, \text{m}$$

This is the range of the first stone.

The second stone is launched from the same place and lands in the same place as the first stone. Therefore its range is the same.

Let the speed of the second stone be $U \, \text{m s}^{-1}$.

Vertically: $u = U \sin 55° \, \text{m s}^{-1}$

$\qquad\qquad a = -9.8 \, \text{m s}^{-2}$

$\qquad\qquad s = 0 \, \text{m}$

Substitute into $s = ut + \dfrac{1}{2}at^2$.

$$0 = Ut\sin 55° + \frac{1}{2} \times (-9.8) \times t^2$$

$$0 = Ut\sin 55° - 4.9t^2$$

$$4.9t^2 - Ut\sin 55° = 0$$

$$t(4.9t - U\sin 55°) = 0$$

Hence $\qquad t = 0 \text{ or } \dfrac{U\sin 55°}{4.9}$

Horizontally: $\quad u = U\cos 55° \, \text{m s}^{-1}$

$$a = 0\,\text{m s}^{-2}$$

$$t = \frac{U\sin 55°}{4.9}\,\text{s}$$

$$s = \frac{28^2 \sin 40° \cos 40°}{4.9}\,\text{m}$$

Substitute into $s = ut + \dfrac{1}{2}at^2$.

$$\frac{28^2 \sin 40° \cos 40°}{4.9} = U\cos 55° \times \frac{U\sin 55°}{4.9}$$

$$\frac{28^2 \sin 40° \cos 40°}{4.9} = \frac{U^2 \sin 55° \cos 55°}{4.9}$$

$$28^2 \sin 40° \cos 40° = U^2 \sin 55° \cos 55°$$

Hence $\qquad U^2 = \dfrac{28^2 \sin 40° \cos 40°}{\sin 55° \cos 55°}$

$$U = \sqrt{\frac{28^2 \sin 40° \cos 40°}{\sin 55° \cos 55°}} = 28.7\,\text{m s}^{-1}$$

Now find the greatest height of the second stone.

Vertically: $\quad u = 28.7\sin 55° \, \text{m s}^{-1}$

$$a = -9.8\,\text{m s}^{-2}$$

$$v = 0\,\text{m s}^{-1}$$

$$s = y\,\text{m}$$

Substitute into $v^2 = u^2 + 2as$.

$$s = \frac{v^2 - u^2}{2a}$$

$$y = \frac{0^2 - (28.7\sin 55°)^2}{2 \times (-9.8)} = 28.1\,\text{m}$$

The greatest height attained by the second stone is 28.1 m.

Alternatively, using the formulae for range and greatest height is more efficient.

The formula for range is given by $x = \dfrac{U^2 \sin 2\theta}{g}$ m.

For the first stone, with a speed of 28 m s^{-1} at an angle of 40°, the range is $\dfrac{28^2 \sin 80°}{9.8}$.

For the second stone, with a speed of $U\,\mathrm{m\,s^{-1}}$ at an angle of 55°, the range is $\dfrac{U^2\sin 110°}{9.8}$.

Since these are equal, $\dfrac{U^2 \sin 110°}{9.8} = \dfrac{28^2 \sin 80°}{9.8}$

Therefore $\qquad\qquad U^2 = \dfrac{28^2 \sin 80°}{\sin 110°}$

The formula for greatest height is given by $y = \dfrac{U^2 \sin^2\theta}{2g}$ m.

For the second stone, the greatest height is

$\dfrac{28^2 \sin 80°}{\sin 110°} \times \dfrac{\sin^2 55°}{2g} = 28.1\,\mathrm{m}$

Exercise 15.5B

Answers page 481

1 A golf ball is struck from ground level and at $U\,\mathrm{m\,s^{-1}}$ and at an angle of 30° to the horizontal. It moves freely under gravity.

   a Show that the golf ball returns to the ground after $\dfrac{U}{g}$ s.

   b Show that the golf ball travels $\dfrac{U^2\sqrt{3}}{2g}$ m horizontally before it returns to the ground.

   c Show that the greatest height achieved by the golf ball is $\dfrac{U^2}{8g}$ m.

2 The range of a particle is given by $x = \dfrac{U^2 \sin 2\theta}{g}$ m, where θ is the angle of projection in degrees and U is the velocity of projection in $\mathrm{m\,s^{-1}}$. Two students, Anna and Johann, were asked to find the maximum range in terms of U, stating the angle for which the maximum range occurs.

Anna differentiated the function with respect to θ and put it equal to zero:

$$\frac{\mathrm{d}x}{\mathrm{d}\theta} = \frac{2U^2 \cos 2\theta}{g} = 0$$

$$\cos 2\theta = 0$$

$$2\theta = 270°$$

$$\theta = 135°$$

Hence the maximum range was $\dfrac{U^2 \sin 270°}{g} = -\dfrac{U^2}{g}$ m.

Johann considered the graph of $y = \sin\theta$, which for $0 \leqslant \theta < 360°$ has a maximum value of 1 when $\theta = 90°$, from which he deduced that the angle was $2\theta = 180°$.

Hence the maximum range was $\dfrac{U^2 \sin 180°}{g} = 0\,\mathrm{m}$.

Neither Anna nor Johann has worked out the correct expression for the maximum range.

Explain what is incorrect for each set of workings and hence find the maximum range in terms of U, stating the angle for which the maximum range occurs.

3 A cannonball is fired from the ground at $20\,\mathrm{m\,s^{-1}}$ and at an angle of θ to the horizontal. It lands $60\,\mathrm{m}$ away on the ground after t s. The ground is assumed to flat and horizontal.

    a Show that $t = \dfrac{3}{2}\sec\theta$.

    b By considering vertical motion, show that $80\sin\theta = gT$.

    c Hence, or otherwise, prove that $\sin 2\theta = \dfrac{3g}{80}$.

 4 A particle is projected at $V\,\mathrm{m\,s^{-1}}$ at an angle of $\arctan 2$ to the horizontal. After t s its vertical displacement is given by y m and its horizontal displacement is given by x m.

    Show that after t s, $y = \dfrac{x(4V^2 - 5gx)}{2V^2}$ m.

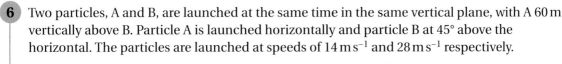 **5** A particle is projected at an angle of $\arcsin\frac{3}{5}$ above the horizontal and at $V\,\mathrm{m\,s^{-1}}$. It returns to the same height after T s.

    a Show that after T s the horizontal distance travelled by the particle is given by $\dfrac{24V^2}{25g}$ m.

    b Prove that the horizontal range of a particle projected at an angle of α above the horizontal is the same as if the particle were to be projected at an angle of $(90° - \alpha)$ above the horizontal.

6 Two particles, A and B, are launched at the same time in the same vertical plane, with A $60\,\mathrm{m}$ vertically above B. Particle A is launched horizontally and particle B at $45°$ above the horizontal. The particles are launched at speeds of $14\,\mathrm{m\,s^{-1}}$ and $28\,\mathrm{m\,s^{-1}}$ respectively.

Show that the particles collide after travelling $40\,\mathrm{m}$ horizontally.

SUMMARY OF KEY POINTS

❯ The position vector of a point is its position relative to the origin.

❯ To find the relative displacement of B from A($\overrightarrow{AB}$), subtract the position vector of A from the position vector B.

❯ A velocity vector shows the direction of travel per unit time.

❯ Position vector (**r**) = initial position vector + t × velocity vector

❯ The relative displacement can be used to find the bearing of one object from another object.

north (000°)/south (180°)	coefficient of **i** equals zero
west (270°)/east (090°)	coefficient of **j** equals zero
northeast (045°)/southwest (225°)	**i** and **j** coefficients are the same
northwest (315°)/southeast (135°)	**i** and **j** coefficients are equal but have opposite signs

❯ By applying Pythagoras' theorem, the relative displacement can be used to find the distance between two particles or the time at which two particles are closest. In order to find the closest two particles get to each other, differentiate or complete the square.

❯ The SUVAT equations of constant acceleration can be used with vectors in two dimensions.

❯ Calculus can be applied to displacement, velocity and acceleration in two dimensions when they are written as functions of time, by differentiating or integrating the horizontal and vertical components individually. When integrating, each component needs its own constant of integration.

❯ A projectile is an object propelled through the air such that its subsequent motion takes place in two dimensions rather than one. The horizontal and vertical components of the motion are considered separately and the equations of constant acceleration are applied to the individual components.

❯ Using the equations of constant acceleration, it is possible to derive specific formulae for calculating the time of flight, the range, the greatest height and the equation of the path for y in terms of x.

EXAM-STYLE QUESTIONS 15   **Answers page 483**

 1 The initial velocity of a particle is $(10\mathbf{i} + 8\mathbf{j})\,\mathrm{m\,s^{-1}}$. Eight seconds later, the displacement of the particle from its starting point is $(-16\mathbf{i} + 224\mathbf{j})$ m.

    a Find the acceleration of the particle.   [4 marks]

    b Find the speed of the particle after 8 s.   [3 marks]

 2 A pebble is launched from the top of a building at a speed of $18\,\mathrm{m\,s^{-1}}$ and at an angle of 20° above the horizontal. The pebble strikes the ground 3 s later.

Find:

a the angle the pebble's velocity makes with the horizontal when it strikes the ground **[5 marks]**

b the speed of the pebble when it strikes the ground **[2 marks]**

c the height of the building. **[3 marks]**

(M) **3** The displacement of a particle after t s is given by $\mathbf{r} = [2t(16 - t)\mathbf{i} + t(t - 1)\mathbf{j}]$ m. Find the speed of the particle, in km h^{-1}, when $t = 5$. **[4 marks]**

(M)(PS) **4** A lion has escaped from a zoo. The zoo is situated at $(17\mathbf{i} - 35\mathbf{j})$ km.

At 1:25 p.m. the lion is spotted in a field at $(-3\mathbf{i} + 13\mathbf{j})$ km.

A zookeeper immediately leaves the zoo and heads in the direction of the lion at 26 km h^{-1}.

At 1:55 p.m. the zookeeper is informed that the original sighting was a false alarm and that the lion is actually at $(6\mathbf{i} - 15\mathbf{j})$ km.

If the zookeeper travels straight there at 30 km h^{-1}, at what time will he reach the lion? **[6 marks]**

(M)(PS) **5** A particle with a mass of 250 g is acted upon by a single force **F**.

After 1.5 s, the displacement of the particle is $\begin{bmatrix} 9 \\ 12 \end{bmatrix}$ m.

After 3.5 s, the displacement of the particle is $\begin{bmatrix} 42 \\ 35 \end{bmatrix}$ m.

a Find the initial velocity of the particle. **[5 marks]**

b Find the magnitude of the force, correct to 3 significant figures. **[3 marks]**

(M)(PS) **6** A radar identifies the initial location of a boat as $(3\mathbf{i} + 8\mathbf{j})$ m.

By tracking the progress of the boat over a 10 s period, the velocity of the boat after t s is shown to be given by $\mathbf{v} = [(0.2t + 0.1)\mathbf{i} + (0.3t - 0.2)\mathbf{j}]$ m s^{-1}.

a Find the position of the boat after 3 s. **[5 marks]**

b Find the magnitude of the acceleration of the boat during the 10 s period. **[3 marks]**

At the end of the 10 s period, the captain of the boat notices that the boat is being tracked and he decelerates uniformly to a stop in 4 s.

c Find the magnitude of the deceleration. **[4 marks]**

(PS)(PF) **7** The HMS *Walton* is docked at $\begin{bmatrix} -6 \\ 63 \end{bmatrix}$ km and the HMS *Barron* is docked at $\begin{bmatrix} 60 \\ -27 \end{bmatrix}$ km. In order to avoid potential disasters, the captains of both ships choose independently to set sail at 3:30 a.m., the HMS *Walton* with a velocity of $\begin{bmatrix} 8 \\ -5 \end{bmatrix}$ km h^{-1} and the HMS *Barron* with a velocity of $\begin{bmatrix} -3 \\ 10 \end{bmatrix}$ km h^{-1}.

a If neither vessel were to alter its direction, show that there would be a collision, stating the time and position vector of the collision. **[6 marks]**

The captain of the HMS *Barron*, having realised the unfortunate coincidence of their identical decisions, takes evasive action. At 7:30 a.m. he alters his course so that the new velocity of the HMS *Barron* is $\begin{bmatrix} -2 \\ 10 \end{bmatrix}$ km h^{-1}.

b Find the new position of the HMS *Barron* at the time at which collision would have happened. **[4 marks]**

c Find the distance between the vessels at this time. **[2 marks]**

(PF) 8 A particle is launched from ground level and returns to the ground $\frac{200}{49}$ s later.

(M)

a Find the greatest height reached by the particle. **[2 marks]**

Given that the particle is launched at an angle of arctan $\frac{3}{4}$:

b show that the range of the particle is just under 109 m. **[5 marks]**

(PS) 9 Two particles, A and B, are projected from a tower block at the same angle and same speed, and in the same vertical plane. Particle A is launched from ground level and lands on the ground 60 m away. Particle B is launched from a window 25 m vertically above particle A and lands on the ground a horizontal distance of 75 m away. Taking g as 10 m s^{-2}, calculate:

(M)

a the angle of projection of each particle **[7 marks]**

b the initial speed of each particle. **[2 marks]**

(PS) 10 Two pirate ships, the *After You* and the *Coathanger*, are out on the ocean waves at 1:49 p.m. The *After You* is at $(20\mathbf{i} + 36\mathbf{j})$ km whereas the *Coathanger* is at $(38\mathbf{i} + 7\mathbf{j})$ km. Their velocities are $(-6\mathbf{i} - 9\mathbf{j})$ km h^{-1} and $(-9\mathbf{i} - 3\mathbf{j})$ km h^{-1} respectively.

(M)

a At what time is the *Coathanger* southeast of the *After You*? **[5 marks]**

b Show that the ships are closest together at 6:53 p.m. **[4 marks]**

c What is the distance between the ships at 6:53 p.m.? **[3 marks]**

(PS) 11 At time t s, a force of $(5\sqrt{t}\ \mathbf{i} + \frac{1}{3\sqrt{t}}\mathbf{j})$ N is being applied to a block of mass $\frac{1}{3}$ kg. Initially, the block is at the origin O travelling at $12\mathbf{j}$ m s^{-1}.

(M)

a Find the time at which the block is moving parallel to the vector $(5\mathbf{i} + \mathbf{j})$. **[6 marks]**

b Find an expression for the displacement of the block in terms of t. **[3 marks]**

(PS) 12 A man projects a ball at an angle of θ at U m s^{-1}. He is aiming to hit a basket 4 m vertically above and 20 m horizontally from its point of projection.

(M)

Given that $U^2 = 25g$:

a show that $2\tan^2\theta - 5\tan\theta + 3 = 0$. **[7 marks]**

b Hence, or otherwise, find both possible angles of projection. **[3 marks]**

16 FORCES

Imagine that you are driving a car up a hill when the engine cuts out. The car will continue to travel up the hill without the driving force exerted by the engine, but it will start to slow down until it comes to instantaneous rest, assuming it does not reach the summit of the hill first. Then the car will roll down the hill backwards.

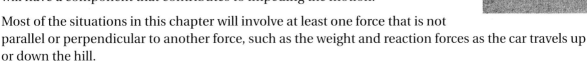

What causes the car to decelerate? There are two main forces acting upon it. Firstly, there is the frictional force, which opposes the motion of the car and which varies according to the texture and angle of the road. Secondly, there is the weight of the car. Since the car is travelling at an angle, the weight (which acts vertically) will have a component that contributes to impeding the motion.

Most of the situations in this chapter will involve at least one force that is not parallel or perpendicular to another force, such as the weight and reaction forces as the car travels up or down the hill.

LEARNING OBJECTIVES

You will learn how to:

> resolve forces in situations where at least one force is not parallel or perpendicular to another (including when a particle is on a slope)

> add forces together in two dimensions using the sine and cosine rules or **i j** notation

> apply the formula $F = \mu R$ which relates the reaction and friction forces

> apply the equations of constant acceleration to problems involving angles and $F = \mu R$

> solve problems involving connected particles involving angles and $F = \mu R$.

TOPIC LINKS

Just like **Chapter 15 Kinematics**, this chapter takes the material from **Book 1** and extends it into two dimensions. It also introduces the relationship between friction and reaction, which will be essential when considering moments (for example, ladders) in **Chapter 17 Moments**.

PRIOR KNOWLEDGE

You should already know how to:

> use **i j** notation and columns to represent vectors

> use Pythagoras' theorem and trigonometry to find the magnitude and direction of a vector

> use the sine and cosine rules

> apply the equations of constant acceleration (SUVAT)

> draw a force diagram using horizontal and/or vertical forces

> select between the forces of weight, tension, thrust, reaction and resistance to motion

> find the sum of the forces in one direction if all forces are horizontal and/or vertical

> apply the formulae $W = mg$ and $F = ma$

> solve problems involving connected particles in which all the forces are horizontal and/or vertical.

You should be able to complete the following questions correctly:

1   In the diagram below, ABD is a triangle and ABC is a straight line. The angle CBD is 58°. AB = 19 cm and BD = 13 cm.

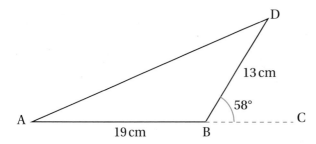

   a Use the cosine rule to find the length of the side AD.

   b Use the sine rule to find the size of the angle BAD.

2   A particle of mass 4 kg experiences a force of $\begin{bmatrix} -16 \\ 28 \end{bmatrix}$ N for 3 s. The initial velocity of the particle is $\begin{bmatrix} 2 \\ -3 \end{bmatrix}$ m s^{-1}.

Find the distance travelled whilst the particle experiences the force.

3   A block of mass 4 500 g is pushed along a rough horizontal table by a horizontal force of magnitude 15 N. The block experiences a resistance to its motion of 6 N.

   a Find the magnitude of the reaction force exerted on the block by the table.

   b Find the acceleration of the block.

4   Two particles of masses 3 kg and 5 kg are connected by a light inextensible string which passes over a smooth pulley. Both particles hang freely. Find the initial acceleration and the tension in the string when the system is released from rest.

16.1 Resolving forces

In **Book 1, Chapter 16 Forces**, you considered forces acting horizontally or vertically. Forces are vector quantities, so if two forces act in opposite directions, they have opposite signs (positive and negative). If a force acts vertically, then it has no horizontal **component** and if a force acts horizontally, it has no vertical component. This is true in general when considering forces acting in parallel and perpendicular directions.

The process of considering the forces acting in a particular direction is called **resolving**. In this chapter the technique is extended to a force acting at an angle to the horizontal or a particle on a slope (so the weight, at least does not act parallel or perpendicular to the slope).

Resolving in a particular direction gives the **resultant force** in that direction.

Consider a force of F N acting at an angle of θ to the horizontal. You can construct a triangle of forces using the triangle law you encountered in **Book 1, Chapter 10 Vectors**.

This diagram shows a right-angled triangle with the force F as its hypotenuse. You can therefore use trigonometry to find the horizontal and vertical components.

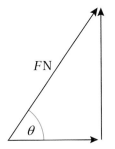

The vertical component is the component opposite to θ. Since $\sin \theta = \dfrac{O}{H}$, you can write $O = H \sin \theta$, and since F is the hypotenuse, the opposite is given by $F \sin \theta$.

Similarly, since $\cos \theta = \dfrac{A}{H}$, you can write $A = H \cos \theta = F \cos \theta$.

Hence the component adjacent to the angle is given by $F \cos \theta$ and the component opposite the angle is given by $F \sin \theta$.

In this diagram, for example, the 17 N force acts at angle of 50° to the horizontal. The horizontal component (adjacent to the angle) is given by $17 \cos 50° = 10.9$ N and the vertical component (opposite the angle) is given by $17 \sin 50° = 13.0$ N.

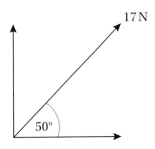

Recall from **Book 1, Chapter 16 Forces** that if a system is in **equilibrium** then the resultant force is zero. Hence the **acceleration** is also zero, because (from $F = ma$) a is given by $\dfrac{F}{m}$ (and the mass will not be zero).

Note that in **Book 1, Chapter 16 Forces**, when a force opposed the motion of a particle it was described as resistance to motion. In this chapter, this force is called **friction**.

> ### KEY INFORMATION
> If a system is in equilibrium, then the resultant force is zero.

Example 1

A box of books is dragged along a rough horizontal floor by a rope at a steady speed. The rope makes an angle of 35° with the floor.

a Draw a force diagram.

b Resolve horizontally.

c Resolve vertically.

Solution

a The box of books is modelled as a particle, which is drawn as a black circle.

Assume that the box is being dragged from left to right along the floor.

There are four forces acting upon the box, each of which is represented as an arrow.

The weight of the box (WN) acts vertically downwards.

The rope has a tension (TN), and since the rope makes an angle of 35° with the floor, so does the tension force.

The reaction force (RN) acts perpendicular to the surface. Since the surface (the floor) is horizontal, the reaction force is vertical.

The frictional force (FN) acts to oppose the motion. Since the box is being dragged to the right, the friction acts to the left.

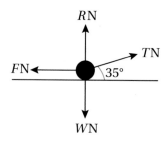

b Since the weight and reaction act vertically, they have no horizontal component. Hence there are only two forces to consider, the tension and the friction.

The tension acts at an angle of 35° to the horizontal, so it has a component of $T\cos 35°$ N horizontally.

The friction acts horizontally.

Because the box is being dragged at a steady speed, there is no acceleration. Hence the resultant force is zero.

Resolve horizontally to find the resultant force in the horizontal direction.

$R(\rightarrow)$

$$T\cos 35° - F = 0$$

c Since the friction acts horizontally, it has no vertical component, and there are three forces to consider – the tension, weight and reaction.

The tension acts at an angle of 35° to the horizontal, so it has a component of $T\sin 35°$ N vertically.

The reaction and weight act vertically.

Because the box is not moving vertically, there is no acceleration. Hence the resultant force is zero.

Resolve vertically to find the resultant force in the vertical direction (you could choose up or down as positive).

$R(\uparrow)$

$$R + T\sin 35° - W = 0$$

> **KEY INFORMATION**
>
> Weight acts vertically downwards, the reaction acts perpendicular to the surface and friction acts to oppose the motion.

> It is useful to consider which forces have any component in the direction you are resolving in, and which do not, before you begin.

> This notation, R(arrow), is used to indicate that you are resolving in the direction of the arrow.

> This component could also be written as $T\cos 55°$.

Stop and think   Why is $T\sin 35°$ the same as $T\cos 55°$?

Example 2

A wetsuit of weight W N hanging from a washing line causes the line to make angles of 22° and 33° with the horizontal, as shown in the diagram.

a Resolve horizontally.   **b** Resolve vertically.

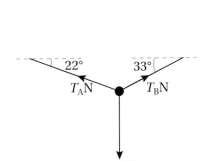

Solution

a Because the angles are different, assume also that the tensions are different. Label these T_A N and T_B N. Also add in the weight of the wetsuit, acting vertically downwards.

Resolve horizontally to find the resultant force in the horizontal direction.

R($\rightarrow$)
$$T_B \cos 33° - T_A \cos 22° = 0$$

b Resolve vertically to find the resultant force in the vertical direction.

R($\uparrow$)
$$T_A \sin 22° + T_B \sin 33° - W = 0$$

This technique applies to a force applied at any angle. The components do not have to be horizontal and vertical. If a particle is on a surface which is itself at an angle, then it is common to resolve parallel and perpendicular to the slope, in which case the weight (and possibly other forces too) will need to be resolved into components. Note that the friction is always parallel to the surface and the reaction force is always perpendicular to the surface.

If a particle is just held in position by a force on a **rough** slope so that the particle is on the point of slipping down the slope, then the frictional force acts up the slope. If the particle is on the point of beginning to move up a rough slope, then the frictional force acts down the slope.

Example 3

A toaster of mass m kg is at rest on a rough slope inclined at θ to the horizontal. The toaster is acted upon by a horizontal force of X N and as a result is on the point of moving up the slope.

Resolve parallel and perpendicular to the slope.

Solution

Start by drawing a force diagram, representing the toaster as a particle.

KEY INFORMATION

Always resolve parallel and perpendicular to the surface.

KEY INFORMATION

If a particle is on the point of slipping down the slope, then the friction acts up. If the particle is on the point of moving up the slope, then the friction acts down.

It is always a good idea to draw a force diagram, even if you have not been asked to.

There are four forces acting upon the toaster.

The weight of the box (mgN) acts vertically downwards.

The horizontal force (XN), which must be acting to the right if it is to keep the particle on the slope.

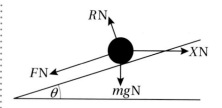

The reaction force (RN) acts perpendicular to the surface. Since the surface is at an angle, so is the reaction force.

The frictional force (FN) acts to oppose the motion. Since the toaster is on the point of moving up the slope, the friction acts down the slope, parallel to the slope.

Start by resolving parallel to the slope. Since the reaction acts perpendicular to the slope, it has no parallel component. Therefore are three forces to consider: X, the weight and the friction.

The friction acts parallel to the slope.

Since XN is horizontal, the component parallel to the slope is equivalent to the adjacent, as shown in the diagram on the right, and hence is given by $X\cos\theta$.

Since mgN is vertical, the component parallel to the slope is equivalent to the opposite, as shown in this diagram, and hence is given by $mg\sin\theta$.

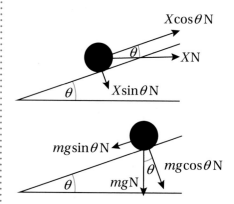

Resolving parallel to the slope, you have:

R($\nearrow$)

$$X\cos\theta - mg\sin\theta - F = 0$$

Now consider resolving perpendicular to the slope. Since the friction acts parallel to the slope, it has no perpendicular component. Again, there are three forces to consider – XN, the weight mgN and the reaction RN.

The reaction acts perpendicular to the slope.

Since XN is horizontal, the component perpendicular to the slope is equivalent to the opposite, as shown in the diagram on the right, and hence is given by $X\sin\theta$.

Since mgN is vertical, the component perpendicular to the slope is equivalent to the adjacent, as shown in this diagram, and hence is given by $mg\cos\theta$.

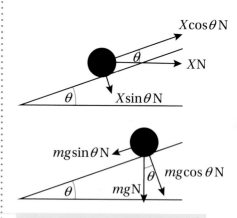

Resolving perpendicular to the slope, you have:

R($\nwarrow$)

$$R - X\sin\theta - mg\cos\theta = 0$$

Alternatively, you could resolve down the slope instead of up the slope.

You could also resolve towards the slope instead of away from the slope.

Exercise 16.1A

Answers page 484

(M) **1** Draw diagrams to represent the following situations:

a A particle is at rest on a rough plane inclined at θ to the horizontal.

b A particle is pulled across a rough horizontal table by a string inclined at 60° to the horizontal.

c A particle is held at rest on a smooth plane inclined at 30° to the horizontal by a string parallel to the plane.

d A car drives up a rough slope inclined at θ to the horizontal.

e A particle is on the verge of moving up a rough slope (inclined at 26° to the horizontal) as a result of being pushed by a horizontal force P.

f A particle is just held in position on a rough slope (inclined at 26° to the horizontal) by a horizontal force P.

g A particle is suspended from a horizontal beam by two unequal strings at angles of 50° and 20° to the beam.

h A particle is at rest on a horizontal surface.

(M) **2** A peg bag of mass $m\,$kg is suspended from a string such that both ends of the string are inclined at θ to the horizontal and the tensions in both ends are identical and given by $T\,$N, as shown in the diagram. The system is in equilibrium.

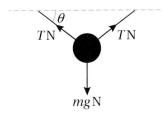

a Resolve vertically.

b Hence find an expression for T in terms of m and θ.

(M) **3** A toy bus of weight $W\,$N is pulled along a rough horizontal carpet at a constant speed by a string. The string makes an angle of 19° with the carpet, as shown in the diagram. The system is in equilibrium.

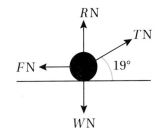

a Resolve parallel to the carpet.

b Resolve perpendicular to the carpet.

(M) **4** A particle of mass $m\,$kg is suspended from two strings, A and B, as shown in the diagram. String A is at an angle of 25° to the horizontal and string B is at an angle of 65° to the horizontal. The system is in equilibrium.

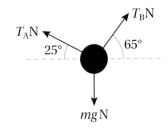

a Resolve horizontally. **b** Resolve vertically.

(M) **5** Resolve each system parallel and perpendicular to the surface:

a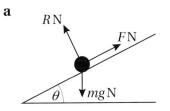

b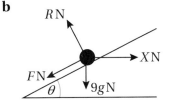

c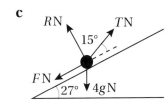

6   Each diagram shows a system in equilibrium. For each diagram, Mary has been asked to resolve in a given direction. Explain what is wrong with Mary's answer in each case and give the correct answer.

a Resolve perpendicular to the slope.

Mary's answer: $R = 3g\cos 24°$

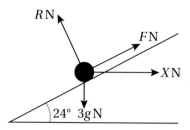

b Resolve parallel to the slope.

Mary's answer: $F + X\cos 24° = 3g\sin 24°$

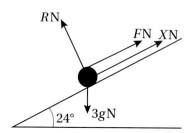

c Resolve parallel to the slope.

Mary's answer: $X\cos 24° = F + 3g\cos 24°$

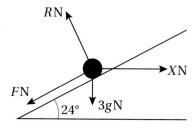

d Resolve perpendicular to the slope.

Mary's answer: $R = X\cos 24° + 3g\cos 24°$

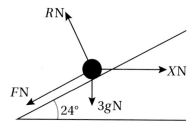

e Resolve parallel to the slope.

Mary's answer: $X\cos 64° = F + 3g\sin 24°$

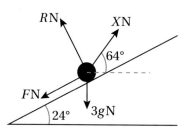

If all the forces are in equilibrium, the sum in any direction will be zero. This means that it is possible to solve problems by setting up and solving equations.

Example 4

A peg bag of mass 2 kg is suspended from two strings, one at 11° to the horizontal and the other at 7° to the horizontal. Find the magnitude of the tension in each string.

Solution

Start by representing the information in a diagram. Let the tensions be T_1 N and T_2 N.

By resolving horizontally and vertically, you can obtain two simultaneous equations in T_1 N and T_2 N.

Resolving horizontally:

R($\rightarrow$)

$$T_2 \cos 7° - T_1 \cos 11° = 0 \qquad ①$$

Resolving vertically:

R($\uparrow$)

$$T_1 \sin 11° + T_2 \sin 7° - 2g = 0 \qquad ②$$

Make T_1 the subject of equation ①.

$$T_1 = \frac{T_2 \cos 7°}{\cos 11°}$$

Substitute for T_1 in equation ②.

$$\frac{T_2 \cos 7°}{\cos 11°} \sin 11° + T_2 \sin 7° - 2g = 0$$

$$T_2 \left(\frac{\cos 7°}{\cos 11°} \sin 11° + \sin 7° \right) = 2g$$

$$T_2 = \frac{2g}{\frac{\cos 7°}{\cos 11°} \sin 11° + \sin 7°}$$

$$= 62.3 \text{ N}$$

$$T_1 = \frac{T_2 \cos 7°}{\cos 11°} = 62.3 \times \frac{\cos 7°}{\cos 11°} = 63.0 \text{ N}$$

The tensions are 62.3 N and 63.0 N.

Because the strings are at different angles to each other, and because you are not told that it is one single string, you need to label the tensions differently.

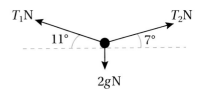

This is an example of solving simultaneous equations by substitution.

An angle is often given in the form of an inverse trigonometric function such as $\theta = \arcsin\left(\frac{a}{b}\right)$. Rather than working out the angle, it is more elegant in this situation to use a right-angled triangle to find the other trigonometric ratios you require, either with a Pythagorean triple or in surd form. This will save you writing down the angle, which will usually be an irrational number and will lead to possible errors if rounded. Note that in this context, the angles will all be acute, so their trigonometric ratios will be positive.

KEY INFORMATION

If an angle is given in the form of an inverse trigonometric function, use Pythagoras' theorem and the tangent ratio to find any other trigonometric ratios.

For example, if $\theta = \arcsin\frac{5}{13}$, sketch a right-angled triangle with an opposite of 5 and a hypotenuse of 13, since $\sin\theta = \frac{O}{H}$.

Using Pythagoras' theorem, the adjacent is given by

$$\sqrt{13^2 - 5^2} = \sqrt{144} = 12$$

Hence $\cos\theta = \frac{A}{H} = \frac{12}{13}$.

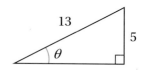

Similarly, if you were told that an angle was $\arctan 0.75$, you could rewrite 0.75 as $\frac{3}{4}$, so the opposite is 3 and the adjacent is 4 (since $\tan\theta = \frac{O}{A}$). This is a 3, 4, 5 triangle, so the hypotenuse is 5.

Hence $\sin\theta = \frac{3}{5}$ and $\cos\theta = \frac{4}{5}$.

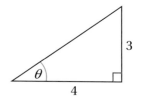

Example 5

A 2.5 kg pulley is suspended from two identical inextensible strings, both at $\arccos\frac{8}{17}$ to the horizontal.

Find the tension in each string.

Solution

Start by drawing the force diagram.

Let each tension be equal to T N and call the angle θ. Note that θ can be placed at either the pulley or the other end of the strings, and the angles are equal since they are alternate.

Resolving horizontally will not provide any additional information because the horizontal components of the tensions are equal and opposite.

Resolving vertically:

R($\uparrow$)

$$T\sin\theta + T\sin\theta - 2.5g = 0$$

Since the tensions are the same, this can simplified to

$$2T\sin\theta - 2.5g = 0$$

Hence, $$T = \frac{2.5g}{2\sin\theta}.$$

In this example, the angle θ has been given as $\arccos\frac{8}{17}$.

Sketch a right-angled triangle with an adjacent of 8 and a hypotenuse of 17, since $\cos\theta = \frac{A}{H}$.

Using Pythagoras' theorem, the opposite is given by

$$\sqrt{17^2 - 8^2} = \sqrt{225} = 15$$

Hence $\sin\theta = \frac{O}{H} = \frac{15}{17}$.

Thus $T = \frac{2.5g}{2 \times \frac{15}{17}} = 13.9$ N.

> Because the strings are identical and at the same angle, the tensions will also be the same.

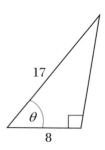

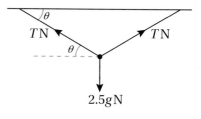

Exercise 16.1B

Answers page 485

 1 A 5 kg chair is at rest on a slope inclined at 28° to the horizontal, as shown in the diagram. The chair is held in position by friction.

Find the magnitude of:

a the reaction force

b the friction force.

 2 An M kg particle is suspended from two identical strings. Each string is inclined at an angle of arcsin $\frac{12}{35}$ to the horizontal.

Given that the tension in each string is 32 N, find the value of M.

  3 A van of mass 800 kg exerts a driving force of 2.4 kN as it travels at constant speed up a hill inclined at 15° to the horizontal. Show that the van experiences a frictional force of 371 N.

4 A particle of mass 3.8 kg, suspended from a rope inclined at 20° to the vertical, is kept in position by a horizontal force XN, as shown in the diagram. Victrix and Colin have been asked to find the value of XN, correct to 3 significant figures.

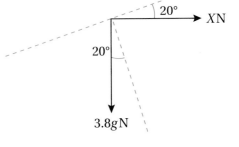

Victrix's solution is as follows:

$R(\uparrow)\ T\cos 20° = 3.8g$

$T = 3.8g \div \cos 20° = 39.630\,\text{N}$

$R(\rightarrow)\ X = T\sin 20° = 39.630 \times \sin 20° = 13.6\,\text{N}\ (3\,\text{s.f.})$

Colin's solution is as follows:

$R(\nearrow)$

$X\cos 20° - 3.8g\sin 20° = 0$

$X\cos 20° = 3.8g\sin 20°$

$X = 3.8g\tan 20° = 13.6\,\text{N}$

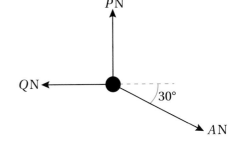

Compare and contrast the two solutions.

 5 A particle at rest on a horizontal table experiences three horizontal forces, PN, QN and AN.

Forces PN and QN are perpendicular.

Force AN acts at 30° to the line of action of Q, as shown in the diagram.

Q is 4 N larger than P.

a Explain why the mass of the particle is not required to find any of the forces P, Q or A.

b Find the magnitude of A.

 6 A case of mass 3 kg is dragged along a rough horizontal floor by a cord inclined at arcsin 0.2 to the horizontal. The case experiences a frictional force of $\sqrt{6}$ N.

Find the reaction exerted by the floor on the case.

(M) **7** A box of bricks weighs 120 N. The box is at rest on a rough plane inclined at an angle of arccos $\frac{40}{41}$ above the horizontal. A 20 N force acts on the particle up the line of greatest slope of the plane.

Find the magnitude and direction of the frictional force.

(M)
(PS) **8** The diagram shows a 1.5 kg peg bag suspended from two unequal strings at 55° and 35° to the horizontal, as shown.

Find each of the tensions.

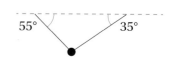

16.2 Adding forces

If two forces act upon a particle in different directions, then it is possible to add them together to find the resultant force.

There are two common methods to add forces.

The first method applies the triangle law you met in **Book 1, Chapter 10 Vectors**. If the forces are perpendicular then you can use Pythagoras' theorem and the tangent ratio. If the forces are not perpendicular then you will need to use the sine and cosine rules instead.

The second method involves resolving the forces into horizontal and vertical (**i** and **j**) components. Because **i** and **j** are always perpendicular, they can then be dealt with using Pythagoras' theorem and the tangent ratio.

Both methods are shown in **Example 6**.

If more than two forces act upon a particle, then it is possible to apply the triangle law as many times as necessary, an additional time for each additional force. However, it is more efficient to resolve all the forces into **i** and **j** components and use the second method.

Note that O$_x$ is an alternative name for the positive x-axis. If a force acts along the positive x-axis, it is said to act along O$_x$.

> **KEY INFORMATION**
>
> Forces can be added using the sine and cosine rules.
>
> Forces can be also be added by writing them using **i j** notation and applying Pythagoras' theorem.
>
> It is more efficient to use the **i j** notation method when there are several forces.

> The positive x-axis can be written as O$_x$.

Example 6

A particle is acted upon by two forces, P and Q. P is a 12 N force acting along O$_x$, whereas Q is a 10 N force which acts at 40° to O$_x$ as shown in the diagram.

Find the magnitude and direction of the resultant force, both correct to 1 decimal place.

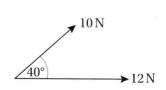

Solution

Method 1

In order to use the triangle law, start by redrawing the forces tip to tail as shown:

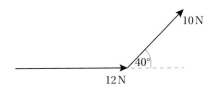

Because the 10 N force acts at 40° to O$_x$, the angle between the two forces is 140° since the two angles lie on a straight line. Thus the resultant force is the longest side in a triangle with two other sides of magnitude 12 and 10, with an angle of 140° between them, as shown. The resultant force can be calculated using the cosine rule.

Remember that angles on a straight line add up to 180°.

Let the resultant force be F N.

Hence, by the cosine rule:

$$F^2 = 12^2 + 10^2 - 2 \times 12 \times 10 \times \cos 140°$$

$$= 144 + 100 - 240 \cos 140°$$

$$= 427.85$$

$$F = \sqrt{427.85} = 20.684 \text{ N}$$

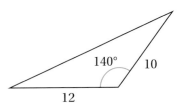

Correct to 1 decimal place, the magnitude of the resultant force is 20.7 N.

Now that the magnitudes of all three sides are known, the sine rule can be applied to find the angle the resultant force makes with O$_x$ (or the cosine rule could be applied again).

In order to apply the cosine rule to find the angle, it would normally be rearranged as $\cos A = \dfrac{b^2 + c^2 - a^2}{2bc}$.

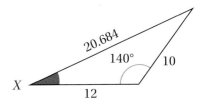

Let the angle the resultant force makes with O$_x$ be X and apply the sine rule:

$$\frac{\sin X}{10} = \frac{\sin 140°}{20.684}$$

$$\sin X = \frac{10 \sin 140°}{20.684} = 0.311$$

$$X = \sin^{-1}(0.311) = 18.105°$$

Correct to 1 decimal place, the direction of the resultant force is 18.1°.

Method 2

This method involves resolving the forces into **i** and **j** components.

The **i** component will consist of the 12 N force (since it acts along O$_x$), plus a component of the 10 N force.

The **j** component will just consist of a component of the 10 N force.

TECHNOLOGY

Make sure you use a sufficiently accurate value for the resultant force, or use the Ans button to use the previously acquired answer.

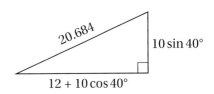

i: $(12 + 10\cos 40°)\,\text{N}$

j: $(10\sin 40°)\,\text{N}$

Apply Pythagoras' theorem to find the magnitude of the resultant force:

Magnitude $= \sqrt{(12 + 10\cos 40°)^2 + (10\sin 40°)^2} = 20.7\,\text{N}$

Apply the tangent ratio to find the angle of the resultant force:

Angle $= \tan^{-1}\left(\dfrac{10\sin 40°}{12 + 10\cos 40°}\right) = 18.1°$

Exercise 16.2A Answers page 485

(M) 1 Forces A and B act along O_x and O_y respectively, with magnitudes $15\,\text{N}$ and $8\,\text{N}$.

   a Find the magnitude of the resultant force.

   b Find the angle the resultant force makes with O_x.

(M) 2 A particle experiences a force of $23\,\text{N}$ south and of $38\,\text{N}$ west.

   a Find the magnitude of the resultant force.

   b Find the bearing of the resultant force.

(M) (PF) 3 Force G of magnitude $9\,\text{N}$ acts along O_x. Force H of magnitude $7\,\text{N}$ acts at $50°$ anticlockwise from O_x.

   a Show that the resultant force is $14.5\,\text{N}$.

   b Show that the resultant force acts at an angle of $21.7°$ to O_x.

(M) 4 Force U of magnitude $11\,\text{N}$ acts along O_x and force V of magnitude $7\,\text{N}$ acts at $110°$ anticlockwise from O_x, as shown in the diagram.

   Find the magnitude and direction of the resultant force.

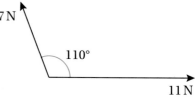

(PS) (CM) 5 Forces P and Q, of magnitudes $3\,\text{N}$ and $4\,\text{N}$, act along O_x and O_y respectively. Force R, of magnitude $5\,\text{N}$, acts at $60°$ to O_x as shown in the diagram.

   Hita has been asked to find the magnitude and direction of the resultant force.

   Hita has written out this solution:

$$a^2 = 3^2 + 5^2 - 2 \times 3 \times 5\,\cos 120° = 49$$

$$a = 7$$

$$\frac{\sin X}{5} = \frac{\sin 120°}{7}$$

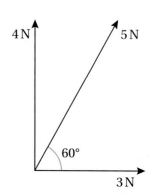

$$X = 38.2°$$

$$a^2 = 7^2 + 4^2 - 2 \times 7 \times 4 \cos 128.2°$$

$$\text{Magnitude} = 9.98 \, \text{N}$$

$$\frac{\sin Z}{4} = \frac{\sin 128.2°}{9.98}$$

$$Z = 18.4°$$

$$\text{Angle} = 56.6°$$

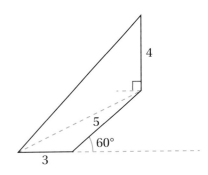

a Explain each part of Hita's solution.

b Write out a more efficient solution for this problem.

6 A boat is travelling north at $15 \, \text{m s}^{-1}$ and experiences wind pushing it at $7 \, \text{m s}^{-1}$ at an acute angle θ of arcsin 0.28 as shown in the diagram.

Show that the speed of the boat is $21.8 \, \text{m s}^{-1}$.

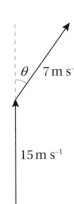

7 Force C of $18 \, \text{N}$ acts along O_x. Force D acts at $52°$ to the horizontal. The magnitude of the resultant force is $28 \, \text{N}$, as shown in the diagram.

a Find the magnitude of D.

b Find the angle that the resultant force makes with O_x.

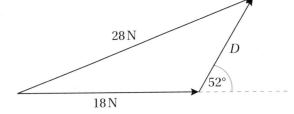

8 A force P acts along O_x. Forces Q, R and S act at $40°$, $130°$ and $220°$ respectively to O_x, each measured anticlockwise to O_x. The forces P and Q both have magnitude $3 \, \text{N}$ and the forces R and S both have magnitude $5 \, \text{N}$.

a Show that the resultant force acts at approximately $124°$ to O_x.

b Show that the magnitude of the resultant force is approximately $3.09 \, \text{N}$.

16.3 Coefficient of friction

Consider a particle at rest on a horizontal surface and what happens as the surface is tilted. Initially, the angle is zero, the particle is stationary and the forces are in equilibrium. As the angle increases, the weight of the particle has a component which should pull the particle down the slope, and the only force preventing the particle from slipping is friction. When the angle of the slope is more than 45°, the particle will inevitably slide or topple down the slope. As the angle increases from 0° to 45°, the frictional force required to prevent slippage increases. At a certain point, the particle will find itself on the point of slipping down the slope, and any further increase in angle will inevitably cause the particle to move down the slope.

There is a relationship between the friction and reaction forces which is dependent upon the angle, as well as on other factors such as the interaction between the particle and the slope (for example, a car and a hill). This relationship is given by $F \leqslant \mu R$, where μ is the **coefficient of friction**. μ always takes a value between 0 and 1.

When a particle is at rest on a horizontal surface with no other external forces, the reaction force is equal and opposite to the weight. The frictional force is zero, which is of course less than the reaction force. As long as F is less than μR, there will be no motion.

When the particle is on the point of slipping or has begun to move, $F = \mu R$. This is called **limiting equilibrium**.

It is possible to prove that for a system in limiting equilibrium, if no forces other than reaction, friction and weight act upon a particle on a rough slope inclined at θ to the horizontal, then $\mu = \tan \theta$.

Resolve parallel to the slope.

R($\nearrow$)

$$F = mg \sin \theta$$

Resolve perpendicular to the slope.

R($\nwarrow$)

$$R = mg \cos \theta$$

Substitute for F and R in $F = \mu R$.

$$mg \sin \theta = \mu \times mg \cos \theta$$

Hence $\qquad \mu = \dfrac{mg \sin \theta}{mg \cos \theta} = \tan \theta$

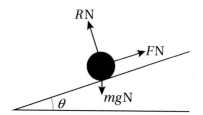

Example 7

A 2 kg particle is held in limiting equilibrium by a horizontal force of 14 N on a rough slope which is inclined at an angle of 17° to the horizontal. The particle is on the point of slipping up the slope. Calculate the coefficient of friction.

Solution

The force diagram has four forces: the weight of $2g$ N, the reaction force, the friction force and the horizontal 14 N force. Since the particle is on the point of slipping up the slope, the friction acts down the slope.

Since the particle is in limiting equilibrium, $F = \mu R$.

Resolve parallel to the slope to find an expression for FN.

$R(\nearrow)$

$$14 \cos 17° - F - 2g \sin 17° = 0$$

Therefore $\qquad F = 14 \cos 17° - 2g \sin 17°$

Resolve perpendicular to the slope to find an expression for RN.

$R(\nwarrow)$

$$R - 2g \cos 17° - 14 \sin 17° = 0$$

Therefore $\qquad R = 2g \cos 17° + 14 \sin 17°$

From $\quad F = \mu R, \; \mu = \dfrac{F}{R}$, so

$$\mu = \frac{14 \cos 17° - 2g \sin 17°}{2g \cos 17° + 14 \sin 17°} = 0.335$$

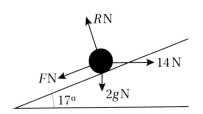

Check that $0 \leqslant \mu \leqslant 1$.

Exercise 16.3A

Answers page 486

M **1** A garden ornament of mass 8 kg is at rest on a concrete path inclined at 28° to the horizontal, as shown in the diagram. The ornament is in limiting equilibrium.

Find the coefficient of friction between the ornament and the path.

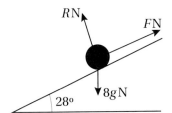

M **2** A car of mass 600 kg is driven up a road at a constant speed. The road is modelled as a rough surface inclined at 8° to the horizontal. Given that the coefficient of friction between the car and the road is 0.12, find the magnitude of the driving force exerted by the engine of the car.

M **PF** **3** A small suitcase of mass 5 kg is pulled along the floor in an airport at a steady speed by a light cord inclined at 20° to the horizontal, as shown in the diagram. The coefficient of friction between the suitcase and the floor is $\frac{1}{4}$.

Show that the tension in the cord is approximately 11.9 N.

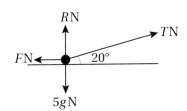

M **PS** **4** A cyclist of mass 65 kg is freewheeling at a steady speed down a hill inclined at an angle α to the horizontal on a bicycle of mass 28 kg.

Given that the coefficient of friction between the bicycle and the hill is 0.3, find the value of α, correct to 1 decimal place.

M **5** A body of mass 3 kg is held in limiting equilibrium on a rough plane inclined at 25° to the horizontal by a horizontal force PN.

Find PN when the body is on the point of slipping down the plane, given that $\mu = 0.15$.

6 A piano of mass m kg is being pushed up a rough slope at a constant speed. The pushing force of 560 N acts on the piano up the line of greatest slope of the plane. The coefficient of friction between the slope and the piano is $\frac{2}{7}$ and the slope is inclined at $\arcsin\frac{1}{10}$ to the horizontal.

Find the mass of the piano.

 7 A shopping trolley is pushed up a rough slope inclined at α to the horizontal. When the trolley is empty it has a mass of 17 kg and requires a pushing force of $4g$ N to travel at a steady speed.

a Show that the coefficient of friction is given by $\dfrac{4 - 17\sin\alpha}{17\cos\alpha}$.

b Given that $\tan\alpha = \frac{13}{84}$, show that the coefficient of friction is $\frac{1}{12}$.

The same trolley, now containing shopping, is pushed up the slope. A pushing force of $10g$ N is required.

c Find the mass of the shopping in the trolley.

 8 A cart is on the point of slipping down a rough lane inclined at an angle of 30° to the horizontal. The coefficient of friction between the cart and the lane is $\frac{2}{9}$. The cart is held in position by a light inextensible cord inclined at 15° to the lane, as shown in the diagram.

The cart is modelled as a particle with a weight of 210 N.

Find the magnitude of the tension in the cord.

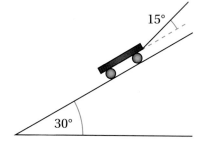

Stop and think   Compare the value of μ for a particle in limiting equilibrium on slopes of different inclinations, such as 10°, 20°, 30° and 40°. Given that $\mu \leqslant 1$, explain why the slope cannot be inclined at 50°.

Consider also a particle on a rough horizontal floor which is pulled by a rope inclined at 15° to the horizontal. How does increasing the value of μ from $\frac{1}{4}$ to $\frac{1}{2}$ affect the frictional force?

Recall Newton's second law, which states that $F = ma$, where F is the resultant force in the direction of motion, m is the mass of the particle and a is the acceleration of the particle. When the forces are in equilibrium in a particular direction, the sum of the forces in that direction is zero. However, if the forces are not in equilibrium, the sum will be equal to ma and the particle will be accelerating in that direction.

KEY INFORMATION

If the forces are not in equilibrium, the sum will be equal to ma and the particle will be accelerating in that direction.

Problems which require the use of the formula $F = ma$ are often combined with the equations of constant acceleration (SUVAT). Additionally, sometimes a force will be removed or the angle will be increased. The value of μ will not change but $F = \mu R$ will then be valid for new values of F and R.

Example 8

A horizontal force of 2 N is just sufficient to prevent a block of mass 1 kg from sliding down a rough plane inclined at an angle of arcsin $\frac{7}{25}$ to the horizontal.

a Find the value of the coefficient of friction.

The horizontal force is removed.

b Hence find the acceleration with which the block will now move.

Solution

a Draw a diagram to represent this information.

This force diagram has four forces: the weight of gN, the reaction force, the friction force and the horizontal 2 N force. Since the block is on the point of sliding down the plane, the friction acts up the plane. Let the angle of inclination be θ.

The phrase 'just sufficient to prevent' indicates that the block is in limiting equilibrium and hence that $F = \mu R$.

Resolve parallel to the slope to find an expression for FN.

$R(\nearrow)$

$$F + 2\cos\theta - g\sin\theta = 0$$

Therefore $\qquad F = (g\sin\theta - 2\cos\theta)\,\text{N}$

Resolve perpendicular to the slope to find an expression for RN.

$R(\nwarrow)$

$$R - g\cos\theta - 2\sin\theta = 0$$

Therefore $\qquad R = (g\cos\theta + 2\sin\theta)\,\text{N}$

$$\mu = \frac{F}{R} = \frac{g\sin\theta - 2\cos\theta}{g\cos\theta + 2\sin\theta}$$

θ is given as arcsin $\frac{7}{25}$. Hence $\sin\theta = \frac{7}{25}$ and you need to find an expression for $\cos\theta$. Sketch a right-angled triangle with an opposite of 7 and a hypotenuse of 25.

Hence $\sqrt{25^2 - 7^2} = \sqrt{576} = 24$ and $\cos\theta = \frac{A}{H} = \frac{24}{25}$.

From which $\mu = \dfrac{g\left(\frac{7}{25}\right) - 2\left(\frac{24}{25}\right)}{g\left(\frac{24}{25}\right) + 2\left(\frac{7}{25}\right)} = \dfrac{7g - 48}{24g + 14} = 0.0827$

b Draw a new diagram, without the horizontal force, to represent the change in information. Add in a double-headed arrow to indicate that the block will now accelerate down the plane.

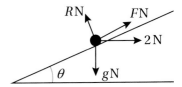

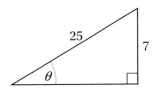

Look out for different ways of expressing the same information for limiting equilibrium, such as 'just sufficient to prevent', 'on the point of slipping' and 'about to move'.

If the original situation is adapted, then it is a good idea to draw a new force diagram.

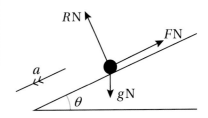

Resolve down the slope in the direction the block will move.

R($\swarrow$)

$$g\sin\theta - F = ma = a$$

In order to find the acceleration it is necessary to find an expression for FN.

Since $F = \mu R$, resolve perpendicular to the plane to find a new expression for RN.

R($\nwarrow$)

$$R - g\cos\theta = 0$$

Therefore $\quad R = g\cos\theta = \frac{24}{25}g$N

Use the value of μ obtained in **part a**, 0.0827.

$$F = 0.0827 \times \tfrac{24}{25}g = 0.778\,\text{N}$$

Hence $a = g\sin\theta - F = \frac{7}{25}g - 0.778 = 1.97\,\text{m s}^{-2}$.

> Note that since the block has a mass of 1 kg, ma can be rewritten as a.

Example 9

A particle of mass 5 kg is just prevented from slipping down a slope by a 3 N force acting parallel to the slope. The slope is inclined at 20° to the horizontal.

Find:

a the coefficient of friction

b the acceleration when the 3 N force is removed

c how long it takes the particle to roll 10 m.

Solution

a Draw the force diagram.

Resolve parallel to the slope.

R($\nearrow$)

$$F + 3 - 5g\sin 20° = 0$$

So $\qquad\qquad F = (5g\sin 20° - 3)\,\text{N}$

Resolve perpendicular to the slope.

R($\nwarrow$)

$$R - 5g\cos 20° = 0$$

Therefore $\qquad R = 5g\cos 20°\,\text{N}$

$$\mu = \frac{F}{R} = \frac{5g\sin 20° - 3}{5g\cos 20°} = 0.299$$

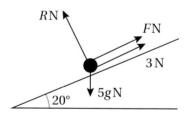

b Draw an updated force diagram.

Resolve parallel to the slope in the direction of motion.

R($\swarrow$)

$$5g\sin 20° - F = 5a$$

The reaction force RN will not have changed, since the 3 N force that was removed was parallel to the slope.

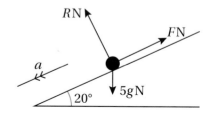

Hence $R = 5g\cos 20°$ and $F = \mu R = 0.299 \times 5g\cos 20°$.

$5a = 5g\sin 20° - 0.299 \times 5g\cos 20° = 3$

$a = \frac{3}{5} = 0.6\,\mathrm{m\,s^{-2}}$

c Since $a = 0.6\,\mathrm{m\,s^{-2}}$, $u = 0\,\mathrm{m\,s^{-1}}$ and $s = 10\,\mathrm{m}$, a SUVAT equation can be used to find the length of time it takes the particle to roll 10 m.

Substitute in $s = ut + \frac{1}{2}at^2$.

$10 = 0 + \frac{1}{2} \times 0.6t^2$

$t^2 = \dfrac{100}{3}$

$t = 5.77\,\mathrm{s}$

Exercise 16.3B

Answers page 486

(M) **1** A ball of mass 3 kg is released from rest on a rough surface inclined at arcsin 0.6 to the horizontal. After 2.5 s the ball has acquired a velocity of $4.9\,\mathrm{m\,s^{-1}}$ down the surface.

Find the coefficient of friction between the ball and the surface.

(M) **2** A body of mass 5 kg is initially at rest at the bottom of a rough inclined plane of length 6.3 metres. The plane is inclined at 30° to the horizontal.

The coefficient of friction between the body and the plane is $\dfrac{1}{2\sqrt{3}}$.

A constant horizontal force of $35\sqrt{3}$ N is applied to the body, causing it to accelerate up the plane.

Find:

a the time taken for the body to reach the top of the plane

b the speed of the body on arrival at the top of the plane.

(M) **3** A box of mass 5 kg is pulled along rough horizontal ground by a rope inclined at arcsin $\frac{44}{125}$ to the ground. When the tension in the rope is 10 N, the box is moving at constant speed.

a Calculate the coefficient of friction between the box and the ground.

b Calculate the acceleration of the box if the tension is increased to 15 N.

(M) (PS) **4** A particle of mass 2 kg rests in limiting equilibrium on a plane inclined at 25° to the horizontal. The angle of inclination is decreased to 20° and a force of magnitude 20 N is applied up a line of greatest slope.

a Find the particle's acceleration.

When the particle has been moving for 2 seconds the force is removed.

b Determine the further distance the particle will move up the plane.

(M) (PF) **5** A body of mass 500 g is placed on a rough plane which is inclined at 40° to the horizontal. The coefficient of friction between the body and the plane is 0.6.

a Find the frictional force.

b Show that motion will occur, stating the acceleration of the body.

6 A horizontal force of 4 N is just sufficient to prevent a block of mass 2 kg from sliding down a rough plane inclined at 23° to the horizontal. If the 4 N force is replaced by a force of X N, then the block is on the point of moving up the plane.

a Find the value of X.

b Find the acceleration if the X N force acts parallel to the slope rather than horizontally.

16.4 Connected particles

This section extends the material from **Book 1, Chapter 16 Forces** on connected particles (such as cars towing trailers or pulleys) by applying the two main new features from this chapter: resolving forces at angles into components and using the formula $F = \mu R$.

Consider a car towing a trailer along a rough horizontal road. A typical force diagram would look like this:

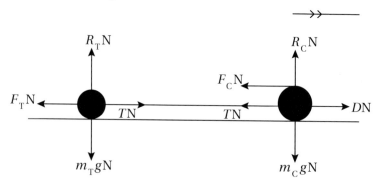

The car is towing the trailer to the right. The engine of the car is exerting a driving force. The tension in the tow-bar is represented in the usual way, with equal and opposite tensions from each of the vehicles. Each vehicle has its own weight, friction and reaction forces labelled with subscripts (C for car and T for trailer), with the friction and reaction usually related by $F = \mu R$.

Note that if the vehicles become uncoupled, they act as if they are no longer connected. In this situation, there are two stages to the motion.

The first stage is when the vehicles are coupled and the particles are accelerating together.

The second stage is when the vehicles are uncoupled and the one that is being pulled is slowed down by friction and/or gravity.

It is important to understand how the two stages differ. When the vehicles are uncoupled, the tension force is removed and, as a result, the vehicle which is producing the driving force will find itself accelerating at a faster rate (since the resultant force is greater), and the vehicle which is being pulled will find itself decelerating as a result of friction and/or its weight. In real life, there will always be a frictional force, however smooth a surface

> **KEY INFORMATION**
>
> If the connection between two particles is impeded, such as vehicles becoming uncoupled or a string going slack or breaking, the particles act as if they are no longer connected.

might appear. It is also worth noting that the final velocity for both vehicles from the first stage will be the initial velocity for both vehicles for the second stage.

Example 10

Two particles, A and B, of masses 2 kg and 3 kg respectively, are connected by a light inextensible 7 m string. A 15 N force is applied to particle B, which accelerates the particles in the direction shown along a rough table. The coefficient of friction between the particles and the table is $\frac{1}{4}$.

a Find:

   i the initial acceleration

   ii the tension in the string.

8 seconds later, particle B hits a wall and the string goes slack.

b How much longer does it take for A to stop?

c How much further does A travel before it stops?

d If the string had been 3 m in length instead, what difference would this have made?

Solution

a **i** Label the forces on the diagram. There are two weights, two tensions, two reactions, two frictions and the 15 N force.

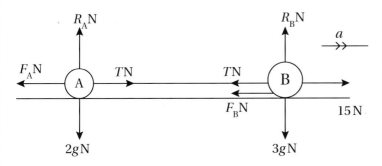

By resolving in the direction of motion for each particle in turn, you obtain simultaneous equations.

Resolve horizontally for B.

$R(\rightarrow)$

$$F = ma$$

$$15 - T - F_B = 3a \qquad \textcircled{1}$$

> Always resolve in the direction of motion when using $F = ma$ (when $a \neq 0\,\mathrm{m\,s^{-2}}$).

Resolve horizontally for A.

R($\rightarrow$)

$$F = ma$$

$$T - F_A = 2a \qquad \textcircled{2}$$

Add equations $\textcircled{1}$ and $\textcircled{2}$ to eliminate T.

$$15 - F_A - F_B = 5a$$

Expressions for F_A and F_B can be determined by resolving vertically and applying $F = \mu R$. Note that there is no acceleration vertically.

Resolve vertically for B.

R($\uparrow$)

$$R_B - 3g = 0$$

$$R_B = 3g\,\text{N}$$

Substitute in $F = \mu R$.

$$F_B = \tfrac{1}{4} \times 3g = \tfrac{3}{4}g\,\text{N}$$

Similarly, $\quad F_A = \tfrac{1}{4} \times 2g = \tfrac{1}{2}g\,\text{N}$

Therefore $\quad 15 - \tfrac{1}{2}g - \tfrac{3}{4}g = 5a$

$$5a = 15 - 4.9 - 7.35 = 2.75$$

$$a = 0.55\,\text{m s}^{-2}$$

The initial acceleration is $0.55\,\text{m s}^{-2}$.

ii From equation $\textcircled{2}$: $\quad T - F_A = 2a$

$$T - \tfrac{1}{2}g = 2 \times 0.55$$

$$T = 2 \times 0.55 + 4.9 = 6\,\text{N}$$

b During the first stage of the motion, $u = 0\,\text{m s}^{-1}$, $a = 0.55\,\text{m s}^{-2}$ and $t = 8\,\text{s}$.

Substitute into $v = u + at$ to find the final velocity for the first stage (and initial velocity for the second stage).

$$v = 0 + 0.55 \times 8$$

$$= 4.4\,\text{m s}^{-1}$$

During the second stage of the motion for A, $u = 4.4\,\text{m s}^{-1}$ and $v = 0\,\text{m s}^{-1}$.

To find the acceleration for the second stage, resolve horizontally again for A but without the tension force.

R($\rightarrow$)

$$F = ma$$

$$-F_A = 2a$$

The reaction force has not changed so F_A is still $\tfrac{1}{2}g\,\text{N}$.

$$-\tfrac{1}{2}g = 2a$$

$$a = -2.45\,\text{m s}^{-2}$$

> Once B hits the wall, the string will go slack. In order to answer the questions in **parts b, c** and **d**, it is necessary to consider the two stages of motion.

> Make sure you remember all the SUVAT equations.

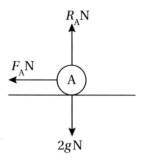

16.4

Substitute into $v = u + at$ to find the time taken for A to stop.

$$t = \frac{v-u}{a}$$

$$= \frac{0 - 4.4}{-2.45}$$

$$= 1.80\,\text{s}$$

c Substitute $u = 4.4\,\text{m s}^{-1}$, $v = 0\,\text{m s}^{-1}$ and $a = -2.45\,\text{m s}^{-2}$ into $v^2 = u^2 + 2as$ to find the displacement for the second stage.

$$s = \frac{v^2 - u^2}{2a}$$

$$= \frac{0 - 4.4^2}{2 \times -2.45}$$

$$= 3.95\,\text{m}$$

d If the string had only been 3 m in length, then A would have hit B and the wall, since 3 m is less than the 3.95 m A would have travelled if the wall had not been there.

Note that you could use $s = ut + \frac{1}{2}at^2$ but this assumes that the answer to **part b** for the time is correct (and accurate). It is usually better to use the information you are given where possible rather than relying on a previous answer.

Now consider a car towing a trailer up a rough hill. The force diagram would look like this:

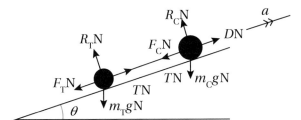

The main difference between this situation and that on a horizontal road is that the weight forces will need to be resolved into components.

Resolve for the car and trailer parallel to the slope.

R(↗) for car

$$D - T - F_C - m_C g\sin\theta = m_C a$$

R(↗) for trailer

$$T - F_T - m_T g\sin\theta = m_T a$$

Resolve for the car and trailer perpendicular to the slope.

R(↖) for car

$$R_C = m_C g\cos\theta$$

R(↖) for trailer

$$R_T = m_T g\cos\theta$$

Example 11

A van of mass 500 kg pulls a trailer of mass 200 kg up a hill inclined at 11° to the horizontal. The coefficient of friction between each vehicle and the hill is $\frac{3}{20}$. Find the driving force exerted by the engine if the van is accelerating at $1.5\,\mathrm{m\,s^{-2}}$.

Solution

Start by drawing the force diagram.

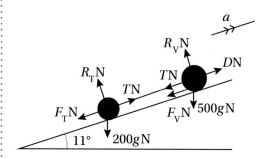

Resolve parallel for the van in the direction of motion.

R($\nearrow$) for van

$$F = ma$$

$$D - T - F_V - 500g\sin 11° = 500 \times 1.5 \qquad ①$$

Resolve parallel for the trailer.

R($\nearrow$) for trailer

$$T - F_T - 200g\sin 11° = 200 \times 1.5 \qquad ②$$

Add equations ① and ②.

$$D - F_V - F_T - 500g\sin 11° - 200g\sin 11° = 700 \times 1.5 \qquad ③$$

Find expressions for F_A and F_B by resolving perpendicular to the hill and applying $F = \mu R$.

Resolve perpendicular for the van and trailer.

R($\nwarrow$)

$$R_V - 500g\cos 11° = 0, \text{ so } F_V = \tfrac{3}{20} \times 500g\cos 11°$$

$$R_T - 200g\sin 11° = 0, \text{ so } F_T = \tfrac{3}{20} \times 200g\cos 11°$$

Substitute in equation ③.

$$D - \tfrac{3}{20} \times 500g\cos 11° - \tfrac{3}{20} \times 200g\cos 11° -$$
$$500g\sin 11° - 200g\sin 11° = 700 \times 1.5$$

$$D - 2319 = 1050$$

$$D = 3370\,\mathrm{N} \text{ (to 3 significant figures)}$$

The driving force exerted by the engine is 3370 N.

> Round answers to mechanics problems to 2 or 3 significant figures unless asked to do otherwise.

In **Book 1, Chapter 16 Forces**, pulleys involved horizontal and vertical strings. This can now be extended to situations with angles and limiting equilibrium.

The force diagrams below show the difference between the situations for a smooth slope and a rough slope for a particle B of mass $2m$ kg suspended freely from a pulley and a particle A of mass m kg on the slope.

Smooth slope

R($\nearrow$) for A

$$T - mg\sin\theta = ma$$

R($\downarrow$) for B

$$2mg - T = 2ma$$

Add.

$$2mg - mg\sin\theta = 3ma$$

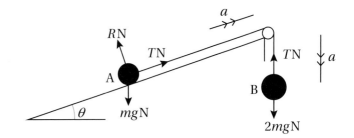

Rough slope

R($\nearrow$) for A

$$T - mg\sin\theta - F = ma$$

R($\nwarrow$) for A

$$R - mg\cos\theta = 0$$

R($\downarrow$) for B

$$2mg - T = 2ma$$

Add.

$$2mg - mg\sin\theta - F = 3ma$$
$$2mg - mg\sin\theta - \mu mg\cos\theta = 3ma$$

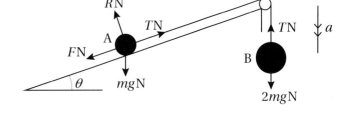

Example 12

Particle A of mass 2 kg lies on a rough plane inclined at an angle of $\arccos\frac{8}{17}$. The coefficient of friction between A and the plane is $\frac{1}{4}$. Particle A is connected to particle B of mass 3 kg by a light inextensible string which lies along a line of greatest slope of the plane and passes over a smooth pulley. The system is initially held at rest with B hanging vertically 2 m above the ground.

When the system is released, find:

a the initial acceleration of B

b the tension

c the total distance travelled by A before it comes to rest.

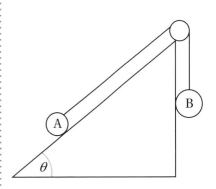

Solution

a Label the diagram with its six forces and two accelerations. Since only one particle lies on the slope, there is no need for subscripts.

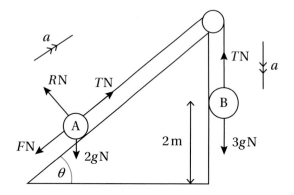

Resolve parallel to the plane in the direction of motion of A to find an equation in terms of TN, FN and a.

R($\nearrow$)

$$T - 2g\sin\theta - F = 2a \qquad \textcircled{1}$$

Resolve vertically downwards in the direction of motion of B to find an equation in terms of TN and a.

R($\downarrow$)

$$3g - T = 3a \qquad \textcircled{2}$$

Add equations $\textcircled{1}$ and $\textcircled{2}$.

$$3g - 2g\sin\theta - F = 5a \qquad \textcircled{3}$$

Resolve perpendicular for A to find the reaction force.

R($\nwarrow$)

$$R - 2g\cos\theta = 0$$

So $\qquad F = \mu R = \tfrac{1}{4} \times 2g\cos\theta$

Substitute in equation $\textcircled{3}$.

$$3g - 2g\sin\theta - \tfrac{1}{4} \times 2g\cos\theta = 5a$$

In **Example 5** it was shown by sketching a right-angled triangle and applying Pythagoras' theorem that if $\cos\theta = \tfrac{8}{17}$, then $\sin\theta = \tfrac{15}{17}$.

So $\quad 3g - 2g \times \tfrac{15}{17} - \tfrac{1}{4} \times 2g \times \tfrac{8}{17} = 5a$

$$3g - \tfrac{30}{17}g - \tfrac{4}{17}g = 5a$$

$$5a = g$$

$$a = \tfrac{1}{5}g = 1.96\,\mathrm{m\,s^{-2}}$$

The initial acceleration of B is $1.96\,\mathrm{m\,s^{-2}}$.

b From equation ②, $3g - T = 3a$

$$T = 3g - 3a$$

$$= 3 \times 9.8 - 3 \times 1.96$$

$$= 23.52$$

The tension is $23.5\,\text{N}$ correct to 3 significant figures.

c There are two stages to A's motion before it comes to rest – the first stage when it is pulled by B until B hits the ground and the second stage when it is slowed down to instantaneous rest by friction and gravity.

During the first stage, $a = 1.96\,\text{m s}^{-2}$, $s = 2\,\text{m}$ and $u = 0\,\text{m s}^{-1}$.

Substitute into $v^2 = u^2 + 2as$ to find the final velocity (and the initial velocity for the second stage).

$$v^2 = 0 + 2 \times 1.96 \times 2$$

$$= 7.84$$

$$v = \sqrt{7.84} = 2.8\,\text{m s}^{-1}$$

For the acceleration during the second stage, resolve again for A but without tension. The reaction, and hence the friction, will be unchanged.

R($\nearrow$)

$$-2g \times \tfrac{15}{17} - \tfrac{1}{4} \times 2g \times \tfrac{8}{17} = 2a$$

$$2a = -2g$$

$$a = -g = -9.8\,\text{m s}^{-2}$$

So during the second stage, $a = -9.8\,\text{m s}^{-2}$, $u = 2.8\,\text{m s}^{-1}$ and $v = 0\,\text{m s}^{-1}$.

Substitute into $v^2 = u^2 + 2as$ to find the displacement during the second stage.

$$s = \frac{v^2 - u^2}{2a}$$

$$= \frac{0 - 2.8^2}{2 \times -9.8}$$

$$= 0.4\,\text{m}$$

Combining the $2\,\text{m}$ travelled during the first stage, the total distance travelled by A is $2.4\,\text{m}$.

Draw a new force diagram because the situation has changed.

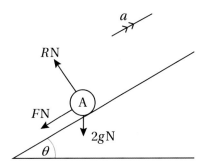

Make sure you read the question carefully so you know whether you need the total distance or the further distance.

Exercise 16.4A
Answers page 487

 1 A car of mass 400 kg pulls a horsebox of mass 600 kg up a rough hill inclined at arcsin 0.2. The resistances to motion are 400 N for the car and 600 N for the horsebox. The car exerts a driving force of 3.2 kN.

   a Find the acceleration.

   The car starts at rest and becomes uncoupled from the horsebox after travelling 75 m.

   b Find the new initial acceleration of the car.

   c Show that the horsebox travels just over 6 m after the uncoupling.

 2 Two particles of masses 7 kg and 4 kg on a rough horizontal table are connected by a light inextensible 50 cm string. The heavier particle is projected away from the lighter particle by a force of 35 N. The numerical value of the coefficient of friction between each particle and the table is one-twentieth of its mass. After 4 seconds, the heavier particle hits a wall.

   a Does the lighter particle collide with the heavier particle?

   b If it does, how much longer does it take the lighter particle to reach the heavier particle; if it does not, how far apart do they end up?

 3 A van pulls a cart along a rough horizontal road. The mass of the van is 900 kg and the mass of the cart is 500 kg. The van and cart experience frictional forces of 800 N and 400 N respectively. The tow-bar coupling the van and cart together is inclined at 12°. The van exerts a driving force of 2.25 kN.

   a Find the acceleration of the van.

   b Show that the magnitude of the tension in the tow-bar is 790 N, correct to 2 significant figures.

 4 A particle J is at rest on a rough horizontal table connected by a light inextensible string via a smooth pulley at the edge of the table to a particle K, which is suspended 3 m above the ground. Particles J and K have masses of 2 kg and 3.2 kg respectively. The coefficient of friction between particle J and the table is 0.3. K is released from rest with the string tense.

   a Find the friction experienced by particle J.

   b Find the acceleration of J and K.

   c Find the tension in the string.

   d Assuming that J does not meet the pulley, how far does it travel altogether before it stops?

 5 A tractor of weight $600g$ N pulls a combine harvester of weight $400g$ N up a slope inclined at 10°. The vehicles experience frictional forces of 500 N and 250 N respectively. The tractor exerts a driving force of 2.6 kN.

   a Find the initial acceleration of each vehicle.

   b Find the tension in the coupling.

   They start from rest, and after 8 s, the vehicles become uncoupled.

   c Show that the combine harvester continues to move for just over half a second before it comes to instantaneous rest.

6 Two particles, X and Y, of masses 700 g and 800 g respectively, are connected by a light inextensible cord. X is at rest on a smooth slope inclined at an angle of arcsin $\frac{1}{4}$ to the horizontal. The cord passes over a smooth light pulley fixed at the top of the slope. Y hangs freely from the pulley.

The system is released from rest with the string taut. Calculate:

a the acceleration of Y

b the tension in the cord.

Y hits the ground after 3 s.

c Find the distance travelled by X before Y hits the ground.

In the subsequent motion, X does not reach the pulley.

d Find the further distance travelled by X before it comes to instantaneous rest.

7 Two particles, A and B, of masses 1.8 kg and 5.2 kg respectively, are connected by a light inextensible string passing over a fixed smooth light pulley. Particle A is placed on a rough plane inclined at 30° to the horizontal. The coefficient of friction between particle A and the plane is $\frac{\sqrt{3}}{9}$. Particle B hangs freely below the pulley. The particles are released from rest with the string taut.

a Show that the acceleration of the particles when released is given by $\frac{4}{7}g\,\text{m s}^{-2}$.

b Calculate the tension in the string.

Particle B strikes the floor after 2 s.

c Find the total time from when the particles were released until A comes to instantaneous rest (assuming it does not reach the pulley).

SUMMARY OF KEY POINTS

> Resolving in a particular direction obtains the resultant force in that direction. It is common to resolve parallel and perpendicular to the surface. The component adjacent to the angle is given by $F\cos\theta$ and the component opposite the angle is given by $F\sin\theta$.

> If a system is in equilibrium, then the resultant force in any direction is zero. If a system is not in equilibrium, then the sum of the forces in the direction it is moving will be equal to ma (i.e. $F = ma$).

> Weight acts vertically downwards.

> The reaction force acts perpendicular to the surface.

> Friction acts to oppose the motion, so if a particle is on the point of slipping down, then friction acts up but if the particle is on the point of moving up the slope, then the friction acts down.

> In general, $F \leqslant \mu R$ (where μ is the coefficient of friction and $0 \leqslant \mu \leqslant 1$). When the particle is on the point of slipping or has begun to move, $F = \mu R$. This is called the limiting equilibrium.

> If an angle is given in the form of an inverse trigonometric function, use Pythagoras' theorem and the tangent ratio to find any other trigonometric ratios.

> If the connection between two particles is impeded, such as vehicles becoming uncoupled, the vehicles act as if they are no longer connected.

> Forces can be added using the sine and cosine rules or by writing them using **i j** notation and applying Pythagoras' theorem. It is more efficient to use the **i j** notation method when there are several forces. The positive x-axis can be written as O_x.

EXAM-STYLE QUESTIONS 16

Answers page 488

(M) **1** A wardrobe is at rest on a rough slope inclined at arcsin $\frac{5}{13}$ to the horizontal. The wardrobe is on the point of slipping down the slope. Find the coefficient of friction between the wardrobe and the slope. **[4 marks]**

(M) (CM) **2** A peg bag of weight 1.8 N hangs in equilibrium from a washing line (made from thick rope). The rope on either side of the bag hangs at angles of 5° and 7° to the horizontal.

    a Describe a suitable model for this situation, including at least two modelling assumptions. **[2 marks]**

    b Find the tension in each part of the rope. **[5 marks]**

(M) (PS) **3** A book of mass 600 g lies on a smooth plane inclined at an angle of $\beta°$ to the horizontal, where $\beta = \arctan\frac{3}{4}$. The book is held in equilibrium by a horizontal force of magnitude G N. Calculate the value of G, correct to 1 significant figure. **[4 marks]**

(M) **4** Force J of magnitude 6 N acts along O_x (the positive x-axis) and force K of magnitude 9 N acts along O_y (the positive y-axis). Force L of magnitude 10 N acts at 50° to O_x as shown in the diagram.

Find:

    a the magnitude of the resultant force **[4 marks]**

    b the angle the resultant force makes with O_x. **[2 marks]**

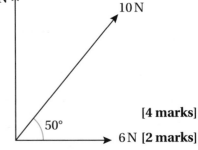

5 A car of mass 750 kg pulls a truck of mass 400 kg along a horizontal track by means of a rope. The driving force exerted by the engine of the car is 5428 N. The coefficient of friction between each vehicle and the track is $\frac{2}{5}$. Initially the vehicles are stationary.

  a Find the acceleration of the car. **[3 marks]**

  b What modelling assumption is made if the accelerations of the car and the truck are considered to be equal? **[1 mark]**

  c Find the tension in the rope. **[2 marks]**

When the car has travelled 90 m:

  d find the speed of the car **[3 marks]**

  e how much time has elapsed? **[2 marks]**

6 A particle of mass m kg is held in position on a rough slope by a horizontal force of 40 N. The coefficient of friction between the particle and the slope is $\frac{1}{3}$. The slope is inclined at an angle of 35° to the horizontal. The particle is on the point of slipping up the slope.

Find the value of m. **[8 marks]**

7 A lorry pulls a cement mixer along a horizontal path. Due to the relative sizes of the vehicles, the coupling is inclined at 9° to the horizontal. The lorry has a mass of 1600 kg and experiences a frictional force of 1000 N. The cement mixer has a mass of 800 kg and experiences a frictional force of 400 N. The lorry sets off from a stationary position with a driving force of 3.2 kN.

  a Find the initial acceleration of the lorry and cement mixer. **[2 marks]**

  b Find the tension in the coupling. **[2 marks]**

After travelling 24 m, the coupling becomes disconnected.

  c Assuming the driving force remains the same, find the new acceleration of the lorry. **[3 marks]**

  d Find the new acceleration of the cement mixer. **[2 marks]**

  e Find the distance travelled before the cement mixer rolls to a stop. **[2 marks]**

8 A particle of mass 250 g is released from rest at the top of a rough plane which is inclined at arccos 0.8 to the horizontal. The coefficient of friction between the particle and the plane is $\frac{3}{7}$. The plane is of length 2.5 m.

Find the speed of the particle on reaching the bottom of the slope (in terms of g). **[8 marks]**

9 A lawnmower of mass M kg is at rest on a grass slope inclined at an angle of θ to the horizontal. The lawnmower cord is also inclined at an angle of θ to the slope. The grass slope is rough and the coefficient of friction between the lawnmower and the grass is given by μ. The lawnmower is on the point of slipping down the grass slope.

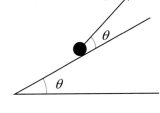

Show that the tension in the cord is given by $\dfrac{Mg(\mu\cos\theta - \sin\theta)}{\mu\sin\theta - \cos\theta}$. **[10 marks]**

10 A parcel C of mass 2 kg is at rest on a smooth slope inclined at an angle of arctan 0.75 to the horizontal. C is connected to another parcel D of mass 3 kg by means of a light inextensible rope and a smooth pulley at the top of the slope. D is suspended above the ground.

a How and why are C and D modelled? [1 mark]

b Find the initial acceleration and tension. [4 marks]

After two seconds the rope snaps.

c How long does it take C to return to its original position after the rope snaps? [3 marks]

11 Particles P and Q, of masses 2.1 kg and 1.5 kg respectively, are attached to the ends of a light inextensible taut string which passes over a smooth pulley fixed at the edge of a rough horizontal table. Initially, P is on the table 2.5 m from the pulley and Q hangs freely below the pulley, 1 m above the ground. The coefficient of friction between P and the table is given by μ. The system is released from rest.

a Show that the initial acceleration of P is given by $\dfrac{(5-7\mu)g}{12}$ m s^{-2}. [6 marks]

After Q hits the ground, P continues to move on the table, coming to rest before reaching the pulley.

b Show that $\frac{1}{5} < \mu < \frac{5}{7}$. [7 marks]

12 A, B and C are horizontal forces acting upon a particle on a horizontal plane. A and B are perpendicular and A makes an angle of α with the vector $-\mathbf{i}$. C acts along the vector $-\mathbf{j}$. The three forces are in equilibrium.

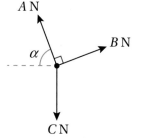

Given that $A = 72$ N and $\alpha = \arcsin \frac{24}{25}$:

a find the magnitude of force C [3 marks]

b find the magnitude of force B. [2 marks]

13 A block of mass 5 kg is resting on a rough slope inclined at an angle of α to the horizontal. There is a horizontal force X N acting upon the block such that the block is on the point of sliding down the slope. The coefficient of friction between the block and the slope is μ.

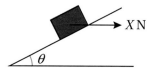

Show that $X = \dfrac{5g(\tan\alpha - \mu)}{1 + \mu\tan\alpha}$. [7 marks]

14 A car is towing a milk float up a hill inclined at an angle of β to the horizontal. The masses of the car and the milk float are 400 kg and 600 kg respectively. The engine of the car exerts a driving force of 3.14 kN. Each vehicle experiences a frictional force of 100g N and they accelerate at 0.2 m s^{-2}.

Find:

a the size of β, correct to 1 decimal place [6 marks]

b the coefficient of friction between the car and the road. [3 marks]

17 MOMENTS

The invention of the wheel is justifiably considered to be one of mankind's greatest achievements, but it was the observation that it could rotate about an axle that allowed technological advancements such as the cart or modern transportation. Similarly, we may take hinges for granted, but they allow us to use rotation to make opening doors easier.

Turning moments are a key feature of many tools because they make work easier. For example, spanners are more efficient than the human hand at tightening nuts. The turning moment is a product of the force applied and the distance from the pivot, so the same force creates a larger turning moment with a spanner, since the force is applied further away than if you turned the nut by hand.

However, turning moments are not only used for solving problems. One of the most enduring of playground activities, the seesaw, is an example of turning moments in action. The closer you sit to the central pivot, the more effort is required to bounce. If an adult and a young child sit at each end, then the adult will inevitably be sitting almost on the ground whilst the child will be several feet up in the air. In order to balance the turning moments, the adult needs to sit closer to the central pivot.

LEARNING OBJECTIVES

You will learn how to:

> take moments about a point

> model a real-life situation

> solve problems using a combination of turning moments and resolving forces

> solve problems involving ladders, hinges and other examples of rigid bodies in equilibrium.

TOPIC LINKS

This is the third and final chapter on mechanics in **Book 2**, following on from **Chapter 15 Kinematics** and **Chapter 16 Forces**. It requires the ability to model a real-life situation involving turning moments produced by forces and the algebraic skills of setting up and solving equations, mainly linear equations. Because forces are vector quantities, this chapter also links with the chapters on vectors in **Books 1** and **2**.

PRIOR KNOWLEDGE

You should already know how to:

> recognise clockwise and anticlockwise turns

> use Pythagoras' theorem and trigonometry to find the magnitude and direction of a vector

> divide a number in a given ratio

› find the sum of the forces in a particular direction

› resolve a force into parallel and perpendicular components

› use the formula $W = mg$ to convert between the mass and the weight of an object

› apply the relationship $F \leq \mu R$ to problems involving friction and reaction (and $F = \mu R$ in the case of limiting equilibrium).

You should be able to complete the following questions correctly:

1 Find the magnitude and direction of the force $(5\mathbf{i} + 7\mathbf{j})$ N.

2 Find the weight, in N, of an apple with a mass of 125 g.

3 Find the horizontal and vertical components of a 9 N force at 52° to the horizontal.

4 Given a particle in limiting equilibrium with a friction force of $7.2g$ N and a reaction force of $18g$ N, find the coefficient of friction.

17.1 Turning moments

In previous chapters, forces have accelerated objects along straight lines but, as you will know from opening a door, sometimes forces will turn (rotate) objects instead. In the case of the door, this is due to hinges attaching the door to its frame.

When a force is applied to an object that is fixed or supported at a point along an axis which does not pass through the point, then that force will generate a **turning moment**.

The formula for the turning moment is given by:

Turning moment = force × perpendicular distance

Using SI units, the force is measured in newtons (N) and the distance in metres (m), giving units for the turning moment of N m. 1 N m is the same as 1 J (joule), the SI unit for energy.

For example, a force of 8 N at a perpendicular distance of 3 m would produce a turning moment of 24 N m.

Note that the formula confirms what you observe when you push a door. Pushing a door requires less effort the further you push from the hinge because, for the same turning moment, the force and distance are inversely proportional (i.e. as one increases, the other decreases).

> **KEY INFORMATION**
>
> Turning moment = force × perpendicular distance
>
> Units are N m.

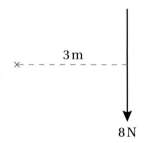

When the distance you are given is not perpendicular to the force, you will need to **resolve** the force into parallel and perpendicular **components**. In the diagram on the right, the horizontal component is $12\cos 50°$; this acts through the origin and so has a perpendicular distance of zero. The vertical component is $12\sin 50°$ and is the component perpendicular to the distance. In general, if the angle between the force and the distance is θ, then the turning moment is given by $Fd\sin\theta$, where F is the force and d is the distance. Note that the $Fd\cos\theta$ component acts through the turning point, so it has no effect because $d = 0$.

For the diagram on the right, the moment is $12 \times 5\sin 50° = 46.0\,\text{N m}$.

Depending upon the direction of the force, the moment may be **clockwise** or **anticlockwise**. This is known as the **sense** of the moment. In this chapter, it will be assumed that anticlockwise moments are positive and clockwise moments are negative.

The overall moment about a point can be found by adding the individual moments. For example, two clockwise moments, of $8\,\text{N}$ at a perpendicular distance of $3\,\text{m}$ and $2\,\text{N}$ at a perpendicular distance of $5\,\text{m}$, have a combined moment of $24 + 10 = 34\,\text{N m}$ clockwise. However, if the first moment was clockwise but the second anticlockwise, then you would find the difference. The moment would be $24 - 10 = 14\,\text{N m}$ clockwise.

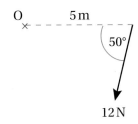

KEY INFORMATION

For a force F acting at a distance d m and at an angle of θ, the turning moment is given by $Fd\sin\theta$.

The sense of a turning moment is whether it is clockwise or anticlockwise.

Example 1

The diagram shows three forces acting about the origin.

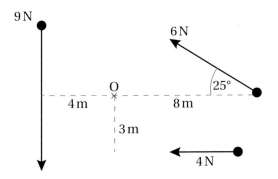

Find the overall turning moment.

Solution

First, appreciate that the moment of the $4\,\text{N}$ force is acting in a clockwise sense whereas the moments of the $9\,\text{N}$ and $6\,\text{N}$ forces are acting in an anticlockwise sense.

The perpendicular distance from the origin to the $9\,\text{N}$ force is $4\,\text{m}$. The moment is $9 \times 4 = 36\,\text{N m}$ anticlockwise.

If the 4 N force is extended along its line of action, then it is easier to see that the perpendicular distance from the origin to the force is 3 m. The moment is $4 \times 3 = 12$ N m clockwise.

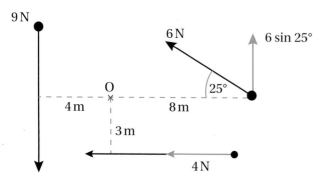

> Make sure the force and distance are perpendicular.

The 6 N force is acting at an angle of 25° to the horizontal, so the 8 m is not perpendicular to the force. If, however, the force is resolved into its horizontal and vertical components, then the vertical component ($6 \sin 25°$) is now perpendicular to the distance. The moment is $6 \sin 25° \times 8 = 20.3$ N m anticlockwise.

The overall moment is $36 + 20.3 - 12 = 44.3$ N m.

Since the answer is positive, the moment is 44.3 N m anticlockwise.

Example 2

The force $(3\mathbf{i} - 8\mathbf{j})$ N acts at the point $(4\mathbf{i} + 5\mathbf{j})$ m. Find the turning moment of the force about the point $(6\mathbf{i} + 2\mathbf{j})$ m.

Solution

Start by representing the information in a diagram. Plot the points (6, 2) and (4, 5) and illustrate the force $(3\mathbf{i} - 8\mathbf{j})$ N by drawing a horizontal 3 N force to the east and a vertical 8 N force to the south.

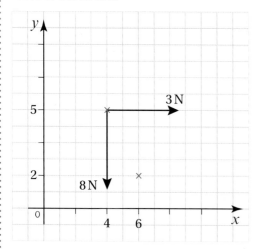

KEY INFORMATION

$\mathbf{i}$ is east, $\mathbf{j}$ is north, $-\mathbf{i}$ is west, $-\mathbf{j}$ is south.

You can then add in the perpendicular distances. The 3 N force is 3 m from the point $(6\mathbf{i} + 2\mathbf{j})$ m and the 8 N force is 2 m away.

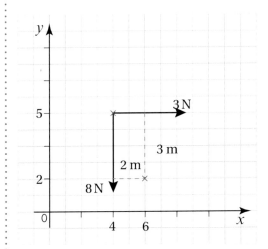

The moment of the 3 N force is acting in a clockwise sense and the moment of the 8 N force is acting in an anticlockwise sense.

The overall moment is $8 \times 2 - 3 \times 3 = 16 - 9 = 7\,\text{N}\,\text{m}$ anticlockwise.

When solving problems, it is important to work out whether each force will produce a clockwise or an anticlockwise moment.

Exercise 17.1A   **Answers page 489**

1 a Find the turning moment, in N m, of:

    i a force of 6 N acting at a perpendicular distance of 5 m

    ii a force of 3.5 N acting at a perpendicular distance of 12 m.

   b Find the force, in N, acting at a perpendicular distance of 4 m, which produces a turning moment of:

    i 28 N m.

    ii 15 N m.

2 Forces P and Q act around a point O. Force P is 12 N and acts at a perpendicular distance of 4 m from O. The moment of P about O is clockwise. Force Q is 17 N and acts at a perpendicular distance of 3 m from O. The moment of Q about O is anticlockwise.

   Find the overall moment about O, stating its sense.

3 Points A, B, C and D lie on a straight line such that AB = 4 m, BC = 5 m and CD = 6 m. There is a 5 N force acting vertically upwards at A and a 3 N force acting vertically downwards at C.

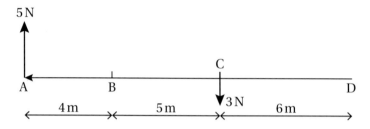

Find the turning moment about

a A **b** B

c C **d** D

4 Find the moment about the origin O for each diagram:

a

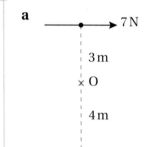

b

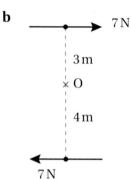

c

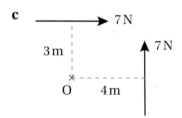

d

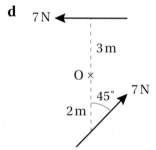

5 Point A has position vector $(\mathbf{i} + 4\mathbf{j})$ m.

A force P of $(6\mathbf{i} - 3\mathbf{j})$ N acts through the point $(-2\mathbf{i} + 6\mathbf{j})$ m.

A force Q of $(-4\mathbf{i} + 7\mathbf{j})$ N acts through the point $(5\mathbf{i} + \mathbf{j})$ m.

Find the overall moment about point A, stating its sense.

PS 6 Sophie and Subash are trying to find the turning moment of a 9 N force about the origin O. The force acts through the point A at 30° to the horizontal as shown in the diagram. The line OA is horizontal and 5 m in length.

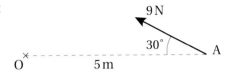

Sophie resolves the force into horizontal and vertical components of $9\cos 30°$ N and $9\sin 30°$ N. She then multiplies $9\sin 30°$ N by 5 m, since $9\sin 30°$ N is the perpendicular component of the force. Her answer is 22.5 N m.

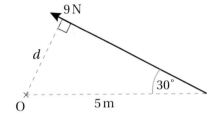

Subash sketches the force as a longer arrow and then marks the perpendicular distance between the origin and the force, which he labels as d. He then uses trigonometry to calculate that $d = 5\sin 30°$ m and multiplies this by 9 N. His answer is 22.5 N m.

Are both methods valid? Whose method is more efficient?

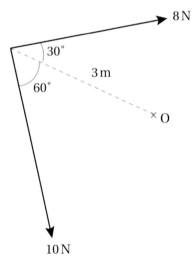

PS 7 An 8 N and a 10 N force act through the same point, 3 m from the origin O. The 8 N force acts at 30° and the 10 N force acts at 60°, as shown in the diagram.

a Find the turning moment about O.

b If the 10 N force acted at 50° instead, how would that affect the turning moment?

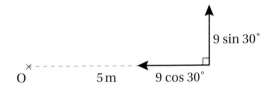

If the overall moment about a point is zero, then the individual clockwise and anticlockwise moments will balance. This means that all the clockwise moments will sum to the same as all the anticlockwise moments.

When an object is in **equilibrium**, the sum of the forces (and hence the **resultant force**) in any direction will be zero, as will the sum of the moments about any point.

KEY INFORMATION

If the overall moment about a point is zero, then the clockwise moments will be equal and opposite to the anticlockwise moments.

Example 3

A force $\mathbf{F}_1$ of $(10\mathbf{i} + 4\mathbf{j})$ N acts at the point $(-\mathbf{i} + 3\mathbf{j})$ m.

A force $\mathbf{F}_2$ of $(-5\mathbf{i} - 6\mathbf{j})$ N acts at the point $(11\mathbf{i} + 13\mathbf{j})$ m.

A force $\mathbf{F}_3$ of $(p\mathbf{i} + 23\mathbf{j})$ N acts at the point $(6\mathbf{i} - 2\mathbf{j})$ m.

Given that the moment about the point $(3\mathbf{i} + 10\mathbf{j})$ m is zero, find the value of p.

Solution

Draw a diagram as accurately as possible. The actual lengths of the forces do not have to correspond to their magnitudes, it is the line of action of each one that is important. Remember that $\mathbf{i}$ and $\mathbf{j}$ are east and north (and that $-\mathbf{i}$ and $-\mathbf{j}$ are west and south). At this stage you do not know whether p is positive or negative.

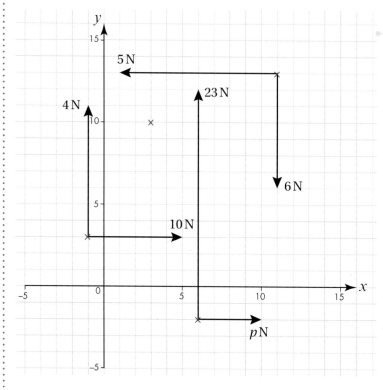

When solving problems, it is helpful to draw a neat diagram.

Add to the diagram the perpendicular distances from each force to the point $(3\mathbf{i} + 10\mathbf{j})$ m.

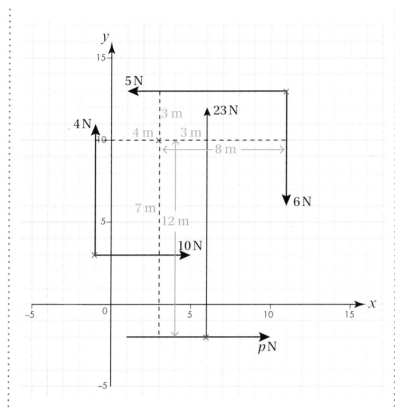

Overall moment $= 5 \times 3 + 10 \times 7 + 23 \times 3 + p \times 12 - (6 \times 8 + 4 \times 4)$
$$= 0$$

$$15 + 70 + 69 + 12p - 48 - 16 = 0$$

$$12p + 90 = 0$$

$$12p = -90$$

$$p = -7.5$$

The negative sign means that the p N force has been drawn in the incorrect direction, but the value is correct.

Exercise 17.1B

Answers page 489

1 Three forces act about a point X. Force A has a magnitude of 13 N and acts at a perpendicular distance of 4 m in a clockwise sense. Forces B and C both act in an anticlockwise sense. Force B has a magnitude of 6 N and acts at a perpendicular distance of 5 m whilst force C has a magnitude of 11 N and acts at a perpendicular distance of d m.

    a If the overall moment is zero, find the value of d.

    b Find the overall moment if force C is increased to 15 N.

M **2** A man and a boy sit on a seesaw which is 4 m long. The pivot of the seesaw is at the centre. The centre of mass of the seesaw is also at the centre, as shown in the diagram.

The boy sits at the end and weighs 36 kg. The man sits on the other side and weighs 60 kg.

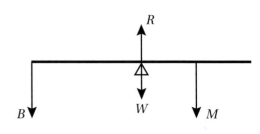

How far must the man sit from the centre of the seesaw to keep the seesaw balanced?

M **3** A 10 m rod is supported by a pivot at point A. There is a 7 N force acting vertically downwards at one end and a 9 N force acting vertically upwards at the other end, as shown in the diagram. The rod is considered to be so light that it has no weight.

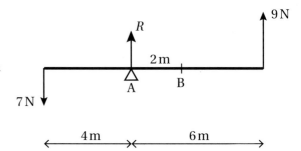

a Find the overall moment about the point A.

Point B is 2 m from A. A vertical force is applied to the rod at B such that the overall moment is now zero.

b **i** Find the magnitude and direction of the force applied at B.

ii By considering all the vertical forces, what is the magnitude of the reaction force at A such that the system is in equilibrium?

PF **4** Point X has position vector $(6\mathbf{i} + 2\mathbf{j})$ m.

A force A of $(3\mathbf{i} - 7\mathbf{j})$ N acts through the point $(\mathbf{i} + 6\mathbf{j})$ m.

A force B of $(-4\mathbf{i} - 2\mathbf{j})$ N acts through the point $(10\mathbf{i} + 4\mathbf{j})$ m.

A force C of $(\mathbf{i} + q\mathbf{j})$ N acts through the point $(p\mathbf{i} - 3\mathbf{j})$ m.

a Show that the horizontal components of the three forces are in equilibrium.

b Given that the vertical components of the three forces are also in equilibrium, find the value of q.

c Given that the moment about X is zero, find the value of p.

PS **5** Georgia was given the following problem.

'The three forces T, U and V shown in the diagram are in equilibrium.

The magnitude of force T is q N and it acts at angle of α to the vertical through a point 8 m from the origin.

Force U acts at an angle of $(\alpha + 90)°$ to the horizontal through a point 3 m from the origin.

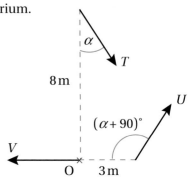

Force V acts through the origin.

a Find the magnitude of U in terms of q.

b Find the value of α.

c Find the magnitude of V in terms of q.

Georgia's solution is shown below.

Resolving vertically:

$$T\cos\alpha = U\cos(90° + \alpha)$$
$$= U\cos 90° + U\cos\alpha.$$
$$q\cos\alpha = 0 + U\cos\alpha$$
$$= U\cos\alpha$$

Hence $U = q\,\text{N}$.

Taking moments about O:

$$U\times 3\sin(\alpha + 90°) - T\times 8\sin\alpha = 0$$
$$q\times 3\sin(\alpha + 90°) - q\times 8\sin\alpha = 0$$
$$3\sin(\alpha + 90°) = 8\sin\alpha$$
$$3\sin\alpha + \sin 90° = 8\sin\alpha$$
$$1 = 5\sin\alpha$$
$$\sin\alpha = 0.2$$
$$\alpha = \sin^{-1}(0.2) = 11.5°$$

Resolving horizontally:

$$V = T\sin\alpha + U\sin(90° + \alpha)$$
$$= q\sin\alpha + q\sin(90° + \alpha)$$
$$= q\sin\alpha + q\sin 90° + q\sin\alpha$$
$$= 2q\sin 11.5° + q$$
$$= 2q\times 0.2 + q$$
$$= 1.4q\,\text{N}$$

Assess Georgia's solution.

Where there are mistakes, provide the correct solution.

17.2 Horizontal rods

Consider a plank of wood balanced upon two logs.

For a system to be in equilibrium, not only do the forces sum to zero in any direction, but the moments must also balance (i.e. all the clockwise forces must balance all the anticlockwise forces).

The plank has a weight force. A standard plank will have its mass spread out evenly so its centre of mass is assumed to act

KEY INFORMATION

For a system to be in equilibrium, not only do the forces sum to zero in any direction, but the moments must also balance (i.e. all the clockwise forces must balance all the anticlockwise forces).

at the middle of the plank. In this case, the plank is described as **uniform**. Alternatively, if the centre of mass is not positioned at the middle of the plank or the mass is not evenly spread out, then it is described as **non-uniform**. However, it is possible for a non-uniform object, for example the dumbbells that weightlifters lift, to have its centre of mass at the middle.

According to Newton's third law, there must be an equal and opposite force to balance the weight. In the example of the plank, this is provided by the reaction forces at the two logs. The points where the logs make contact with the plank are analogous to the point where the ground makes contact with a particle, where there is a normal reaction force acting perpendicular to the surface.

In situations of this type, the plank is modelled as a **rod**, which is defined as rigid (and hence one-dimensional), no matter what are the magnitudes of the forces that act upon it. All the forces will be vertical. There will be weight forces from the rod (acting at its centre of mass) and from any objects placed upon the rod, acting vertically downwards. There will also be reaction forces from the supports (or tensions from strings that the rod is suspended from), acting vertically upwards. The force diagram for the plank and logs would be drawn as shown on the right, with reactions at A and B and a weight at the middle.

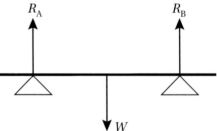

Consider a uniform rod AD of length 10 m, with a centre of mass at C, supported at B and D. AB = 3 m. Assume that the weight of the rod is 42 N and that you need to find the magnitudes of the reaction forces at B and D.

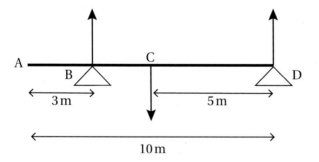

Since you are told that the rod is uniform, the 42 N weight is modelled as a single force acting at the middle of the rod (and since the rod is 10 m long, this will be 5 m from each end).

You do not know either reaction force. You could take moments about two points chosen at random (assuming these are neither B nor D) and solve two simultaneous equations. However, if you take moments about B or D then the moment of the reaction force about that point will be zero (since the perpendicular distance is zero). Eliminating a force from the equation means that there will only be one unknown rather than two. Note that is also easier to take moments about the end of a rod than about a point in the middle, because all the forces with moments of the same sense will be pointing in the same direction (upwards or downwards).

Since there is a reaction force at D, and D is at the end of the rod, this is a sensible choice of point to take moments about.

$$\text{Turning moment} = \text{force} \times \text{perpendicular distance}$$

The reaction at B is acting clockwise with a moment of $R_B \times 7$.

The weight at C is acting anticlockwise with a moment of 42×5.

$$\text{Hence, the overall moment is } 42 \times 5 - R_B \times 7 = 0$$

$$210 = R_B \times 7$$

$$R_B = 30\,\text{N}$$

To find the reaction force at D, you could now take moments about any other point, since you know the value of R_B. However, the simplest approach is to resolve vertically. Resolving vertically, the sum of all the upward forces equals the sum of all the downward forces.

$$R_B + R_D - 42 = 0$$

$$R_D = 42 - R_B$$

$$= 42 - 30$$

$$= 12\,\text{N}$$

Example 4

A uniform 8 m rod AB in equilibrium rests upon supports at C and B. $AC : CB = 1 : 3$.

The reaction at B is $2g\,\text{N}$ and there is a mass of 5 kg at A.

Find:

a   the mass of the rod   b   the reaction at C.

Solution

a   Start by drawing a diagram to represent the information. Because AB is 8 m with C dividing it in the ratio 1 : 3, AC is 2 m and CB is 6 m. Since the rod is uniform, the weight force W will act at the middle of the rod. There will be a reaction force at each support, including $2g\,\text{N}$ at B and a second weight force of $5g\,\text{N}$ at A.

It helps to take moments about the position at which an unknown force is acting because then the unknown force will not feature in the equation and there will be fewer variables. The unknown force in this example is at C.

If the rod is non-uniform and you have not been given the position of the centre of mass, assign a position yourself.

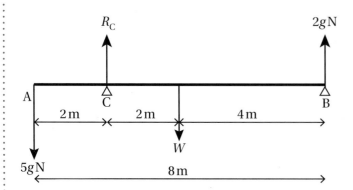

Since you do not know the reaction at C, eliminate the reaction force at C by taking moments about C. W is 2 m from C, the reaction at B is 6 m from C and the 5 kg mass is 2 m from C.

$$2g \times 6 + 5g \times 2 - W \times 2 = 0$$
$$2W = 22g$$
$$W = 11g \, \text{N}$$

Hence the mass of the rod is 11 kg.

Resolving vertically:

$$R_C + R_B - 5g - W = 0$$
$$R_C + 2g - 5g - 11g = 0$$
$$R_C = 14g = 137 \, \text{N}$$

The reaction force at C is 137 N.

Round to 3 significant figures unless otherwise stated.

Example 5

A non-uniform tree trunk AB of length 36 m rests in equilibrium on a support at C where $AC = 9$ m. A woman of mass 48 kg sits at A and a boy of mass 12 kg sits at B.

a   Find the position of the centre of mass if the tree trunk weighs 6 kg.

b   Find the mass of the tree trunk if the centre of mass is 12 m from A.

State any assumptions you have made in answering this question.

Solution

a   Draw a diagram, representing the tree trunk as a rod and the woman and boy as particles (with their mass concentrated at a single point). Because the tree trunk is non-uniform, the centre of mass of the trunk does not have to be in the middle. For **part b**, you are told its

position. For **part a**, however, just draw an arrow for the
weight force at a distance of x m from A. Note that the
weight force is $6g$ N for **part a**.

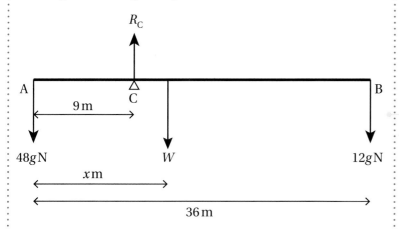

> Drawing a force diagram
> will help you to visualise the
> problem.

Take moments about C to eliminate the reaction force.

$$48g \times 9 - 6g \times (x - 9) - 12g \times 27 = 0$$

$$432g = 6g(x - 9) + 324g$$

$$108g = 6g(x - 9)$$

Divide by $6g$.

$$x - 9 = 18$$

$$x = 27$$

The centre of mass is 27 m from A.

b Use the same diagram, but the centre of mass is 12 m
from A and it is now the weight force that is unknown.

Again, take moments about C to eliminate the reaction
force.

$$48g \times 9 - W \times 3 - 12g \times 27 = 0$$

$$108g = 3W$$

$$W = 36g$$

The mass of the tree trunk is 36 kg.

The assumptions made in this question are that the tree
trunk can be modelled as a rod, and the woman and boy as
particles.

Stop and think   Why is there no reaction force acting on the woman or the boy?

Answers page 490

(M) **1** Draw a diagram to represent each situation:

   a An 8 m uniform pole weighs 30 kg. It has one sack loaded at each end, of masses 20 kg and 40 kg. It rests on one support.

   b A rod AB of mass 9 kg and length 15 m has its centre of mass 6 m from B. It is suspended horizontally from the ceiling by two vertical strings, one attached at B and the other at C, where BC = 12 m.

   c A non-uniform rod AB of length 6 m is supported horizontally on two supports, one at A and the other at B. The reactions at these supports are 5gN and 3gN respectively.

   d A 12 m uniform pole AB of mass 20 kg has supports at C and D. AC : CD : DB = 1 : 3 : 2.

   e A playground seesaw consists of a uniform beam of length 4 m supported at its midpoint. A girl of mass 25 kg sits at one end of the seesaw. Her brother of mass 40 kg sits on the other side and the seesaw is horizontal.

   f A pole vaulter uses a uniform pole of length 5 m and mass 7 kg. He holds the pole horizontally by placing one hand at one end of the pole and the other hand at a position 80 cm away.

(M) **2** A uniform log AB of mass 60 kg has loads of 30 kg and 50 kg at A and B respectively, as shown in the diagram. The log is held by two supports at C and D, with AC = 2 m, CD = 2 m and DB = 1 m.

   a Explain why the log is modelled as a uniform rod.

   b Find the magnitudes of the reaction forces at C and D.

(M) **3** AB is a non-uniform 5 m rod of mass 20 kg. It rests upon a single support at C, where CB = 4 m. There is a 30 kg rock at A and a 2 kg stone at B.

   Find:

   a the distance of the centre of mass from A

   b the reaction at C.

(M) **4** A pole vaulter uses a uniform pole of length 4 m and mass 5 kg. He holds the pole horizontally by placing one hand at one end of the pole and the other hand at a position 80 cm away.

   Find the vertical forces exerted by his hands.

(M)
(PF) **5** A uniform beam JK of length 12 m and mass 16 kg has a sack of mass 24 kg attached at J and a sack of mass 40 kg attached at K. The beam is balanced horizontally on a support at L.

   a By taking moments about L, prove that JL is 7.2 m.

   b Explain why the reaction at L is 784 N.

   c What assumptions have you made in answering this question?

6 Three uniform rods, each of length 25 cm and of masses 3, 5 and 7 kg, are joined rigidly to make one long 75 cm rod, with the 5 kg rod in the middle. The rod is suspended from a vertical string attached to the rod at a point that is x cm from its centre. The rod is horizontal.

Find:

a the value of x **b** the tension in the string.

7 A lumberjack has laid a tree AC of mass 64 kg and 6 metres in length on two supports, one at A and one at B. When a 16 kg boulder is placed on the tree at C, the reaction force at B is three times as large as the reaction force at A. The tree is assumed to be uniform.

a What is the reaction force at B?   **b** Find the position of B.

The 16 kg boulder is removed and replaced by a boulder of mass m kg. The reaction force at B is now five times as large as the reaction force at A.

c Show that m is 32.

d What assumption have you made about the boulders and why did you need to make this assumption?

8 A 10 m non-uniform rod CD is supported at A and B, where AC = 3 m and BD = 2 m. When a 10 kg object is placed at D, the ratio of the reactions at A and B is 1 : 4, but when a 5 kg object is placed at C instead, the ratio of the reactions at A and B is 4 : 3.

a Find the weight of the rod.

b Find the distance of the centre of mass of the rod from C.

Tilting

Consider a plank of wood balanced upon two logs. If you were to stand on one end of the plank, unless you were very light or the plank was very heavy, the plank would be likely to tilt.

The diagram shows a rod resting on two supports, P and Q.

If the rod rotates about P then the rod will no longer be in contact with Q. At the instant that the rod starts to rotate, the rod is described as being on the point of **tilting** about P, and the reaction force at Q is zero.

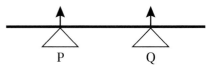

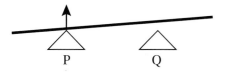

Example 6

A uniform 18 m rod AB rests on supports at C and D, where AC : CD : DB = 1 : 3 : 2.

It is on the point of tilting about D. There is a mass of 3 kg at A and a mass of 8 kg at B.

Find the weight of the rod and the reaction at D.

KEY INFORMATION

If a rod is supported at P and Q and the rod is on the point of tilting about P, then $R_Q = 0$.

Solution

Dividing 18 m in the ratio $1:3:2$, AC = 3 m, CD = 9 m and DB = 6 m. The rod is uniform, so the weight of the rod is 9 m from A. Because the rod is on the point of tilting about D, the reaction force at C is zero.

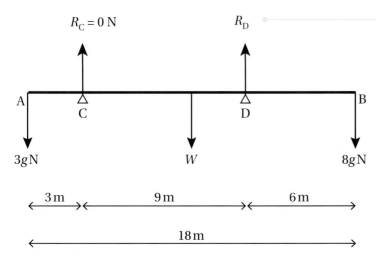

Even though the reaction at C is zero, it is still advisable to include the force in the diagram.

Take moments about D to eliminate the reaction force at D. D is 3 m from the centre of mass.

$$3g \times 12 + W \times 3 - 8g \times 6 = 0$$
$$36g + 3W = 48g$$
$$3W = 12g$$
$$W = 4g = 39.2 \, \text{N}$$

The weight of the rod is 39.2 N.

Resolving vertically:

$$R_D + R_C - 3g - W - 8g = 0$$
$$R_D + 0 - 3g - 4g - 8g = 0$$
$$R_D = 15g = 147 \, \text{N}$$

The reaction force at D is 147 N.

Exercise 17.2B

Answers page 491

1 A uniform beam AB of mass 6 kg is suspended from two vertical strings attached at points J and K, where AJ is d m. The beam is $4d$ m in length and horizontal. A particle of mass 9 kg is attached to the beam and, as a result, the beam is now on the point of tilting about J.

How far is the particle from A, in terms of d?

2 A non-uniform rod AB has a mass of 10 kg and length of 2.1 m. It rests on supports at C and D, where AC, CD and DB are all the same distance. The rod is horizontal and in equilibrium. When a particle of mass 4 kg is attached to the rod at B, the rod is on the point of tilting about D.

Show that the centre of mass is 1.12 m from A.

3 A uniform rod CF has a mass of 70 kg. It rests on supports D and E, 3 m and 7 m from C respectively, as shown in the diagram. The rod is on the point of tilting about D. There is a 50 kg particle at C and an m kg particle at F.

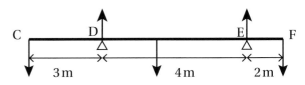

a State the magnitude of the reaction force at E.

b Find the value of m.

4 A telegraph pole AB has length 12 m and mass 20 kg. It is resting in equilibrium in a horizontal position on supports at points P and Q, where AP = 4 m and QB = 3 m. The telegraph pole is modelled as a uniform rod.

a Find the reaction forces at P and Q.

A crate of mass M kg is placed at point X, where AX = 10 m. The rod is now on the point of tilting about Q.

b Find the value of M.

5 The diagram shows a uniform plank XY of length 10 m and weight W N, balanced on supports at U and V. There is a 70 kg load 1 m from X and a 5 kg load at Y. XU = 2 m and XV = 7 m.

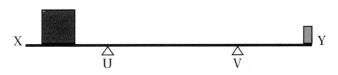

a Given that the plank is on the point of tilting about U, find the value of W.

b Another load is added at Y so that the rod is now on the point of tilting about V. Show that the mass of the load that has been added can be written in the form $\frac{a}{3}$ kg, where a is an integer.

6 A non-uniform 8 m rod AB resting on supports at E and F has a weight of Mg. AE = 2 m and AF = 5 m. In this situation, the reaction at F is $18g$ N, but when a 72 kg block is placed 1.5 m from B, the rod is on the point of tilting about F.

Find the value of M and the distance of the rod's centre of mass from A.

17.3 Equilibrium of rigid bodies

In the previous section, all the forces acted vertically on a horizontal rod. In this section, the forces can act in any direction.

Ladders

When considering a ladder, all the forces are either horizontal or vertical. The ladder rests against a wall and experiences a horizontal reaction from the wall. If the wall is rough, there will also be a friction force vertically upwards. The ladder rests on the ground and again there will be a reaction force, but this time it will be vertical. There must be a force keeping the ladder from slipping – if the floor is rough, this will be a horizontal friction force acting towards the wall. However, it could be a tense string keeping it in place. The ladder will also have a weight.

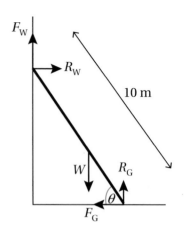

The diagram on the right shows the weight and both of the friction and reaction forces for a 10 m length rod inclined at an angle of θ to the horizontal.

Problems on ladders are solved by a combination of resolving forces and taking moments.

For the diagram above:

Resolving horizontally, $R_W - F_G = 0$

Resolving vertically, $R_G + F_W - W = 0$

Taking moments about the point of contact of the ladder with the ground,

$$W \times 5\cos\theta - F_W \times 10\cos\theta - R_W \times 10\sin\theta = 0$$

The three diagrams below show the perpendicular distances for the friction force at the wall, the reaction force at the wall and the weight.

> **KEY INFORMATION**
>
> Solve ladder problems using simultaneous equations. Generate the equations using resolving, moments and $F \leqslant \mu R$.

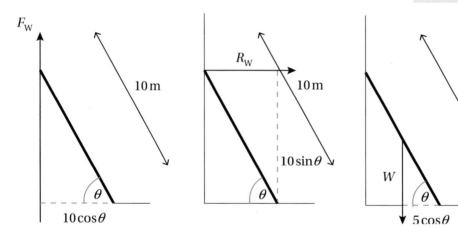

Example 7

A 30 kg uniform ladder rests against a smooth wall at an angle of arctan 3 to the horizontal. The ground is rough. Given that the ladder is in limiting equilibrium, find the coefficient of friction at the ground.

Solution

Draw a diagram. Let the point of contact with the wall be W and the point of contact with the ground be G. The ground is rough so it has both a reaction and a friction force whereas the wall is smooth so it only has a reaction force. The ladder is

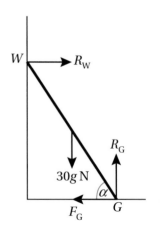

uniform so its weight acts at the middle. The ladder is inclined at an angle of arctan 3 to the horizontal. Let the angle be α, where $\tan \alpha = 3$.

The length of the ladder is not given. Let the length of the ladder equal $2a$ m, so the weight acts at a m from each end.

> If a length is not given, then it must not be needed. What is important, however, is the ratio.

Since there are two forces acting at G and only one at W, take moments about G.

The perpendicular distance between G and the line of action of R_W is $2a \sin \alpha$ m.

The perpendicular distance between G and the line of action of $30g$ N is $a \cos \alpha$ m.

Hence $30g \times a \cos \alpha - R_W \times 2a \sin \alpha = 0$.

Dividing by $a \cos \alpha$, $30g - R_W \times 2 \tan \alpha = 0$.

Since $\tan \alpha = 3$, $30g = R_W \times 2 \times 3$

$$R_W = 5g \, \text{N}$$

Resolving horizontally and vertically:

$$R_W - F_G = 0$$
$$R_G - 30g = 0$$

The question asks for the coefficient of friction at the ground. Since the ladder is in limiting equilibrium, you have $F_G = \mu R_G$, from which $\mu = \dfrac{F_G}{R_G}$.

> Use $F = \mu R$ for limiting equilibrium.

$$\mu = \frac{F_G}{R_G} = \frac{R_W}{30g} = \frac{5g}{30g} = \frac{1}{6}$$

Exercise 17.3A

Answers page 491

(M) **1** A ladder of mass M kg is modelled as a uniform rod. The wall is smooth but the floor is rough. The coefficient of friction acting on the ladder at the floor is 0.4. The ladder is inclined at an angle of α to the vertical.

Find:

a the reaction at the floor in terms of M

b the friction at the floor in terms of M

c the reaction at the wall in terms of M

d the angle α.

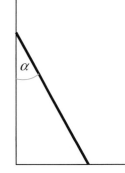

(M) **2** A non-uniform ladder AB of mass 50 kg and length 8 m stands on rough ground against a smooth wall at A. The angle between the ground and the ladder is 70°. The coefficient of friction between the ladder and the ground is $\frac{1}{4}$.

Find the distance of the centre of mass of the ladder from B.

3 A uniform ladder GW is 6 m long, weighs 36 kg and makes an angle of 30° with the vertical.
G is the point of contact of the ladder with the ground and W is the point of contact of the
ladder with the wall. A 45 kg boy stands on the ladder 2 m from G. The wall is smooth.

Show that the coefficient of friction between the ladder and the ground is given by $\frac{11\sqrt{3}}{81}$.

 4 A uniform 54 kg ladder CD rests against a rough wall, where C is on the ground. The length of
the ladder is $2a$ metres. The ladder is at an angle of α to the horizontal, where $\sin \alpha = 0.8$. The
coefficient of friction between the ladder and the wall is $\frac{1}{6}$.

Show that the coefficient of friction between the ladder and the ground can be written as $\frac{k}{17}$,
where k is a constant to be found.

 5 A ladder AB of mass 60 kg and length 3.5 m leans against a smooth wall. The ladder is inclined
at an angle α to the horizontal, where $\tan \alpha = \frac{20}{13}$. The foot of the ladder B is on rough
horizontal ground. The coefficient of friction between the ladder and the ground is 0.4. The
ladder is modelled as a uniform rod. A man of mass 70 kg climbs the ladder.

Find the maximum distance he can climb safely.

 6 A uniform 40 kg ladder CD rests against a rough wall, where C is on the ground. The length of
the ladder is $2a$ metres. The ladder is at an angle of α to the horizontal, where $\sin \alpha = 0.8$. The
coefficient of friction between the ladder and the wall, μ, is the same as the coefficient of
friction between the ladder and the ground.

Find the value of μ.

> **Stop and think**   What is the minimum angle a ladder can make with the horizontal? Why is this?

Other bodies

You will now consider other rigid bodies in which one or more
forces will act at an angle (i.e. not all the forces will be either
parallel or perpendicular). In this case, it will be necessary to
resolve a force into its components.

Often there will still be a reaction force and a friction force acting
at the point of contact and related by the relationship $F \leqslant \mu R$ (or
$F = \mu R$ for limiting equilibrium).

Alternatively, where a rod is connected to a surface by a hinge,
there will be a reaction force at the hinge. Since the hinge is a
point rather than a surface, the angle cannot be considered to
be perpendicular. In this case, the reaction force will need to be
resolved into X and Y components. The magnitude of the reaction
force can then be deduced using Pythagoras' theorem and the
angle by trigonometry.

KEY INFORMATION
Use X and Y for the
components of the reaction
force at a hinge, since the
angle is unknown.

Example 8

The diagram shows a uniform pole AB of mass 8 kg and length 2.5 m freely hinged at A to a vertical wall. The pole is held horizontally in position by a light inextensible rope BC with C vertically above A. The string is inclined at α to the horizontal such that $\tan \alpha = \frac{2}{3}$.

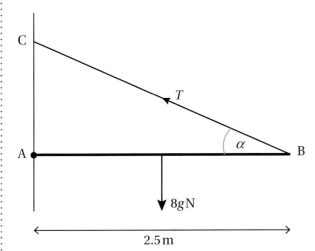

a Show that the tension in the rope is given by $T = 2\sqrt{13}\,g\,$N.

A load of mass 5 kg is now attached to the pole at D, where BD = 0.5 m.

b Find the magnitude and direction of the force exerted by the hinge on the pole at A.

Solution

a Start by completing the diagram with the reaction force at the hinge. Since you do not know what angle the reaction acts at, represent it as two perpendicular components, X for the horizontal and Y for the vertical.

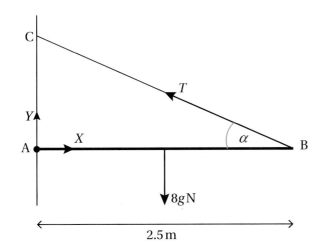

Take moments about A to eliminate both X and Y from the equation. This requires you to resolve T into components. The perpendicular component is $T\sin \alpha$.

$$8g \times 1.25 = T\sin \alpha \times 2.5$$

Given that $\tan \alpha = \frac{2}{3}$ (and that α is acute), you can find the value of $\sin \alpha$ using a right-angled triangle and trigonometry. For any right-angled triangle, $\tan \alpha = \frac{O}{A}$. If the opposite is <u>2 and</u> the adjacent is 3, then the hypotenuse is $\sqrt{2^2 + 3^2} = \sqrt{13}$.

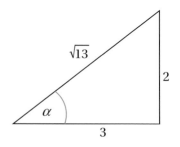

Hence $\sin \alpha = \frac{O}{H} = \frac{2}{\sqrt{13}}$.

$$8g \times 1.25 = T \times \frac{2}{\sqrt{13}} \times 2.5$$

$$10g = T \times \frac{5}{\sqrt{13}}$$

$$T = 10g \times \frac{\sqrt{13}}{5}$$

$$= 2\sqrt{13}\,g\,\text{N}$$

b A load of mass 5 kg has been added at D, so add this to the diagram.

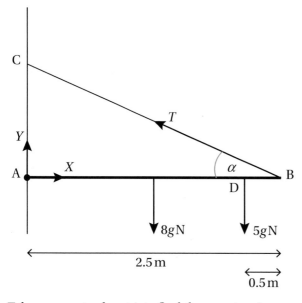

Take moments about A to find the new tension.

$$T\sin \alpha \times 2.5 - 8g \times 1.25 - 5g \times 2 = 0$$

$$T \times \frac{5}{\sqrt{13}} = 20g$$

$$T = 4\sqrt{13}\,g\,\text{N}$$

Resolve horizontally to find the magnitude of X.

$$X - T\cos \alpha = 0$$

You can use the same right-angled triangle as before to find the value of $\cos \alpha$.

$$\cos \alpha = \frac{A}{H} = \frac{3}{\sqrt{13}}$$

$$X = 4\sqrt{13}\,g \times \frac{3}{\sqrt{13}} = 12g\,\text{N}$$

Resolve vertically to find the magnitude of Y.

$$Y + T\sin \alpha - 13g = 0$$

$$Y + 4\sqrt{13}\,g \times \frac{2}{\sqrt{13}} = 13g$$

$$Y + 8g = 13g$$

$$Y = 5g\,\text{N}$$

Using the triangle law from **Book 1, Chapter 10 Vectors**, you can add the vectors X and Y. Since X and Y are perpendicular, the magnitude of the reaction force R can be calculated using Pythagoras' theorem.

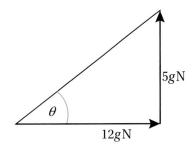

$$R = \sqrt{(5g)^2 + (12g)^2}$$

$$= \sqrt{25g^2 + 144g^2}$$

$$= \sqrt{169g^2}$$

$$= 13g = 127\,\text{N}$$

Similarly, the direction of the reaction force can be found using the same triangle and trigonometry.

$$\tan \theta = \frac{5g}{12g} = \frac{5}{12}$$

Hence $\theta = \arctan \frac{5}{12} = 22.6°$ to the horizontal.

Example 9

The diagram shows a non-uniform beam AB of mass 5 kg and length 4 m resting against a smooth peg at C, where AC = 3 m. The end A of the beam lies on a rough horizontal floor. The angle between the beam and the floor is 30°. The reaction force exerted on the beam by the peg is $14\sqrt{3}$ N. The beam is in limiting equilibrium. The centre of mass of the beam is x m from A.

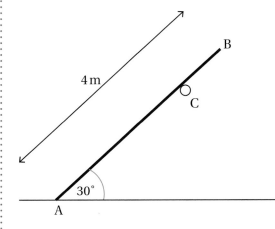

a Find the value of x.

b Show that the coefficient of friction at A is given by $k\sqrt{3}$, where k is a constant to be found.

Solution

a Start by adding extra information to the diagram. The peg is 1 m from B (3 m from A). The reaction force at the peg is $14\sqrt{3}$ N, which acts perpendicular to the beam. There will be a reaction force, R_A, and a friction force, F, at A.

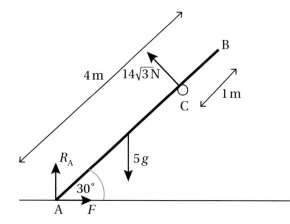

Take moments about A to eliminate R_A and F.

$$14\sqrt{3} \times 3 - 5g \times x\cos 30° = 0$$

$$x = \frac{14\sqrt{3} \times 3}{5g\cos 30°} = \tfrac{12}{7} = 1.71 \text{ m}$$

b Resolve horizontally to find the friction force at A.

$$F - 14\sqrt{3} \times \sin 30° = 0$$

$$F = 14\sqrt{3} \times \tfrac{1}{2}$$

$$= 7\sqrt{3} \text{ N}$$

Resolve vertically to find the reaction force at A.

$$R_A + 14\sqrt{3} \times \cos 30° - 5g = 0$$

$$R_A + 14\sqrt{3} \times \frac{\sqrt{3}}{2} = 5g$$

$$R_A + 21 = 49$$

$$R_A = 28 \text{ N}$$

Since the beam is in limiting equilibrium, $F = \mu R$ and $\mu = \dfrac{F}{R}$.

$$\mu = \frac{7\sqrt{3}}{28} = \tfrac{1}{4}\sqrt{3}$$

Hence $k = \tfrac{1}{4}$.

Answers page 492

1 A uniform rod AB of mass 12 kg and length 1.8 m rests against a smooth peg at C, where AC : CB = 2 : 1. The rod is inclined at an angle of 25° to the horizontal, with A at rest on rough horizontal ground.

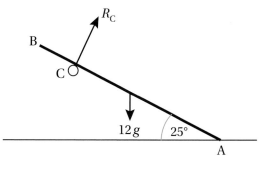

a Find the reaction force exerted on the rod by the peg.

b Find the frictional force at A.

c Find the reaction force at A.

d Find the magnitude and direction of the force exerted on the rod by the ground at A.

2 A thin metal pole AB is modelled as a uniform rod with a mass of 8 kg and a length of 1.6 m. The pole is freely hinged at A to a vertical wall and held horizontally by a light inextensible cable at B. The cable makes an angle of 60° with the horizontal as shown in the diagram.

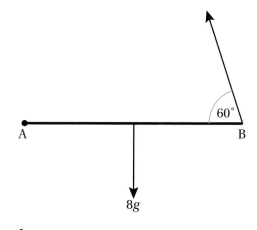

a Find the tension in the cable.

b Find the magnitude and direction of the force exerted by the hinge on the rod.

3 A non-uniform pole AB of mass M kg and length $12a$ m is held horizontally in position with A at rest on a rough vertical wall by a light inextensible string BC of length $13a$ m, with C vertically above A. The pole is in limiting equilibrium. The coefficient of friction between the wall and the pole is given by μ. The centre of mass of the pole is $4a$ m from A.

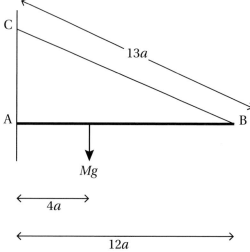

a Show that the tension in the string is $\frac{13}{15} Mg$ N.

b Show that $\mu \geqslant \frac{5}{6}$.

M **4** ABCD is a uniform rectangular
PF sheet of metal of mass 6 kg. AB is
$2a$ m and BC is $10a$ m. The lamina
is hinged at A and is able to swing
freely. The sheet is held in place
with AD horizontal by a force of
$2F$ N acting along BC and a force
of F N acting along CD.

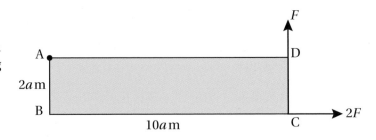

Show that the reaction force exerted on the sheet by the hinge acts at an angle of arctan 0.9 to
the horizontal.

M **5** A plank AB of mass 7.5 kg is supported at rest
by a smooth peg at C. The reaction force
exerted on the plank by the peg is 49 N. The
end A rests on rough horizontal ground. The
plank is in limiting equilibrium and the
coefficient of friction between the ground and
the plank is μ. AB is 3 m in length. The peg is
2.5 m from A and the centre of mass of the
plank is 2 m from A. The plank is inclined at α
to the horizontal.

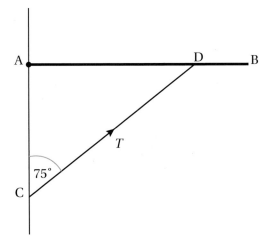

a Find the size of the angle α, correct to 1 decimal place.

b Find the value of μ.

M **6** A plastic bar AB of mass 0.6 kg is hinged at
PF point A on a vertical wall. The bar is held in a
horizontal position by a solid strut CD, where C
is vertically below A. The angle between the
wall and the strut is 75°. The length of AD is $\frac{3}{4}$
of the length of AB. The bar is modelled as a
uniform rod.

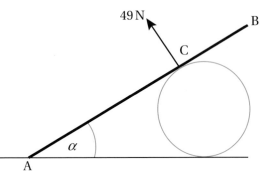

a Show that the thrust in the strut is 15.1 N,
correct to 3 significant figures.

b Find the magnitude of the force exerted on
the bar by the hinge.

M **7** A 2 kg rod AB of length 3 m is held
PS horizontally as shown in the diagram by
two light inextensible ropes. The rope at
A is inclined at 20° to the horizontal and
the rope at B is inclined at 50° to the
horizontal. The centre of mass of the rod
is x m from A.

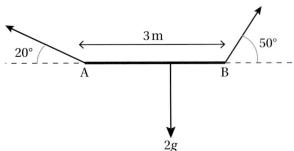

a Find the magnitude of the tension in
each rope.

b Find the value of x.

SUMMARY OF KEY POINTS

> Turning moment = force × perpendicular distance

> The units for turning moment are N m.

> The sense of a turning moment is whether it is clockwise or anticlockwise.

> For a force F acting at a distance d m and at an angle of θ, the turning moment is given by $Fd\sin\theta$.

> If the overall moment about a point is zero, then the clockwise moments will be equal and opposite to the anticlockwise moments. For a system to be in equilibrium, not only do the forces sum to zero in any direction, but the moments must also balance (i.e. all the clockwise forces must balance all the anticlockwise forces).

  > For a horizontal rod, it is assumed that the weight force acts through its centre of mass.

  > A uniform rod has its centre of mass at the middle of the rod.

  > For a non-uniform rod, the centre of mass might be at the middle, but it is usually somewhere else.

  > A rod is modelled as rigid and straight.

  > Reaction forces act vertically on the other side of the rod from the support.

  > Weight forces act vertically downwards.

> If a rod is supported at P and Q and the rod is on the point of tilting about P, then $R_Q = 0$.

> Ladder problems can be solved using simultaneous equations. Generate the equations using resolving, moments and $F \leqslant \mu R$.

  > Use $F = \mu R$ for limiting equilibrium.

  > Use X and Y for the components of the reaction at a hinge since the angle is unknown.

EXAM-STYLE QUESTIONS 17

Answers page 492

 1 The diagram shows a non-uniform rod AB of length 10 m and mass 90 kg with its centre of mass x m from A. There are loads of 60 kg and 40 kg at A and B. C is 1 m from A, and D is 3 m from B. The reaction force at D is $95g$ N.

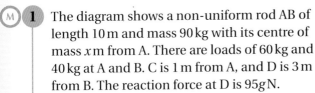

  a Find the magnitude of the reaction force at C. **[2 marks]**

  b Find the value of x. **[3 marks]**

 2 A broom consists of a wooden broomstick of length 90 cm and mass 700 g with a broom head of mass 1.1 kg attached at one end. The broom rests horizontally on a support.

  a State suitable models for the broomstick and broom head. **[2 marks]**

  b Find the distance of the support from the broom head. **[3 marks]**

(M) **3** A non-uniform plank AB has length $5d$ m and mass $8m$ kg. It is in equilibrium in a horizontal position resting on supports at the points U and, V where AU = $2d$ m and AV = $4d$ m. A parcel of mass $6m$ kg is placed on the plank at B. The plank is on the point of tilting about V.

   a Calculate the distance of the centre of mass of the plank from A in terms of d. **[3 marks]**

   b Comment upon any assumptions you have made. **[2 marks]**

(M) (PS) **4** A uniform rod WZ of length 14 m and mass 5 kg rests horizontally on two supports at X and Y. WX : XZ = 1 : 6.

   There is a particle of mass 7 kg attached at W and a particle of mass 9 kg attached at Z.

   The reaction at X is half the magnitude of the reaction at Y.

   Given that the rod is in equilibrium, find the distance WY. **[5 marks]**

(M) (PF) **5** A uniform log AB has a length of 12 m and a weight of W newtons, as shown in the diagram. It is held in equilibrium in a horizontal position by two vertical ropes attached at C and D, where AC is 4 m and DB is 1 m. A backpack of weight 28 N is suspended from the rod at A.

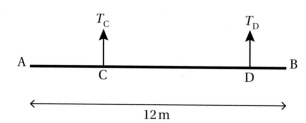

   a How are the ropes modelled in this question? **[1 mark]**

   b Show that the tension in the rope at C is given by $\left(\frac{5}{7}W + 44\right)$ N. **[4 marks]**

   The tension in the rope at C is 13 times the tension in the rope at D.

   c Hence, or otherwise, find the value of W. **[6 marks]**

(M) (PS) (CM) **6** Romain has been set this problem to solve:

   Anais is a pole vaulter. Her 4.8 metre pole is a 2300 g uniform rod which she holds with one hand at the end and the other hand 0.5 m from the same end. Calculate each of the reactions exerted by Anais' hands to keep the pole horizontal.

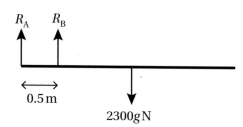

   Romain's solution is as follows:

   Moments about A: $R_B \times 0.5 - 2300g \times 2.4 = 0$

   $$R_B = 108\,000\,\text{N (3 s.f.)}$$

   Moments about B: $R_A \times 0.5 - 2300g \times 1.9 = 0$

   $$RA = 85\,700\,N\,(3\,s.f.)$$

   a Explain what mistakes Romain has made. **[2 marks]**

   b Write out a correct solution. **[5 marks]**

(PF) **7** Point A has position vector $(5\mathbf{i} - 2\mathbf{j})$ m.

   A force P of $(4\mathbf{i} + 10\mathbf{j})$ N acts through the point $(2\mathbf{i} - 3\mathbf{j})$ m.
   A force Q of $(-3\mathbf{i} - 2\mathbf{j})$ N acts through the point $(-\mathbf{i} + 9\mathbf{j})$ m.

   Show that the overall turning moment about point A is 19 N m, stating its sense. **[4 marks]**

8 A large beam AD is 10 m long and rests horizontally on supports at B and C which are 2 m from each end of the beam, as shown in the diagram. When a 1800 N force is applied vertically upwards at A, the beam is on the point of tilting about C. When a 1575 N force is applied vertically upwards at D instead, the beam is on the point of tilting about B.

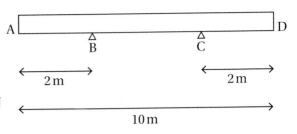

The beam is modelled as a non-uniform rod with its centre of mass x m from A.

Find:

a the weight of the beam   [6 marks]

b the value of x.   [2 marks]

9 Harry's uniform ladder of mass 50 kg and length 3.5 m is perched against a smooth wall at an angle of arctan 2 to the floor. Harry weighs 70 kg. The coefficient of friction between the ladder and the floor is $\frac{1}{3}$.

Find the farthest up the ladder that Harry can climb before the ladder slips.   [8 marks]

10 A wooden beam AB has a length of 7.2 m and a mass of 2.5 kg. A is at rest on a rough horizontal platform with B attached to a cable. The cable makes an angle β with the horizontal and the tension in the cable is T N. The beam is inclined at an angle α to the horizontal as shown in the diagram, such that $\sin \alpha = 0.6$. The beam is in limiting equilibrium and the coefficient of friction between the ground and the beam is $\frac{1}{2}$. The beam is uniform.

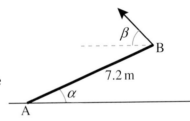

a Find the magnitude of the vertical reaction force on the beam at A.   [5 marks]

b State any modelling assumptions you have made.   [2 marks]

c Given that $\sin \beta = 0.28$, show that T is 17.5.   [2 marks]

11 The non-uniform 6 m rod AB is supported by two rests at C and D as shown in the diagram.

Particles are attached to the rod at A and B.

When the particle at A is 100 kg and the particle at B is 20 kg, the rod is on the point of tilting about C.

When the particle at A is 20 kg and the particle at B is 250 kg, the rod is on the point of tilting about D.

Find the mass of the rod and the position of its centre of mass.   [10 marks]

12 Two students, Colin and Nevita, were asked to find the position of the centre of mass of a 7 m rod which was held at its ends by two strings at 40° and 60° to the horizontal. The weight of the rod was $3g$ N. The students were given this diagram.

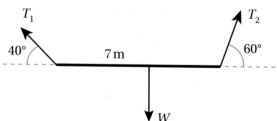

Colin resolved horizontally and vertically and used simultaneous equations to show that $T_2 = 22.9$ N.

He then took moments about the left-hand end of the rod, writing $3gx = 22.9 \sin 60° \times 7$, from which he deduced that x was 4.72 m.

Nevita annotated the diagram as shown and wrote down $x \tan 40° = (7 - x) \tan 60°$, from which she deduced that x was 4.72 m.

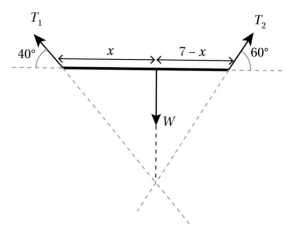

a Explain why $x \tan 40° = (7 - x) \tan 60°$. **[3 marks]**

b Write down Colin's method for finding T_2. **[4 marks]**

c Whose method is more efficient? Explain your answer. **[2 marks]**

13 A ladder of mass M kg and length $8a$ m stands against a rough wall with its foot on rough ground. The ladder is inclined at an angle of θ to the horizontal. The reaction force exerted by the ground on the ladder is given by R_G. The coefficient of friction at the wall is given by μ_W and coefficient of friction at the ground is given by μ_G. The centre of mass of the ladder is $6a$ m from the foot of the ladder.

a By resolving, find R_G in terms of M, μ_W and μ_G. **[4 marks]**

b By taking moments about the point of contact of the ladder with the wall, find R_G in terms of M, μ_G and θ. **[4 marks]**

c Hence show that $\mu_W = \dfrac{3 - 4\mu_G \tan\theta}{\mu_G}$. **[1 mark]**

14 A uniform ladder of mass M kg and length $4a$ m rests on rough ground against a smooth vertical wall. The ladder is inclined at an angle of 60° to the horizontal. The coefficient of friction between the ladder and the ground is $\dfrac{5\sqrt{3}}{24}$. A boy of mass $\frac{1}{2}M$ kg wants to climb up the ladder. Find, in terms of a, the maximum distance the boy can climb up the ladder before it slips. **[10 marks]**

EXAM-STYLE EXTENSION QUESTIONS

Chapter 1 Algebra and functions 1: Functions

(PF) 1 The function f is defined by $f : x \mapsto \dfrac{x}{x^2 - 4} - \dfrac{1}{x+2}$, $x > 2$

     a Show that $f(x) = \dfrac{2}{x^2 - 4}$.   **[3 marks]**

     b Find the range of f.   **[2 marks]**

     c Find the inverse function $f^{-1}(x)$, stating its domain.   **[4 marks]**

     The function g is defined by

     $g : x \mapsto \ln(1 - 2x)$, $x < \dfrac{1}{2}$

     d Find the exact value of gf(3).   **[2 marks]**

     (From Core 3 Examination Questions)

(M) 2 The function f is defined by

     $f(x) = \dfrac{3}{x-1}$ $\{x \in R, \ x \neq 1\}$

     and the function g is defined by

     $g(x) = \dfrac{4}{x}$ $\{x \in R, \ x \neq 0\}$

     a Find $f^{-1}(x)$ **[2 marks]**

     b Find $g^{-1}(x)$ **[1 mark]**

     c Solve $gf^{-1} = fg^{-1}$ **[4 marks]**

(M) 3 The function f is defined by

     $f(x) = \dfrac{1 - 3x}{4x + 5}$ $\left\{x \in R, \ x \neq -\dfrac{5}{4}\right\}$

     a Find $f^{-1}(x)$ **[3 marks]**

     b Find the domain and range of $f^{-1}(x)$ **[2 marks]**

     c Solve $f^{-1}(x) = -6$ **[2 marks]**

(M) 4 The function f is defined as

     $f(x) = 6 - x$, $x \leqslant 2$

     $ = 2x + 5$, $x > 2$

     a Sketch the graph of f(x) for $-3 \leqslant x \leqslant 4$.   **[3 marks]**

     b Find the value(s) of x for which $f(x) = 10$.   **[3 marks]**

(M) 5 **a** Given that $f(x) = |3x - 5|$ and $g(x) = 2x$, sketch, on the same pair of axes, the graphs of $f(x)$ and $g(x)$.   **[2 marks]**

     b Use your graphs to solve $|3x - 5| > 2x$.   **[1 mark]**

     c On the same pair of axes, sketch the graphs of $2f(x)$ and $|g(x)|$.   **[2 marks]**

     d Use your graphs to solve $2f(x) > |g(x)|$.   **[1 mark]**

6 The function f is defined by

$$f(x) = e^x \qquad\qquad \{x \in R, x > 0\}$$

and the function g is defined by

$$g(x) = 2x^2 - 3 \qquad\qquad \{x \in R\}$$

a Find $f^{-1}(x)$ **[2 marks]**

b Find $fg(x)$ **[1 mark]**

c Solve $fg(x) = 8$. Give your answer(s) to 1 decimal place.   **[3marks]**

(PS) 7 The function f is defined by

$$f(x) = \frac{5}{x-2} - \frac{2x-1}{x^2-4}$$

a Write $f(x)$ in the form $\dfrac{Ax + B}{(x-2)(x+2)}$ and state the domain.   **[3 marks]**

b Find $f^{-1}(x)$ **[4 marks]**

(PS) 8 The function f is defined by

$$f(x) = e^{x+1} \qquad\qquad \{x \in R, x > -1\}$$

and the function g is defined by

$$g(x) = x^2 - x - 6 \qquad\qquad \{x \in R\}$$

Solve $gf(x) = 10$. Give you answer(s) to 3 decimal places.   **[8 marks]**

(CM) 9 The function f is defined by

$$f(x) = 2x \qquad\qquad \{x \in R\}$$

and the function g is defined by

$$g(x) = \sin x \qquad\qquad \left\{x \in R, \ -\frac{\pi}{2} \leqslant x \leqslant \frac{\pi}{2}\right\}$$

a Find $fg(x)$ and $gf(x)$ **[2 marks]**

b On the same pair of axes sketch the graphs of $fg(x)$ and $gf(x)$. Describe the transformations that map $g(x)$ to $fg(x)$ and $g(x)$ to $gf(x)$ **[2 marks]**

c Find $g^{-1}(x)$ and state its domain.   **[2 marks]**

d On a separate pair of axes sketch the graph of $g^{-1}(x)$ **[2 marks]**

(PS) 10 The function f is defined by

$$f(x) = x^3 + 17 \qquad\qquad \{x \in R\}$$

and the function g is defined by

$$g(x) = \frac{8}{x^2 - 7} \qquad\qquad \left\{x \in R, \ x \neq \sqrt{7}\right\}$$

a Find $f^{-1}(x)$.   **[2 marks]**

b Solve $f^{-1}g(x) = 7$. Give your answer(s) to 3 decimal places.   **[7 marks]**

Chapter 2 Algebra and functions 2: Partial fractions

1 **a** Express $f(x) = \dfrac{3(x+1)}{(x+2)(x-1)}$ in partial fraction form. **[3 marks]**

b Hence, or otherwise, find a series expansion for f(x) in ascending powers, up to and including the term in x^2. For what interval of x is your series expansion valid? **[6 marks]**

c Use your partial fractions in (a) to find the exact value of $\displaystyle\int_2^5 \dfrac{3(x+1)}{(x+2)(x-1)}\,dx$, giving your answer as a single logarithm. **[5 marks]**

(From Core 4 Examination Questions)

(PF) **2** **a** Given that $f(x) = 2x^3 + 11x^2 + 10x - 8$ show that $\dfrac{5}{f(x)}$ can be expressed as partial fractions. **[6 marks]**

(PS) **b** Hence, or otherwise, find the exact value of the gradient of $\dfrac{5}{f(x)}$ when $x = -3$. **[4 marks]**

(PS) **3** **a** Express $f(x) = \dfrac{3x^2 - 4x + 1}{(x-1)(x^2 - 4)}$ in partial fraction form. **[3 marks]**

(M) **b** Determine if there are any real solutions to $f'(x) = 0$. **[6 marks]**

(M) **4** **a** Express $f(x) = \dfrac{56}{(9x^2 - 4)}$ as partial fractions in the form $= \dfrac{A}{(3x - 2)} + \dfrac{B}{(3x + 2)}$. **[3 marks]**

b Hence, or otherwise, find the exact value of $\displaystyle\int_2^3 \dfrac{56}{(9x^2 - 4)}\,dx$, giving your answer as a single logarithm. **[7 marks]**

(PF) **5** **a** Express $f(x) = \dfrac{3 - 5x}{(x-3)(x-7)}$ in partial fraction form. **[3 marks]**

b Hence, or otherwise, find a series expansion for f(x) in ascending powers, and show that the coefficient of x^2 is $-\dfrac{271}{3087}$. **[6 marks]**

(M) **6** Express $\dfrac{7 + 4x - x^2}{-x^3 + 3x^2 + 4x - 12}$ in partial fraction form. **[6 marks]**

(M) **7** **a** Express $f(x) = \dfrac{3x^2 + 4x - 7}{(x-2)^3}$ in partial fraction form. **[4 marks]**

b Hence, or otherwise, find $f'(x)$ **[3 marks]**

(PF) **8** **a** Show that $f(x) = \dfrac{4x^2 + 7x - 3}{(x-1)(x-2)^2}$ cannot be split into partial fractions in the form

(CM) $\dfrac{A}{x-1} + \dfrac{B}{(x-2)} + \dfrac{C}{(x-2)}$.

Clearly state the reasoning for your conclusion. **[4 marks]**

b Express $f(x) = \dfrac{4x^2 + 7x - 3}{(x-1)(x-2)^2}$ in partial fraction form. **[4 marks]**

(PS) **9** Find the gradient of the curve C given by $f(x) = \dfrac{x^4 - x^3 + x^2 + 1}{x^2 - 3x - 4}$ when $x = 1$. **[8 marks]**

(M) **10** **a** Express $f(x) = \dfrac{8x + 17}{(x+1)^2}$ in partial fraction form. **[3 marks]**

b Hence, or otherwise, find a series expansion for f(x) in ascending powers, up to and including the term in x^3. **[4 marks]**

Chapter 3 Coordinate geometry: Parametric equations

1 A curve has parametric equations

$$x = 3\sin t, \ y = \cos 2t, \qquad 0 \le t \le \pi$$

  a Find an expression for $\dfrac{dy}{dx}$ in terms of the parameter t.   **[4 marks]**

  b Find the equation of the tangent at the point where $t = \dfrac{\pi}{3}$ **[4 marks]**

  c Find a Cartesian equation of the curve in the form $y = f(x)$.   **[4 marks]**

    (From Core 4 Examination Questions)

(PS) 2 The curve C shown is given by the parametric equations $x = t^2 + 3$ and $y = \sin t$. C intersects the x-axis at points A and B.

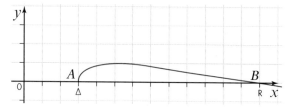

  a Find the exact coordinates of points A and B.   **[3 marks]**

  b Find the exact coordinates of the local maximum between A and B.   **[4 marks]**

  c Find the area under the curve between A and B.   **[6 marks]**

(PS) (M) 3 A curve C is given by the parametric equations $x = t^2 - 1$ and $y = 3t - t^2$.

  a Draw a graph of the curve C where $0 \le t \le 3$.   **[4 marks]**

  b Find the equation of the normal to the curve at $(8, 0)$. Give your answer in the form $ax + by + c = 0$.   **[6 marks]**

(PS) (M) 4 An object moves with constant velocity and the parametric equations for the position of the object are $x = 15 - 7t$ and $y = 9t + 17$.

  a Draw a graph of the position of the object where $0 \le t \le 5$.   **[4 marks]**

  b Find the distance travelled by the object in the first 5 seconds of its journey to the nearest unit.   **[3 marks]**

  c Find the speed of the object.   **[3 marks]**

  d Find the bearing, to the nearest degree, on which the object is moving.   **[2 marks]**

(M) 5 The equation of a line is given by the parametric equations $x = \dfrac{4 + 5t}{3 - t}$ and $y = \dfrac{8 - 7t}{1 + t}$.

  a Find the exact coordinates of the x-intercept.   **[2 marks]**

  b Find the Cartesian equation of the line.   **[6 marks]**

(CM) (PS) 6 The parametric equations for an ellipse are $x = 5\cos t + 1$ and $y = 3\sin t - 2$. Will the ellipse fit inside a rectangle that is 11 units by 5.5 units? You must show all your working and provide justification for your answer.   **[6 marks]**

(PS) (M) 7 A curve C is given by the parametric equations $x = t^2$ and $y = t^3 - 1$. Find the y-intercept of the tangent to the curve at $(9, 26)$.   **[5 marks]**

(CM) (M) **8** Find the Cartesian equation of the curve C given by the parametric equations $x = \sin 3t - 5$ and $y = \cos 3t + 7$. Describe the shape of the curve. **[5 marks]**

(PS) **9** A curve C is given by the parametric equations $x = 2 - t - t^2$ and $y = t^3$. A line L is given by the equation $4x + y + 8 = 0$.

Find the area A under the curve C, bounded by the line L and the x-axis. **[8 marks]**

(PS) **10** A curve C is given by the parametric equations $x = t^3 - 2$ and $y = \sqrt{t}$. Find the area bounded by the curve and the x-axis from $x = -2$ to $x = 5$ by evaluating $\int y \frac{dx}{dt} dt$. Show that your answer is correct by finding and integrating the Cartesian equation of C. **[marks]**

11 Find the Cartesian equation for each of the curves given by the following parametric equations:

a $x = 4\sin t$ and $y = 5\cos t$

b $x = 3\cos t + 1$ and $y = 7\sin t - 4$

c $x = \sin 2t$ and $y = \cos 2t$

Which one(s) of a, b and c are circles? For those that are, state the radius and coordinates of the centre of the circle.

12 Find the Cartesian equation for each of the curves given by the following parametric equations:

a $x = \sin t$ and $y = \cos 2t$

b $x = 2\cos t$ and $y = 3\cos 2t$

c $x = \tan 2t$ and $y = \sin 2t$

(PF) **13** Show that the Cartesian equation for the curve given by the parametric equations $x = \cos t$ and $y = \sin 2t$ can be written as $y = 2x\sqrt{1-x}$.

Chapter 4 Sequences and series

(PS) **1** On her first birthday, an uncle gives his niece £50.

On each subsequent birthday, he gives her £5 more than he gave her on the previous birthday.

How old is the niece when the **total** amount given first exceeds £1000? **[4 marks]**

(PF) **2** The nth term of a sequence is given by the formula $a_n = \dfrac{2n + 3}{n + 1}$ $n = 1, 2, 3, \ldots$

a Prove that this is a decreasing sequence. **[2 marks]**

b Prove that the sequence approaches a limit and state its value. **[2 marks]**

(PS) **3** The first term of a geometric series is 5. The sum to infinity is 100.

Work out how many terms are needed before the sum of the series is more than 50. **[4 marks]**

4 In January, a saver puts £40 in a savings account.

Each subsequent month, the saver deposits £5 more than the previous month.

How many months does it take to save £2000? **[4 marks]**

(M) (MR) **5** Some years ago, the following 'chain-letter scheme' was in operation.

Carol Black	Glasgow
John Brown	London
Alice Green	Birmingham
Peter White	Leeds

A person received a postcard with a list of four names and addresses on it.

The receiver was asked to send £1 to the name at the top of the list. In this example, the money goes to Carol Black.

He or she then sends **four** postcards to four different people. The list of names is the same but with the top name removed and their own name and address at the bottom.

John Brown	London
Alice Green	Birmingham
Peter White	Leeds
Mick Scarlet	Swansea

The scheme starts with one person sending four postcards. This is stage 1.

Then those four people each send £1 and four postcards. This is stage 2.

Assume that everyone does what they are asked and no-one receives more than one postcard.

a Theoretically, how much money will be sent to you if you receive a card and carry out the instructions. **[1 mark]**

b Work out the total number of £1 payments made after stage 10. **[2 marks]**

c Work out the total number of postcards sent after stage 10. **[1 mark]**

d The person who started this chain claimed that everyone would make money from it. Explain why this is impossible, even if everyone follows the instructions. **[1 mark]**

(PF) **6** **a** Prove that the sum of the first n odd numbers is n^2 **[3 marks]**

b Find a formula for the sum of the first n multiples of 3 **[2 marks]**

(MR) **7** **a** Find a formula in terms of n for $\sum_{i=1}^{n} 80 - 4i$. **[2 marks]**

b What can you say about the value of n if the sum in part (a) is negative? **[2 marks]**

(M) **8** A man borrows £2835 from a friend. He makes a repayment each month.

In the first month he pays £200, in the second he pays £195, in the third he pays £190, and so on.

How many months does he take to repay the loan? **[4 marks]**

9 **a** Write down a cubic approximation for $\dfrac{1}{x+1}$ for small values of x. **[1 mark]**

b Use your answer to part (a) to work out a cubic approximation to $\dfrac{1}{x+2}$ for small values of x. **[2 marks]**

c Use your answers to parts (a) and (b) to find a cubic approximation to $\dfrac{1}{x^2+3x+2}$ for small values of x. **[5 marks]**

10 **a** Find the first four terms of a binomial expansion of $\sqrt[3]{8 + x^2}$, stating the range of values of x for which the expansion is valid. **[4 marks]**

b Show that the expression in part (a) gives the value of $\sqrt[3]{9}$ correct to 4 decimal places. **[1 mark]**

 11 **a** The first term of an arithmetic sequence is a and the common ratio is r where $|r| < 1$

Prove that the sum of the first n terms of the sequence is $\dfrac{a\left(1 - r^n\right)}{1 - r}$ **[3 marks]**

b The first three terms of a geometric series are $10 + 8 + 6.4 + \cdots$

Find the number of terms required before the difference between the sum of the series and the sum to infinity is less than 0.1 **[4 marks]**

 12 A UN website says that the world population in 2016 was 7.4 billion. They estimate that the world population will be 11.2 billion in 2100.

Assume that the population each year from 2016 to 2100 can be modelled as the terms of a geometric sequence.

a Find the first year in which the world population will exceed 10 billion. **[5 marks]**

b Explain why a geometric sequence is a better model that an arithmetic sequence. **[1 mark]**

 13 Find the exact value of $\displaystyle\sum_{r=1}^{20} \frac{4}{(r + 2)(r + 3)}$ **[5 marks]**

14 **a** Find a quadratic approximation to $\sqrt{1 - x}$ for small values of x. **[2 marks]**

b Find a quadratic approximation to $\dfrac{\sqrt{1 - x}}{1 + x}$ for small values of x. **[4 marks]**

15 **a** The first term of an arithmetic sequence is a and the common difference is d.

Show that the sum of n terms is $\dfrac{n}{2}\{2a + (n - 1)d\}$ **[2 marks]**

b The sum of the first 20 terms of an arithmetic sequence is 1070.

The sum of the first 30 terms is 2055.

Find a and d. **[2 marks]**

Chapter 5 Trigonometry

PS 1

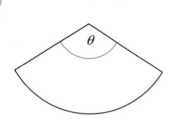
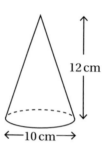

A sector is cut out of paper and the straight edges are joined to make a cone.

The height of the cone is 12 cm and the diameter of the base is 10 cm.

Calculate the angle, θ, of the sector, in radians. **[4 marks]**

MR 2 Use standard small angle approximations for sine and cosine to show that, if x is small,

$$\frac{\sin^2 2x}{(\cos 0.5x - 1)} \approx -32$$

[3 marks]

3 Sketch the graph of $y = 4\arcsin 2x$ and state the domain and range of the function.

Angles are in radians. **[3 marks]**

PS 4

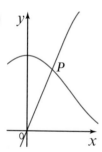

The graphs $y = 2\sin x$ and $y = \cos^2 x$ cross at the point P and x is in radians.

Work out the coordinates of P. **[3 marks]**

5 OA and OB are rods of length 6 units and 4 units.

Angle OAB is a right angle and the rods can rotate about the origin O.

The angle between OA and the positive x-axis is θ radians.

a Show that the x-coordinate of B is $6\cos\theta + 4\sin\theta$ **[1 mark]**

b Write the x-coordinate of B in the form $r\cos(\theta - \alpha)$ where r and α are constants. **[3 marks]**

c Find the possible values of θ if the x-coordinate of B is 5 **[3 marks]**

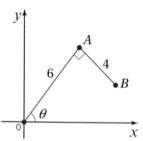

(M) **6** The depth, y m, of water in a harbour t hours after midnight on one day is given by the formula $y = 3.5 + 2.8\sin(15t + 60)°$ $0 \leqslant t \leqslant 24$

    a Sketch a graph of the height through the 24-hours. [2 marks]

    b Find the times when the depth of water is more than 4 m. [4 marks]

7 **a** Prove that $\sin x + \cos x \equiv \sqrt{2}\sin\left(x + \dfrac{\pi}{4}\right)$ [2 marks]

    b Find the exact value of $\sin\dfrac{7\pi}{12}$ [2 marks]

(MR) **8** **a** Show that $\sin(x + y) + \sin(x - y) = 2\sin x \cos y$ [2 marks]

    b Show that $\sin a + \sin b = 2\sin\dfrac{a + b}{2}\cos\dfrac{a - b}{2}$ [2 marks]

    c Find a similar expression for $\cos a + \cos b$ [2 marks]

(PF) **9** Show that $\cos 4x = 8\cos^4 x - 8\cos^2 x + 1$ [3 marks]

10 Solve the equation $10\sin x - 8\cos x = 6$

    when x is in radians and $0 < x < 2\pi$ [5 marks]

(PS) **11** **a** Find the exact value of $\sin 75°$ [2 marks]

(PF)   **b** Show that $\tan 75° = 2 + \sqrt{3}$ [3 marks]

(MR) **12** Show that $\dfrac{\sin 3\theta}{\sin\theta} = \cos 2\theta + 2\cos^2\theta$ [4 marks]

(MR) **13** **a** Solve the equation $\tan x + \cot x = 4$ $0° < x < 360°$ [4 marks]

    b Find the range of values of a for which the equation $\tan x + \cot x = a$ has no solution. [2 marks]

(PF) **14** Prove that $\sec 2x = \dfrac{\sec^2 x}{2 - \sec^2 x}$ [3 marks]

(PF) **15** Show that $\tan 3x = \dfrac{3\tan x - \tan^3 x}{1 - 3\tan^2 x}$ [3 marks]

Chapter 6 Differentiation

(M) **1** The number of apes in an African country is declining.

    A model for the size of the population (y apes) in t years' time is $y = 4600e^{-0.077t}$

    a In how many years will the population be half its present size? [3 marks]

    b Find the rate of decline of the population at that time. [3 marks]

(MR) **2** The equation of a curve is $y = x^4 - 8x^3 + 20$

    Show that the curve has one minimum point and find its coordinates. [4 marks]

3 **a** $f(x) = \sin x - \sin^2 x$

      Find the stationary points of the graph of $y = f(x)$ in the interval $0 < x < \pi$ [4 marks]

    b Sketch the graph for the interval $0 < x < \pi$ [1 mark]

4 Differentiate $\sqrt[3]{1 - 2x^2}$ with respect to x. [3 marks]

5 Find the coordinates of any stationary points of the curve with the equation
$y = (x^2 - 10x + 30)^3$. **[3 marks]**

6 Differentiate with respect to x:

a $\cos 2x^2$ **[2 marks]**

b $\cos^2 2x$ **[2 marks]**

7 **a** $y = e^{ax} - e^{-ax}$, where a is a positive constant.

Show that $\dfrac{d^2 y}{dx^2} = a^2 y$. **[2 marks]**

(MR) **b** Write down and prove a similar result for $y = \cos ax - \sin ax$. **[3 marks]**

8 $y = \ln(1 + 2x)$ $x > -\dfrac{1}{2}$

Show that $\dfrac{d^2 y}{dx^2} = -\left(\dfrac{dy}{dx}\right)^2$. **[3 marks]**

9 $f(x) = 2^{\sqrt{x}}$ $x > 0$

a Show that $(1, 2)$ and $(4, 4)$ are on the graph of $y = f(x)$. **[1 mark]**

b Show that the tangents to the curve at $(1, 2)$ and $(4, 4)$ are parallel. **[3 marks]**

10 **a** Differentiate $\cos^2 x$ with respect to x. **[1 mark]**

b Differentiate $\dfrac{1}{2}\cos 2x$ with respect to x. **[1 mark]**

(MR) **c** Show that the answers to parts (a) and (b) are identical and explain why this is the case. **[2 marks]**

11 Differentiate $\sin 2x$ with respect to x from first principles. **[5 marks]**

12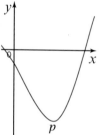

This is a sketch of the curve $y = \cos\left(e^x + \dfrac{\pi}{4}\right)$.

Find the coordinates of the minimum point P. **[4 marks]**

13 If $y = \ln\dfrac{2x^2}{1 + x^2}$, show that $\dfrac{dy}{dx} = \dfrac{2}{x(1 + x^2)}$. **[4 marks]**

14 $y = \cos^2 x$

Show that $\dfrac{d^2 y}{dx^2} + 4y = 2$. **[4 marks]**

15 The equation of a curve is $y = x^3 + x^2 + cx$ where c is a constant.

The curve has a stationary point that is a point of inflection.

a Find the value of c. **[4 marks]**

(PS) **b** Find the coordinates of the stationary point. **[1 mark]**

Chapter 7 Further differentiation

1 The equation of a curve is $x^2 - 2y^2 = 28$.

Find the equation of the tangent at the point $(6, 2)$. **[4 marks]**

2 The parametric equation of a curve is $x = t - \dfrac{1}{t}$ and $y = t + \dfrac{4}{t}$. $\quad t > 0$

Show that the curve has a minimum point and find its coordinates. **[4 marks]**

3

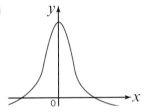

The equation of this curve is $y = \dfrac{4 - x^2}{1 + 2x^2}$.

Find the gradient of the curve at each point where it crosses the x-axis. **[5 marks]**

(MR) **4** $y = \sec x$

Show that $\dfrac{\mathrm{d}^2 y}{\mathrm{d}x^2} = 2\sec^3 x - \sec x$. **[4 marks]**

5

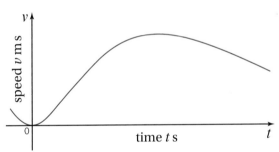

A car is accelerating from rest. The speed of the car, $v\,\mathrm{m\,s^{-1}}$, after t seconds is given by the formula $v = t^2 e^{-0.2t}$.

a Find the acceleration after 5 seconds. **[3 marks]**

(M) **b** Find the maximum speed of the car. **[2 marks]**

6 The equation of a curve is $y = \dfrac{e^x}{1 + x^2}$ and it has a stationary point.

a Find the coordinates of the stationary point. **[2 marks]**

b State the type of stationary point and justify your answer. **[2 marks]**

7 $y = \arctan 0.5x$

Find the coordinates of any point where the gradient is 0.1. **[3 marks]**

8 **a** Differentiate $\sqrt{x^2 + 3}$ with respect to x. **[1 mark]**

b Sajiv uses a different method.

He writes $y^2 = x^2 + 3$ and differentiates this implicitly.

Show that this method gives the same answer. **[2 marks]**

c Olivia uses another method.

She writes x as a function of y and finds $\dfrac{dx}{dy}$ and hence $\dfrac{dy}{dx}$.

Show that this method gives the same answer. **[3 marks]**

9 $y = e^{3x} \sin 2x$

Find $\dfrac{d^2y}{dx^2}$. **[4 marks]**

10

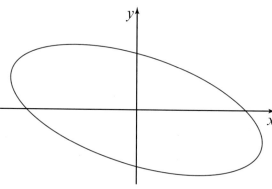

The equation of this ellipse is $x^2 + 2xy + 4y^2 = 36$.

a Find the gradient where the curve crosses the y-axis. **[4 marks]**

b Find the exact coordinates of the maximum point. **[2 marks]**

11

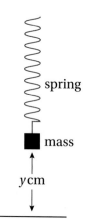

A mass on the end of a spring is pulled down and released.

The graph shows how the vertical height (y cm) varies with time.

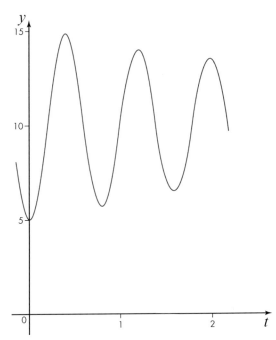

The height after t seconds is given by the formula $y = 10 - 5e^{-0.2t}\cos 8t$.

a Work out the speed of the mass after 5 seconds. **[4 marks]**

b Find the time when the mass gains its maximum height. **[3 marks]**

12 $x = \dfrac{2t}{1+t^2}$ and $y = \dfrac{1-t^2}{1+t^2}$

Show that $\dfrac{\mathrm{d}y}{\mathrm{d}x} = -\dfrac{x}{y}$ **[5 marks]**

13

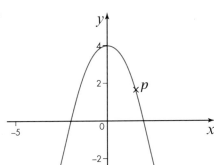

P is a point on the graph of $y = 4 - x^2$ and O is the origin.

Find the minimum possible length of OP. **[5 marks]**

14 Differentiate x^x with respect to x. **[4 marks]**

15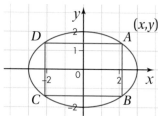

The equation of the ellipse is $\dfrac{x^2}{9} + \dfrac{y^2}{4} = 1$.

a Show that $\dfrac{dy}{dx} = -\dfrac{4x}{9y}$. **[1 mark]**

b $ABCD$ is a rectangle and the coordinates of A are (x, y).

Explain why the area of the rectangle is $4xy$. **[1 mark]**

c Show that the maximum possible area of the rectangle is 12 square units. **[4 marks]**

Chapter 8 Integration

1 Find the area enclosed by the curve $y = x^2 + 4$ and the line $y = 12 - 2x$. **[5 marks]**

2 Work out $\displaystyle\int \dfrac{10}{(4x - 1)^3}\, dx$ **[3 marks]**

3 **a** Solve the differential equation $\dfrac{dy}{dx} = -2xy$ **[3 marks]**

b Sketch the graph of the particular solution that passes through the point $(0, 100)$. **[2 marks]**

4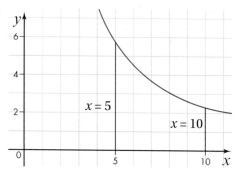

Find the area between the curve $y = \dfrac{40}{2x - 3}$ and the x-axis between $x = 5$ and $x = 10$. Give your answer to 2 decimal places. **[4 marks]**

5 Find $\displaystyle\int \ln 2x\, dx$ **[4 marks]**

6

a Find the area between the graph of $y = 2^{-x}$ and the x-axis between $x = 0$ and $x = a$. **[3 marks]**

b What is the limit of the area as $a \to \infty$? **[1 mark]**

7 Find $\int x\sqrt{x+4}\,dx$ **[4 marks]**

8 Find $\int x\sin x\cos x\,dx$ **[4 marks]**

9 Find:

    a $\int \dfrac{1}{x^2+4}\,dx$ **[2 marks]**

    b $\int \dfrac{x}{x^2+4}\,dx$ **[2 marks]**

    c $\int \dfrac{x^2}{x^2+4}\,dx$ **[3 marks]**

10

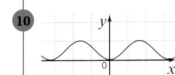

This is a graph of $y=\sin^2 x$.

Find the area between one wave and the x-axis. **[4 marks]**

11 Find $\int e^{-0.5x}\sin x\,dx$ **[6 marks]**

(MR) 12 Show that $\displaystyle\int_0^{\frac{\pi}{2}} \tan 0.5x\,dx = \ln 2$ **[4 marks]**

13 An infectious disease is spreading through a city.

7 days after the infection is first reported, 100 people are infected.

t days after the first report, x people are infected and the rate of increase of x is modelled by the differential equation $\dfrac{dx}{dt} = 0.08x$.

    a Find a formula for the number of people infected after t days. **[4 marks]**

    b Explain why the model is no longer appropriate if t is large. **[1 mark]**

14 Find $\int \dfrac{x+5}{x^2+4x+3}\,dx$ **[5 marks]**

15 The origin of a pair of axes is at the centre of a sphere of radius r.

The sphere is approximately equal to a large number of circular discs. The side view of one disc is shown in the diagram.

    a If the thickness of the disc is δx, show that the volume of the disc is $\pi(r^2 - x^2)\,\delta x$. **[1 mark]**

    b Use integration to prove that the volume of the sphere is $\dfrac{4}{3}\pi r^3$. **[4 marks]**

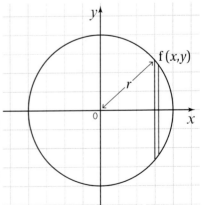

Chapter 9 Numerical methods

 1 **a** If $y = 5^x$ complete this table giving the values of 5^x to 3 decimal places.

x	0	0.2	0.4	0.6	0.8	1
y	1		1.904			5

[2 marks]

b Use the trapezium rule with 5 strips to find an approximation for the value of $\int_0^1 5^x dx$ **[4 marks]**

(From Core 2 Examination Questions)

 2 A function is given by $f(x) = x^3 - 2x^2 - 3$

a Show that the equation $f(x) = 0$ has a root in the interval $[2, 3]$. **[2 marks]**

b Show that $f(x) = 0$ can be rearranged as

$$x = \sqrt{\left(2x + \frac{3}{x}\right)}, \qquad\qquad x \neq 0$$

The root is estimated using the iterative formula

$$x_{n+1} = \sqrt{\left(2x_n + \frac{3}{x_n}\right)} \text{ with } x_1 = 2$$

[3 marks]

c Use this iterative formula to find the values of x_2, x_3 and x_4 correct to 3 decimal places. **[3 marks]**

d Prove that the root is 2.486 correct to 3 decimal places. **[2 marks]**

(From Core 3 Examination Questions)

 3 The trapezium rule is to be used to approximate the integral

$$\int_0^2 e^{-x^2} dx$$

a Complete the table with the missing values of y.

x	0	0.5	1.0	1.5	2.0
y	1		e^{-1}		e^{-4}

[2 marks]

b Use the trapezium rule with five y values to approximate the integral

$$\int_0^2 e^{-x^2} dx$$

[4 marks]

(From Core 4 Examination Questions)

 4 **a** Show that the equation $x^2 - 4x - 7 = 0$ has a root in the interval $5 < x < 6$. **[2 marks]**

b Using an iteration formula in the form $x_{n+1} = (ax_n + b)^{\frac{1}{2}}$, find an estimate for the root of $x^2 - 4x - 7 = 0$ to 3 decimal places. **[4 marks]**

c Calculate the percentage error of your answer in part (b). **[3 marks]**

(M) **5** $f(x) = 2^{x-1} - 3x$

    a Use the iteration formula $\frac{1}{6}(2^{x_n}) = x_{n+1}$ with $x_0 = 1$ to find an approximation to 3 decimal places for the root near to 1. **[4 marks]**

    b Use the iteration formula $\frac{1}{3}(2^{x_n-1}) = x_{n+1}$ with $x_0 = 1$ to find an approximation to 3 decimal places for the root near to 1. **[4 marks]**

    c Comment on the two different formulae used in parts (a) and (b). **[1 mark]**

(M) **6** $\text{Area} = \int_0^{\frac{\pi}{4}} \tan x \, dx$

    a Find the exact value of the area. **[3 marks]**

    b Use the trapezium rule with 2 strips to find an approximation for the area. **[2 marks]**

    c Use the trapezium rule with 4 strips to find an approximation for the area. **[3 marks]**

    d Find the percentage error for the more accurate result in part (b) or (c). **[2 marks]**

7 $f(x) = x + e^x$

    a Show that $f(x)$ has a root in the interval $-1 < x < 0$. **[2 marks]**

    b Use the Newton–Raphson method to find the root in the interval $-1 < x < 0$, to 3 decimal places. **[4 marks]**

    c Show graphically the first two iterations for the formula. **[3 marks]**

8 **a** Sketch the graphs of $y = x^2 - 3$ and $y = \ln x$ on the same pair of axes. **[3 marks]**

    b Show that $x^2 - \ln x - 3 = 0$ has two solutions in the interval $0 < x < 2$. **[2 marks]**

    c **i** Use the iteration formula $e^{x_n^2 - 3} = x_{n+1}$ with $x_0 = 1$ four times to find an approximation for the root near to 1. **[4 marks]**

        ii Use the iteration formula $(\ln x_n + 3)^{\frac{1}{2}} = x_{n+1}$ with $x_0 = 1$ four times to find an approximation for the root near to 1. **[4 marks]**

    d Comment on your findings in part (c). **[1 mark]**

9 $f(x) = x^3 - x^2 + 2x - 5$

    a Show graphically that the equation $x^3 = x^2 - 2x + 5$ has one root. **[3 marks]**

    b Show that this root is in the interval $0 < x < 2$. **[2 marks]**

    c Use the Newton-Raphson method to find the root, to 2 decimal places, that is in the interval $0 < x < 2$. **[4 marks]**

(PS) **10** Use the Newton–Raphson process to find an approximation of $\ln 13$ to four decimal places. **[7 marks]**

Chapter 10 Three-dimensional vectors

(PF) (M) 1 The rooms in a hemispherical dome are modelled using vectors.

There are offices at position vectors $(7\mathbf{i} + 9\mathbf{j} + 12\mathbf{k})$ m, $(11\mathbf{i} + \mathbf{j} + 12\mathbf{k})$ m and $(3\mathbf{i} + \mathbf{j} + 12\mathbf{k})$ m.

a Show that the offices lie on the circumference of a circle with centre $(7\mathbf{i} + 4\mathbf{j} + 12\mathbf{k})$. **[3 marks]**

The main reception has position vector $(7\mathbf{i} + 4\mathbf{j})$ m. There are escalators from the main reception to each of the offices.

b Find the length of each escalator. **[2 marks]**

(PF) 2 Points A, B and C have position vectors $\begin{bmatrix} 10 \\ -4 \\ 32 \end{bmatrix}$, $\begin{bmatrix} 8 \\ p \\ q \end{bmatrix}$ and $\begin{bmatrix} 4 \\ 14 \\ 2 \end{bmatrix}$, respectively.

a Find the values of p and q such that A, B and C are collinear. **[3 marks]**

Point D has position vector $\begin{bmatrix} 20 \\ 11 \\ 25 \end{bmatrix}$.

b Show that BD is perpendicular to AC. **[4 marks]**

(PF) 3 W, X, Y and Z have position vectors $(8\mathbf{i} + 3\mathbf{j} + 9.5\mathbf{k})$, $(22\mathbf{i} + 21\mathbf{j} + 72.5\mathbf{k})$, $(116.5\mathbf{i} + 42\mathbf{j} + 45.5\mathbf{k})$ and $(102.5\mathbf{i} + 24\mathbf{j} - 17.5\mathbf{k})$.

a Verify that $WY = XZ$. **[4 marks]**

b Verify that $WX = \frac{2}{3}WZ$. **[2 marks]**

c Hence state what type of quadrilateral $WXYZ$ is. **[1 mark]**

(PF) 4 The vector $\overrightarrow{XY}$ is given by $\begin{bmatrix} 8 \\ -9 \\ 10 \end{bmatrix}$. The vector $\overrightarrow{XZ}$ is given by $\begin{bmatrix} 1 \\ 5 \\ -4 \end{bmatrix}$.

M divides $\overrightarrow{YZ}$ in the ratio $4:3$.

a Find $\overrightarrow{XM}$. **[3 marks]**

b Given that $\overrightarrow{OM} = \begin{bmatrix} -1 \\ 3 \\ 5 \end{bmatrix}$

find $\overrightarrow{OZ}$. **[2 marks]**

(PF) 5 Points P and R have position vectors $(-63\mathbf{i} + 102\mathbf{j} - 26\mathbf{k})$ and $(25\mathbf{i} - 52\mathbf{j} + 30\mathbf{k})$, respectively.

T is the midpoint of the line segment PR.

a Find the position vector of T. **[2 marks]**

Point Q has position vector $(58\mathbf{i} + 53\mathbf{j} + a\mathbf{k})$.

b Given that PT is perpendicular to QT,

find the value of a. **[4 marks]**

c Find the area of the triangle PQR. **[3 marks]**

(PF) **6** Quadrilateral *STUV* has vertices with position vectors $S(9\mathbf{i} + \mathbf{j} - 9\mathbf{k})$, $T(2\mathbf{i} + 11\mathbf{j} + 6\mathbf{k})$, $U(-11\mathbf{i} + 14\mathbf{j} + 20\mathbf{k})$ and $V(-4\mathbf{i} + 4\mathbf{j} + 5\mathbf{k})$. Prove that *STUV* is a rhombus. **[8 marks]**

(PF) **7** Points *A* and *C* have position vectors $(17\mathbf{i} + 16\mathbf{j} - \mathbf{k})$ and $(29\mathbf{i} + 22\mathbf{j} + 11\mathbf{k})$, respectively.

  a Find the position vector of E such that AE:EC = 1:1. **[2 marks]**

  Points *B* and *D* have position vectors $(27\mathbf{i} + 15\mathbf{j} + 3\mathbf{k})$ and $(9\mathbf{i} + 33\mathbf{j} + 12\mathbf{k})$, respectively.

  b Show that *E* divides *BD* in the ratio 2:7. **[2 marks]**

  c Prove that *ABCD* is a kite. **[3 marks]**

  d Given that the area of *ABC* is 54 units², find the area of *ABCD*. **[2 marks]**

(PF) (PS) **8** Points *P*, *Q*, *R* and *S* are such that $\overrightarrow{OP} = \begin{bmatrix} 5 \\ 6 \\ 2 \end{bmatrix}$, $\overrightarrow{OQ} = \begin{bmatrix} 9 \\ 14 \\ 10 \end{bmatrix}$, $\overrightarrow{OR} = \begin{bmatrix} 13 \\ 10 \\ 12 \end{bmatrix}$ and $\overrightarrow{OS} = \begin{bmatrix} 15 \\ -4 \\ 7 \end{bmatrix}$.

  a Show that *PS* and *QR* are parallel. **[3 marks]**

  b State the ratio *PS*:*QR*. **[1 mark]**

  c Verify that $\angle QPS$ is 90°. **[3 marks]**

  d Find the area of *PQRS*. **[3 marks]**

(PS) **9** *ABCD* is a square such that *A* has position vector $\begin{bmatrix} 11 \\ -7 \\ 4 \end{bmatrix}$, *B* has position vector $\begin{bmatrix} 5 \\ 2 \\ 22 \end{bmatrix}$ and *C* has position vector $\begin{bmatrix} 23 \\ -4 \\ 31 \end{bmatrix}$.

  a Find the position vector of *D*. **[2 marks]**

  b Find the position vector of *E*, such that *BE*:*ED* = 2:1. **[3 marks]**

  c What type of quadrilateral is *ABCE*? **[1 mark]**

  d Find the area of *ABCE*. **[4 marks]**

(PF) **10** *A*, *B* and *C* have position vectors $(-2.6\mathbf{i} - 8\mathbf{j} - 13\mathbf{k})$, $(16.6\mathbf{i} + 4\mathbf{j} + 19\mathbf{k})$ and $(47\mathbf{i} - 26\mathbf{j} - 12\mathbf{k})$.

  M is the midpoint of *AB*.

  a Show that *MC* is perpendicular to *AB*. **[5 marks]**

  D is the reflection of *C* in the line segment *AB*.

  b Find the position vector of *D*. **[2 marks]**

Chapter 11 Proof

(PF) (M) **1** Show that if *n* is even then $(n - 1)^3$ is always odd. **[4 marks]**

(PF) (M) **2** Prove by contradiction that there do not exist integers *x* and *y* such that $7x + 28y = 11$. **[4 marks]**

(PF) (M) **3** Prove by contradiction that every non-zero real number does not have more than one reciprocal. **[4 marks]**

(PF) (4) Show that the quadratic formula for the equation $ax^2 - bx - c = 0$ is $x = \dfrac{b \pm \sqrt{b^2 + 4ac}}{2a}$
(CM)
(M) where $a, b, c \in \mathbb{R}$ and $a, b, c > 0$. Hence, or otherwise, state what can be deduced about quadratic equations of this form. **[4 marks]**

(PF) (5) Show that if n is odd then $n^2 + 2n + 1$ is even. **[4 marks]**
(M)

(PF) (6) Show using proof by contradiction that $\sqrt[3]{11}$ is irrational. **[5 marks]**

(PF) (7) Prove by contradiction that if x and y are positive real numbers, then $x + y \geqslant \sqrt{xy}$. **[6 marks]**

(PF) (8) Show that $2\sqrt{2}$ is irrational. **[6 marks]**

(PF) (9) Show using proof by contradiction that if $p, q \in \mathbb{Z}$ then $p^2 - 4q - 2 \neq 0$ **[6 marks]**

(PF) (10) Show that if $a^2 + b^2 = c^2$ where $a, b, c \in \mathbb{Z}$ then either a or b must be even. **[5 marks]**

Chapter 12 Probability

(M) (1) A group of 5 year old boys were asked what they got for Christmas. 65% of them got building bricks and 35% were given a computer console. There are 13% of the boys who got both. Calculate the probability that a boy chosen at random has building bricks or a computer console but not both. **[3 marks]**

(2) J and W are two events and $P(J) = 0.48$, $P(W) = 0.64$ and $P(J \cap W) = 0.13$. Find
   a $P(J \cup W)$ **[1 mark]**
   b $P(W')$ **[1 mark]**
   c $P(J \cap W')$ **[2 marks]**
   d $P(J \cup W')$ **[2 marks]**

(3) H and K are two events such that $P(H) = 0.55$ and $P(K) = 0.22$ and $P(H|K) = 0.43$.
   Find
   a $P(K|H)$ **[3 marks]**
   b $P(H' \cap K')$ **[1 mark]**
   c $P(K' \cap H)$ **[2 marks]**

(4) For events F and G, $P(F) = 0.51$, $P(G|F) = 0.32$ and $P(G'|F') = 0.46$.
   a Draw a tree diagram representing the events and the corresponding probabilities. **[4 marks]**
   b Find $P(G)$. **[3 marks]**
   c Find $P(F'|G')$. **[3 marks]**

(5) John Paul is a 37-year-old man, living and born in England. In 2016, gun crime in America was at 23.4% for 30 to 39 year olds, and there were approximately 12.9 million men between 30 and 39 living in America at that time. What is the probability that John Paul will get accused of a gun crime?

   Critically analyse your answer, stating any other assumptions you would make and the effect they would have on the probability. **[3 marks]**

6 Tom visits the post office every Monday morning to send off his online orders. Items he may send off are boxes, packages or letters.

The probability, P(B), that he sends off a box on any given Monday is 0.37.

The probability, P(P), that he sends off a package on any given Monday is 0.18.

The probability, P(L), that he sends off a letter on any given Monday is 0.45.

Calculate the probability that, on any given Monday, Tom sends off:

i none of those items [3 marks]

ii exactly two or three items. [4 marks]

7 In a drama school of 50 girls, 17 girls dance only, 23 girls have both singing and dancing lessons and 3 girls do neither. Each girl must do at least one discipline from dancing and singing.

a Find the number of girls who sing or dance. [1 mark]

b One performing girl is selected at random.

i Find the probability that she does either singing or dancing but not both. [2 marks]

ii Given that the girl selected does only one discipline, find the probability that she dances. [2 marks]

Let D be the event that a girl dances and S be the event that a girls sings.

c Explain why D and S are not mutually exclusive. [2 marks]

d Show that D and S are not independent. [3 marks]

8 The events A and B are such that $P(A) = \frac{3}{7}$, $P(B) = \frac{1}{5}$ and $P(A|B') = \frac{1}{3}$.

a i Find $P(A \cap B')$,

ii $P(A \cap B)$ [2 marks]

iii $P(A \cup B)$ [3 marks]

iv $P(A|B)$ [3 marks]

b State, with a reason, whether or not A and B are:

i mutually exclusive [2 marks]

ii independent. [2 marks]

9 The probability of two events is given as $P(F) = f$ and $P(N) = n$.

a Write down an expression for $P(F \cup N)$ in terms of f and n, stating any assumptions. [2 marks]

It turns out that $P(F) = 0.13$ and $P(N) = 0.33$ and $P(F|N) = 0.08$

Find the value of

b $P(F \cap N)$ [3 marks]

c $P(F \cup N)$ [3 marks]

d $P(F')$ [1 mark]

10 A game consists of rolling a tetrahedral die and then flipping a coin. The die has four fair sides labelled 2, 4, 6 and 8.

When the die lands on a square number, a biased coin with a probability of $\frac{2}{7}$ of landing on heads is flipped.

When the die does not land on a square number, a biased coin with a probability of $\frac{1}{3}$ of landing on heads is flipped.

a Complete a tree diagram. [3 marks]

Aleena plays the game.

b Find the probability that she obtains tails. [2 marks]

c Given that Heather played the game and obtained a head when she flipped the coin, find the probability that her die landed on a square number. [4 marks]

Chapter 13 Statistical distributions

1 A footballer found that the time taken, M minutes, to score a goal in training can be modelled by a normal distribution with mean 54 minutes and standard deviation 6.2 minutes.

Find

a $P(M > 60)$ [3 marks]

b $P(55 \leqslant M \leqslant 65)$ [4 marks]

c the value of m, to 1 decimal place, such that $P(M \leqslant m) = 0.12$. [4 marks]

2 The hand spans of a group of rugby players are modelled by a normal distribution with mean 29 cm and standard deviation 2.8 cm. The heights of this group of players are modelled by a normal distribution with mean 181 cm and standard deviation 5.7 cm.

Find the probability that a randomly chosen player,

a has a hand span smaller than 26 cm [2 marks]

b is taller than 185 cm. [3 marks]

c Assuming that for these athletes hand span and height are independent, find the probability that a randomly chosen athlete is taller than 185 cm and has a hand span smaller than 26 cm. [3 marks]

d Comment on the assumption that height and hand span are independent. [1 mark]

3 The random variable X is normally distributed with mean μ and variance σ^2.

a Write down 3 properties of the distribution of X. [3 marks]

b Given that $\mu = 5$ and $\sigma = 0.1$

find $P(4.9 < X < 5.1)$. [4 marks]

4 Brand new trainers in their boxes are normally distributed with mean 950 g and standard deviation 7.7 g. The weight of an empty box is normally distributed with mean 42.5 g and standard deviation 3.18 g. The weight of a box is independent of the weight of the trainers it contains.

Find the probability that a randomly selected box weighs less than 42 g and the trainers are heavier than 952 g. Give your answer to 2 significant figures. [8 marks]

5 The random variable $X \sim N(\mu, \sigma^2)$.

It is known that

$P(X \leq 10) = 0.0122$ and $P(X \geq 12) = 0.4013$.

  a Use a clearly labelled sketch to represent these probabilities on a normal distribution curve. **[1 mark]**

  b **i** Show that the value of σ is 0.8.

      ii Find the value of μ. **[8 marks]**

  c Find $P(10 \leq X \leq 14)$. **[3 marks]**

6 A children's farm, is open 10 am to 4 pm each day. Visitors are able to use its facilities for as long as they wish once they have paid to go in. The farm's records suggest that the length of a visit can be modelled by a normal distribution with mean 180 minutes. Only 5% of visitors stay for less than 60 minutes.

  a Find the standard deviation of the normal distribution. **[4 marks]**

  b Find the probability that a visit lasts longer than 200 minutes. **[3 marks]**

7 The random variable X is normally distributed with mean 84 and variance 16.

Find

  a $P(X < 79)$ **[3 marks]**

  b $P(80 < X < 88)$. **[3 marks]**

8 The shelf life of powdered soup has a mean of 68 days and a standard deviation of 2 days. Shelf life may be assumed to be normally distributed.

Find the shelf life, d days, of a powdered soup such that 1 packet in 8 will have a lifetime longer than d. **[6 marks]**

9 An A-level technology group design a soap dispenser to put soap onto your hands. It is designed so that the dispenser releases a mean of 5 ml of soap. Another group of students monitor the amount actually dispensed and notice that it has a mean amount of 4.76 ml per hand wash and a standard deviation of 0.46.

What percentage of dispensers are below the advertised amount of soap? **[4 marks]**

10 Cat food is sold in pouches containing a stated weight of 200 g. The pouches are filled by a machine. The actual weight of the pouch is normally distributed with mean 202 g and standard deviation 2.3 g.

  a Find the probability that a pouch contains less than the stated weight. **[2 marks]**

  b In a box of 44 pouches, find the expected number of pouches containing less than the stated weight. **[2 marks]**

Chapter 14 Statistical hypothesis testing

1 A teacher wishes to see if a student who is good at biology is also good at statistics. She thinks that the two subjects are linked so if a student does well in biology, they do well in statistics. She collects the data from the end of year exams for 10 students who are studying both biology and statistics.

Student	A	B	C	D	E	F	G	H	I	J
Biology/100	34	56	74	88	18	91	58	29	75	82
Statistics/100	45	78	73	76	23	79	66	21	68	72

a Draw a scatter graph to represent the data and comment upon the correlation. **[2 marks]**

b State the null and alternate hypothesis. **[2 marks]**

The correlation coefficient is calculated to be 0.897.

c Using the correlation coefficient as a test statistic, carry out the test at the 1% significance level. **[5 marks]**

2 A school rugby team decide to change their training strategy and do more weights rather than cardio based training. The captain thinks this will not work but the coach insists that the players will get fitter and stronger and therefore score more tries. The points scored for the next six matches are shown in the table below:

	Match A	Match B	Match C	Match D	Match E	Match F
Points scored	17	3	32	19	8	21
Points conceded	6	7	22	33	13	3

a Draw a scatter graph and comment upon your findings. **[2 marks]**

b State the null and alternate hypothesis. **[2 marks]**

The correlation coefficient is calculated to be 0.377.

c Using the correlation coefficient as a test statistic, carry out the test at the 10% significance level. **[5 marks]**

3 The temperature of a pan of spaghetti bolognaise is monitored over a period of time to see how long it will take to cool, before feeding it to a small child.

Time (m)/ minutes	0	2	4	6	8	10	12	14	16
Temperature (T)/°C	140	100	85	78	75	70	61	55	50

a Draw a graph of temperature against time. **[2 marks]**

b By drawing a suitable graph, test whether the relationship between time and temperature is of the form $T = ka^m$, where k and a are constants. **[4 marks]**

c Calculate the value of a and k. **[4 marks]**

4 The results of an experiment are plotted on a graph of $\log_{10} y$ against $\log_{10} x$, producing a straight line with gradient -0.31 and intercept 0.5.

The relationship between y and x is either $y = px^q$ or $y = pq^x$.

a Write the equation of the straight line in terms of the data values. **[3 marks]**

b Write the relationship between y and x. **[4 marks]**

5 A worker at a children's play barn wishes to know if age has anything to do with how fast a child goes down a slide. She decided to collect data on the next 8 children who go down the slide.

Age/months	Time/seconds
181	18.4
66	19.8
154	17.4
88	20.3
96	21.7
163	18.9
77	22.4
149	19.6

a Plot the data and comment on your findings. **[2 marks]**

b State the null and alternate hypotheses. **[3 marks]**

c The correlation coefficient is found to be -0.754. Carry out the test at a 5% significance level. **[5 marks]**

d Comment upon the validity of the test. **[1 mark]**

6 The speed limit on a motorway for vans is 60 mph. A traffic officer thinks that the speeds of vans are increasing, as the drivers are less aware of their speed limit compared with car drivers. He records the speeds of 10 vans and the results in mph are:

65, 72, 77, 54, 66, 68, 72, 57, 61, 71.

It is known that speeds on the motorway are normally distributed with a standard deviation of 5.2 mph.

Use this data set to test, at the 10% significance level, the traffic officer's belief that the speeds of vans are increasing. **[8 marks]**

7 New LED TVs have had a mean lifetime of 8 years. An electrician puts forward a theory that the new LED TVs are better and will have a longer lifetime compared with older types of TV.

A random sample of 1000 LED TVs is taken from the population of those sold.

The raw scores, x, are collected and are as follows.

$n = 1000$, $\Sigma x = 9200$, $\Sigma x^2 = 86\,023$

Carry out a suitable hypothesis test on the electrician's claim at the 1% significance level.

State your assumptions carefully. **[8 marks]**

8 A machine is designed to put screen wash into the washer bottle of a car. The car owner electronically controls the mean volume of screen wash that goes into the washer bottle. The standard deviation is fixed at 0.37 litres. The mean volume is suspected to be lower than it should be. If a random sample of 108 cars has a total volume of 514.08 litres, is there evidence at the 5% significance level to support this suspicion if the mean volume is set to 5 litres? **[8 marks]**

9 A mobile phone battery lasts on standby for 36 hours. The supplier of LITHICELL batteries claims that the mean life of their batteries on standby is 50 hours. The manufacturer of a phone buys 1500 of their batteries to test this claim, and they find the mean to be 42.9 hours, with a sample standard deviation of 14 hours. Is there evidence at a 5% level that the mean is higher than 36 hours? **[8 marks]**

10 A researcher wishes to find out if there is any connection between the amount of sleep a teenager gets and their IQ. She collects data on 50 eighteen-year-olds and generates a product moment correlation coefficient value of $r = 0.2743$

Carry out a hypothesis test at the 5% level to determine whether there is any connection between the amount of sleep a teenager gets and their IQ. **[6 marks]**

Chapter 15 Kinematics

1 A particle is travelling in a straight line along the x-axis at a constant deceleration of $3\,\text{m}\,\text{s}^{-2}$. Initially the particle starts at the origin with a velocity of $9\,\text{m}\,\text{s}^{-1}$. Find the times at which the particle is displaced 10 m from the origin. **[5 marks]**

M 2 A diver is preparing to dive from the 10 m board. He runs towards the edge and jumps off at an angle of 40° at a speed of $4\,\text{m}\,\text{s}^{-1}$.

Find

a the time the diver spends in the air once he jumps from the board, **[4 marks]**

b the speed with which the diver strikes the water. **[2 marks]**

M 3 Two particles are moving on a horizontal plane with constant velocities. At time $t = 0$ s, particle A has position $(-\mathbf{i} + 6\mathbf{j})$ m and particle B has position $(31\mathbf{i} + 11\mathbf{j})$ m. The velocity of particle A is $(3\mathbf{i} + \mathbf{j})\,\text{m}\,\text{s}^{-1}$.

a Find the position of particle A when $t = 3$ s. **[2 marks]**

At time $t = 3$ s, particle B has position $(16\mathbf{i} + 5\mathbf{j})$ m.

b Find the velocity of particle B in $\text{m}\,\text{s}^{-1}$. **[2 marks]**

c Find the time at which particle B is due south of particle A. **[3 marks]**

d Find the distance between the particles when A is north-east of B. **[4 marks]**

M 4 A cannonball is fired from a ship. It lands in the water 140 m from the ship after travelling for 5 s.

a Find the horizontal component of the cannonball's velocity. **[2 marks]**

Given that the cannonball is fired at an angle of arcsin 0.6,

b find the height from which the cannonball is fired **[5 marks]**

c find the maximum height the cannonball reaches above the sea. **[3 marks]**

(PF) **5** At time t s, the position vector of a particle is given by $\mathbf{r} = [(12t^2 - 3t + 5)\mathbf{i} + (1 + \frac{7}{3}t^3 - 13t)\mathbf{j}]$ m.

    a Find the time at which the particle is travelling on a bearing of 135°. **[5 marks]**

    b Show that the particle is accelerating at 25 m s^{-2} at time $t = \frac{1}{2}$ s. **[4 marks]**

(M) **6** An island has two ports, X and Y with position vectors $(-5\mathbf{i} + 80\mathbf{j})$ km and $(46\mathbf{i} - 7\mathbf{j})$ km respectively. Ship U leaves Port X at 15:40, travelling in the direction $(4\mathbf{i} - 3\mathbf{j})$ km at a speed of 20 km h^{-1}.

    a Find the velocity of ship U in the form $(a\mathbf{i} + b\mathbf{j})$ km h^{-1}. **[3 marks]**

    Ship V leaves Port Y at 16:40 at a velocity of $(2\mathbf{i} + 18\mathbf{j})$ km h^{-1}.

    b Show that if the ships were to continue on their courses, they would collide, stating the time and position where this collision would take place. **[6 marks]**

    The captain of Ship U changes direction at 18:40 so that the ship is travelling at a new velocity of $(12\mathbf{i} - 9\mathbf{j})$ km h^{-1}.

    c Find the distance between the ships at the time the collision would have taken place. **[3 marks]**

(M) **7** In this question, $\mathbf{i}$ is horizontal and $\mathbf{j}$ is vertical.

    A ball is thrown at $(12\mathbf{i} + 9\mathbf{j})$ m s^{-1} from the point with position vector $8\mathbf{j}$ m.

    a Find the position vector of each point at which the speed of the ball is 13 m s^{-1}. Write the answers in terms of g. **[6 marks]**

    b Find the angle the direction of motion makes to the horizontal when the ball lands on the ground. Give the answer correct to the nearest degree. **[4 marks]**

8 A particle is moving on a horizontal plane with a constant acceleration. At time $t = 5$ s the particle has a velocity of $(\mathbf{i} + 8\mathbf{j})$ m s^{-1} and at $t = 9$ s the particle has a velocity of $(-3\mathbf{i} + 20\mathbf{j})$ m s^{-1}.

    a Find the acceleration of the particle. **[2 marks]**

    b Find the initial velocity of the particle. **[2 marks]**

    c Find the time at which the particle is moving parallel to $\mathbf{i}$. **[3 marks]**

    d Find the distance of the particle from its initial position at $t = 7$ s. **[4 marks]**

(M) **9** A lorry is moving along the straight line $WXYZ$ at a constant acceleration. It takes the lorry 4 s to travel the 88 m from W to X and 2 s to travel the 54.5 m from X to Y. At Y the acceleration is increased by 0.5 m s^{-2}. The lorry then travels from Y to Z in 8 s.

    a Sketch a velocity–time graph. **[3 marks]**

    b Find the total distance from W to Z. **[5 marks]**

(M) **10** A tennis ball, A, is dropped from the top of a building. At the same time a second tennis ball, B, is projected vertically upwards at 15 m s^{-1} from the ground at the point directly below where A was dropped. The tennis balls collide after 2.4 s.

    a Find the height of the building. **[5 marks]**

    The experiment is repeated, except that A is dropped 0.2 s before B is projected.

    b Find the height above the ground where the balls collide. **[6 marks]**

11 Two carts, P and Q, are in the desert. Initially the carts are at the points $(-9\mathbf{i} + \mathbf{j})$ km and $(-5\mathbf{i} - 2\mathbf{j})$ km, respectively. The two carts start driving simultaneously, cart P at $(5\mathbf{i} + 2\mathbf{j})$ km h^{-1} and cart Q at $(2\mathbf{i} + 3\mathbf{j})$ km h^{-1}.

   a Find the position vector for cart P after t hours. Give the answer in the form $(a\mathbf{i} + b\mathbf{j})$ km, where a and b are given in terms of t. **[2 marks]**

   b Find the time at which the carts are closest together. **[6 marks]**

   c Show that the carts are approximately 1.58 km apart at this time. **[2 marks]**

12 A rocket is launched from ground level at $20\,\mathrm{m\,s^{-1}}$ at an angle of β to the horizontal. The rocket lands at a point on the ground that is d m away.

   a Show that $d = \dfrac{400\sin 2\beta}{g}$. **[4 marks]**

   b Given that the maximum height reached by the rocket is $\dfrac{100}{g}$ m find the value of d, giving the answer in terms of g. **[6 marks]**

13 For $0 \leqslant t \leqslant 6$ s, a car accelerates at a constant $(2\mathbf{i} + \mathbf{j})\,\mathrm{m\,s^{-2}}$ from an initial starting point of $(-9\mathbf{i} - 5\mathbf{j})$ m.

   Given that the car starts from rest,

   a find the position of the car after 6 s. **[4 marks]**

   For $t > 6$, the velocity is given by $[(t + 6)\mathbf{i} + t\mathbf{j}]\,\mathrm{m\,s^{-1}}$.

   b Find the position of the car at $t = 10$ s. **[6 marks]**

Chapter 16 Forces

1 A box of mass 600 g is at rest on a rough plane inclined at $\arctan\frac{1}{3}$ to the horizontal. The box is acted upon by a horizontal force X N such that it is about to move up the plane. The coefficient of friction between the box and the plane is $\frac{1}{17}$.

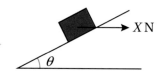

   Prove that X is given by $\frac{6}{25}g$ N. **[7 marks]**

2 A passenger is pulling a suitcase of mass 12 kg through an airport. The floor of the airport is rough and horizontal. The cord is inclined at 60° to the floor. The tension in the cord is 40 N and the passenger is accelerating at $0.1\,\mathrm{m\,s^{-2}}$.

   Find the coefficient of friction between the suitcase and the floor. **[6 marks]**

3 Three forces act upon a particle of mass 2 kg. Force $\mathbf{F}_1 = 6\mathbf{j}$ N and force $\mathbf{F}_2 = 7\mathbf{i}$ N. Force $\mathbf{F}_3$ has a magnitude of 9 N and acts on a bearing of 060°.

Find:

a the magnitude of the acceleration [4 marks]

b the bearing of the resultant force. [3 marks]

(M) **4** An object of mass $\frac{5}{2}$ kg is acted upon by two forces, $\begin{bmatrix} 7 \\ -19 \end{bmatrix}$ N and $\begin{bmatrix} 3 \\ 4 \end{bmatrix}$ N.

The object has an initial velocity of $\begin{bmatrix} 9 \\ 12 \end{bmatrix}$ m s^{-1}.

Find the distance travelled after $\frac{1}{2}$ s. [7 marks]

(M) (PF) **5** A girl is pulling her toboggan up a snowy hill using a thick rope. The hill is at 14° to the horizontal and the rope is parallel to the slope. The toboggan has a mass of 30 kg. The coefficient of friction between the toboggan and the hill is 0.05.

Given that the girl is pulling the toboggan at a steady speed,

a show that the tension in the rope is about 85 N. [6 marks]

b How is the thick rope being modelled? Explain why this model is necessary. [2 marks]

(M) **6** A truck is towing a cart along a straight horizontal road. The truck is six times heavier than the cart and the tow rope is inclined at 20° to the horizontal as shown in the diagram. The truck and cart experience resistive forces of 300 N and 50 N respectively. The engine of the truck exerts a driving force of 560 N. The truck accelerates from rest at 0.15 m s^{-2}.

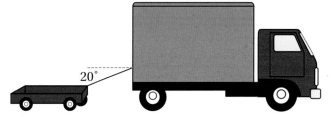

a Find the mass of the truck. [4 marks]

b Find the tension in the rope. [2 marks]

After 8 s, the tow rope snaps.

c Find the time taken for the cart to stop. [5 marks]

(M) (PF) **7** A particle of mass M kg is held on a rough slope inclined at θ to the horizontal. The coefficient of friction between the particle and the slope is given by μ.

The particle is released and starts to accelerate down the slope.

a Show that $a = g(\sin\theta - \mu\cos\theta)$. [5 marks]

b Given that $\mu = \frac{\sqrt{5}}{10}$, $\theta = \arcsin\frac{2}{3}$,

find the time taken for the particle to travel 50 m. [4 marks]

8 Two spheres P and Q, of masses 3 kg and 5 kg respectively, are connected by a light inextensible string across a smooth pulley. The pulley is at the top of a rough block in the shape of an equilateral triangle as shown in the diagram. The base of the block is 4 m wide. The coefficient of friction between each sphere and the block is given by μ. Initially the spheres are both 3 m from the pulley and resting on the block.

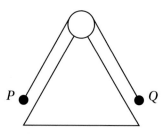

a Show that the acceleration of each sphere, when the system is released, is given by

$$a = \frac{\left(\sqrt{3} - 4\mu\right)g}{8} \text{ m s}^{-2}.$$

[6 marks]

b Given that $\mu = \frac{\sqrt{3}}{8}$,

find the total distance travelled by sphere P before it comes to instantaneous rest. [6 marks]

9 A particle of mass 4 kg is at rest on a rough slope inclined at an angle α to the horizontal. It is acted upon by a force P which is at an angle of α below the horizontal, as shown in the diagram. The particle is on the point of moving up the slope. The coefficient of friction between the particle and the slope is $\frac{1}{8}$.

a Show that P is given by $\dfrac{4g(\cos\alpha + 8\sin\alpha)}{8\cos 2\alpha - \sin 2\alpha}$ N. [6 marks]

b Given that $\alpha = \arcsin \frac{1}{2}$,

show that $P = 4g\dfrac{8 + \sqrt{3}}{8 - \sqrt{3}}$ N. [3 marks]

10 A peg bag, of mass M kg, hangs from a washing line in equilibrium, as shown in the diagram. The washing line is attached to points A and B and makes angles of $(x + 10)°$ and $x°$, respectively, at each end. The tension in the line attached to A is 35 N and in the line attached to B is 32 N.

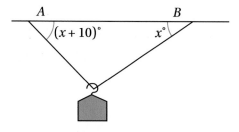

a Find the value of x. [6 marks]

b Find the value of M. [3 marks]

11 A car of mass 0.5 tonnes is pulling a broken down van of mass 300 kg up a hill. The hill makes an angle of α with the horizontal, such that $\tan\alpha = \frac{7}{24}$. The hill is modelled as a rough slope and the coefficient of friction between each vehicle and the hill is $\frac{1}{8}$. The driving force exerted by the engine of the car is 3.4 kN.

Given that the car and van are initially stationary, find the velocity of the car when it has pulled the van 40 m. [10 marks]

12 Two particles X and Y of masses 3 kg and 4 kg, respectively, are connected by a light inextensible cord which passes over a smooth pulley. X is at rest on a rough ramp inclined at an angle of arcsin $\frac{3}{5}$. The coefficient of friction between X and the ramp is μ. Y is suspended from the pulley as shown in the diagram.

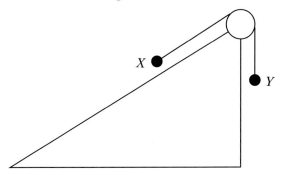

    a Show that $\mu < \dfrac{11}{12}$.   **[7 marks]**

    The tension in the cord is given by T N.

    b Show that the force on the pulley is given by $\frac{4}{5}\sqrt{5}T$.   **[5 marks]**

Chapter 17 Moments

 1 A uniform girder AB of mass 40 kg and length 10 m is supported on two logs at C and D where $AC = 2$ m and $DB = 3$ m.

    a Find the reaction force exerted on the girder by the log at C.   **[3 marks]**

    A block of mass M kg is now placed at B so that the girder is on the point of tilting about D.

    b Find the value of M.   **[3 marks]**

    c How have you modelled the girder to answer this question?   **[1 mark]**

2 Point A has position vector $\begin{bmatrix} 5 \\ -3 \end{bmatrix}$.

A force F_1 of $\begin{bmatrix} -7 \\ -3 \end{bmatrix}$ N acts about the point with position vector $\begin{bmatrix} 9 \\ -2 \end{bmatrix}$.

A force F_2 of $\begin{bmatrix} 10 \\ 6 \end{bmatrix}$ N acts about the point with position vector $\begin{bmatrix} 2 \\ -8 \end{bmatrix}$.

Find the turning moment of the two forces about A.   **[4 marks]**

 3 A steel cable WZ of mass 8 kg is suspended horizontally from two light inextensible wires at X and Y, where $WX = x$ m and $YZ = 2x$ m. $WZ = 5$ m. The centre of mass of the cable acts 2 m from W. The tensions at X and Y are in the ratio $4:3$.

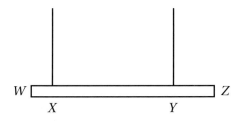

Find the value of x.   **[6 marks]**

(M) **4** A non-uniform horizontal plank *AB* of mass 5 kg and length 8 m is supported at point *C* at the centre of the plank. A force of 30 N is applied to the plank at A at 45° to the horizontal and a force of *X* N is applied to the plank at D at 60° to the horizontal, as shown in the diagram. *D* is 2 m from *B*. The plank is in equilibrium.

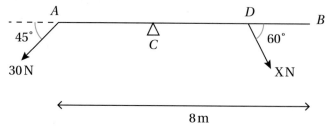

   a Find the value of *X*.   **[3 marks]**

   b Find the magnitude of the reaction force at *C*.   **[2 marks]**

   c Given that the centre of mass is *d* m from A,

      find the value of *d*.   **[3 marks]**

(M) (PF) **5** A uniform beam *PQ* of mass $3\sqrt{3}$ kg and length 7 m is balanced in limiting equilibrium against a smooth peg such that the beam is inclined at 30° to the rough horizontal ground as shown in the diagram. The peg is $\frac{49}{8}$ m from *P*. A load of mass $2\sqrt{3}$ kg is attached at *Q*.

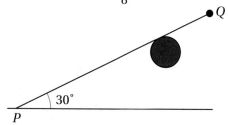

   a Show that the reaction force exerted by the peg on the beam is given by $6g$ N.   **[6 marks]**

   b Show that the coefficient of friction between the beam and the ground is

      given by $\frac{\sqrt{3}}{2}$.   **[4 marks]**

(M) (PS) **6** A 15 m pole *AD* has a mass of 44 kg. It is supported at points *B* and *C* where $AB:BC:CD = 1:2:2$. The centre of mass of the pole is halfway between *B* and *C*. There is a load of 10 kg at *A* and a load of *X* kg at *D*. The reaction force at *C* is 1.5 times as large as the reaction force at *B*.

A ▪——————————————————▪ D
       △   △
       B   *C*

   a Find the magnitude of the reaction force at *B*.   **[5 marks]**

   b Find the value of *X*.   **[2 marks]**

7
 A non-uniform pillar SV of length 6 m rests horizontally in equilibrium, supported at points T and U where $ST = 1$ m and $UV = 2$ m.

When a rock of mass 36 kg is placed at V, the pillar is on the point of tilting about U.

When the first rock is removed and a rock of mass 18 kg is placed at S, the pillar is on the point of tilting about T.

Given that the mass of the pillar is M kg and the centre of mass is x m from S,

a find the value of M [6 marks]

b find the value of x. [2 marks]

 8 A horizontal rod AB of mass 6 kg is hinged to a wall at A. B is attached to a taut string of length $13a$ m. The other end of the string is attached to the wall at C where C is $5a$ m vertically above A. The centre of mass of the rod acts at D where $AD = 4a$ m.

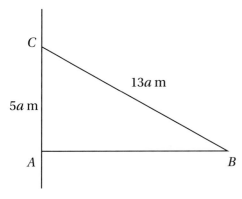

a Find the magnitude of the tension in the string in terms of g. [3 marks]

b i Find the magnitude of the reaction force at the hinge. [4 marks]

   ii Find the angle the reaction force makes with the horizontal. [2 marks]

 9 A non-uniform ladder XZ of length $6a$ m rests with Z against a rough wall and X on a rough floor. The ladder is inclined at an angle of α to the horizontal. The centre of mass of the ladder acts at Y where $XY : YZ = 1 : 2$. The coefficients of friction at X and Z are given by μ_X and μ_Z, respectively.

Show that $\tan\alpha = \dfrac{1 - 2\mu_X\mu_Z}{3\mu_X}$. [10 marks]

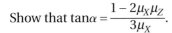 **10** A uniform ladder AB, of length 8 m, is at rest against a smooth wall with A against the wall and B on rough ground. The coefficient of friction between the ladder and the ground is $\dfrac{3}{10}$. The ladder is inclined at an angle of $\arctan\left(\dfrac{15}{8}\right)$ to the horizontal. Given that it is safe a for man of mass 75 kg to climb 6 m up the ladder but no further, find the mass of the ladder. [8 marks]

Answers

Short answers are given in this book, with full worked solutions available to teachers by emailing: education@harpercollins.co.uk

1 Algebra and functions 1: Functions

Prior knowledge   **page 1**

1 **a**

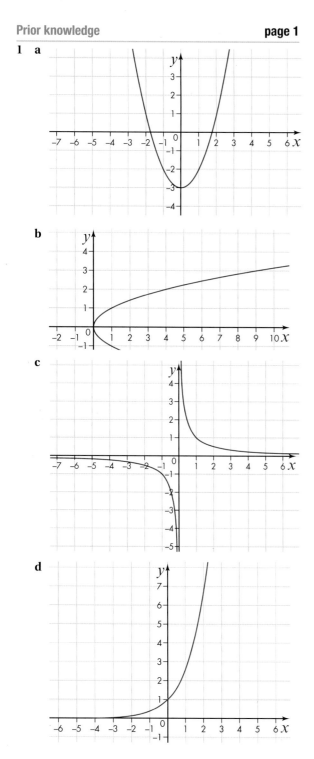

e

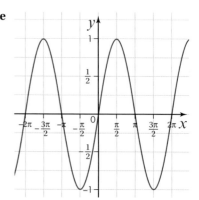

f

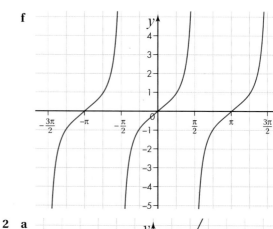

2 **a**

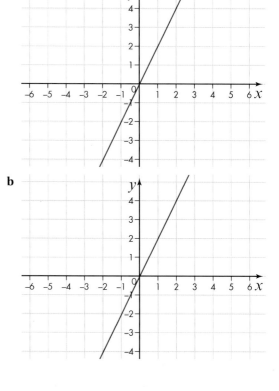

b

c

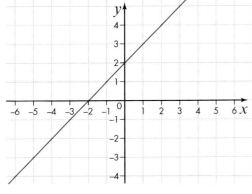

d

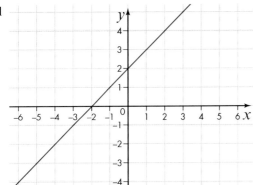

3 **a** $y = 2$ **b** $y = 1$
    c $y = 5$ **d** $y = \frac{3}{2}$

Exercise 1.1A   **page 7**

1 **a** $f(x) = 5 - 3x$ is a one-to-one function.
    b $g(x) = x^2 + x$ is a many-to-one function.
    c $h(x) = \pm\sqrt{x+1}$ is not a function.
    d This is not a function.
    e This is not a function.
    f This is not a function.

2 **a** The range of $f(x)$ is {−15, −11, −7, −3, 1, 5}.
    b This is a one-to-one function.

3 **a** 5
    b

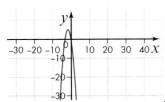

    Range of $g(x)$ is $-72 \leqslant g(x) \leqslant \frac{49}{8}$
    c $b = -\frac{3}{2}$ or $b = -1$
    d This is a many-to-one function.

4 $d = 3$

5 **a**

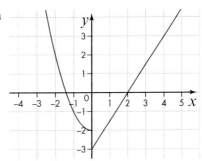

    Range of k(x) is k(x) ≥ −3.
    b 5
    c This is a many-to-one function.

6 The relationship could become a function if the domain was limited to either $\{x < -a,\, x \in -\mathbb{R}\}$ or $\{x > 0,\, x \in +\mathbb{R}\}$

7 m(x) = x when $x = -\frac{1}{2}$

Exercise 1.2A   **page 11**

1 **a** 7
    b 49
    c 16
    d −17

2 **a** $\dfrac{1}{x^2 - 1}$
    b $\dfrac{1 - x^2}{x^2}$
    c $x^4 - 2x^2$
    d x

3 $a = 3$ or $a = 7$

4 $fg(x) = 5(x^2 + 3) - 2 = 5x^2 + 13$
    $gf(x) = (5x - 2)^2 + 3 = 25x^2 - 20x + 7$
    So fg ≠ gf.

5 **a** $x + 2 = ff(x)$
    b $x^2 + 2x - 6 = gf(x)$

6 $hg(x) = \dfrac{1}{x^3 + 1}$
    $fhg(x) = \dfrac{2}{x^3 + 1} - 3$
    $ffhg(x) = \dfrac{4}{x^3 + 1} - 6 - 3$
    $f^2hg = \dfrac{4}{x^3 + 1} - 9$

7 1.778 (3 d.p.)

8 **a** $h^2fg(x)$
    b $ghf(x)$
    c $gh^2f(x)$

421

Exercise 1.3A   **page 14**

1   **a** $f^{-1}(x) = \dfrac{5-x}{4}, \{x \in \mathbb{R}\}$

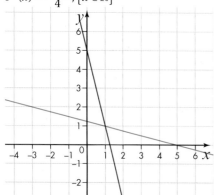

  b $f^{-1}(x) = 3 + \sqrt{x}, \{x \geqslant 0, x \in \mathbb{R}\}$ and for all $+\sqrt{x}$

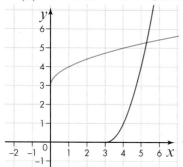

  c $f^{-1}(x) = \sqrt[3]{x}, \{x \in \mathbb{R}\}$

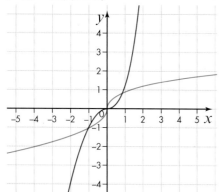

2   Let $y = 7x - 2$.

$$\dfrac{y+2}{7} = x$$

$$f^{-1}(x) = \dfrac{x+2}{7}$$

$$f^{-1}(2) = \dfrac{2+2}{7} = \dfrac{4}{7}$$

3   **a** $f(x) = \dfrac{5-x}{4}, \{x \in \mathbb{R}\}$

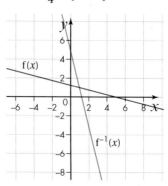

  b $f(x) = (x-4)^2, \{x \in \mathbb{R}, x \geqslant 4\}$

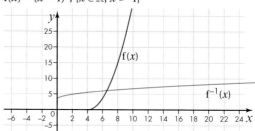

  c $f(x) = \ln x, \{x > 0, x \in \mathbb{R}\}$

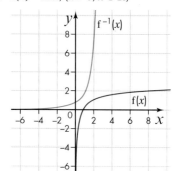

4 $f^{-1}(x) = \sqrt{x + \dfrac{93}{4}} - \dfrac{7}{2}$

5   **a** $f^{-1}(x) = \dfrac{2}{x-5} + 1$, where $\{x \in \mathbb{R}, x > 5\}$

  b

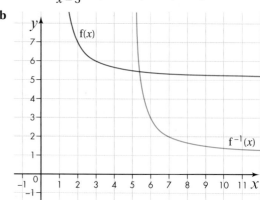

  c $x = 3 \pm \sqrt{6}$

6 Let $y = 2x^2 + 5x - 13$.

$$y = 2\left[\left(x + \frac{5}{4}\right)^2 - \frac{129}{16}\right]$$

$$x = \sqrt{\frac{8y + 129}{16}} - \frac{5}{4}$$

$$f^{-1}(x) = \sqrt{\frac{8x + 129}{16}} - \frac{5}{4}$$

$$ff^{-1}(x) = 2\left(\sqrt{\frac{8x + 129}{16}} - \frac{5}{4}\right)^2 + 5\left(\sqrt{\frac{8x + 129}{16}} - \frac{5}{4}\right) - 13$$

$$= \frac{8x + 129}{8} - 5\sqrt{\frac{8x + 129}{16}} + \frac{25}{8} +$$

$$5\sqrt{\frac{8x + 129}{16}} - \frac{25}{4} - \frac{52}{4}$$

$$= x + \frac{129}{8} + \frac{25}{8} - \frac{50}{8} - \frac{104}{8}$$

$$= x$$

Exercise 1.4A   **page 17**

1 **a**

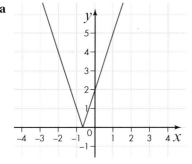

b

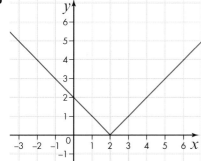

c

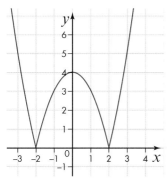

d

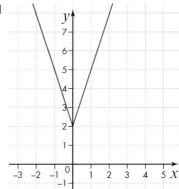

e

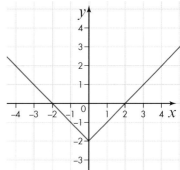

f

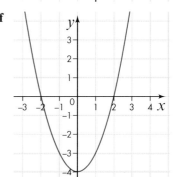

2 $y = \left|\frac{1}{x}\right|$ is similar to $y = \frac{1}{x}$. However, the modulus turns any negative y values positive.

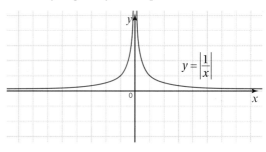

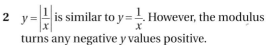

3 a

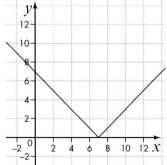

b

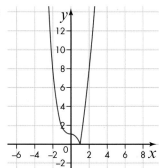

c

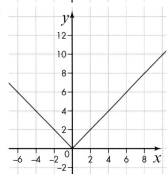

d

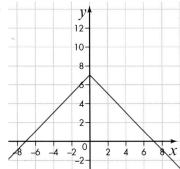

e

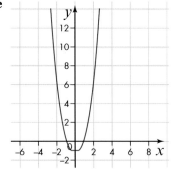

f

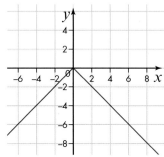

4 $y = |\sin x|$ is symmetrical about the y-axis.

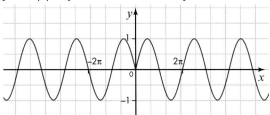

$y = \sin |x|$ is symmetrical about the y-axis.

5　a

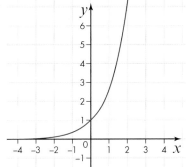

b

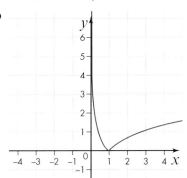

c

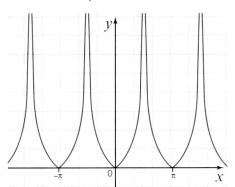

d

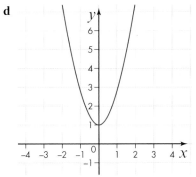

e

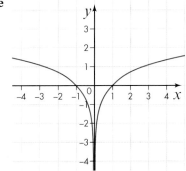

f

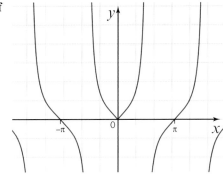

1　a

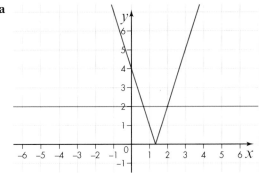

b $x = \dfrac{2}{3}$ or $x = 2$

2　a

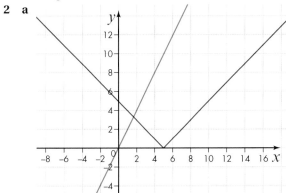

b $x < 1.67$

425

3

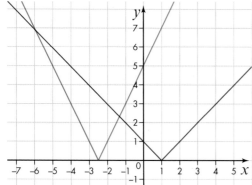

$$x = -6$$

$$x = -\frac{4}{3}$$

4 $0 < x < 2.322$

5 $x = 4.828$

$x = 3.236$

6 $-\frac{\pi}{6} \leqslant x \leqslant \frac{\pi}{6}$

Exercise 1.5A   **page 22**

1 **a**

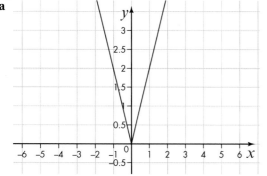

b

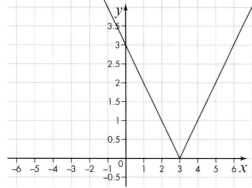

c

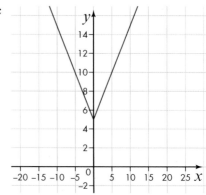

2 **a** D

b A

c B

d C

3 **a**

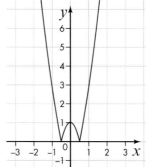

b

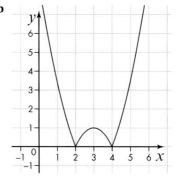

c

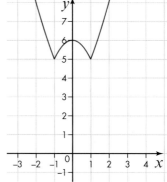

4

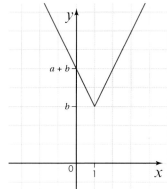

5 a

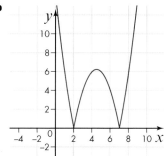

b

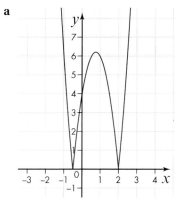

c

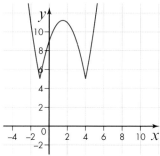

6

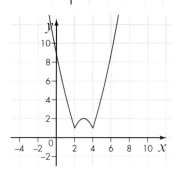

1 a 1

b When $t = 10$, $N = 2^{10} = 1024$

When $t = 28$, $N = 2^{28} = 268\,435\,456$

c

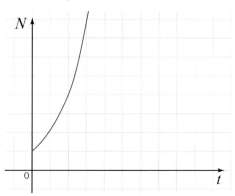

2 a £600

b When $t = \frac{1}{2}$, $V = £535.37$

When $t = 2$, $V = £383.59$

When $t = 4$, $V = £252.33$

c

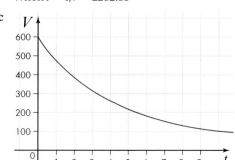

d Yes, it is a suitable model because the value of the phone will never become negative at some point in the future.

3 a 220 000

b When $t = 6$, $S = 236\,442$

When $t = 16$, $S = 299\,060$

c 2022

d 2031

e The model assumes that the number of people wanting to go to university will continue to increase and that the number of students that the universities can accommodate will also increase. This won't necessarily happen.

Exam-style questions 1   **page 27**

1 a 11

b
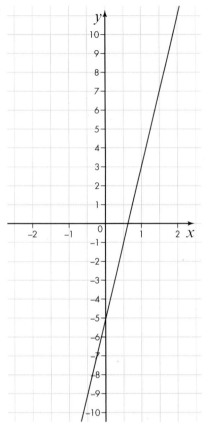

Range of f(x) is {−29, −21, −13, −5, 3, 11, 19}

c $b = -1$

d This is a one-to-one function.

2 a $2(x^2 - 4)$

b $(2x)^2 - 4$

c $(x^2 - 4)^2 - 4$

d $\dfrac{x}{2}$

e $\sqrt{x + 4}$

3 a £250000

b £575243

c
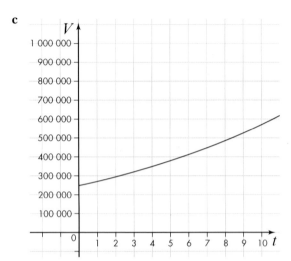

d No, because the value of a house does not increase at the same rate all the time and is subject to many external forces.

4 a
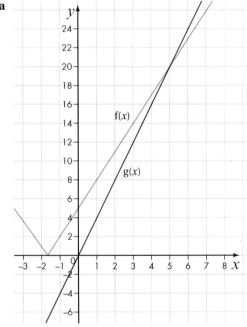

b $x < 5$

c

d $x < -5$ or $x > -1$

5 a

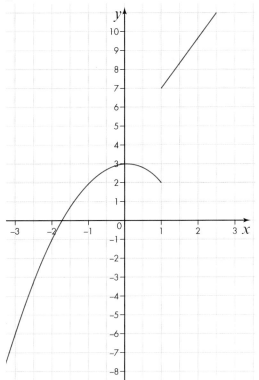

The range is k(x) ⩽ 3 or k(x) ⩾ 7.

b $c = -3$

c This is a many-to-one function.

6 a $\dfrac{1}{x^2 + 5x - 13}$

b x

c $\sqrt{x + \dfrac{77}{4}} - \dfrac{5}{2}$

7 a $-2.924 < x < 3$

b f(x) blue, |f(x)| red, f(|x|) green.

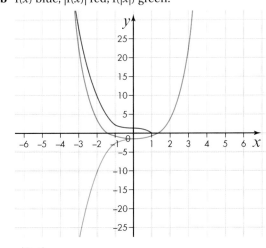

c $-|f(x)|$

8 a $f^{-1}(x) = \dfrac{11 - x}{3}, \{x \in \mathbb{R}, x \leqslant 11\}$

b $\left(\dfrac{11}{4}, \dfrac{11}{4}\right)$

c Reflection in the line $y = x$

9 a transformation is $-f(x)$

b transformation could be $2f(x)$ or $f(2x)$

c transformation is $f(x - 1)$

d transformation is $f(x) + 3$

10 $\{x > 2, x \in +\mathbb{R}\}$ or $\{x < 2, x \in -\mathbb{R}\}$

11 a $f^{-1}(x) = \sqrt{\dfrac{8x + 137}{16}} - \dfrac{7}{4}$

b $ff^{-1}(x) = x$

12 Although scientist A's model will show growth, it is exponential growth and so the population of rabbits will never be limited in this model and, in theory, could be infinite, which is reality is not possible.

The rabbit population using scientist B's model can never exceed 1 million. A limit is more realistic than infinity, but it is unknown whether the limit of 1 000 000 is correct.

13 a there are no solutions for $-\dfrac{\pi}{2} \leqslant x \leqslant \dfrac{\pi}{2}$

b $f(x) = |\cos x|$ blue, $f\left(\dfrac{x}{2}\right)$ red

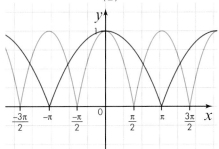

14 a $\dfrac{8}{3x - 3} + 3$

b $\dfrac{1}{6x + 3}$

c $x = 0$ or -0.333

2 Algebra and functions 2: Partial fractions

Prior knowledge   **page 31**

1 $x^2 + 4x + 3$

2 $x^2 + 9x + 38$ remainder 120

3 $x^2 - 2x + 1$

Exercise 2.1A   **page 32**

1 a 5

   b $\dfrac{1}{x-2}$

   c $\dfrac{1}{x-4}$

2 a $\dfrac{x^2}{1+x}$

   b $\dfrac{4(x-3)}{(x+3)^2}$

   c $\dfrac{x}{x+7}$

3 a $\dfrac{x(x^2-8)}{x^2-4}$

   b $\dfrac{2(x-6)}{x+5}$

   c $\dfrac{2x(x-2)(x+2)}{(x+4)(x-7)}$

Exercise 2.2A   **page 36**

1 a i and ii $\dfrac{-9}{x+2}+\dfrac{11}{x+3}$

   b i and ii $\dfrac{-1}{x-1}+\dfrac{7}{x+5}$

   c i and ii $\dfrac{3}{x-3}-\dfrac{8}{x-7}$

2 $\dfrac{11-2x}{(x-2)(x+5)}=\dfrac{A}{x-2}+\dfrac{B}{x+5}$

   $11-2x=A(x+5)+B(x-2)$

   Substitute $x=-5$.

   $21=-7B$

   $B=-3$

   Substitute $x=2$.

   $7=7A$

   $A=1$

   $\dfrac{11-2x}{(x-2)(x+5)}=\dfrac{1}{x-2}-\dfrac{3}{x+5}$

3 a $\dfrac{1}{2x+1}-\dfrac{1}{3-x}$

   b $\dfrac{11}{x}-\dfrac{9}{x+2}$

   c $\dfrac{-5}{x-5}+\dfrac{9}{x-3}$

4 $\dfrac{1}{8(x-4)}-\dfrac{1}{8(x+4)}$

   Notice that $\dfrac{1}{x^2-a^2}=\dfrac{1}{2a(x-a)}-\dfrac{1}{2a(x+a)}$

5 a $\dfrac{3}{x-1}-\dfrac{8}{x-2}+\dfrac{7}{x-3}$

430

b $\dfrac{-2}{x}+\dfrac{5}{x-1}+\dfrac{2}{1-2x}$

c $\dfrac{1}{x-1}-\dfrac{13}{3(x-2)}-\dfrac{5}{3(x+4)}$

6 $\dfrac{1}{x^2-a^2}=\dfrac{A}{x-a}+\dfrac{B}{x+a}$

   $1=A(x+a)+B(x-a)$

   Substitute $x=a$.

   $1=2aA$

   $A=\dfrac{1}{2a}$

   Substitute $x=-a$.

   $1=-2aB$

   $B=-\dfrac{1}{2a}$

   $\dfrac{1}{x^2-a^2}=\dfrac{1}{2a(x-a)}-\dfrac{1}{2a(x+a)}$

7 $\dfrac{5+3x-x^2}{-x^3+3x^2+4x-12}$

   First factorise the denominator using the factor theorem.

   Substitute $x=2$.

   $-8+12+8-12=0$

   So $(x-2)$ is a factor.

   Substitute $x=-2$.

   $8+12-8-12=0$

   So $(x+2)$ is a factor.

   Substitute $x=3$.

   $-27+27+12-12=0$

   So $(3-x)$ is a factor as coefficient of x^3 is negative.

   Rewrite the expression as partial fractions.

   $\dfrac{5+3x-x^2}{-x^3+3x^2+4x-12}=\dfrac{A}{x+2}+\dfrac{B}{x-2}+\dfrac{C}{3-x}$

   $5+3x-x^2=A(x-2)(3-x)+B(x+2)(3-x)$
   $\qquad\qquad\qquad +C(x+2)(x-2)$

   Substitute $x=2$.

   $5+6-4=B(4)(1)$

   $B=\tfrac{7}{4}$

   Substitute $x=3$.

   $5+9-9=C(5)(1)$

   $C=1$

   Substitute $x=-2$.

   $5-6-4=A(-4)(5)$

   $A=\tfrac{1}{4}$

   $\dfrac{5+3x-x^2}{-x^3+3x^2+4x-12}=\dfrac{1}{4(x+2)}+\dfrac{7}{4(x-2)}+\dfrac{1}{3-x}$

1 $\dfrac{x^2 + 8x + 4}{x^2(x - 2)} = \dfrac{A}{x} + \dfrac{B}{x^2} + \dfrac{C}{x - 2}$

$x^2 + 8x + 4 = A(x)(x - 2) + B(x - 2) + C(x^2)$

Substitute $x = 0$.

$4 = -2B$

$B = -2$

Substitute $x = 2$.

$4 + 16 + 4 = 4C$

$C = 6$

Substitute $x = 1$.

$1 + 8 + 4 = A(1)(-1) + (-2)(-1) + (6)(1)$

$13 = -A + 2 + 6$

$A = -5$

$\dfrac{x^2 + 8x + 4}{x^2(x - 2)} = \dfrac{-5}{x} - \dfrac{2}{x^2} + \dfrac{6}{x - 2}$

2 $p = 5$

3 $\dfrac{7}{x - 4} + \dfrac{25}{(x - 4)^2}$

4 $\dfrac{2x^2 - x - 6}{x^3 + 4x^2 + 4x} = \dfrac{2x^2 - x - 6}{x(x + 2)^2}$

$\equiv \dfrac{A}{x} + \dfrac{B}{(x + 2)} + \dfrac{C}{(x + 2)^2}$

$\equiv \dfrac{A(x + 2)^2 + Bx(x + 2) + Cx}{x(x + 2)^2}$

$2x^2 - x - 6 \equiv A(x + 2)^2 + B(x)(x + 2) + C(x)$

Substitute $x = 0$.

$-6 = 4A$

$A = -\dfrac{3}{2}$

Substitute $x = -2$.

$8 + 2 - 6 = -2C$

$C = -2$

Substitute $x = 1$.

$2 - 1 - 6 = \left(-\dfrac{3}{2}\right)(9) + B(1)(3) + (-2)(1)$

$-5 = -\dfrac{27}{2} + 3B - 2$

$B = \dfrac{7}{2}$

$\dfrac{2x^2 - x - 6}{x^3 + 4x^2 + 4x} \equiv -\dfrac{3}{2(x)} + \dfrac{7}{2(x + 2)} - \dfrac{2}{(x + 2)^2}$

5 $\dfrac{1}{(x + 1)(x - 2)^2} = \dfrac{A}{x + 1} + \dfrac{B}{x - 2} + \dfrac{C}{(x - 2)^2}$

$1 = A(x - 2)^2 + B(x + 1)(x - 2) + C(x + 1)$

Substitute $x = 2$.

$1 = 3C$

$C = \dfrac{1}{3}$

Substitute $x = -1$.

$-1 = 9A$

$A = -\dfrac{1}{9}$

Substitute $x = 0$.

$1 = \left(-\dfrac{1}{9}\right)(-2) + B(1)(-2) + \left(\dfrac{1}{3}\right)(1)$

$B = -\dfrac{1}{9}$

$\dfrac{1}{(x + 1)(x - 2)^2} = \dfrac{1}{9(x + 1)} - \dfrac{1}{9(x - 2)} + \dfrac{1}{3(x - 2)^2}$

Yes, $\dfrac{1}{(x + 1)(x - 2)^2}$ can be split into partial fractions.

6 $\dfrac{2x^2 + 6x + 5}{(x - 2)^3} = \dfrac{A}{x - 2} + \dfrac{B}{(x - 2)^2} + \dfrac{C}{(x - 2)^3}$

$2x^2 + 6x + 5 = A(x - 2)^2 + B(x - 2) + C$

Substitute $x = 2$.

$8 + 12 + 5 = C$

$C = 25$

Substitute $x = 0$.

$5 = 4A - 2B + 25$ ①

Substitute $x = 1$.

$13 = A - B + 25$ ②

Multiply ② by 2.

$-24 = 2A - 2B$

$26 = 2A - 2B$ ③

Subtract ① from circled ③.

$-4 = -2A$

$A = 2$

Substitute into ②.

$B = 14$

$\dfrac{2x^2 + 6x + 5}{(x - 2)^3} = \dfrac{2}{(x - 2)} + \dfrac{14}{(x - 2)^2} + \dfrac{25}{(x - 2)^3}$

7 The two different methods to split a rational function with linear terms in its denominator into partial fractions are *substitution* and *equating coefficients*.

For partial fractions without repeated terms, the substitution method relies only on basic mathematical operations without the need for any algebraic manipulation, whereas the equating coefficients method results in the solving of simultaneous equations.

For partial fractions with repeated terms, the substitution method relies on a combination of basic mathematical operations and the solving of simultaneous equations, whereas the equating coefficients method results in the solving of simultaneous equations with three unknowns.

Exercise 2.4A   page 40

1 **a** Proper; numerator < denominator

b Improper; numerator = denominator

c Mixed; whole + proper fraction

d Improper; numerator > denominator

e Improper; numerator = denominator

f Proper; numerator < denominator

g Mixed; whole + proper fraction

h Improper; numerator > denominator

i Proper; numerator < denominator

2 **a** $6 + \dfrac{18x - 12}{(x^2 - 3x + 2)}$

b $6 - \dfrac{6}{x - 1} + \dfrac{24}{x - 2}$

3

$$x^2 + 6x + 9 \overline{\smash{\big)}\ x^3 + 0\ \ \ + 0\ \ \ + 6} \quad \overset{\displaystyle x - 6}{}$$

$$\begin{array}{r} x^3 + 6x^2 + 9x \\ \hline 0 - 6x^2 - 9x\ +6 \\ -6x^2 - 36x - 54 \\ \hline 27x + 60 \end{array}$$

$$\frac{27x + 60}{(x + 3)^2} = \frac{C}{x + 3} + \frac{D}{(x + 3)^2}$$

$$27x + 60 = C(x + 3) + D$$

Substitute $x = -3$.

$$-21 = D$$

Substitute $x = 0$.

$$60 = 3C - 21$$

$$C = 27$$

$$51 = 3A + 132$$

$$C = -27$$

$$\frac{x^3 + 6}{(x + 3)^2} = x - 6 + \frac{27}{x + 3} - \frac{21}{(x + 3)^2}$$

$$A = 1, B = -6, C = 27, D = -21$$

4 $-2 + \dfrac{3}{4(2 - 5x)} + \dfrac{19}{4(x + 2)}$

5 $-\dfrac{27}{2}$

6 $(x + 2)^3 = x^3 + 6x^2 + 12x + 8$

$$x^3 + 6x^2 + 12x + 8 \overline{\smash{\big)}\ x^4 - x^3\ + x^2\ \ +0\ \ \ +0} \quad \overset{\displaystyle x - 7}{}$$

$$\begin{array}{r} x^4 + 6x^3 + 12x^2 + 8x \\ \hline 0 - 7x^3 - 11x^2 - 8x \\ -7x^3 - 42x^2 - 84x - 56 \\ \hline 31x^2 + 76x + 56 \end{array}$$

$$\frac{31x^2 + 76x + 56}{(x + 2)^3} = \frac{C}{x + 2} + \frac{D}{(x + 2)^2} + \frac{E}{(x + 2)^3}$$

$$31x^2 + 76x + 56 = C(x + 2)^2 + D(x + 2) + E$$

Substitute $x = -2$.

$$124 - 152 + 56 = E$$

$$E = 28$$

Substitute $x = 0$.

$$28 = 4C + 2D \qquad ①$$

$$56 = 4C + D \qquad ①$$

Substitute $x = -1$.

$$31 - 76 + 56 = C + D + 28$$

$$-17 = C + D \qquad ②$$

Subtract ② from ①.

$$2C = 62$$

$$C = 31$$

$$D = -48$$

$$\frac{x^4 - x^3 + x^2}{(x + 2)^3} = x - 7 + \frac{31}{x + 2} - \frac{48}{(x + 2)^2} + \frac{28}{(x + 2)^3}$$

$$A + B + C + D + E = 1 - 7 + 31 - 48 + 28 = 5$$

Exam-style questions 2   page 43

1 $\dfrac{x + 3}{x + 9}$ cannot be simplified further.

2 $(x - 1)(x - 2) = x^2 - 3x + 2$

$$x^2 - 3x + 2 \overline{\smash{\big)}\ 2x^2 + 0\ \ +0} \quad \overset{\displaystyle 2}{}$$

$$\begin{array}{r} 2x^2 - 6x \\ \hline 6x - 4 \end{array}$$

$$\frac{6x - 4}{(x - 1)(x - 2)} = \frac{A}{x - 1} + \frac{B}{x - 2}$$

$$6x - 4 = A(x - 2) + B(x - 1)$$

Substitute $x = 2$.

$$B = 8$$

Substitute $x = 1$.

$$A = -2$$

$$\frac{2x^2}{(x - 1)(x - 2)} = 2 - \frac{2}{x - 1} + \frac{8}{x - 2}$$

3 If x is a positive integer then $2x$ is an even number and $2x + 1$ is an odd number.

$$\frac{1}{2x} \times \frac{1}{2x+1} = \frac{1}{2x(2x+1)}$$

$$\frac{1}{2x(2x+1)} = \frac{A}{2x} + \frac{B}{2x+1}$$

$$1 = A(2x+1) + B(2x)$$

Substitute $x = 0$.

$$A = 1$$

Substitute $x = -\frac{1}{2}$.

$$B = -1$$

$$\frac{1}{2x(2x+1)} = \frac{1}{2x} - \frac{1}{2x+1}$$

4 $\dfrac{4}{x-3} - \dfrac{1}{3x-2}$

5 $\dfrac{18x-11}{(x-3)(x-2)} = \dfrac{A}{x-3} + \dfrac{B}{x-2}$

$$18x - 11 = A(x-2) + B(x-3)$$

Substitute $x = 2$.

$$B = -25$$

Substitute $x = 3$.

$$A = 43$$

$$\frac{18x-11}{(x-3)(x-2)} = \frac{43}{x-3} - \frac{25}{x-2}$$

6 $\dfrac{6}{x+1} + \dfrac{1}{(x+1)^2}$

7 $f(x) = \dfrac{9}{x+1} - \dfrac{9}{x+2}$

$$f'(x) = -\frac{9}{(x+1)^2} + \frac{9}{(x+2)^2}$$

8 $\dfrac{5-9x-2x^2}{6x^2+27x-15} = -\dfrac{(2x-1)(x+5)}{3(2x-1)(x+5)} = -\dfrac{1}{3}$

9 $\dfrac{7}{2x-3} - \dfrac{7}{2x+3}$

10 $5 + \dfrac{5}{x-1} + \dfrac{20}{x-2} + \dfrac{40}{(x-2)^2}$

A, B, C and D are all multiples of 5.

11 $-\dfrac{2}{x-1} + \dfrac{5}{x-4} - \dfrac{3}{x+4}$

12 a $\dfrac{A}{px+q} + \dfrac{B}{(px+q)^2}$

b $\dfrac{A}{mx+n} + \dfrac{B}{(px+q)}$

c cannot be split into partial fractions.

13 $a = 2$

14 $f(x) = \dfrac{7}{4(x+2)} + \dfrac{5}{4(x-2)}$

$$f'(x) = 1 - \frac{7}{4(x+2)^2} - \frac{5}{4(x-2)^2}$$

It is easy to forget to differentiate the single x as it is not involved in the partial fractions part of the question.

15 $-\dfrac{2}{9x} + \dfrac{2}{3x^2} + \dfrac{2}{9(x+3)}$

16 a $\dfrac{1}{3x-1} + \dfrac{5}{2x+5}$

b $-\dfrac{187}{196}$

17 $\dfrac{1}{6(x+1)} - \dfrac{1}{6(x+7)}$

18 $2(3x-1)$

19 $\dfrac{3x^2+2x+2}{(x-2)(x-3)^2} = \dfrac{A}{x-2} + \dfrac{B}{x-3} + \dfrac{C}{x-3}$

$$3x^2 + 2x + 2 = A(x-3)^2 + B(x-2)(x-3) + C(x-2)(x-3)$$

Substitute $x = 2$.

$$b + 4 + 2 = A$$

$$A = 18$$

Substitute $x = 0$.

$$-160 = 6B + 6C \quad \text{①}$$

Substitute $x = 1$.

$$-65 = 2B + 2C \quad \text{②}$$

Multiply ② by 3.

$$-195 = 6B + 6C \quad \text{③}$$

Equations ① and ③ are contradictory and consequently cannot be solved. Consequently,

$\dfrac{3x^2+2x+2}{(x-2)(x-3)^2}$ cannot be split into partial fractions

in the form $\dfrac{A}{x-2} + \dfrac{B}{x-3} + \dfrac{C}{x-3}$.

20 $\dfrac{3}{(x+4)} + \dfrac{2}{(2x+1)} - \dfrac{1}{(2x+1)^2}$

21 $(x^2-4)(2x+1) = 2x^3 + x^2 - 8x - 4$

$$
\begin{array}{r}
2x^2 - x + \frac{17}{2} \\
2x^3 + x^2 - 8x - 4 \overline{\smash{\big)}\, 4x^5 + 0 \ + \ 0 \ + 0 \ - 3x + 2} \\
\underline{4x^5 + 2x^4 - 16x^3 - 8x^2} \\
-2x^4 + 16x^3 + 8x^2 - 3x + 2 \\
\underline{-2x^4 - \ x^3 + 8x^2 + 4x} \\
17x^3 - 7x + 2 \\
\underline{17x^3 + \frac{17}{2}x^2 - 68x - 34} \\
-\frac{17}{2}x^2 + 61x + 36
\end{array}
$$

$$\frac{-\frac{17}{2}x^2 + 61x + 36}{(x-2)(x+2)(2x+1)} = \frac{D}{x-2} + \frac{E}{x+2} + \frac{F}{2x+1}$$

$$-\tfrac{17}{2}x^2 + 61x + 36 = D(x+2)(2x+1) + E(x-2)(2x+1)$$
$$+ F(x-2)(x+2)$$

Substitute $x = 2$.

$$D = \frac{31}{5}$$

Substitute $x = -2$.

$$E = -10$$

Substitute $x = -\frac{1}{2}$.

$$F = -\frac{9}{10}$$

$$\frac{4x^5 - 3x + 2}{(x^2 - 4)(2x+1)} = 2x^2 - x + \frac{17}{2} + \frac{31}{5(x-2)} - \frac{10}{x+2} - \frac{9}{10(2x+1)}$$

The expression $\dfrac{4x^5 - 3x + 2}{(x^2 - 4)(2x+1)}$ can be split into partial fractions.

The coefficient of $(x+2)^{-1}$ is -10.

22 $x(2-3x)^2 = 9x^3 - 12x^2 + 4x$

$$9x^3 - 12x^2 + 4x \overline{)27x^4 + 0 - 9x^2 +0 +5}$$

$$\begin{array}{r} 3x+4 \\ \hline 27x^4 - 36x^3 + 12x^2 \\ \hline 0 + 36x^3 - 21x^2 + 5 \\ 36x^3 - 48x^2 + 16x \\ \hline 27x^2 - 16x + 5 \end{array}$$

$$\frac{27x^2 - 16x + 5}{x(2-3x)^2} = \frac{C}{x} + \frac{D}{2-3x} + \frac{E}{(2-3x)^2}$$

$$27x^2 - 16x + 5 = C(2-3x)^2 + D(x)(2-3x) + Ex$$

Substitute $x = 0$.

$$C = \frac{5}{4}$$

Substitute $x = \frac{2}{3}$.

$$E = \frac{19}{2}$$

Substitute $x = 1$.

$$D = -\frac{21}{4}$$

$$\frac{27x^4 - 9x^2 + 5}{x(2-3x)^2} = 3x + 4 + \frac{5}{4x} - \frac{21}{4(2-3x)} + \frac{19}{2(2-3x)^2}$$

$$A + B + C + D + E = 3 + 4 + \frac{5}{4} - \frac{21}{4} + \frac{19}{2} = \frac{25}{2}$$

23 $\dfrac{x^4 - x^3 + x^2 + 1}{(x-2)^3} = x + 5 + \dfrac{19}{x-2} + \dfrac{24}{(x-2)^2} + \dfrac{13}{(x-2)^3}$

$$f'(6) = -\frac{279}{256}$$

24 $f(-2) = -16 + 44 - 20 - 8$ so $(x+2)$ is a factor of $f(x)$.

$$\begin{array}{r} 2x^2 + 7x - 4 \\ x+2 \overline{)\,2x^3 + 11x^2 + 10x - 8} \\ \underline{2x^3 + 4x^2} \\ 7x^2 + 10x \\ \underline{7x^2 + 14x} \\ -4x - 8 \\ \underline{-4x - 8} \\ 0 \end{array}$$

$$2x^2 + 7x - 4 = (x+4)(2x-1)$$

$$\frac{8}{2x^3 + 11x^2 + 10x - 8} = \frac{A}{x+2} + \frac{B}{x+4} + \frac{C}{2x-1}$$

$$8 = A(x+4)(2x-1) + B(x+2)(2x-1) + C(x+2)(x+4)$$

Substitute $x = -4$.

$$B = \frac{4}{9}$$

Substitute $x = -2$.

$$A = \frac{-4}{5}$$

Substitute $x = \frac{1}{2}$.

$$C = \frac{32}{45}$$

$$\frac{8}{2x^3 + 11x^2 + 10x - 8} = -\frac{4}{5(x+2)} + \frac{4}{9(x+4)} + \frac{32}{45(2x-1)}$$

25 $\dfrac{1}{c^2 - x^2} = \dfrac{A}{c+x} + \dfrac{B}{c-x}$

$$1 = A(c-x) + B(c+x)$$

Substitute $x = c$.

$$B = \frac{1}{2c}$$

Substitute $x = -c$.

$$A = \frac{1}{2c}$$

$$\frac{1}{c^2 - x^2} = \frac{1}{2c(c+x)} + \frac{1}{2c(c-x)}$$

$$= \frac{1}{2c(x+c)} - \frac{1}{2c(x-c)}.$$

26 $\dfrac{x^2 + 5x + 4}{(x-1)(x-4)(x+3)} = \dfrac{A}{x-1} + \dfrac{B}{x-4} + \dfrac{C}{x+3}$

$$x^2 + 5x + 4 = A(x-4)(x+3) + B(x-1)(x+3)$$
$$+ C(x-1)(x-4)$$

Substitute $x = 4$.

$$B = \frac{40}{21}$$

Substitute $x = 1$.

$$A = \frac{-5}{6}$$

Substitute $x = -3$.

$$C = \frac{-1}{14}$$

$$\frac{x^2 + 5x + 4}{(x-1)(x-4)(x+3)} = -\frac{5}{6(x-1)} + \frac{40}{21(x-4)} - \frac{1}{14(x+3)}$$

$$A + B + C = -\frac{5}{6} + \frac{40}{21} - \frac{1}{4} = \frac{-35 + 80 - 3}{42} = \frac{42}{42} = 1$$

3 Coordinate geometry: Parametric equations

Prior knowledge page 45

1

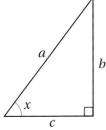

For the right-angled triangle above,

$$a^2 = b^2 + c^2$$

Rearrange.

$$1 = \frac{b^2 + c^2}{a^2} = \frac{b^2}{a^2} + \frac{c^2}{a^2} \qquad ①$$

$$\sin x = \frac{b}{a} \quad \text{so} \quad \sin^2 x = \frac{b^2}{a^2}$$

$$\cos x = \frac{c}{a} \quad \text{so} \quad \cos^2 x = \frac{c^2}{a^2}$$

Substitute into ①.

$$\sin^2 x + \cos^2 x = 1$$

2 $f^{-1}(x) = \sqrt{x+9}$

3 $8x - 1$

4 $f^{-1}(x) = \sqrt{\frac{1}{x} - 7}$

5 $f^{-1}(x) = \sqrt{x + \frac{93}{4}} - \frac{7}{2}$

Exercise 3.1A page 49

1

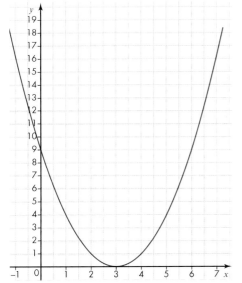

The coordinates of the y-intercept are $(0, 9)$.

2

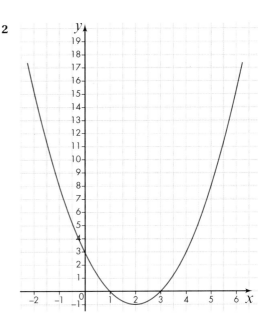

The coordinates of the turning point are $(2, -1)$.

3 **a** The x-intercept is at $(2, 0)$.

  The y-intercept is at $(0, -6)$.

  b The x-intercept is at $(-1, 0)$.

  The y-intercept is at $(0, 1)$.

  c The x-intercepts are at $(-1, 0)$ and $(1, 0)$.

  The y-intercept is at $(0, -4)$.

4 $(1, 4)$ and $(4, 1)$.

5

t	0	$\frac{\pi}{6}$	$\frac{\pi}{3}$	$\frac{\pi}{2}$	$\frac{2\pi}{3}$	$\frac{5\pi}{6}$	π	$\frac{7\pi}{6}$	$\frac{4\pi}{3}$	$\frac{3\pi}{2}$	$\frac{5\pi}{3}$	$\frac{11\pi}{6}$	2π
x	1	1.5	1.87	2	1.87	1.5	1	0.5	0.13	0	0.13	0.5	1
y	−1	−1.13	−1.5	−2	−2.5	−2.87	−3	−2.87	−2.5	−2	−1.5	−1.13	−1

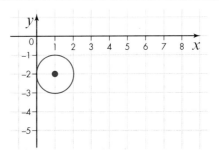

The curve is a circle, radius 1 and centre $(1, -2)$.
The coordinates of the centre of the circle are the same as the values in the parametric equations.

6 **a** The x-intercepts are at $(0.189, 0)$ and $(-2.189, 0)$

  The y-intercept is at $(0, -1)$.

  b The x-intercepts are at $\left(-\sqrt{\frac{2}{3}}, 0\right)$ and $\left(\sqrt{\frac{2}{3}}, 0\right)$.

  There is no y-intercept.

  c The x-intercepts are at $(1, 0)$ and $(4, 0)$.

  The y-intercept is at $(0, 2)$.

7 The points of intersection are $(1, -1)$ and $(-1, -1)$.

8 $p = \pm 3$

the parametric equations are $x = 3t - 1$ and $y = \dfrac{3}{t}$ or

$x = -3t - 1$ and $y = -\dfrac{3}{t}$.

9 The x-intercepts are at $(3, 0)$ and $(-1, 0)$

The y-intercepts are at $\left(0, -\dfrac{\sqrt{3}}{2}\right)$ and $\left(0, \dfrac{\sqrt{3}}{2}\right)$.

10 $q = -0.681$ or $q = 24.68$

$x = -0.681t + 2$ and $y = t^3 - 0.681$

or

$x = 24.68t + 2$ and $y = t^3 + 24.68$.

Exercise 3.2A   page 51

1 a $y = (x + 1)^2$

   b $y = x^2 - 4x + 3$

   c $y = 4x^2 - 4$

   All of these are quadratic curves.

2 $x = t + 2$

        $x - 2 = t$

        $y = 3(x - 2)$

        $\quad = 3x - 6$

   $3x - y - 6 = 0$

   So $a = 3$, $b = -1$, $c = -6$.

3 This is a circle centred on the origin with a radius of 1.

4 a $y = (x + 1)^4 - 2$

   This is a quartic curve.

   b $y = \dfrac{2}{x^2} - 3$

   None of these.

   c $y = x - 3\sqrt{x} + 2$

   None of these.

5 $x = 4\sin t - 3$

   $4(x - 3) = \sin t$ should be $\dfrac{x + 3}{4} = \sin t$

   $y = \dfrac{\csc t}{5}$

   $5y = \csc t$

   $\csc t = \dfrac{1}{\cos t}$ should be $\csc t = \dfrac{1}{\sin t}$

   $5y = \dfrac{1}{\cos t}$ should be $5y = \dfrac{1}{\sin t}$

   $\cos t = \dfrac{1}{5y}$ should be $\sin t = \dfrac{1}{5y}$

   $\sin^2 t + \cos^2 t = 1$ redundant

$(4x - 12)^2 + \left(\dfrac{1}{5y}\right)^2 = 1$ redundant

$\dfrac{x + 3}{4} = \dfrac{1}{5y}$

$y = \dfrac{4}{5(x + 3)}$

Exercise 3.3A   page 56

1 diameter $= 22$

2 So $a = 14$, $b = 6$

3 $x = t + 2t^2$

        $= 2\left(t^2 + \dfrac{t}{2}\right)$

        $= 2\left[\left(t + \dfrac{1}{4}\right)^2 - \dfrac{1}{16}\right]$

   $\dfrac{x}{2} + \dfrac{1}{16} = \left(t + \dfrac{1}{4}\right)^2$

   $\dfrac{8x + 1}{16} = \left(t + \dfrac{1}{4}\right)^2$

   $\dfrac{\sqrt{8x + 1}}{4} = t + \dfrac{1}{4}$

   $t = \dfrac{-1 + \sqrt{8x + 1}}{4}$

   $y = 5t$

   $t = \dfrac{y}{5}$

   $\dfrac{-1 + \sqrt{8x + 1}}{4} = \dfrac{y}{5}$

   $y = \dfrac{5(-1 + \sqrt{8x + 1})}{4}$

C

4 $x = t + 1$

   $y = \dfrac{4}{3}t + 5$

5 When $x = a \sin t$, then the width is $2a$

   When $y = b \sin 2t$, then the height is $2b$

   For a frame twice the size:

   $x = 8 \sin t$

   $y = 3 \sin 2t$

6 $x = \dfrac{3}{4}t + 2$

   $y = \dfrac{5}{2}t + 7$

7 $v = \dfrac{2\sqrt{13}}{21}$

   $x = \dfrac{4t}{21}$

   $y = -\dfrac{2}{7}t + 30$

   the toy should be placed at $\left(\dfrac{80}{7}, \dfrac{90}{7}\right)$.

8 The lengths of the axes are unaffected by the location of the centre of the ellipse, consequently this can be ignored and the parametric equations used in a form without the centre of the ellipse, i.e $x = 2\sin t$ and $y = 5\cos t$.

$x = 2\sin t$

$\dfrac{x}{2} = \sin t$

$y = 5\cos t$

$\dfrac{y}{5} = \cos t$

$\sin^2 t + \cos^2 t = 1$

$\left(\dfrac{x}{2}\right)^2 + \left(\dfrac{y}{5}\right)^2 = 1$

9 10 units

10 $v = \sqrt{1762}$

On a bearing of $168°$

Exam-style questions 3 **page 58**

1 y-intercept is at $(0, 7)$

x-intercepts are $(-7, 0)$ and $(-1, 0)$

2 $(4, 1)$ and $(1, 4)$

3

$x = \sin 2t$

$y = \cos 2t$

$\sin^2 2t + \cos^2 2t = 1$

$x^2 + y^2 = 1$

This is a circle centred on the origin with a radius of 1.

4 $p = 0$ and 2

The equations are:

$x = \dfrac{1}{2}(t + 3)$ and $y = 4t^2$

5

$x = \sec t$

$y = \tan t$

$1 + \tan^2 t = \sec^2 t$

$1 + y^2 = x^2$

$x = \operatorname{cosec} t$

$y = \cot t$

$1 + \cot^2 t = \operatorname{cosec}^2 t$

$1 + y^2 = x^2$

6 $x = -11\sin t$

$\dfrac{x}{-11} = \sin t$

$y = 5\cos t$

$\dfrac{y}{5} = \cos t$

$\sin^2 t + \cos^2 t = 1$

$\left(\dfrac{x}{-11}\right)^2 + \left(\dfrac{y}{5}\right)^2 = 1$

The lengths of the axes are 22 and 10 so the ellipse will not fit inside a rectangle that is 20 units by 12 units.

7 $x = \dfrac{4}{3}t - 1$

$y = 3 - \dfrac{4}{3}t$

$\left(\dfrac{49}{3}, -\dfrac{43}{3}\right)$

8 a $(x + 2)^2 + (y - 5)^2 = 36$

b

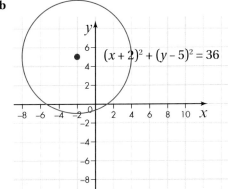

$(x + 2)^2 + (y - 5)^2 = 36$

9 a $y^2 = 4x^2(1 - x^2)$

b $t = \pi$ and $t = 0$

10 84

11 $v = 25$ on a bearing of $074°$

12 $x = 3\cos^2 t$

$\dfrac{x}{3} = \cos^2 t$

$y = 7\cos^2 t$

$\dfrac{y}{7} = \cos^2 t$

$\dfrac{y}{7} = \dfrac{x}{3}$

$y = \dfrac{7x}{3}$

This is a linear graph but only for positive x and y, since $\cos^2 t$ can never be negative.

D

13 a $x = 5\sin t + 2$, $y = 5\cos t - 3$ or vice versa.

b y-intercepts are at $(0, 1.6)$ and $(0, -7.6)$.

x-intercepts are at $(-2, 0)$ and $(6, 0)$.

14 a $y = 2 \pm \sqrt{x^{-1}}$

b When any curve intersects the y-axis, $x = 0$. As 0^{-1} is undefined, the curve will never intersect the y-axis.

15 Rearrange the parametric equation for x.

$$t = \frac{x}{1 - 2x}$$

Rearrange the parametric equation for y.

$$t = \frac{y + 1}{1 - y}$$

Equate these expressions for t.

$$\frac{x}{1 - 2x} = \frac{y + 1}{1 - y}$$

Rearrange.

$$y = \frac{1 - 3x}{x - 1}$$

16 $y = 2x^2 - 1$

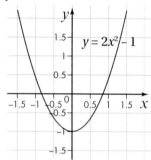

4 Sequences and series

Prior knowledge
page 62

1 a 15

b If $\frac{1}{2}n(n+1) = 45$ then $n^2 + n - 90 = 0$.

$$(n + 10)(n - 9) = 0$$

$n = 9$ is a solution so the 9th triangle number is 45.

If $\frac{1}{2}n(n+1) = 54$ then $n^2 + n - 108 = 0$.

This cannot be factorised with whole numbers so 54 is *not* a triangle number.

2 160

3 £30 630.76

Exercise 4.1A
page 64

1 a 16

b $x_n = n^2$

c $x_n = n^2 + 1$

2 a $x_{10} = 55$

$x_{20} = 210$

b $x_{n+1} = x_n + n + 1$

3 a $x_1 = 6, x_2 = 1.5874..., x_3 = 2.0338..., x_4 = 1.9971...,$
$x_5 = 2.0002...,$ approaching the limit 2.

b $\sqrt[3]{10 - 2} = \sqrt[3]{8} = 2$ so the limit satisfies the equation.

4 $t_1 = \tan 60° = 1.732$

$t_2 = \tan 120° = -\tan 60° = -1.732$

$t_3 = \tan 180° = 0$

$t_4 = \tan 240° = \tan 60° = 1.732$, etc.

The sequence is periodic of order 3.

5 a x_0 is the initial debt in pounds. Adding 2% is done by multiplying by 1.02 and then £50 is paid so 50 is subtracted from the product.

b £876.35.

c If $x_n = 0$ then $2500 - 1500 \times 1.02^n = 0$.

Rearrange.

$$1.02^n = \frac{2500}{1500} = 1.6667$$

$$n \ln 1.02 = \ln 1.6667$$

$$n = \frac{\ln 1.6667}{\ln 1.02} = 25.79$$

so the debt reduces to 0 when $n = 26$.

d The debt would increase every month

6 a 1.414 214

b 1.414 214

c The limit is $\sqrt{2}$.

d $\sqrt{3}$.

e Yes, the limit is the same.

f You can try any positive number and you will find that the limit is $\sqrt{a}$.

7 a a periodic sequence of order 2

b a periodic sequence of order 2

c a periodic sequence of order 2

d a periodic sequence of order 2

8 a The sequence becomes cyclic between 1 and −1, with order 2.

b The sequence is increasing and divergent.

c As it continues, the sequence approximately forms a cycle of order 3.

Exercise 4.2A
page 67

1 a 38

b 40

c 44

2 68.5

3 0

4 a 91

   b 333 495 150

5 a 3825

   b 2500

6 9260

7 a $\displaystyle\sum_{i=1}^{34}(3i-2)$

   b $\displaystyle\sum_{i=1}^{10}(i^2+2)$

   c $\displaystyle\sum_{i=1}^{20}\frac{2^i}{2i+1}$

8 a $t_{100}=\frac{1}{2}\times100\times101=5050$

   b 35

   c $\displaystyle\sum_{i=1}^{n}\tfrac{1}{2}i(i+1)=\tfrac{1}{2}\sum_{i=1}^{n}(i^2+i)$

$$=\tfrac{1}{2}\sum_{i=1}^{n}i^2+\tfrac{1}{2}\sum_{i=1}^{n}i$$

$$=\tfrac{1}{4}n(n+1)+\tfrac{1}{12}n(n+1)(2n+1)$$

$$=\tfrac{1}{12}n(n+1)(3+(2n+1))$$

$$=\tfrac{1}{12}n(n+1)(2n+4)$$

$$=\tfrac{1}{12}n(n+1)\times2\times(n+2)$$

$$=\tfrac{1}{6}n(n+1)(n+2)$$

Exercise 4.3A   page 70

1 a 83

   b 515

   c 195

2 a 425

   b 846

   c 397.5

3 800

4 a 51 mm

   b 231 mm

5 £2580

6 17.1

7 a 330

   b $S_n=\dfrac{n}{2}(12+(n-1)\times6)$

$$=\dfrac{n}{2}(12+6n-6)=\dfrac{n}{2}(6n+6)=3n(n+1)$$

8 a $a=1$ and $d=2$ so

$$S_n=\frac{n}{2}(2+(n-1)\times2)$$

$$=\frac{n}{2}(2+2n-2)=\frac{n}{2}\times2n=n^2$$

   b $S_n=n^2+n$

9 18

Exercise 4.4A   page 73

1 81 920

2 a 59 048

   b 569.533 to 3 d.p.

   c −5462.7

   d 1145.5

3 £137 566.55

4 a 2

   b $\dfrac{3}{2}$

   c $\dfrac{3}{4}$

   d 4

   e $\dfrac{3}{5}$

5 a 1 048 576 p, which is £10 485.76.

   b 2 097 151p or £20 971.51.

   c On his 65th birthday he would receive 2^{63} p, which is more than £9.2×10^{16}. This is an absurdly large amount.

6 a £3125

   b 9 years

7 a $\frac{5}{16}$

   b $\frac{21}{64}$

   c $\frac{1}{3}$

8 a $266\frac{2}{3}$ m north of the starting point.

   b 800 m.

9 $\dfrac{5}{11}$

10 11 terms

Exercise 4.5A   page 77

1 a $1-2x+3x^2$

   b $1+2x+3x^2$

2 a $\frac{5}{16}$

b $-\frac{5}{128}$

3 a $1+\frac{1}{3}x-\frac{1}{9}x^2$

b $-1<x<1$ or $|x|<1$

c 1.032 to 3 d.p.

4 a $1+\frac{1}{2}x-\frac{1}{8}x^2+\frac{1}{16}x^3$

b $1+x-\frac{1}{2}x^2+\frac{1}{2}x^3$

c $-0.5<x<0.5$ or $|x|<0.5$

5 a $1-x+x^2-x^3$

b $1-x^2+x^4-x^6$

6 a $1-2y+3y^2-4y^3$

b $\frac{1}{4}-\frac{3}{4}x+\frac{27}{16}x^2-\frac{27}{8}x^3$

The range of validity is $\left|\frac{3}{2}x\right|<1$ so $|x|<\frac{2}{3}$

7 $\frac{5}{2}$

8 $4.01^{\frac{3}{2}}=(4\times1.0025)^{\frac{3}{2}}$

$\qquad =4^{\frac{3}{2}}\times(1.0025)^{\frac{3}{2}}$

$\qquad =8\times(1+0.0025)^{\frac{3}{2}}$

So you need an expansion of $(1+x)^{\frac{3}{2}}$.

$\qquad =1+\frac{3}{2}x+\frac{\frac{3}{2}\times\frac{1}{2}}{2}x^2+\frac{\frac{3}{2}\times\frac{1}{2}\times\left(-\frac{1}{2}\right)}{6}x^3+\dots$

$\qquad =1+\frac{3}{2}x+\frac{3}{8}x^2-\frac{1}{16}x^3+\dots$

So when $x=0.0025$ then

$4.01^{\frac{3}{2}}=8+12x+3x^2-\frac{1}{2}x^3+\dots$

$\qquad =8+0.03+0.00001875-\dots=8.03002$ to 5 d.p.

Exam-style questions 4   **page 78**

1 a 14

b $x_n=n^2+2.$

2 16

3 a 67.2

b 447.282

4 a £19 020

b In the second month he earns

$1200\times1.05=£1260$, which is less than $1200+70=£1270$.

In the year he earns $\frac{1200(1.05^{12}-1)}{1.05-1}=£19\,100.55$, which is more than £19 020.

5 2.4

6 £5180.94

7 a $a=20$ and $d=4$

$\frac{n}{2}\big(40+4(n-1)\big)=504$

$n(4n+36)=1008$

$n^2+9n=252$

b 12 terms

8 a $1+x+x^2+x^3+\dots$

b $\frac{1}{2}+\frac{x}{4}+\frac{x^2}{8}+\frac{x^3}{16}+\dots$

The expansion is valid if $\left|\frac{x}{2}\right|<1$ so $|x|<2$

9 $0\leqslant x_1\leqslant0.5$

10 a $1-x-\frac{1}{2}x^2-\frac{1}{2}x^3$

b it is accurate to 8 significant figures.

11 a $1+3+5+\dots$ (n terms) is an arithmetic series with $a=1$ and $d=2$.

$S_n=\frac{n}{2}\{2a+(n-1)d\}=\frac{n}{2}\{2+(n-1)\times2\}$

$\qquad =\frac{n}{2}(2+2n-2)=\frac{n}{2}\times2n=n^2$

b $n(n+1)$

12 a $2-2x-x^2-x^3$

b $|x|<\frac{1}{2}$

13 a 20 units

b $(16, 8)$

14 33 terms

15 35 terms

5 Trigonometry

Prior knowledge
page 81

1

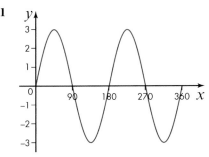

2 $x = 46.4°$ or $273.6°$

3 $RHS = (1 + \cos x)(1 - \cos x)$

Multiply out the brackets.

$$= 1 - \cos^2 x = \sin^2 x = LHS$$

Exercise 5.1A
page 82

1 **a** $28.6°$

   b $68.8°$

   c $5.7°$

   d $229.2°$

   e $315.1°$

2 **a** $\dfrac{\pi}{6}$

   b $\dfrac{5\pi}{6}$

   c $\dfrac{3\pi}{2}$

   d $\dfrac{7\pi}{4}$

   e 4π

3 **a** $36°$

   b $210°$

   c $292.5°$

   d $630°$

   e $54°$

4 **a** $27\,cm$

   b $64.8\,cm$

   c $26\,cm$

5 **a** $57\,cm$

   b $112.8\,cm$

   c $39\,cm$

6 **a** $202.5\,cm^2$

   b $777.6\,cm^2$

   c $84.5\,cm^2$

7 $1140\,cm^2$

8 **a** $100\,cm^2$

   b $50\,cm$

9 **a** $50\,cm$

   b 0.4 radians

10 a $2 \times 10^{-4}\,cm$

   b The angle can be any size.

Exercise 5.2A
page 85

1 **a** 0.5 **b** $\dfrac{\sqrt{3}}{2}$ **c** $\dfrac{1}{\sqrt{3}}$

2 **a** $\dfrac{\sqrt{3}}{2}$ **b** 0.5 **c** $\sqrt{3}$

3 **a** $\dfrac{1}{\sqrt{2}}$ **b** $-\dfrac{1}{\sqrt{2}}$ **c** -1

4 **a** $\dfrac{\sqrt{3}}{2}$ **b** -0.5 **c** $-\sqrt{3}$

5 **a** $\theta = \dfrac{\pi}{3}$ or $\dfrac{2\pi}{3}$

   b $\theta = \dfrac{\pi}{6}$ or $\dfrac{7\pi}{6}$

6 $\dfrac{7}{4}$

Exercise 5.3A
page 87

1 **a** Period π and amplitude 4

   b

2 **a** Period $\dfrac{\pi}{2}$ and amplitude 10

   b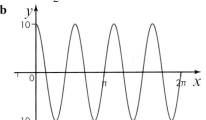

3 **a** Period 4π and amplitude 2

b (green curve) and **c** (red)

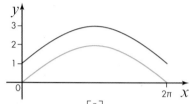

d A translation of $\begin{bmatrix} 0 \\ 1 \end{bmatrix}$

4 $y = 3\sin\frac{3}{2}x$.

5

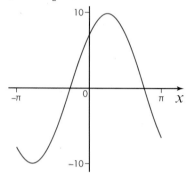

Exercise 5.4A **page 89**

1 **a** Period $= \dfrac{2\pi}{2.1} = 2.991... = 3.0$ to 1 d.p.

b

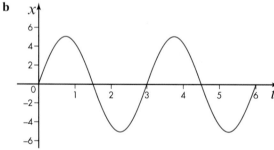

c 5 cm

2 **a** 0.004

b 250

c $\dfrac{2\pi}{0.004}$ or 1571

3 **a** 101.25 m

b 135 m

c 30 minutes

d

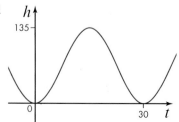

e 10.2 minutes

4 **a** 2 times each second

b

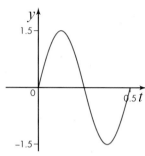

c 0.134 s

5 **a** 2.9 m

b h is a maximum when $30(t + 1) = 90$

This is the case when $t = 2$, which is at 2 a.m.

c 8 a.m.

d

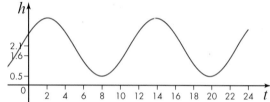

e The boat can enter before 05:44, between 10:16 and 17:44, or after 22:16.

Exercise 5.5A **page 92**

1 **a** 1.666×10^{-4}

b 0.17%

2 $\cos^2\theta \approx \left(1 - \frac{1}{2}\theta^2\right)^2 = 1 - \theta^2 + \frac{1}{4}\theta^4$

If θ is small you can ignore the last term because it is much smaller than the others, and the result follows.

3 **a** $1 - 2x^2$

b $1 - 2x^2$

c $\cos 2x$ and $1 - 2\sin^2 x$ are (approximately) equal if x is small.

4 $\frac{1}{8}$

5 **a** $\sin^2\theta + \cos^2\theta \approx \theta^2 + \left(1 - \frac{1}{2}\theta^2\right)^2$

$$= \theta^2 + 1 - \theta^2 + \frac{1}{4}\theta^4 = 1 + \frac{1}{4}\theta^4$$

b The error is less than $\frac{1}{4} \times 0.5^4 = 0.015625$.

6 **a** The base is $\cos x \approx 1 - \frac{1}{2}x^2$ and the height is $\sin x \approx x$.

Area $= \frac{1}{2} \times$ base $\times$ height $\approx \frac{1}{2}\left(1 - \frac{1}{2}x^2\right)x = \frac{1}{2}x - \frac{1}{4}x^3$

b 0.13%

7 $x + x^3$

8 **a** $-4x(1 - 2x^2)$

Exercise 5.6A page 95

1 $\cos 75° = \cos(45° + 30°)$

$$= \cos 45° \cos 30° - \sin 45° \sin 30°$$

$$= \frac{1}{\sqrt{2}} \times \frac{\sqrt{3}}{2} - \frac{1}{\sqrt{2}} \times \frac{1}{2} = \frac{\sqrt{3} - 1}{2\sqrt{2}}$$

2 $\dfrac{\sqrt{3} - 1}{\sqrt{3} + 1}$

3 $\sin 5A$

4 $\dfrac{\sqrt{3}}{2}$

5 RHS $= \sqrt{2} \sin\left(A + \dfrac{\pi}{4}\right) = \sqrt{2}\left(\sin A \cos \dfrac{\pi}{4} + \cos A \sin \dfrac{\pi}{4}\right)$

$$= \sqrt{2}\left(\sin A \times \frac{1}{\sqrt{2}} + \cos A \times \frac{1}{\sqrt{2}}\right)$$

$$= \sin A + \cos A = \text{LHS}$$

6 $\tan(A + B) = \dfrac{\tan A + \tan B}{1 - \tan A \tan B}$

When $A + B = \dfrac{\pi}{2}$, the tangent is indeterminate so the denominator of the fraction is zero

$$1 - \tan A \tan B = 0$$

Therefore $\tan A \tan B = 1$

7 a LHS $= \tan A + \tan B = \dfrac{\sin A}{\cos A} + \dfrac{\sin B}{\cos B}$

$$= \frac{\sin A \cos B + \cos A \sin B}{\cos A \cos B} = \frac{\sin(A + B)}{\cos A \cos B} = \text{RHS}$$

b $\dfrac{\sin(A - B)}{\cos A \cos B}$

8 a $\sin(A + B) + \sin(A - B) = (\sin A \cos B + \cos A \sin B)$
$$+ (\sin A \cos B - \cos A \sin B)$$
$$= 2 \sin A \cos B = \text{RHS}$$

b Let $A + B = x$ and $A - B = y$.

Add the equations to get $2A = x + y$

Therefore $\qquad A = \dfrac{x + y}{2}$

Subtract the equations to get $2B = x - y$

Therefore $\qquad B = \dfrac{x - y}{2}$.

Substitute these expressions into the identity in **part a** and the result follows.

9 a $\tan(A + B) = \dfrac{\tan A + \tan B}{1 - \tan A \tan B} = \tan \dfrac{\pi}{4} = 1$

Therefore $\tan A + \tan B = 1 - \tan A \tan B$

Rearrange: $\tan A + \tan A \tan B = 1 - \tan B$

Factorise: $\tan A(1 + \tan B) = 1 - \tan B$

Hence $\qquad \tan A = \dfrac{1 - \tan B}{1 + \tan B}$

b $\dfrac{1 - \tan A}{1 + \tan A}$

Exercise 5.6B page 97

1 $\sin 2A = 2 \sin A \cos A$

Let $x = 2A$ so $\dfrac{x}{2} = A$ and $\sin x = 2 \sin \dfrac{x}{2} \cos \dfrac{x}{2}$

So $\qquad \frac{1}{2} \sin x = \sin \dfrac{x}{2} \cos \dfrac{x}{2}$

2 a $\cos 2\theta = 2 \cos^2 \theta - 1$

Rearrange: $2 \cos^2 \theta = 1 + \cos 2\theta$

Hence $\qquad \cos^2 \theta = \frac{1}{2}(1 + \cos 2\theta)$

b $\cos 2\theta = 1 - 2 \sin^2 \theta$

$$2 \sin^2 \theta = 1 - \cos 2\theta$$

$$\sin^2 \theta = \frac{1}{2}(1 - \cos 2\theta)$$

3 $2 \cos^2 x = 1 + \cos 2x$ so $2 \cos^3 x = \cos x + \cos x \cos 2x$

RHS $= \frac{1}{2}\left(\sin x \sin 2x + 2 \cos^3 x\right)$

$$= \frac{1}{2}\left(\sin x \sin 2x + \cos x + \cos x \cos 2x\right)$$

$$= \frac{1}{2}\left(\cos x + \cos(2x - x)\right) = \cos x = \text{LHS}$$

4 LHS $= \cos 2\theta + 2 \cos \theta + 1$

$$= 2 \cos^2 \theta - 1 + 2 \cos \theta + 1 = 2 \cos^2 \theta + 2 \cos \theta$$

$$= 2 \cos \theta(\cos \theta + 1) = \text{RHS}$$

5 a If $x = \dfrac{\pi}{8}$ then $2x = \dfrac{\pi}{4}$ so $\tan 2x = 1$

$$\tan 2x = \frac{2 \tan x}{1 - \tan^2 x} = 1$$

Hence $\qquad 2 \tan x = 1 - \tan^2 x$

b $\sqrt{2} - 1$

6 LHS $= \sin(2A + A) = \sin 2A \cos A + \cos 2A \sin A$

$$= 2 \sin A \cos^2 A + (1 - 2 \sin^2 A) \sin A$$

$$= 2 \sin A (1 - \sin^2 A) + \sin A - 2 \sin^3 A$$

$$= 2 \sin A - 2 \sin^3 A + \sin A - 2 \sin^3 A$$

$$= 3 \sin A - 4 \sin^3 A = \text{RHS}$$

Exercise 5.7A page 98

1 a $2 \sin\left(\theta + \dfrac{\pi}{6}\right)$

b

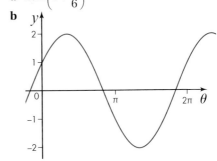

2 $5 \sin(\theta + 0.927)$

3 a $\sqrt{2} \sin\left(\theta + \dfrac{\pi}{4}\right)$

b

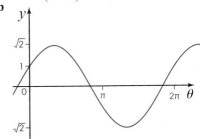

c $\sqrt{2}\cos\left(\theta+\dfrac{\pi}{4}\right)$

d

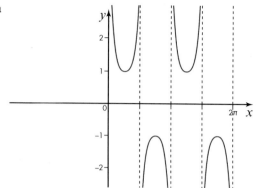

4 $10\sin(\theta-0.644)$

5 a $13\sin(\theta+0.395)$

b 13

c $\begin{bmatrix} -0.395 \\ 0 \end{bmatrix}$

d $\theta=0.268$

6 a $\sqrt{244}\cos(\theta+0.876)$

b $\theta=0.369$ or 4.162

7 a Vertical distance from R to OB is $6\sin\theta$ and vertical distance from P to R is $4\cos\theta$ so the sum of these is the height of P.

b $7.21\,\text{m}$

c 0.740 or 1.225

8 a $\theta=2.149$ and 5.926

b If $0.640\sin(\theta+0.675)=0.7$ then

$\sin(\theta+0.675)=1.094$

This has no solution because

$-1\le\sin(\theta+0.675)\le1$

Exercise 5.8A page 102

1 a $\sqrt{2}$ **b** $\sqrt{3}$

   c $\dfrac{2}{\sqrt{3}}$ **d** $\dfrac{2}{\sqrt{3}}$

2

θ	$\sin\theta$	$\cos\theta$	$\tan\theta$	$\sec\theta$	$\operatorname{cosec}\theta$	$\cot\theta$
$\dfrac{\pi}{6}$	$\dfrac{1}{2}$	$\dfrac{\sqrt{3}}{2}$	$\dfrac{1}{\sqrt{3}}$	$\dfrac{2}{\sqrt{3}}$	2	$\sqrt{3}$
$\dfrac{\pi}{3}$	$\dfrac{\sqrt{3}}{2}$	$\dfrac{1}{2}$	$\sqrt{3}$	2	$\dfrac{2}{\sqrt{3}}$	$\dfrac{1}{\sqrt{3}}$

3 a $x=0.253$ or 2.889

b $x=1.911$ or 4.373

4 a 2 **b** $\dfrac{2}{\sqrt{3}}$ **c** $\sqrt{3}$

5 a

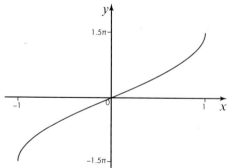

b $f(x)\ge1$ and $f(x)\le-1$

6 a $f(x)\ge0.5$ and $f(x)\le-0.5$

b $f(x)\ge1.5$ and $f(x)\le-0.5$

c $f(x)\ge1$ and $f(x)\le-1$

7 $\tan\theta+\cot\theta=\dfrac{\sin\theta}{\cos\theta}+\dfrac{\cos\theta}{\sin\theta}=\dfrac{\sin^2\theta+\cos^2\theta}{\cos\theta\sin\theta}$

$\qquad\qquad\quad =\dfrac{1}{\cos\theta\sin\theta}=\sec\theta\operatorname{cosec}\theta$

8 $\operatorname{cosec}2x=\dfrac{1}{\sin2x}=\dfrac{1}{2\sin x\cos x}=\tfrac{1}{2}\operatorname{cosec}x\sec x$

9 $\cot2x=\dfrac{1}{\tan2x}=\dfrac{1-\tan^2x}{2\tan x}$

Divide the numerator and the denominator by

$\tan^2x$ to get $\dfrac{\dfrac{1}{\tan^2x}-1}{2\times\dfrac{1}{\tan x}}=\dfrac{\cot^2x-1}{2\cot x}$

10 $x=\dfrac{\pi}{6}$ or $\dfrac{5\pi}{6}$ or $\dfrac{3\pi}{2}$

Exercise 5.9A page 104

1 a $-1\le x\le1$

b $-0.5\le x\le0.5.$

c $4\le x\le6$

2 a

b $-1.5\pi\le y\le1.5\pi$

3 $\left(\dfrac{1}{\sqrt{2}},\dfrac{\pi}{4}\right)$.

4 a

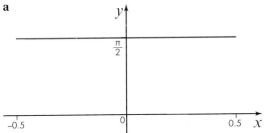

b If $x = \sin\theta$ and $x = \cos\varphi$ then $\theta + \varphi = \dfrac{\pi}{2}$

So the graph of $y = \theta + \varphi$ will be a horizontal straight line as shown.

5 $x = \pm\sqrt{0.479} = \pm 0.692$

6 a $x \geqslant 1$ or $x \leqslant -1$

b $-\dfrac{\pi}{2} \leqslant f(x) < 0$ and $0 < f(x) \leqslant \dfrac{\pi}{2}$, with 0 excluded.

Exam-style questions 5   **page 105**

1 $\dfrac{7\pi}{6}$

2 $288\,\text{cm}^2$

3 $\tan 75° = \tan(45° + 30°)$

$$= \dfrac{\tan 45° + \tan 30°}{1 - \tan 45° \tan 30°} = \dfrac{1 + \dfrac{1}{\sqrt{3}}}{1 - \dfrac{1}{\sqrt{3}}} = \dfrac{\sqrt{3} + 1}{\sqrt{3} - 1}$$

4 a $\sin 2x = 2\sin x \cos x$

Let $\theta = 2x$ so $x = \dfrac{\theta}{2}$ and then $\sin\theta = 2\cos\dfrac{\theta}{2}\sin\dfrac{\theta}{2}$

b In the same way, $\sin\dfrac{\theta}{2} = 2\cos\dfrac{\theta}{4}\sin\dfrac{\theta}{4}$

So $\sin\theta = 2\cos\dfrac{\theta}{2} \times 2\cos\dfrac{\theta}{4}\sin\dfrac{\theta}{4}$

$$= 4\cos\dfrac{\theta}{2}\cos\dfrac{\theta}{4}\sin\dfrac{\theta}{4}$$

5 a

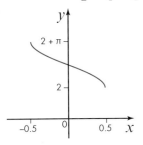

b Domain $-\dfrac{1}{2} \leqslant x \leqslant \dfrac{1}{2}$ and range $2 \leqslant f(x) \leqslant 2 + \pi$

6 $2\sin^2\theta - \sin\theta = 0$

Factorise.

$\sin\theta\,(2\sin\theta - 1) = 0$

Either $\sin\theta = 0$ or $2\sin\theta - 1 = 0$

If $\sin\theta = 0$ then $\theta = 0$ or π

If $2\sin\theta - 1 = 0$ then $\sin\theta = \dfrac{1}{2}$ so $\theta = \dfrac{\pi}{6}$ or $\dfrac{5\pi}{6}$

There are four values for θ.

7 a $10\,\text{m}$

b $8\,\text{s}$

c $4.7\,\text{s}$ and $7.3\,\text{s}$

8 $\dfrac{\cos 2\theta}{\cos^2\theta} + \tan^2\theta = \dfrac{\cos^2\theta - \sin^2\theta}{\cos^2\theta} + \tan^2\theta$

$$= 1 - \dfrac{\sin^2\theta}{\cos^2\theta} + \tan^2\theta$$

$$= 1 - \tan^2\theta + \tan^2\theta = 1$$

9 Area of triangle $= \dfrac{1}{2}a^2 \sin 60°$

Area of sector $= \dfrac{1}{2}a^2\theta$

So $\theta = \sin 60° = 0.866$

The perimeter of the triangle is $3a$ and that of the sector is $0.866a + 2a = 2.866a$

The perimeter of the sector is $\dfrac{2.866}{3} \times 100 = 95.53\%$ of the perimeter of the triangle, so is about 4.5% less.

10 $\theta = 35.3°$ or $324.7°$

11 $\cos\left(\theta + \dfrac{\pi}{4}\right) = \cos\theta\cos\dfrac{\pi}{4} - \sin\theta\sin\dfrac{\pi}{4}$

$\cos\dfrac{\pi}{4} = \sin\dfrac{\pi}{4} = \dfrac{1}{\sqrt{2}}$ and $\sin\theta = \sqrt{1 - \left(\dfrac{4}{5}\right)^2} = \dfrac{3}{5}$

So $\cos\left(\theta + \dfrac{\pi}{4}\right) = \dfrac{4}{5} \times \dfrac{1}{\sqrt{2}} - \dfrac{3}{5} \times \dfrac{1}{\sqrt{2}} = \dfrac{1}{5\sqrt{2}} = \dfrac{\sqrt{2}}{10}$

12 a $r = 17.21$

$\alpha = 0.951$

b $\theta = 0.108$ or 1.132

13 a $\cos 2A = 2\cos^2 A - 1$ so

$\cos^2 2A = (2\cos^2 A - 1)^2$

$$= 4\cos^4 A - 4\cos^2 A + 1$$

b $\cos 4A = 2\cos^2 2A - 1$

$$= 2(4\cos^4 A - 4\cos^2 A + 1) - 1$$

$$= 8\cos^4 A - 8\cos^2 A + 1$$

14 a $x = 1.37$ or 0.206

b $k \geqslant 2$ or $k \leqslant -2$

15 $x = \dfrac{\pi}{3}$ or $\dfrac{5\pi}{3}$.

$x = 1.91$ or 4.37

6 Differentiation

Prior knowledge
page 108

1 (0, 0) a maximum point

   (4, –32) a minimum point

2 If θ is small, $\cos\theta \approx 1 - \frac{1}{2}\theta^2$

   Let $\theta = 2\alpha$ and then $\cos 2\alpha \approx 1 - \frac{1}{2}(2\alpha)^2 = 1 - 2\alpha^2$

3 a $6x^2 - 2$

   b $\dfrac{x^2 + 1}{2x^2}$

   c $-\frac{1}{2}x^{-\frac{3}{2}}$

Exercise 6.1A
page 109

1 (5, 41)

   $\dfrac{d^2y}{dx^2} = -2 < 0$, which means it is a maximum point.

2 a $\dfrac{dy}{dx} = 3x^2 - 6x + 3$

      If $3x^2 - 6x + 3 = 0$ then $x^2 - 2x + 1 = 0$ so $(x-1)^2 = 0$ and the only solution is $x = 1$.

   b (1, –19) is the stationary point.

      $\dfrac{d^2y}{dx^2} = 6x - 6$, which is 0 when $x = 1$.

      When $x = 0$, $\dfrac{dy}{dx} = 3$ and when $x = 2$, $\dfrac{dy}{dx} = 3$, so the gradient goes 3, 0, 3.

      This is a point of inflection.

3 (0, 0) a maximum point.

   (1, –1) a minimum point.

   (–1, –1) a minimum point.

4 a $\dfrac{dy}{dx} = 3x^2$ and $\dfrac{d^2y}{dx^2} = 6x$

      If $\dfrac{dy}{dx} = 0$ then $x = 0$, $y = 0$ and $\dfrac{d^2y}{dx^2} = 0$

      (0, 0) is the stationary point.

      when $x = -1$, $\dfrac{dy}{dx} = 3$ and when $x = 1$ $\dfrac{dy}{dx} = 3$ so the gradient goes 3, 0, 3.

      This is a point of inflection.

   b This curve is a translation of $y = x^3$ by $\begin{bmatrix} 10 \\ 20 \end{bmatrix}$ so it has a point of inflection at (10, 20).

5 $\dfrac{dy}{dx} = 2ax + b$.

   The curve has a stationary point when $2ax + b = 0$ or $x = -\dfrac{b}{2a}$.

   Since $a \neq 0$ this will always exist. $\dfrac{d^2y}{dx^2} = 2a$, which can be positive or negative but never zero. It will have a minimum or a maximum point but not a point of inflection (which would require $\dfrac{d^2y}{dx^2} = 0$).

6 a The total length is $8x + 4(9 - 2x) = 8x + 36 - 8x = 36$

   b The volume $V = x^2(9 - 2x) = 9x^2 - 2x^3$

      $\dfrac{dV}{dx} = 18x - 6x^2$

      When the volume is a maximum, $18x - 6x^2 = 0$.

      Therefore $3x - x^2 = 0$, so $x(3 - x) = 0$ and $x = 0$ or 3.

      $x = 0$ is clearly a minimum volume.

      Now $\dfrac{d^2V}{dx^2} = 18 - 12x$

      When $x = 3$, $\dfrac{d^2V}{dx^2} = -18 < 0$ so the volume is a maximum.

      When $x = 3$, $9 - 2x = 3$ so all the sides are 3 cm and the shape is a cube.

   c The surface area $A = 2x^2 + 4x(9 - 2x)$

      $\qquad = 2x^2 + 36x - 8x^2 = 36x - 6x^2$

      $\dfrac{dA}{dx} = 36 - 12x$ and if $36 - 12x = 0$ then $x = 3$.

      This means A has a stationary value when $x = 3$.

      $\dfrac{d^2A}{dx^2} = -12 < 0$ so the surface area is also a maximum when $x = 3$ and the shape is a cube.

Exercise 6.2A
page 111

1 a $16(2x + 1)$

   b $12(4x + 2)^2$

   c $20(4x + 2)^4$

2 a $-10(8 - x)^9$

   b $60x(1 + 3x^2)^9$

   c $60(6x - 3x^2)^9(1 - x)$

3 a $\dfrac{1}{2\sqrt{x - 3}}$

   b $\dfrac{1}{2\sqrt{x - 3}}$

   c $\dfrac{x}{\sqrt{x^2 - 3}}$

4 a -1.25

   b (8, 2)

5 a If $x = 10$, $y = \dfrac{600}{150} = 4$

   b $-\dfrac{1200x}{(x^2 + 50)^2}$

   c $y = -\frac{8}{15}x + 9\frac{1}{3}$

      or $y = -0.533x + 9.333$

Exercise 6.3A
page 113

1 a $2e^{2x}$ b $-e^{-x}$

   c $0.4e^{0.4x}$ d $4e^{4x + 2}$

2 a $2e^{0.5x}$

   b $-10e^{-0.1x}$

   c $100e^{2x - 10}$

3 **a** $e^{(\ln 1.5)x}$

b $(\ln 1.5)1.5^x$

4 **a** $8e^{2x}$

b $16e^{2x}$

5 **a** $-4xe^{-0.5x^2}$

b 1.08

6 **a** The multiplier for an increase of 3% is $1 + 0.03 = 1.03$ so after t years the value is multiplied by 1.03^t.

b $5000e^{(\ln 1.03)t}$

c £161.50 per year

7 $f'(x) = 2e^{2x} - 2e^{-2x}$

$f''(x) = 4e^{2x} - 2e^{-2x} \times -2 = 4e^{2x} + 4e^{-2x} = 4(e^{2x} + e^{-2x})$

$= 4f(x)$

8 $\dfrac{dy}{dx} = e^x - 4e^{-x}$

At a stationary point $e^x - 4e^{-x} = 0$.

Multiply by e^x.

$e^{2x} - 4 = 0$ so $e^{2x} = 4$

Take logs.

$2x = \ln 4$ so $x = \frac{1}{2}\ln 4 = \ln\left(4^{\frac{1}{2}}\right) = \ln 2$

$\dfrac{d^2y}{dx^2} = e^x + 4e^{-x}$ which equals y.

When $x = \ln 2$ then $y = e^{\ln 2} + 4e^{-\ln 2} = 2 + 4 \times \frac{1}{2} = 4$

This means $\dfrac{d^2y}{dx^2} = 4 > 0$ so the stationary point is a minimum.

The coordinates are $(\ln 2, 4)$.

Exercise 6.4A page 115

1 **a** $\dfrac{1}{x}$

b $\dfrac{3}{x}$

c $\dfrac{3x^2}{x^3 + 2}$

2 **a** $\dfrac{4}{x}$ **b** $\dfrac{1}{x}$ **c** $\dfrac{4}{x}$

3 **a**

b $\begin{bmatrix} 0 \\ \ln 2 \end{bmatrix}$

c $f(x) = \ln kx = \ln k + \ln x$ and since $\ln k$ is a constant, $f'(x) = \dfrac{1}{x}$

d The translation is parallel to the y-axis so the gradient for any particular x value is unchanged.

4 **a** $y = x + 4$

b $y = \frac{1}{5}x + \ln 5$

5 **a** If $x = e$, then $y = \ln e = 1$ so $(e, 1)$ is on the curve.

b $\dfrac{dy}{dx} = \dfrac{1}{x}$ and if $x = e$ then $\dfrac{dy}{dx} = \dfrac{1}{e}$

The equation of the tangent is $y - 1 = \dfrac{1}{e}(x - e)$ which simplifies to $y = \dfrac{1}{e}x$, which passes through the origin.

c The gradient of the normal is $-e$ and the equation is $y - 1 = -e(x - e)$.

Where it crosses the x-axis, $y = 0$ so $-1 = -e(x - e)$, which rearranges to $\dfrac{1}{e} = x - e$ and so $x = e + \dfrac{1}{e}$

6 **a** $-\dfrac{1}{x}$

b $-\dfrac{2}{x}$

c $\dfrac{dy}{dx} = -\dfrac{n}{x}$

7 $y = \log_{10} x$ therefore $x = 10^y = (e^{\ln 10})^y = e^{(\ln 10)y}$

Take logs. $(\ln 10)^y = \ln x$

$$y = \frac{1}{\ln 10}\ln x$$

$$\frac{dy}{dx} = \frac{1}{x \ln 10}$$

Exercise 6.5A page 118

1 **a** $2\cos 2x$

b $-5\sin(5x - 2)$

c $2x\cos(x^2 + 1)$

2 **a** $5\cos 0.5x$

b $3\cos 3x - 6\sin 6x$

c $-(2x - 3)\sin(x^2 - 3x - 4)$

3 $f'(x) = 4\cos 4x - 4\sin 4x$

$f''(x) = -16\sin 4x - 16\cos 4x$

$= -16(\sin 4x + \cos 4x) = -16f(x)$

Hence $f''(x) + 16f(x) = 0$

4 **a** $y = (\sin x)^2$

Use the chain rule.

$$\frac{dy}{dx} = 2\sin x \times \cos x = 2\sin x \cos x = \sin 2x$$

b $-2\sin x \cos x = -\sin 2x$

c $\sin^2 x + \cos^2 x = 1$ so the derivative of $\sin^2 x + \cos^2 x$ is 0.

Hence the two derivatives are identical apart from the sign.

5 a 1

b At a stationary point, $\cos x\, e^{\sin x} = 0$.

Since $e^{\sin x}$ is always positive, $\cos x = 0$.

One solution is $x = \dfrac{\pi}{2}$ and then $y = e^{\sin\frac{\pi}{2}} = e$ so $\left(\dfrac{\pi}{2}, e\right)$ is a stationary point.

6 a $0.177\,\mathrm{m\,s^{-1}}$

b the displacement is $0.1\,\mathrm{m}$

c The bob is in the central position.

Exam-style questions 6   page 119

1 a $\dfrac{dy}{dx} = 3(x-10)^2$ and $\dfrac{d^2y}{dx^2} = 6(x-10)$

b At a stationary point, $3(x-10)^2 = 0$ so $x = 10$ and $y = 15$.

When $x = 10$, $\dfrac{d^2y}{dx^2} = 0$

When $x = 9$, $\dfrac{dy}{dx} = 3$ and when $x = 11$, $\dfrac{dy}{dx} = 3$

The gradient goes $+$, 0, $+$ so $(10, 15)$ is a point of inflection.

2 $y = (1+x^2)^{\frac{1}{2}}$ so $\dfrac{dy}{dx} = \dfrac{1}{2}(1+x^2)^{-\frac{1}{2}} \times 2x = \dfrac{x}{(1+x^2)^{\frac{1}{2}}} = \dfrac{x}{y}$

3 $f(x) = (\cos x)^{-1}$ so $f'(x) = -1 \times (\cos x)^{-2} \times (-\sin x)$
$$= \dfrac{\sin x}{\cos^2 x} = \dfrac{\tan x}{\cos x}$$

4 a $-2e^{-2x}$ **b** $-2xe^{-x^2}$

5 $(4, \ln 16)$

6 Student A is incorrect because it is not true that $\sin(x+2) = \sin x + \sin 2$

Student B is correct. The student's answer is an expansion of $\cos(x+2)$ and this is a more concise way to give the answer. It can be found by differentiating $\sin(x+2)$ directly and it is not necessary to use an addition formula.

7 a After 40 minutes, $t = \frac{2}{3}$ and $y = 10 \times 2^1 = 20$, which is double the initial area.

b $226.3\,\mathrm{mm^2}$

c $235\,\mathrm{mm^2}$ per hour

d The model will no longer be valid if the bacteria run out of space to expand or nutrient to feed on. This is likely to happen as the area increases.

8 $(1.073, 0.877)$

9 $y = 2.773x + 8$

10 a $\dfrac{dy}{dx} = 3x^2 + 6x + 4$

At a stationary point, $3x^2 + 6x + 4 = 0$

Use the quadratic formula.

$$x = \dfrac{-6 \pm \sqrt{36 - 48}}{6} = \dfrac{-6 \pm \sqrt{-12}}{6}$$

Since $\sqrt{-12}$ is not a real number, the equation has no solution, so there are no stationary points.

b $y = x + 4$

11 $k = 0.125$

$a = 5.73$

12 a $y = \sin^3 x$ so $\dfrac{dy}{dx} = 3\sin^2 x \times \cos x$
$$= 3\cos x \sin^2 x = 3\cos x\,(1 - \cos^2 x)$$
$$= 3\cos x - 3\cos^3 x$$

b $\dfrac{d^2y}{dx^2} = -3\sin x - 9\cos^2 x \times (-\sin x)$
$$= -3\sin x + 9\sin x \cos^2 xt$$
$$= -3\sin x + 9\sin x\,(1 - \sin^2 x)$$
$$= -3\sin x + 9\sin x - 9\sin^3 x$$
$$= 6\sin x - 9\sin^3 x$$

13 the maximum and minimum values of the gradient are $\pm\dfrac{\pi}{18}$

14 If $y = e^{-\frac{1}{2}x^2}$ then $\dfrac{dy}{dx} = e^{-\frac{1}{2}x^2} \times (-x) = -xe^{-\frac{1}{2}x^2}$

When $x = a$, $y = e^{-\frac{1}{2}a^2}$ and $\dfrac{dy}{dx} = -ae^{-\frac{1}{2}a^2}$

The equation of the tangent at P is
$$y - e^{-\frac{1}{2}a^2} = -ae^{-\frac{1}{2}a^2}(x - a)$$

Where this crosses the x-axis, $y = 0$ and
$$-e^{-\frac{1}{2}a^2} = -ae^{-\frac{1}{2}a^2}(x - a)$$

Hence $1 = a(x - a)$, which you can rearrange to get $x = a + \dfrac{1}{a}$.

15 $y = \sin 2x$. If δx is a small change in x and δy is the corresponding change in y, then

$y + \delta y = \sin 2(x + \delta x) = \sin(2x + 2\delta x)$
$$= \sin 2x \cos 2\delta x + \cos 2x \sin 2\delta x$$

$\dfrac{\delta y}{\delta x} = \dfrac{(y + \delta y) - y}{\delta x} = \dfrac{\sin 2x \cos 2\delta x + \cos 2x \sin 2\delta x - \sin 2x}{\delta x}$
$$= \dfrac{\sin 2x(\cos 2\delta x - 1)}{\delta x} + \dfrac{\cos 2x \sin 2\delta x}{\delta x}$$

If δx is small then $\cos 2\delta x \approx 1 - \frac{1}{2}(2\delta x)^2 = 1 - 2(\delta x)^2$ and $\sin 2\delta x \approx 2\delta x$.

Therefore $\dfrac{\delta y}{\delta x} \approx \dfrac{\sin 2x \times \left(-2(\delta x)^2\right)}{\delta x} + \dfrac{\cos 2x \times 2\delta x}{\delta x}$
$$= -2\sin 2x \times \delta x + 2\cos 2x$$

Therefore $\dfrac{dy}{dx} = \lim\limits_{\delta x \to 0} \dfrac{\delta y}{\delta x} = 2\cos 2x$

7 Further differentiation

Prior knowledge   page 122

1 **a** $2e^{2x} + 2e^{-2x}$ **b** $12\cos 4x - 10\sin 2x$

  c $\dfrac{1}{x}$

2 at $(9, 6)$ and at $(9, -6)$.

3 $\sec^2 A = 1 + \tan^2 A$

Therefore $\tan^2 A = \sec^2 A - 1 = (\sec A - 1)(\sec A + 1)$

Hence $\dfrac{\tan^2 A}{\sec A + 1} + 1 = \dfrac{(\sec A - 1)(\sec A + 1)}{\sec A + 1} + 1$

$\qquad\qquad = \sec A - 1 + 1 = \sec A$

Exercise 7.1A   page 124

1 **a** $\sin x + x\cos x$

  b $\cos x - x\sin x$

  c $2x\sin 2x + 2x^2\cos 2x$

2 **a** $e^x + xe^x$

  b $-xe^{-x}$

  c $2x(x+1)e^{2x}$

3 **a** $\cos^2 x - \sin^2 x$

  b $\cos 2x$

  c The identity $\cos 2x = \cos^2 x - \sin^2 x$ means that the two answers are identical.

4 **a** $e^x\sin x + e^x\cos x$

  b $2e^x\cos x$

5 **a** $1 + \ln x$

  b 1

  c (e, e)

  d $(e^{-1}, -e^{-1})$

6 **a** $\dfrac{dy}{dx} = e^{-x} - xe^{-x}$

At a stationary point, $e^{-x} - xe^{-x} = 0$ so $e^{-x}(1 - x) = 0$.

e^{-x} cannot be 0 so the only solution is $x = 1$.

Then $y = 1 \times e^{-1}$ and the coordinates are $(1, e^{-1})$.

  b $\dfrac{dy}{dx} = 2xe^{-x} - x^2e^{-x}$

At a stationary point, $2xe^{-x} - x^2e^{-x} = 0$ so

$\qquad\qquad x(2 - x)e^{-x} = 0$

Either $x = 0$ or $x = 2$

When $x = 0$, $y = 0$

When $x = 2$, $y = 4e^{-2}$

The points are $(0, 0)$ and $(2, 4e^{-2})$.

7 $(0, 0)$, $(1, 10e^{-1})$ and $(-1, 10e^{-1})$

8 $\dfrac{dy}{dx} = -0.2e^{-0.2x}\sin 2x + 2e^{-0.2x}\cos 2x$

At a stationary point,

$-0.2e^{-0.2x}\sin 2x + 2e^{-0.2x}\cos 2x = 0$

$\qquad 2e^{-0.2x}\cos 2x = 0.2e^{-0.2x}\sin 2x$

$\qquad\qquad 2\cos 2x = 0.2\sin 2x$

$\qquad\qquad \dfrac{\sin 2x}{\cos 2x} = \dfrac{2}{0.2}$

$\qquad\qquad \tan 2x = 10$

$2x = \arctan 10 = 1.4711$ or $0.4711 + \pi$ or $0.4711 + 2\pi$

or ...

$2x = 1.4711 + n\pi$ where $n = 0, 1, 2, 3, \ldots$

So $x = \dfrac{n\pi}{2} + 0.736$ where $n = 0, 1, 2, 3, \ldots$

9 $28.77\,\text{m s}^{-1}$

Exercise 7.2A   page 126

1 **a** $\dfrac{1}{(x+1)^2}$

  b $\dfrac{1 - 2x - x^2}{(x^2 + 1)^2}$

  c $\dfrac{4x^3 - 3x^2}{(2x - 1)^2}$

2 **a** $\dfrac{x\cos x - \sin x}{x^2}$

  b $\dfrac{\cos x + (x + 1)\sin x}{\cos^2 x}$

  c $\dfrac{2x\sin 2x - 2x^2\cos 2x}{\sin^2 2x}$

3 **a** $-2e^{-2x}\sin x + e^{-2x}\cos x$

  b $\dfrac{dy}{dx} = \dfrac{e^{2x}\cos x - \sin x \times 2e^{2x}}{e^{4x}}$

Divide numerator and denominator by e^{2x}.

$\dfrac{\cos x - 2\sin x}{e^{2x}} = \dfrac{\cos x}{e^{2x}} - \dfrac{2\sin x}{e^{2x}}$

which is the same answer.

4 **a** $\dfrac{1 - \ln x}{x^2}$

  b $\dfrac{2\ln x - 3}{x^3}$

  c The coordinates of the stationary point are $\left(e, \dfrac{1}{e}\right)$.

When $x = e$, then $\dfrac{d^2y}{dx^2} = \dfrac{2\ln e - 3}{e^3} = -\dfrac{1}{e^3}$ which is negative so the stationary point is a maximum point.

5 **a** $\dfrac{2e^{-x}}{(1+e^{-x})^2}$

b $\frac{1}{2}$

c At a stationary point, $\dfrac{2e^{-x}}{(1+e^{-x})^2} = 0$ but since e^{-x} is always positive, this can never happen. Therefore there is no turning point.

6 **a** If $\quad v = \sqrt{x^2 + 1} = (x^2 + 1)^{\frac{1}{2}}$ then

$$\dfrac{dv}{dx} = \dfrac{1}{2}(x^2+1)^{-\frac{1}{2}} \times 2x = \dfrac{x}{\sqrt{x^2+1}}$$

So if $y = \dfrac{x}{\sqrt{x^2+1}}$ then $\dfrac{dy}{dx} = \dfrac{\sqrt{x^2+1} - x \times \dfrac{x}{\sqrt{x^2+1}}}{x^2+1}$

The numerator is

$$\sqrt{x^2+1} - x \times \dfrac{x}{\sqrt{x^2+1}} = \dfrac{x^2+1-x^2}{\sqrt{x^2+1}} = \dfrac{1}{\sqrt{x^2+1}}$$

So $\dfrac{dy}{dx} = \dfrac{1}{(x^2+1)\sqrt{x^2+1}} = \dfrac{1}{\sqrt[3]{x^2+1}}$

b $-\dfrac{1}{x^2\sqrt{x^2+1}}$

7 Use the quotient rule.

If $y = \tan x = \dfrac{\sin x}{\cos x}$ then

$$\dfrac{dy}{dx} = \dfrac{\cos x \times \cos x - \sin x \times (-\sin x)}{\cos^2 x} = \dfrac{\cos^2 x + \sin^2 x}{\cos^2 x}$$

$$= \dfrac{1}{\cos^2 x}$$

Exercise 7.3A　　　　　　　　　　page 130

1 **a** $3\sec^2 3x$　　**b** $-3\csc^2 3x$　　**c** $2\sec^2(2x+3)$

2 **a** $2\sec 2x \tan 2x$

b $2\sec^2 x \tan x$

c $4\sec^2 2x \tan 2x$

3 **a** $\dfrac{2}{\sqrt{1-4x^2}}$

b $\dfrac{2x}{\sqrt{1-x^4}}$

c $\dfrac{2}{4+x^2}$

4 **a** $\cot x - x\csc^2 x$

b $2x\csc x - x^2 \csc x \cot x$

c $\dfrac{x\sec x \tan x - \sec x}{x^2}$

5 $y = 2\sqrt{3}x - \dfrac{2\sqrt{3}\pi}{3} + 2$

6 $2\sec^3 x - \sec x$

7 $\left(\dfrac{1}{2}, \dfrac{\pi}{4}\right)$ and $\left(-\dfrac{1}{2}, -\dfrac{\pi}{4}\right)$

8 **a** $2\tan x \sec^2 x$

b $2\sec^2 x \tan x$

c $\sec^2 x = 1 + \tan^2 x$. If you differentiate both sides then the derivative of $\sec^2 x$ is the same as the derivative of $\tan^2 x$, because the derivative of 1 is 0.

Exercise 7.4A　　　　　　　　　　page 132

1 **a** $2t$

b $4t^3$

c $1 - \dfrac{3}{2t}$

2 $\dfrac{dy}{dx} = \dfrac{dy}{dt} \div \dfrac{dx}{dt} = \dfrac{2t}{1} = 2t$

When $t = 1$, $x = 3$, $y = 0$ and $\dfrac{dy}{dx} = 2$

The equation of the tangent is $y - 0 = 2(x - 3)$ or $y = 2x - 6$ or $2x - y - 6 = 0$

3 $\dfrac{dy}{dx} = \dfrac{dy}{dt} \div \dfrac{dx}{dt} = \dfrac{2t}{\frac{1}{2}} = 4t$

When $t = -1$, $x = -\frac{1}{2}$ and $y = -3$, and the gradient of the normal is $-\dfrac{1}{4t} = \frac{1}{4}$.

The equation of the normal is $y + 3 = \frac{1}{4}\left(x + \frac{1}{2}\right)$

therefore $\qquad\qquad 4y + 12 = x + \frac{1}{2}$

So $\qquad\qquad\qquad 8y + 24 = 2x + 1$

So $\qquad\qquad\qquad 8y - 2x + 23 = 0$

4 **a** $-\dfrac{5}{4}\tan t$

b $6\cos t$

c $\cos^3 2t$

5 $\dfrac{dy}{dx} = \dfrac{dy}{dt} \div \dfrac{dx}{dt} = \dfrac{7\cos t}{-3\sin t} = -\frac{7}{3}\cot t$

When $t = \dfrac{\pi}{4}$, the gradient is $-\frac{7}{3}\cot\dfrac{\pi}{4} = -\frac{7}{3} \times 1 = -\frac{7}{3}$

6 $x = \sin 2t$

$\dfrac{dx}{dt} = 2\cos 2t$

$y = \cos 2t$

$\dfrac{dy}{dt} = -2\sin 2t$

$\dfrac{dy}{dx} = -2\sin 2t \div 2\cos 2t = -\tan 2t$

When $t = \dfrac{\pi}{3}$, $\dfrac{dy}{dx} = -\tan\dfrac{2\pi}{3} = \sqrt{3}$

So the gradient of the normal is $-\dfrac{1}{\sqrt{3}}$.

7 $\dfrac{3}{4}$ and $-\dfrac{3}{4}$

8 $\dfrac{dy}{dx} = \dfrac{dy}{dt} \div \dfrac{dx}{dt} = \dfrac{2\sin t}{2 - 2\cos t}$

When $= \dfrac{\pi}{4}$, $\dfrac{dy}{dx} = \dfrac{2\sin\dfrac{\pi}{4}}{2 - 2\cos\dfrac{\pi}{4}} = \dfrac{2\times\dfrac{1}{\sqrt{2}}}{2 - 2\times\dfrac{1}{\sqrt{2}}} = \dfrac{\sqrt{2}}{2 - \sqrt{2}}$

$= \dfrac{\sqrt{2}}{2 - \sqrt{2}} \times \dfrac{2 + \sqrt{2}}{2 + \sqrt{2}} = \dfrac{2\sqrt{2} + 2}{4 - 2} = \sqrt{2} + 1$

So the gradient of the tangent is $\sqrt{2} + 1$.

The gradient of the normal is

$-\dfrac{1}{\sqrt{2} + 1} = -\dfrac{1}{\sqrt{2} + 1} \times \dfrac{\sqrt{2} - 1}{\sqrt{2} - 1} = -\dfrac{\sqrt{2} - 1}{2 - 1} = 1 - \sqrt{2}$

9 $(0, \sqrt{2}b)$

10 a $\dfrac{-0.4\sin 2t}{\cos t}$

   b $t = 0$

   c $36°$

Exercise 7.5A   page 136

1 a $-\dfrac{2x}{y}$

   b $-\dfrac{3}{2}$

   c $\dfrac{3}{2}$

2 a If $x = 2$ and $y = 1$, then $x^2 y = 4$; also, $4(2 - y) = 4$ so the point is on the curve.

   b $-\dfrac{1}{2}$

3 a If $x = 5$ and $y = 0$, then $x^2 + 4xy + y^2 = 25 + 0 + 0 = 25$

   If $x = 0$ and $y = 5$, then $x^2 + 4xy + y^2 = 0 + 0 + 25 = 25$

   b -0.5

   c $2x + 4y + (4x + 2y)\dfrac{dy}{dx} = 0$

   $(4x + 2y)\dfrac{dy}{dx} = -(2x + 4y)$

   $\dfrac{dy}{dx} = \dfrac{-(2x + 4y)}{4x + 2y} = -\dfrac{x + 2y}{2x + y}$

4 a If $x = 3$ and $y = 5$, then $y^2 - x^2 = 25 - 9 = 16$

   b $5y - 3x = 16$

5 Differentiate.

$2(x^2 + y^2) \times \left(2x + 2y\dfrac{dy}{dx}\right) = 50\left(2x + 2y\dfrac{dy}{dx}\right)$

At a stationary point, $\dfrac{dy}{dx} = 0$ and so

$2(x^2 + y^2) \times (2x) = 50(2x)$

Therefore $2(x^2 + y^2) = 50$

$x^2 + y^2 = 25$

This is a circle of radius 5.

6 a $\dfrac{dy}{dx} = \dfrac{x}{y}$

   b $\dfrac{dy}{dx} = \dfrac{\sec t}{\tan t}$

   c If $x = \sec t$ and $y = \tan t$, then $\dfrac{x}{y} = \dfrac{\sec t}{\tan t}$ and the expressions are equivalent.

   d $\left(\dfrac{2}{\sqrt{3}}, \dfrac{1}{\sqrt{3}}\right)$ and $\left(-\dfrac{2}{\sqrt{3}}, -\dfrac{1}{\sqrt{3}}\right)$.

Exercise 7.6A   page 139

1 a $\dfrac{dy}{dx} = cx^2$, where c is a constant.

   b If $y = x^3 + 4$, then $\dfrac{dy}{dx} = 3x^2$

   This is a solution with $c = 3$.

2 a $\dfrac{dy}{dx} = \dfrac{c}{y}$

   b If $y = \sqrt{x} = x^{\frac{1}{2}}$, then $\dfrac{dy}{dx} = \dfrac{1}{2}x^{-\frac{1}{2}} = \dfrac{1}{2\sqrt{x}}$

   This is a solution with $c = \dfrac{1}{2}$.

   c If $y = a\sqrt{x} = ax^{\frac{1}{2}}$, then $\dfrac{dy}{dx} = \dfrac{a}{2}x^{-\frac{1}{2}} = \dfrac{a}{2\sqrt{x}}$

   This is a solution with $c = \dfrac{a}{2}$.

3 a $\dfrac{dy}{dx} = cxy$

   b If $y = e^{\frac{1}{2}x^2}$ then $\dfrac{dy}{dx} = e^{\frac{1}{2}x^2} \times x = xe^{\frac{1}{2}x^2} = xy$

   This is a solution with $c = 1$

4 a $\dfrac{dy}{dt} = 0.005y$

   b If $y = ae^{0.005t}$, then $\dfrac{dy}{dt} = a \times 0.005e^{0.005t} = 0.005y$

   c 55.3 million

5 a $\dfrac{dx}{dt} = \dfrac{c}{x}$

   When $x = 5$, $\dfrac{dx}{dt} = 0.4$ so $0.4 = \dfrac{c}{5}$ and $c = 2$

   Therefore $\dfrac{dx}{dt} = \dfrac{2}{x}$ or $x\dfrac{dx}{dt} = 2$

   b $-0.08\,\mathrm{m\,s^{-2}}$

6 a $\dfrac{dr}{dt} = \dfrac{c}{r^2}$

b $0.5 = \dfrac{c}{100}$ so $c = 50$ and $\dfrac{dr}{dt} = \dfrac{50}{r^2}$ or $r^2 \dfrac{dr}{dt} = 50$

c $r = at^{\frac{1}{3}}$ so $\dfrac{dr}{dt} = \frac{1}{3} at^{-\frac{2}{3}}$ and

$$r^2 \dfrac{dr}{dt} = a^2 t^{\frac{2}{3}} \times \frac{1}{3} at^{-\frac{2}{3}} = \frac{1}{3} a^3$$

This is a solution if $\frac{1}{3} a^3 = 50$, so $a^3 = 150$ and

$a = \sqrt[3]{150}$.

Exam-style questions 7　　　　　　　**page 140**

1 $2x \cos x - x^2 \sin x$

2 $\dfrac{4x}{(x^2 + 1)^2}$

3 a $-2 \operatorname{cosec}^2 2x$

b $-4 \cot 2x \operatorname{cosec}^2 2x$

4 $\dfrac{dx}{dt} = 0.6 \cos 2t + 0.8 \sin 2t$ and

$\dfrac{d^2 x}{dt^2} = -1.2 \sin 2t + 1.6 \cos 2t$

$4x = 1.2 \sin 2t - 1.6 \cos 2t = -\dfrac{d^2 x}{dt^2}$ so $\dfrac{d^2 x}{dt^2} + 4x = 0$

5 a There is a sign error: it should say

$\dfrac{dy}{dx} = 2(\sin x \times (-\sin x) + \cos x \times \cos x)$

and continue

$\dfrac{dy}{dx} = 2(-\sin^2 x + \cos^2 x) = 2 \cos 2x$

b You could write $y = 2 \sin x \cos x = \sin 2x$ and

differentiate to get $\dfrac{dy}{dx} = 2 \cos 2x$

6 a $\dfrac{\sin \theta}{1 - \cos \theta}$

b $\dfrac{\sin \theta}{1 - \cos \theta} = 0.5$ so $\sin \theta = 0.5(1 - \cos \theta)$

$2 \sin \theta = 1 - \cos \theta$

$2 \sin \theta + \cos \theta = 1$

7 a $-e^{-x} \sin 10x + 10 e^{-x} \cos 10x$

b At a stationary point, $-e^{-x} \sin 10x + 10 e^{-x} \cos 10x = 0$

Therefore　　　　　　$-\sin 10x + 10 \cos 10x = 0$

$\sin 10x = 10 \cos 10x$

$\tan 10x = 10$

$10x = 1.471$ or $1.471 + \pi$ or ...

$x = 0.147$ is the smallest value and is the x-coordinate of the first maximum point.

c $x = 0.461$ at the first minimum point.

8 a $k = 0.1$

b $10 \, \text{m s}^{-2}$

c $14.8 \, \text{m s}^{-1}$

9 $\dfrac{dy}{dx} = 3 \sin^2 x \cos x - 3 \cos^2 x \sin x$

At a stationary point, $\dfrac{dy}{dx} = 0$ so

$3 \sin^2 x \cos x - 3 \cos^2 x \sin x = 0$ and therefore

$$\sin^2 x \cos x = \cos^2 x \sin x$$

Either $\sin x = 0$ or $\cos x = 0$ or $\sin x = \cos x$

If $\sin x = 0$, then $x = 0, \pm\pi, \pm 2\pi, \pm 3\pi, \ldots$

If $\cos x = 0$, then $x = \pm\dfrac{\pi}{2}, \pm\dfrac{3\pi}{2}, \pm\dfrac{5\pi}{2}, \ldots$

If $\sin x = \cos x$, then $\tan x = 1$ and $x = \dfrac{\pi}{4}, \dfrac{5\pi}{4}, \dfrac{9\pi}{4}, \ldots$

or $-\dfrac{3\pi}{4}, -\dfrac{7\pi}{4}, -\dfrac{11\pi}{4}, \ldots$

Putting these together, $\dfrac{k\pi}{4}$ is a turning point if k is an integer with a remainder of 0, 1 or 2 when it is divided by 4, but not if the remainder is 3.

10 a $\dfrac{e^x}{(e^x + 1)^2}$

b $y(1 - y) = \dfrac{e^x}{e^x + 1}\left(1 - \dfrac{e^x}{e^x + 1}\right) = \dfrac{e^x}{e^x + 1}\left(\dfrac{e^x + 1 - e^x}{e^x + 1}\right)$

$= \dfrac{e^x}{e^x + 1} \times \dfrac{1}{e^x + 1} = \dfrac{e^x}{(e^x + 1)^2} = \dfrac{dy}{dx}$

11 a $(3, 3)$

b $y + x = 3$ and $y + x = -3$

12 $\dfrac{dx}{dt} = \dfrac{2(1 + t^2) - 2t \times 2t}{(1 + t^2)^2} = \dfrac{2 + 2t^2 - 4t^2}{(1 + t^2)^2} = \dfrac{2 - 2t^2}{(1 + t^2)^2}$

$\dfrac{dy}{dt} = \dfrac{-2t(1 + t^2) - 2t(1 + t^2)}{(1 + t^2)^2} = \dfrac{-2t - 2t^3 - 2t + 2t^3}{(1 + t^2)^2}$

$= \dfrac{-4t}{(1 + t^2)^2}$

Then $\dfrac{dy}{dx} = \dfrac{-4t}{(1 + t^2)^2} \div \dfrac{2 - 2t^2}{(1 + t^2)^2} = \dfrac{-4t}{2 - 2t^2}$

$= \dfrac{-2t}{1 - t^2} = \dfrac{-x}{y}$

13 $(10, 20)$ and $(-10, -20)$

8 Integration

Prior knowledge　　　　　　　　　　**page 144**

1 a $\cos(x^2 + 1) \times 2x = 2x \cos(x^2 + 1)$

b $-e^{-x} \cos 2x - 2e^{-x} \sin 2x$

c $\dfrac{3}{3x + 2}$

2 a $4 \times 2x^{\frac{1}{2}} + c = 8\sqrt{x} + c$

b $\frac{1}{4} e^{4x} + c$

c $\tan x + c$

3 20.83

4 $\dfrac{2}{x} - \dfrac{2}{x+2}$

Exercise 8.1A   **page 146**

1   **a** $\frac{3}{4}x^{\frac{4}{3}} + c$

    b $\frac{3}{16}(4x+3)^{\frac{4}{3}} + c$

    c $\frac{3}{4}(2x+5)^{\frac{2}{3}} + c$

2   **a** $-2e^{-2x} + c$

    b $2e^{0.5x} + 2e^{-0.5x} + c$

    c $x^2 + \frac{3}{4}e^{4x+5} + c$

3   **a** $-\frac{1}{2}\cos 2x - \frac{1}{2}\sin 2x + c$

    b $40\sin(0.1t + 1.3)$

    c $0.1\cos(2-\theta) + c$

4   **a**   4

    b

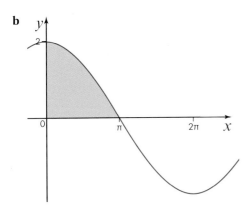

5   **a** $\frac{1}{18}(3x-2)^6 + c$

    b $-\dfrac{1}{5(5x+1)} + c$

    c $\dfrac{20}{3}(0.1x + 0.3)^{\frac{3}{2}} + c$

6 $e^{2.5}$ or 12.18

7   **a** $2\ln(x+1) + c$

    b $\frac{3}{2}\ln(2x+1) + c$

    c $\frac{1}{4}x^2 - 3\ln x + c$

8   1.264

9   **a** $2x\cos(x^2)$

    b $\frac{1}{2}\sin(x^2) + c$

    c $-\frac{1}{2}\cos(x^2) + c$

10 Both are correct.

    $\ln 2x = \ln 2 + \ln x$ so Bhaskar's answer is equivalent to $\frac{1}{2}\ln 2 + \frac{1}{2}\ln x + c$, which differs from Alice's only by the constant, $\frac{1}{2}\ln 2$.

Exercise 8.2A   **page 149**

1   **a** $\tan x + c$

    b $\frac{1}{2}\tan 2x + c$

    c $\frac{1}{2}\tan(2x-1) + c$

2   **a** $\sec x + c$

    b $\frac{1}{4}\sec 4x + c$

3   **a** $-3\operatorname{cosec} x + c$

    b $-\frac{3}{5}\operatorname{cosec} 5x + c$

4 $-\frac{1}{2}\cot(2x-0.4) + c$

5 $\sec x + c$

6 $\cos 2x = 2\cos^2 x - 1$ so $2\cos^2 x = 1 + \cos 2x$

    Therefore $\cos^2 x = \frac{1}{2} + \frac{1}{2}\cos 2x$

    Integrate.

$$\int \cos^2 x\, \mathrm{d}x = \int \left(\frac{1}{2} + \frac{1}{2}\cos 2x\right)\mathrm{d}x$$
$$= \frac{1}{2}x + \frac{1}{2}\times\frac{1}{2}\sin 2x + c$$
$$= \frac{1}{2}x + \frac{1}{4}\sin 2x + c$$

7   **a** $\dfrac{2}{\sqrt{1-4x^2}}$

    b $\frac{1}{2}\arcsin 2x + c$

8   **a** $-\tan x$

    b $-\ln|\cos x| + c$

    c $\ln|\sin x| + c$

9   **a** $\frac{1}{2}x + \frac{1}{8}\sin 4x + c$

    b $\frac{1}{2}x - \sin x + c$

10 $-\cot x - x + c$

Exercise 8.3A   **page 152**

1   **a** $(-2, 4)$ and $(1, 1)$

    b   4.5

2   2.607

3   **a** $\frac{1}{3}$

    b $\dfrac{n-1}{n+1}$

    c   When n is large the area is close to, but less than, 1.

4 $8 - 2\pi$

5 Where the curves cross, $4x - x^2 = x^2 - 4x + 6$

Therefore
$$2x^2 - 8x + 6 = 0$$
$$x^2 - 4x + 3 = 0$$

Factorise.
$$(x - 3)(x - 1) = 0$$

The curves cross where $x = 1$ or 3.

Area between $= \int_1^3 (4x - x^2) - (x^2 - 4x + 6)dx$

$$= \int_1^3 (8x - 2x^2 - 6)dx$$

$$= \left[4x^2 - \tfrac{2}{3}x^3 - 6x \right]_1^3$$

$$= \left[36 - 18 - 18 \right] - \left[4 - \tfrac{2}{3} - 6 \right]$$

$$= \left[0 \right] - \left[-\tfrac{8}{3} \right] = \tfrac{8}{3} \text{ or } 2\tfrac{2}{3}$$

6 a When $x = 1$, $2^{2-x} = 2$ and $2x = 2$ so they cross at $(1, 2)$.

b 1.885

7 a If $x = \dfrac{\pi}{3}$, $\sin x = \sin \dfrac{\pi}{3} = \dfrac{\sqrt{3}}{2}$ and

$\sin 2x = \sin \dfrac{2\pi}{3} = \dfrac{\sqrt{3}}{2}$

So the curves cross where $x = \dfrac{\pi}{3}$

b $\tfrac{1}{4}$

c $2\tfrac{1}{4}$

8 a $\pi y^2 \delta x$

b The volume of the cone $= \displaystyle\lim_{\delta x \to 0} \sum \pi y^2 \delta x = \int_0^h \pi y^2 dx$

The gradient of the diagonal line representing the side of the cone is $\dfrac{r}{h}$

So $y = \dfrac{r}{h}x$

So the volume is $\displaystyle\int_0^h \pi y^2 \, dx = \int_0^h \pi \left(\dfrac{r}{h}x\right)^2 dx$

$$= \int_0^h \dfrac{\pi r^2}{h^2}x^2 \, dx = \left[\dfrac{\pi r^2}{h^2} \times \tfrac{1}{3}x^3 \right]_0^h$$

$$= \left[\dfrac{\pi r^2}{h^2} \times \tfrac{1}{3}h^3 \right] - \left[0 \right] = \tfrac{1}{3}\pi r^2 h$$

Exercise 8.4A page 156

1 $\tfrac{1}{5}(x + 2)^5 - \tfrac{1}{2}(x + 2)^4 + c$

2 $u^{\frac{1}{2}} + c = \sqrt{x^2 + 4} + c$

3 $\tfrac{1}{30}(2x^3 + 5)^5 + c$

4 $-\tfrac{1}{2}\cos x^2 + c$

5 a $\ln|x + 2| + c$

b $x + 2 - 2\ln|x + 2| + c$

c $\tfrac{1}{2}(x + 2)^2 - 4(x + 2) + 4\ln|x + 2| + c$

6 a $\tfrac{1}{2} \times \tfrac{2}{3}u^{\frac{3}{2}} + c = \tfrac{1}{3}(x^2 + 3)^{\frac{3}{2}} + c$

b $\tfrac{2}{5}(x + 3)^{\frac{5}{2}} - 2(x + 3)^{\frac{3}{2}} + c$

c $\tfrac{2}{7}(x + 3)^{\frac{7}{2}} - \tfrac{12}{5}(x + 3)^{\frac{5}{2}} + 6(x + 3)^{\frac{3}{2}} + c$

7 a $\ln|\sec x| + c$

b $-\ln|\operatorname{cosec} x| + c$

8 $\arcsin\dfrac{x}{10} + c$

9 a $\tfrac{1}{2}\arctan\dfrac{x}{2} + c$

b $\dfrac{1}{a}\arctan\dfrac{x}{a} + c$

Exercise 8.5A page 159

1 a $x\sin x + \cos x + c$

b $(x + 1)\sin x + \cos x + c$

2 a $-\tfrac{1}{2}x\cos 2x + \tfrac{1}{4}\sin 2x + c$

b $\tfrac{1}{4}x\sin 4x + \tfrac{1}{16}\cos 4x + c$

3 a $xe^x - e^x + c$

b $-xe^{-x} - e^{-x} + c$

c $2xe^{0.5x} - 4e^{0.5x} + c$

4 $x^2e^x - 2xe^x + 2e^x + c$

5 $-(2x^2 + 4x + 3)e^{-x} + c$

6 $\tfrac{1}{2}e^x(\cos x + \sin x) + c$

7 a $x\ln x - x + c$

b $x\ln 2x - x + c$

c $2(x\ln x - x) + c$

d $(x + 1)\ln(x + 1) - (x + 1) + c$

8 a $-\tfrac{1}{5}e^{-x}(\sin 2x + 2\cos 2x) + c$

b 0.483

9 a $\tfrac{2}{5}(x + 1)^{\frac{5}{2}} - \tfrac{2}{3}(x + 1)^{\frac{3}{2}} + c$

b $\tfrac{2}{3}x(x + 1)^{\frac{3}{2}} - \tfrac{4}{15}(x + 1)^{\frac{5}{2}} + c$

c The answer to **part a** can be rewritten.

$$\tfrac{2}{5}(x + 1)^{\frac{5}{2}} - \tfrac{2}{3}(x + 1)^{\frac{3}{2}} = (x + 1)^{\frac{3}{2}}\left\{\tfrac{2}{5}(x + 1) - \tfrac{2}{3}\right\}$$
$$= (x + 1)^{\frac{3}{2}}\left(\tfrac{2}{5}x - \tfrac{4}{15}\right)$$

The answer to **part b** can be rewritten.

$$\tfrac{2}{3}x(x + 1)^{\frac{3}{2}} - \tfrac{4}{15}(x + 1)^{\frac{5}{2}} = (x + 1)^{\frac{3}{2}}\left\{\tfrac{2}{3}x - \tfrac{4}{15}(x + 1)\right\}$$
$$= (x + 1)^{\frac{3}{2}}\left\{\tfrac{2}{3}x - \tfrac{4}{15}x - \tfrac{4}{15}\right\}$$
$$= (x + 1)^{\frac{3}{2}}\left(\tfrac{2}{5} - \tfrac{4}{15}\right)$$

So the two are the same.

Exercise 8.6A page 162

1 **a** $\frac{1}{2}\ln|x|+c$

   b $\frac{1}{2}\ln|2x+5|+c$

2 **a** $-\dfrac{1}{x+2}+c$

   b $\ln(x+2)+\dfrac{2}{x+2}+c$

3 $x+3-6\ln|x+3|+c$

4 $\arctan x+\frac{1}{2}\ln\left|1+x^2\right|+c$

5 $\frac{1}{2}\ln|u|+c=\frac{1}{2}\ln\left|x^2-4x+1\right|+c$

6 **a** $x-\dfrac{x}{x+1}=\dfrac{x(x+1)-x}{x+1}=\dfrac{x^2+x-x}{x+1}=\dfrac{x^2}{x+1}$

   b $\frac{1}{2}x^2-(x+1)+\ln|x+1|+c$

7 **a** $2\arctan u+c=2\arctan 2x+c$

   b $\ln\left|\dfrac{2x-1}{2x+1}\right|+c$

8 **a** $-\ln|x+1|+2\ln|x+2|+c$

   or $\ln\left|\dfrac{(x+2)^2}{x+1}\right|+c$

   b $\ln\left|\dfrac{(x+1)^2}{x}\right|+c$

Exercise 8.7A page 167

1 $2x-0.1x^3+c$

2 $x=-\dfrac{10}{y}+c$

3 $y=\left(\dfrac{x+c}{2}\right)^2$

4 $y=ke^{x^3}$

5 **a** $|xy|=k$

   b $xy=24$

6 **a** Separate the variables.

$$3y\,dy=-2x\,dx$$
$$\int 3y\,dy=-\int 2x\,dx+c$$
$$\tfrac{3}{2}y^2=-x^2+c$$
$$x^2+\tfrac{3}{2}y^2=c$$

or $x^2+\frac{3}{2}y^2=a$, replacing c with a, and $n=\frac{3}{2}$.

   b $(7, 0)$ and $(-7, 0)$.

7 **a** $y=1000e^{1.5t}$

   b 36.3 million

   c The colony cannot increase indefinitely as it will run out of space or nutrients.

   The solution will no longer be valid for larger values of t.

8 $y=2e^{-\frac{1}{2}x^2}$

9 **a** $v=20-20e^{-0.5t}$

   b As t increases, $e^{-0.5t}\to 0$

   Therefore the speed approaches a limiting value of $20-0=20\,\text{m s}^{-1}$.

   c The solution is only valid while the stone is in the air.

   If the building is not very high then the stone could hit the ground before it gets close to the limiting value of $20\,\text{m s}^{-1}$.

10 **a** $\dfrac{dt}{dx}=-\dfrac{1}{\lambda x}$

$$=-\int\dfrac{1}{\lambda x}\,dx$$
$$t=-\dfrac{1}{\lambda}\ln x+c$$
$$\ln x=\lambda c-\lambda t$$
$$x=e^{\lambda c-\lambda t}$$

which you can write as $x=ae^{-\lambda t}$ where $a=e^{-\lambda c}$.

When $t=0$, $x=a$ so a is the initial mass of the substance.

   b When half the substance remains, $x=\frac{1}{2}a$

so $\frac{1}{2}a=ae^{-\lambda t}$

$$e^{-\lambda t}=\tfrac{1}{2}$$
$$-\lambda t=\ln\tfrac{1}{2}$$
$$-\lambda t=-\ln 2$$

so $t=\dfrac{\ln 2}{\lambda}$

Exam-style questions 8 page 169

1 $0.2\ln|5x+10|+c$

2 **a** $2\tan 0.5x+c$

   b $2\tan 0.5x-x+c$

3 5.410

4 $\frac{1}{5}(2x+1)^{\frac{5}{2}}-\frac{1}{3}(2x+1)^{\frac{3}{2}}+c$

5 **a** $-2x\cos 0.5x+4\sin 0.5x+c$

   b 4π

6 **a** $\frac{1}{3}u+c=\frac{1}{3}\arctan\dfrac{x}{3}+c$

   b $\frac{1}{2}\ln(x^2+9)+c$

   c $x-3\arctan\dfrac{x}{3}+c$

7 **a** $v=\dfrac{100}{t+5}$

   b In this formula the speed is never actually 0, however large t is.

   In practice the car will stop at some point and the solution will no longer be valid.

8 11.05

9 **a** The product of the gradients of the tangent and the normal at any point is −1 because they are perpendicular.

$$\frac{x}{2y} \times -\frac{2y}{x} = -1$$

b 6.25

10 $\ln\frac{(x-3)^2}{|x+2|} + c$

11 $-2x^2 e^{-0.5x} - 8x e^{-0.5x} - 16 e^{-0.5x} + c$

12 **a** The acceleration is $\frac{dv}{dt}$ and the difference between v and 20 is $20 - v$.

$\frac{dv}{dt} \propto (20 - v)$ and this can be written as the equation $\frac{dv}{dt} = k(20 - v)$ for some k.

b $v = 20 - e^{c-kt}$

c As t increases, $ae^{-kt} \to 0$ and the speed approaches a limit of 20 m s^{-1}.

9 Numerical methods

Prior knowledge

page 172

1

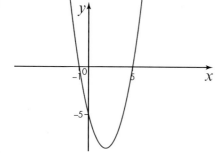

2

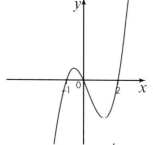

3

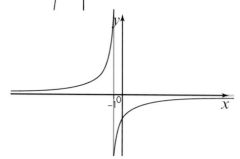

Asymptotes at $x = -1$ and $y = 0$

4

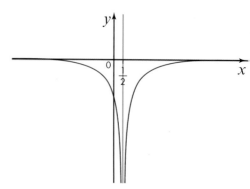

Asymptotes at $x = \frac{1}{2}$ and $y = 0$

Exercise 9.1A

page 174

1 **a** $f(x) = 3x^2 + 4x - 11$

$f(1.2) = 4.32 + 4.8 - 11 = -1.88$

$f(1.4) = 5.88 + 5.6 - 11 = 0.48$

There is a sign change between f(1.2) and f(1.4), therefore there is a root of f(x) in the interval $1.2 < x < 1.4$.

b $f(x) = x^3 + 6x^2 + 11x + 6$

$f(-1.4) = -2.744 + 11.76 - 15.4 + 6 = -0.384$

$f(-0.8) = -0.512 + 3.84 - 8.8 + 6 = 0.528$

There is a sign change between f(−1.4) and f(−0.8), therefore there is a root of f(x) in the interval $-1.4 < x < -0.8$.

c $f(x) = 8x^4 + 2x^3 - 53x^2 + 37x - 6$

$f(1) = 8 + 2 - 53 + 37 - 6 = -12$

$f(3) = 648 + 54 - 477 + 111 - 6 = 330$

There is a sign change between f(1) and f(3), therefore there is a root of f(x) in the interval $1 < x < 3$.

2 $f(x) = 1 - e^x - \ln x$

$f(0.2) = 1.3888$

$f(0.7) = -0.657$

There is a sign change between f(0.2) and f(0.7), therefore there is a root of f(x) in the interval $0.2 < x < 0.7$.

3 $f(x) = \dfrac{\sin x}{e^x}$

$f(-0.8) = -1.597$

$f(-0.7) = -1.297$

There is not a sign change between f(−0.8) and f(−0.7)

therefore there is not a root of f(x) in the interval $-0.8 < x < -0.7$.

4 a

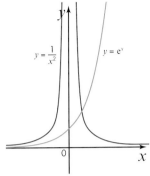

There is one point of intersection of the two graphs, therefore $\dfrac{1}{x^2} = e^x$ has one root.

b $f(x) = \dfrac{1}{x^2} - e^x$

$f(0.6) = 0.956$

$f(0.8) = -0.663$

There is a sign change between $f(0.6)$ and $f(0.8)$, therefore there is a root of $f(x)$ in the interval $0.6 < x < 0.8$.

5 $f(x) = \sin^2 x - \cos^2 x$

$f\left(\dfrac{\pi}{8}\right) = -0.707$

$f\left(\dfrac{3\pi}{8}\right) = 0.707$

$x = \dfrac{\pi}{4}$ is in the interval $\dfrac{\pi}{8} < x < \dfrac{3\pi}{8}$.

There is a sign change between $f\left(\dfrac{\pi}{8}\right)$ and $f\left(\dfrac{3\pi}{8}\right)$, therefore there is a root of $f(x)$ in the interval $\dfrac{\pi}{8} < x < \dfrac{3\pi}{8}$, which is near to $\dfrac{\pi}{4}$.

6 a

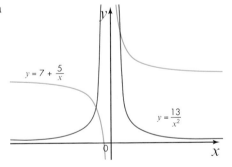

There are two points of intersection, therefore the equation $7 + \dfrac{5}{x} = \dfrac{13}{x^2}$ has two roots.

b $f(x) = 7 + \dfrac{5}{x} - \dfrac{13}{x^2}$

$f(0.95) = -2.141$

$f(1.1) = 0.802$

There is a sign change between $f(0.95)$ and $f(1.1)$, therefore there is a root of $f(x)$ in the interval $0.95 < x < 1.1$.

c Rearranging $7 + \dfrac{5}{x} = \dfrac{13}{x^2}$ gives $7x^2 + 5x - 13 = 0$.

Consequently, by solving the quadratic, the roots to the original equation are also found.

$a = 7,\, b = 5,\, c = -13$

$x = \dfrac{-b \pm \sqrt{b^2 - 4ac}}{2a}$

$x = \dfrac{-5 \pm \sqrt{5^2 - (4)(7)(-13)}}{14}$

$= 1.052 \text{ or} - 1.766$

Exercise 9.2A page 177

1 $f(x) = 6x^2 - x - 2$

Root in the interval $-1 < x < 0$:

$f(-1) = 5$

$f(0) = -2$

There is a sign change between $f(-1)$ and $f(0)$, therefore there is a root of $f(x)$ in the interval $-1 < x < 0$.

Root in the interval $0 < x < 1$:

$f(0) = -2$

$f(1) = 3$

There is a sign change between $f(0)$ and $f(1)$, therefore there is a root of $f(x)$ in the interval $0 < x < 1$.

2 $f(x) = \dfrac{1}{x - 3}$

$f(2.4) = -\dfrac{5}{3}$

$f(3.3) = \dfrac{10}{3}$

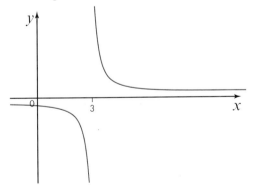

There is a discontinuity in the interval $2.4 < x < 3.3$, so although there is a sign change between $f(2.4)$ and $f(3.3)$ there isn't a root, as shown on the sketch graph.

3 Functions in the form $f(x) = (x \pm a)^2$ do not intersect the x-axis. Instead the curve 'sits on' the x-axis (the discriminant equals zero and so the function has one repeated root). Consequently in the case of $f(x) = (x \pm a)^2$, $f(a-1)$ and $f(a+1)$ are both positive.

4

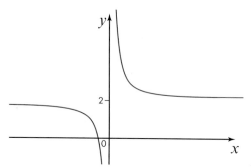

a $f(x) = \dfrac{1}{x} + 2$

Root in the interval $-1 < x < -0.2$:

$f(-1) = 1$

$f(-0.2) = -3$

There is a sign change between $f(-1)$ and $f(-0.2)$, therefore there is a root of $f(x)$ in the interval $-1 < x < -0.2$.

b $f(x) = \dfrac{1}{x} + 2$

Asymptote in the interval $-0.2 < x < 0.2$:

$f(-0.2) = -3$

$f(0.2) = 7$

There is a sign change between $f(-0.2)$ and $f(0.2)$, therefore there is an asymptote of $f(x)$ in the interval $-0.2 < x < 0.2$.

5 Asymptote in the interval $\dfrac{\pi}{4} < x < \dfrac{3\pi}{4}$:

$f\left(\dfrac{\pi}{4}\right) = 1$

$f\left(\dfrac{3\pi}{4}\right) = -1$

There is a sign change between $f\left(\dfrac{\pi}{4}\right)$ and $f\left(\dfrac{3\pi}{4}\right)$, therefore there is an asymptote of $f(x)$ in the interval $\dfrac{\pi}{4} < x < \dfrac{3\pi}{4}$.

Root in the interval $\dfrac{3\pi}{4} < x < \dfrac{5\pi}{4}$:

$f\left(\dfrac{3\pi}{4}\right) = -1$

$f\left(\dfrac{5\pi}{4}\right) = 1$

There is a sign change between $f\left(\dfrac{3\pi}{4}\right)$ and $f\left(\dfrac{5\pi}{4}\right)$, therefore there is a root of $f(x)$ in the interval $\dfrac{3\pi}{4} < x < \dfrac{5\pi}{4}$.

6 Discontinuity in the interval $-1 < x < -0.1$:

$f(-1) = 0.333$

$f(-0.1) = -0.490$

There is a sign change between $f(-1)$ and $f(-0.1)$, therefore there is a discontinuity of $f(x)$ in the interval $-1 < x < -0.1$.

Discontinuity in the interval $0.1 < x < 1$:

$f(0.1) = -0.543$

$f(1) = 0.2$

There is a sign change between $f(0.1)$ and $f(1)$, therefore there is a discontinuity of $f(x)$ in the interval $0.1 < x < 1$.

Exercise 9.3A   **page 184**

1   **a** $x^2 = 11 - 5x$

Square root both sides.

$x = \sqrt{11 - 5x}$

b $x^2 = 11 - 5x$

Divide both sides by x.

$x = \dfrac{11}{x} - 5$

c $x^2 = 11 - 5x$

$x^2 - 11 = -5x$

$x = \dfrac{x^2 - 11}{-5}$

2 $x_3 = 2.408$

3   **a** $f(x) = x^2 + 4x - 7$

$f(-6) = 5$

$f(-5) = -2$

There is a sign change between $f(-6)$ and $f(-5)$, therefore there is a root of $f(x)$ in the interval $-6 < x < -5$.

b $x^2 + 4x - 7 = 0$

$x^2 = 7 - 4x$

$x = \dfrac{7}{x} - 4$; so $a = -4$ and $b = 7$

c -5.32

d

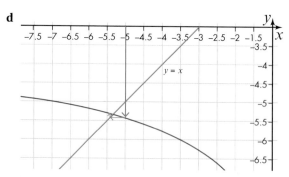

4   **a** $f(-2) = -5$

$f(-1) = 4$

There is a sign change between $f(-2)$ and $f(-1)$, therefore there is a root of $f(x)$ in the interval $-2 < x < -1$.

b $x_{n+1} = \dfrac{2x_n^3 + 1}{5}$

$x_0 = -1$

$x = 0.203$

c

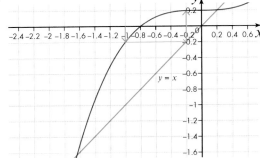

5 a $x = 1.21$

b

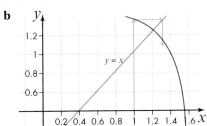

6 $x = 0.25$ and 1.86

Exercise 9.4A — page 186

1 a $f'(x) = 2x + 5$

b $x = 1.7$

2 $x = 1.21$

3 $x = 2.219$

4 $x = -1.67$

5 $x = -5.32$ and 1.32

6 a $x = -2.11$

b $x = -2.11$

c Both the iterative method and the Newton–Raphson method require a similar number of iterations to find the result to the desired number of decimal places. However, it is possible to make a number of false starts using the iterative method, depending on the iterative formula used. Consequently, it seems more efficient to choose the Newton–Raphson method in this case.

Exercise 9.5A — page 189

1 a When finding x_2 need to find the square root of a negative number, so the iterations stop.

b Converges to a different root, $x = -6.65$ to 2 d.p.

c Converges to this root, $x = 1.65$ to 2 d.p.

2 $x_4 = 5.477\,702$

This combination of iteration formula and starting value does not result in a convergent sequence of results, and eventually the iterations stop because one iteration results in finding the square root of a negative number.

3 $x_{n+1} = \dfrac{x_n^2 - 7}{4}$

$x_0 = -5$

This form converges to the root $x = 1.32$

$x_{n+1} = \dfrac{7}{x_n} - 4$

This form converges to a different root, $x = -5.32$

4 No value of x_0 in the interval $-100 < x < 100$ produces a divergent set of results.

5 Yes: $x_{n+1} = \dfrac{7 - x_n^5}{3x_n}$

6 No value of x_0 in the interval $1.9 < x < 1\,000\,000$ produces a divergent set of results.

Exercise 9.6A — page 192

1

x	1	$\dfrac{3}{2}$	2	$\dfrac{5}{2}$	3
y	1	$\dfrac{2}{3}$	$\dfrac{1}{2}$	$\dfrac{2}{5}$	$\dfrac{1}{3}$

Approximate area is 1.117

2 a 20.4

b

x	0	$\dfrac{1}{2}$	1	$\dfrac{3}{2}$	2
y	7	7.0625	8	12.0625	23

Approximate area is 21.0625

c In this case, the trapezium rule has a percentage error of 3.2%.

3 a 0.102

b This answer is an overestimate. The curve is concave between 1 and 3 and consequently the trapezia will include extra areas outside the curve.

4 a 1

b 0.987

c 0.997

d The more strips used, the more accurate the result.

5 0.169%

6 Percentage error is 1.3% with four strips

1   3.317

2   **a** 29.359

   b 29.25

   c 0.373%

3 $f(x) = x^2 - a$

   $f'(x) = 2x$

   $$x_{n+1} = x_n - \frac{x_n^2 - a}{2x_n}$$

   $$= \frac{2x_n^2 - x_n^2 + a}{2x_n}$$

   $$= \frac{x_n^2 + a}{2x_n}$$

   $$= \frac{1}{2}\left(x_n + \frac{a}{x_n}\right)$$

4   6.683 and 14.37 units.

5   67.677 g

6   0.53

1   **a** $f(2.1) = -0.037$

   $f(2.3) = 0.046$

   There is a sign change between $f(2.1)$ and $f(2.3)$ so $f(x)$ has a root in the interval $2.1 < x < 2.3$.

   b $x_3 = 2.17$

2   **a** $f'(x) = 2x + 6$

   b $x = 1.7$

3   **a** 6

   b 6

   c It is not necessary to use the trapezium rule with eight strips as the exact answer has already been found with four strips.

   d No comparison necessary, as the exact answer has been found with four strips.

4   2.648

5   **a** $f(x) = x^5 + 2x - 7$

   $f(1) = -4$

   $f(2) = 29$

   There is a sign change between $f(1)$ and $f(2)$ so a root of the equation $x^5 + 2x - 7 = 0$ lies in the interval $1 < x < 2$.

   b $x^5 + 2x - 7 = 0$

   $x^5 = 7 - 2x$

   $x = (7 - 2x)^{\frac{1}{5}}$

   $x_{n+1} = (7 - 2x_n)^{\frac{1}{5}}$

   $a = 7, b = 2$

c $x = 1.34$

d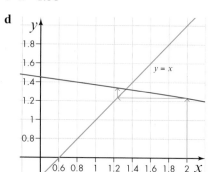

6 $(0.74, 0.74)$

7   3.529%

8 $x = -1.83$ and 3.83

9   **a** $x = 1.890$

   b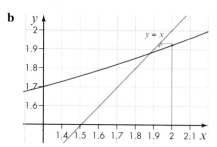

10 $r = 1.79$

11   a minimum of 6 strips is required for <3% error.

12   16% per year

10 Three-dimensional vectors

1   **a** 145

   b $\frac{1}{145}(143\mathbf{i} - 24\mathbf{j})$

2   **a** $\begin{bmatrix} 2 \\ 5 \end{bmatrix}$

   b 26

3   **a** Write as position vectors.

   $$\mathbf{a} = \begin{bmatrix} 31 \\ 81 \end{bmatrix}, \mathbf{b} = \begin{bmatrix} 86 \\ 33 \end{bmatrix}, \mathbf{c} = \begin{bmatrix} 38 \\ -22 \end{bmatrix} \text{ and } \mathbf{d} = \begin{bmatrix} -17 \\ 26 \end{bmatrix}$$

   $$\overrightarrow{AB} = \mathbf{b} - \mathbf{a} = \begin{bmatrix} 86 \\ 33 \end{bmatrix} - \begin{bmatrix} 31 \\ 81 \end{bmatrix} = \begin{bmatrix} 55 \\ -48 \end{bmatrix}$$

   $$\overrightarrow{BC} = \mathbf{c} - \mathbf{b} = \begin{bmatrix} 38 \\ -22 \end{bmatrix} - \begin{bmatrix} 86 \\ 33 \end{bmatrix} = \begin{bmatrix} -48 \\ -55 \end{bmatrix}$$

   $$\overrightarrow{CD} = \mathbf{d} - \mathbf{c} = \begin{bmatrix} -17 \\ 26 \end{bmatrix} - \begin{bmatrix} 38 \\ -22 \end{bmatrix} = \begin{bmatrix} -55 \\ 48 \end{bmatrix}$$

$$\overrightarrow{DA} = \mathbf{a} - \mathbf{d} = \begin{bmatrix} 31 \\ 81 \end{bmatrix} - \begin{bmatrix} -17 \\ 26 \end{bmatrix} = \begin{bmatrix} 48 \\ 55 \end{bmatrix}$$

Length of $\overrightarrow{AB} = \sqrt{55^2 + (-48)^2} = 73$

Length of $\overrightarrow{BC} = \sqrt{(-48)^2 + (-55)^2} = 73$

Length of $\overrightarrow{CD} = \sqrt{(-55)^2 + 48^2} = 73$

Length of $\overrightarrow{DA} = \sqrt{48^2 + 55^2} = 73$

All the sides are the same length.

The gradient of $\overrightarrow{AB} = \dfrac{33 - 81}{86 - 31} = -\dfrac{48}{55}$

The gradient of $\overrightarrow{BC} = \dfrac{-22 - 33}{38 - 86} = \dfrac{55}{48}$

$\overrightarrow{AB}$ and $\overrightarrow{BC}$ are perpendicular because
$-\dfrac{48}{55} \times \dfrac{55}{48} = -1$

Since the sides are all the same length and there is a right angle, the quadrilateral is a square.

b Area $= 73^2 = 5329$

Exercise 10.1A   **page 202**

1 a i 17

   ii 69

   iii $5\sqrt{2}$

   iv $7\sqrt{3}$

b i $\dfrac{1}{29}(12\mathbf{i} + 21\mathbf{j} - 16\mathbf{k})$

  ii $\dfrac{1}{97}(-72\mathbf{i} + 33\mathbf{j} + 56\mathbf{k})$

  iii $\dfrac{1}{\sqrt{354}}(11\mathbf{i} + 13\mathbf{j} + 8\mathbf{k})$

  iv $\dfrac{1}{3\sqrt{5}}(5\mathbf{i} - 4\mathbf{j} - 2\mathbf{k})$

2 a Since the second toilet block is not further than 9 units from Room 168, the manager's rule is satisfied for Room 168.

b It has been assumed that the units are equally spaced for all three axes.

3 a i $2\mathbf{i} + \dfrac{43}{2}\mathbf{j} + 5\mathbf{k}$

  ii 39

b i $13\mathbf{i} - 2\mathbf{k}$

  ii 14

c i $\begin{bmatrix} 6 \\ \frac{123}{2} \\ 7 \end{bmatrix}$

  ii 93

4 a not parallel

 b not parallel

 c not parallel

 d $\begin{bmatrix} 12 \\ -8 \\ 32 \end{bmatrix} = 4\begin{bmatrix} 3 \\ -2 \\ 8 \end{bmatrix}$, $\begin{bmatrix} 21 \\ -14 \\ 56 \end{bmatrix} = 7\begin{bmatrix} 3 \\ -2 \\ 8 \end{bmatrix}$: parallel, as they have a common factor.

5 $20\mathbf{i} - 40\mathbf{j} + 5\mathbf{k}$

6 a B lies outside the surface of S_1.

 b Centre of S_2, X $= \dfrac{1}{2}\left(\begin{bmatrix} -4 \\ 6 \\ -1 \end{bmatrix} + \begin{bmatrix} -2 \\ 2 \\ 17 \end{bmatrix} \right) = \begin{bmatrix} -3 \\ 4 \\ 8 \end{bmatrix}$

For radius of S_2,
$$d^2 = (-3 - (-4))^2 + (4 - 6)^2 + (8 - (-1))^2$$
$$= 1^2 + (-2)^2 + 9^2 = 86$$
Radius of $S_2 = \sqrt{86}$

For $\overrightarrow{LX}$, $\begin{bmatrix} -3 \\ 4 \\ 8 \end{bmatrix} - \begin{bmatrix} 2 \\ 10 \\ 3 \end{bmatrix} = \begin{bmatrix} -5 \\ -6 \\ 5 \end{bmatrix}$

Length of $\overrightarrow{LX} = \sqrt{(-5)^2 + (-6)^2 + 5^2} = \sqrt{86}$

For $\overrightarrow{MX}$, $\begin{bmatrix} -3 \\ 4 \\ 8 \end{bmatrix} - \begin{bmatrix} 4 \\ -2 \\ 7 \end{bmatrix} = \begin{bmatrix} -7 \\ 6 \\ 1 \end{bmatrix}$

Length of $\overrightarrow{MX} = \sqrt{(-7)^2 + 6^2 + 1^2} = \sqrt{86}$

Hence both L and M lie on the surface of S_2.

7 a $\sqrt{3^2 + 4^2 + 12^2} = 13$. No other sets of three non-zero integers satisfy this rule.

Total $= 6 \times 8 = 48$ sets of values

 b If zero is permitted then you also have 0, 5, 12. There are six permutations of the numbers but only four choices for positive and negative because 0 is neither. Hence $6 \times 4 = 24$ more sets, making 72 sets in total.

8 $(1, -3, -1)$ or $(-11, 5, -5)$.

9 a $\overrightarrow{FG} = \mathbf{g} - \mathbf{f} = \begin{bmatrix} 9 \\ 32 \\ 5 \end{bmatrix} - \begin{bmatrix} -7 \\ 10 \\ 1 \end{bmatrix} = \begin{bmatrix} 16 \\ 22 \\ 4 \end{bmatrix}$

Magnitude of $\overrightarrow{FG} = \sqrt{16^2 + 22^2 + 4^2} = \sqrt{756}$

$\overrightarrow{GH} = \mathbf{h} - \mathbf{g} = \begin{bmatrix} 3 \\ 20 \\ 29 \end{bmatrix} - \begin{bmatrix} 9 \\ 32 \\ 5 \end{bmatrix} = \begin{bmatrix} -6 \\ -12 \\ 24 \end{bmatrix}$

Magnitude of $\overrightarrow{GH} = \sqrt{(-6)^2 + (-12)^2 + 24^2} = \sqrt{756}$

Since FG and GH have the same magnitude, FG = GH.

b (–2, 15, 15).

c Since FG = GH, the triangle is isosceles and the area is given by $\frac{1}{2} \times$ FH $\times$ GM.

$$\overrightarrow{MG} = \mathbf{g} - \mathbf{m} = \begin{bmatrix} 9 \\ 32 \\ 5 \end{bmatrix} - \begin{bmatrix} -2 \\ 15 \\ 15 \end{bmatrix} = \begin{bmatrix} 11 \\ 17 \\ -10 \end{bmatrix}$$

Magnitude of $\overrightarrow{MG} = \sqrt{11^2 + 17^2 + (-10)^2} = \sqrt{510}$

$$\overrightarrow{FH} = \mathbf{h} - \mathbf{f} = \begin{bmatrix} 3 \\ 20 \\ 29 \end{bmatrix} - \begin{bmatrix} -7 \\ 10 \\ 1 \end{bmatrix} = \begin{bmatrix} 10 \\ 10 \\ 28 \end{bmatrix}$$

Magnitude of $\overrightarrow{FH} = \sqrt{10^2 + 10^2 + 28^2} = \sqrt{984}$

Area $= \frac{1}{2} \times \sqrt{984} \times \sqrt{510} = \sqrt{125\,460}$

Exercise 10.2A
page 206

1 $\overrightarrow{DE} = \mathbf{e} - \mathbf{d} = \begin{bmatrix} -11 \\ 3 \\ -7 \end{bmatrix} - \begin{bmatrix} 2 \\ -8 \\ 5 \end{bmatrix} = \begin{bmatrix} -13 \\ 11 \\ -12 \end{bmatrix}$

Magnitude of $\overrightarrow{DE} = \sqrt{(-13)^2 + 11^2 + (-12)^2} = \sqrt{434}$

$$\overrightarrow{DF} = \mathbf{f} - \mathbf{d} = \begin{bmatrix} -6 \\ -1 \\ 2 \end{bmatrix} - \begin{bmatrix} 2 \\ -8 \\ 5 \end{bmatrix} = \begin{bmatrix} -8 \\ 7 \\ -3 \end{bmatrix}$$

Magnitude of $\overrightarrow{DF} = \sqrt{(-8)^2 + 7^2 + (-3)^2} = \sqrt{122}$

$$\overrightarrow{EF} = \mathbf{f} - \mathbf{e} = \begin{bmatrix} -6 \\ -1 \\ 2 \end{bmatrix} - \begin{bmatrix} -11 \\ 3 \\ -7 \end{bmatrix} = \begin{bmatrix} 5 \\ -4 \\ 9 \end{bmatrix}$$

Magnitude of $\overrightarrow{EF} = \sqrt{5^2 + (-4)^2 + 9^2} = \sqrt{122}$

Since EF = DF ≠ DE, the triangle is isosceles.

2 a $\overrightarrow{TU} = \mathbf{u} - \mathbf{t} = (13\mathbf{i} + 12\mathbf{j} - 14\mathbf{k}) - (10\mathbf{i} + 8\mathbf{j} - 3\mathbf{k})$
$= 3\mathbf{i} + 4\mathbf{j} - 11\mathbf{k}$

Magnitude of $\overrightarrow{TU} = \sqrt{3^2 + 4^2 + (-11)^2} = \sqrt{146}$

$\overrightarrow{TV} = \mathbf{v} - \mathbf{t} = (3\mathbf{i} + 27\mathbf{j} + 2\mathbf{k}) - (10\mathbf{i} + 8\mathbf{j} - 3\mathbf{k})$
$= -7\mathbf{i} + 19\mathbf{j} + 5\mathbf{k}$

Magnitude of $\overrightarrow{TV} = \sqrt{(-7)^2 + 19^2 + 5^2} = \sqrt{435}$

$\overrightarrow{UV} = \mathbf{v} - \mathbf{u} = (3\mathbf{i} + 27\mathbf{j} + 2\mathbf{k}) - (13\mathbf{i} + 12\mathbf{j} - 14\mathbf{k})$
$= -10\mathbf{i} + 15\mathbf{j} + 16\mathbf{k}$

Magnitude of $\overrightarrow{UV} = \sqrt{(-10)^2 + 15^2 + 16^2} = \sqrt{581}$

Since TU ≠ TV ≠ UV, the triangle is scalene.

b 146 + 435 = 581 so TU² + TV² = UV²; the triangle is right-angled.

3 a $\overrightarrow{EF} = \mathbf{f} - \mathbf{e} = \begin{bmatrix} 26 \\ 20 \\ 2 \end{bmatrix} - \begin{bmatrix} 12 \\ 27 \\ 16 \end{bmatrix} = \begin{bmatrix} 14 \\ -7 \\ -14 \end{bmatrix}$

Magnitude of $\overrightarrow{EF} = \sqrt{14^2 + (-7)^2 + (-14)^2} = 21$

$$\overrightarrow{FG} = \mathbf{g} - \mathbf{f} = \begin{bmatrix} 12 \\ 6 \\ -5 \end{bmatrix} - \begin{bmatrix} 26 \\ 20 \\ 2 \end{bmatrix} = \begin{bmatrix} -14 \\ -14 \\ -7 \end{bmatrix}$$

Magnitude of $\overrightarrow{FG} = \sqrt{(-14)^2 + (-14)^2 + (-7)^2} = 21$

$$\overrightarrow{HG} = \mathbf{g} - \mathbf{h} = \begin{bmatrix} 12 \\ 6 \\ -5 \end{bmatrix} - \begin{bmatrix} -2 \\ 13 \\ 9 \end{bmatrix} = \begin{bmatrix} 14 \\ -7 \\ -14 \end{bmatrix}$$

Magnitude of $\overrightarrow{HG} = \sqrt{14^2 + (-7)^2 + (-14)^2} = 21$

$$\overrightarrow{EH} = \mathbf{h} - \mathbf{e} = \begin{bmatrix} -2 \\ 13 \\ 9 \end{bmatrix} - \begin{bmatrix} 12 \\ 27 \\ 16 \end{bmatrix} = \begin{bmatrix} -14 \\ -14 \\ -7 \end{bmatrix}$$

Magnitude of $\overrightarrow{EH} = \sqrt{(-14)^2 + (-14)^2 + (-7)^2} = 21$

$$\overrightarrow{EG} = \mathbf{g} - \mathbf{e} = \begin{bmatrix} 12 \\ 6 \\ -5 \end{bmatrix} - \begin{bmatrix} 12 \\ 27 \\ 16 \end{bmatrix} = \begin{bmatrix} 0 \\ -21 \\ -21 \end{bmatrix}$$

Magnitude of $\overrightarrow{EG} = \sqrt{(-21)^2 + (-21)^2} = 21\sqrt{2}$

$$\overrightarrow{FH} = \mathbf{h} - \mathbf{f} = \begin{bmatrix} -2 \\ 13 \\ 9 \end{bmatrix} - \begin{bmatrix} 26 \\ 20 \\ 2 \end{bmatrix} = \begin{bmatrix} -28 \\ -7 \\ 7 \end{bmatrix}$$

Magnitude of $\overrightarrow{FH} = \sqrt{(-28)^2 + (-7)^2 + (-7)^2} = 21\sqrt{2}$

Since all the sides are the same length and the diagonals are the same length, the quadrilateral is a square.

b $\overrightarrow{JK} = \mathbf{k} - \mathbf{j} = \begin{bmatrix} 5 \\ 11 \\ -2 \end{bmatrix} - \begin{bmatrix} 6 \\ -1 \\ 3 \end{bmatrix} = \begin{bmatrix} -1 \\ 12 \\ -5 \end{bmatrix}$

$$\overrightarrow{ML} = \mathbf{l} - \mathbf{m} = \begin{bmatrix} 3 \\ 9 \\ -7 \end{bmatrix} - \begin{bmatrix} 4 \\ -3 \\ -2 \end{bmatrix} = \begin{bmatrix} -1 \\ 12 \\ -5 \end{bmatrix}$$

$\overrightarrow{JK}$ and $\overrightarrow{ML}$ are equal vectors.

Magnitude of $\overrightarrow{JK} = \sqrt{(-1)^2 + 12^2 + (-5)^2} = \sqrt{170}$

$$\overrightarrow{KL} = \mathbf{l} - \mathbf{k} = \begin{bmatrix} 3 \\ 9 \\ -7 \end{bmatrix} - \begin{bmatrix} 5 \\ 11 \\ -2 \end{bmatrix} = \begin{bmatrix} -2 \\ -2 \\ -5 \end{bmatrix}$$

$$\overrightarrow{JM} = \mathbf{m} - \mathbf{j} = \begin{bmatrix} 4 \\ -3 \\ -2 \end{bmatrix} - \begin{bmatrix} 6 \\ -1 \\ 3 \end{bmatrix} = \begin{bmatrix} -2 \\ -2 \\ -5 \end{bmatrix}$$

$\overrightarrow{KL}$ and $\overrightarrow{JM}$ are equal vectors.

Magnitude of $\overrightarrow{JM} = \sqrt{(-2)^2 + (-2)^2 + (-5)^2} = \sqrt{33}$

$$\overrightarrow{JL} = \mathbf{l} - \mathbf{j} = \begin{bmatrix} 3 \\ 9 \\ -7 \end{bmatrix} - \begin{bmatrix} 6 \\ -1 \\ 3 \end{bmatrix} = \begin{bmatrix} -3 \\ 10 \\ -10 \end{bmatrix}$$

Magnitude of $\overrightarrow{JL} = \sqrt{(-3)^2 + 10^2 + (-10)^2} = \sqrt{209}$

$$\overrightarrow{KM} = \mathbf{m} - \mathbf{k} = \begin{bmatrix} 4 \\ -3 \\ -2 \end{bmatrix} - \begin{bmatrix} 5 \\ 11 \\ -2 \end{bmatrix} = \begin{bmatrix} -1 \\ -14 \\ 0 \end{bmatrix}$$

Magnitude of $\overrightarrow{KM} = \sqrt{(-1)^2 + (-14)^2} = \sqrt{197}$

Since the sides are two pairs of equal vectors and the diagonals are different lengths, the quadrilateral is a parallelogram.

4 (3, 3, 3)

5 a $\overrightarrow{AB} = \mathbf{b} - \mathbf{a} = (3\mathbf{i} - \mathbf{j} - \mathbf{k}) - (6\mathbf{i} - 11\mathbf{j} + 4\mathbf{k})$
$= -3\mathbf{i} + 10\mathbf{j} - 5\mathbf{k}$

Magnitude of $\overrightarrow{AB} = \sqrt{(-3)^2 + 10^2 + (-5)^2} = \sqrt{134}$

$\overrightarrow{BC} = \mathbf{c} - \mathbf{b} = (10\mathbf{i} + \mathbf{j} + 8\mathbf{k}) - (3\mathbf{i} - \mathbf{j} - \mathbf{k}) = 7\mathbf{i} + 2\mathbf{j} + 9\mathbf{k}$

Magnitude of $\overrightarrow{BC} = \sqrt{7^2 + 2^2 + 9^2} = \sqrt{134}$

$\overrightarrow{CD} = \mathbf{d} - \mathbf{c} = (23\mathbf{i} - 17\mathbf{j} + 27\mathbf{k}) - (10\mathbf{i} + \mathbf{j} + 8\mathbf{k})$
$= 13\mathbf{i} - 18\mathbf{j} + 19\mathbf{k}$

Magnitude of $\overrightarrow{CD} = \sqrt{13^2 + (-18)^2 + 19^2} = \sqrt{854}$

$\overrightarrow{AD} = \mathbf{d} - \mathbf{a} = (23\mathbf{i} - 17\mathbf{j} + 27\mathbf{k}) - (6\mathbf{i} - 11\mathbf{j} + 4\mathbf{k})$
$= 17\mathbf{i} - 6\mathbf{j} + 23\mathbf{k}$

Magnitude of $\overrightarrow{DA} = \sqrt{17^2 + (-6)^2 + 23^2} = \sqrt{854}$

Since adjacent sides are the same length, the quadrilateral is a kite.

b $8\mathbf{i} - 5\mathbf{j} + 6\mathbf{k}$

6 the quadrilateral is a square.

7 $\frac{1}{2}\sqrt{65}$

8 a $p = 25$

$q = -32$

b $\begin{bmatrix} -\frac{1}{2} \\ \frac{61}{2} \\ -\frac{9}{2} \end{bmatrix}$

c 35 937

Exam-style questions 10   page 207

1 a 2037 m

b This is the distance in a straight line. It is unlikely that the climb will be in a straight line.

2 a $\overrightarrow{PQ} = \mathbf{q} - \mathbf{p} = \begin{bmatrix} 43 \\ 36 \\ -26 \end{bmatrix} - \begin{bmatrix} 59 \\ -19 \\ 13 \end{bmatrix} = \begin{bmatrix} -16 \\ 55 \\ -39 \end{bmatrix}$

Magnitude of $\overrightarrow{PQ} = \sqrt{(-16)^2 + 55^2 + (-39)^2}$
$= \sqrt{4802}$

$\overrightarrow{PR} = \mathbf{r} - \mathbf{p} = \begin{bmatrix} 19 \\ -4 \\ -11 \end{bmatrix} - \begin{bmatrix} 59 \\ -19 \\ 13 \end{bmatrix} = \begin{bmatrix} -40 \\ 15 \\ -24 \end{bmatrix}$

Magnitude of $\overrightarrow{PR} = \sqrt{(-40)^2 + 15^2 + (-24)^2}$
$= \sqrt{2401} = 49$

$\overrightarrow{QR} = \mathbf{r} - \mathbf{q} = \begin{bmatrix} 19 \\ -4 \\ -11 \end{bmatrix} - \begin{bmatrix} 43 \\ 36 \\ -26 \end{bmatrix} = \begin{bmatrix} -24 \\ -40 \\ 15 \end{bmatrix}$

Magnitude of $\overrightarrow{QR} = \sqrt{(-24)^2 + (-40)^2 + 15^2}$
$= \sqrt{2401} = 49$

Since PR = QR ≠ PQ, the triangle is isosceles.

b $PR^2 + QR^2 = 2401 + 2401 = 4802 = PQ^2$

Since Pythagoras' theorem is satisfied, the triangle is right-angled.

3 $\overrightarrow{AX} = \mathbf{x} - \mathbf{a} = \begin{bmatrix} 2 \\ -10 \\ 8 \end{bmatrix} - \begin{bmatrix} -12 \\ 7 \\ 5 \end{bmatrix} = \begin{bmatrix} 14 \\ -17 \\ 3 \end{bmatrix}$

Radius $= \sqrt{14^2 + (-17)^2 + 3^2} = \sqrt{494}$

Midpoint of $\overrightarrow{BC} = \frac{1}{2}\left(\begin{bmatrix} 23 \\ -6 \\ -7 \end{bmatrix} + \begin{bmatrix} -5 \\ 8 \\ -13 \end{bmatrix} \right) = \begin{bmatrix} 9 \\ 1 \\ -10 \end{bmatrix}$

Let the midpoint be M.

$\overrightarrow{MX} = \mathbf{x} - \mathbf{m} = \begin{bmatrix} 2 \\ -10 \\ 8 \end{bmatrix} - \begin{bmatrix} 9 \\ 1 \\ -10 \end{bmatrix} = \begin{bmatrix} -7 \\ -11 \\ 18 \end{bmatrix}$

Length of $\overrightarrow{MX} = \sqrt{(-7)^2 + (-11)^2 + 18^2} = \sqrt{494}$

Hence M lies on the surface of S.

4 $(20\mathbf{i} - 30\mathbf{j} + 12\mathbf{k})\,\text{m s}^{-1}$

5 a Jayanie has only shown that opposite sides are the same length, so ABCD could just be a parallelogram. She also needs to show that the diagonals are the same length.

b For AC,
$d^2 = (-12 - 69)^2 + (35 - 13)^2 + (-97 - 103)^2$
$= (-81)^2 + (22)^2 + (-200)^2 = 47\,045$

For BD,
$d^2 = (36 - 21)^2 + (101 - (-53))^2 + (79 - (-73))^2$
$= (15)^2 + (154)^2 + (152)^2 = 47\,045$

Since $AC^2 = BD^2$, the diagonals are the same length and ABCD is a rectangle.

6 $\overrightarrow{LM} = \mathbf{m} - \mathbf{l} = \begin{bmatrix} 113 \\ 69 \\ 173 \end{bmatrix} - \begin{bmatrix} 50 \\ -75 \\ 61 \end{bmatrix} = \begin{bmatrix} 63 \\ 144 \\ 112 \end{bmatrix}$

Magnitude of $\overrightarrow{LM} = \sqrt{63^2 + 144^2 + 112^2} = 193$

$$\overrightarrow{MN} = \mathbf{n} - \mathbf{m} = \begin{bmatrix} -31 \\ 181 \\ 110 \end{bmatrix} - \begin{bmatrix} 113 \\ 69 \\ 173 \end{bmatrix} = \begin{bmatrix} -144 \\ 112 \\ -63 \end{bmatrix}$$

Magnitude of $\overrightarrow{MN} = \sqrt{(-144)^2 + 112^2 + (-63)^2} = 193$

$$\overrightarrow{ON} = \mathbf{n} - \mathbf{o} = \begin{bmatrix} -31 \\ 181 \\ 110 \end{bmatrix} - \begin{bmatrix} -94 \\ 37 \\ -2 \end{bmatrix} = \begin{bmatrix} 63 \\ 144 \\ 112 \end{bmatrix}$$

Magnitude of $\overrightarrow{ON} = \sqrt{63^2 + 144^2 + 112^2} = 193$

$$\overrightarrow{LO} = \mathbf{o} - \mathbf{l} = \begin{bmatrix} -94 \\ 37 \\ -2 \end{bmatrix} - \begin{bmatrix} 50 \\ -75 \\ 61 \end{bmatrix} = \begin{bmatrix} -144 \\ 112 \\ -63 \end{bmatrix}$$

Magnitude of $\overrightarrow{LO} = \sqrt{(-144)^2 + 112^2 + (-63)^2} = 193$

The sides of the quadrilateral are all the same length.

$$\overrightarrow{LN} = \mathbf{n} - \mathbf{l} = \begin{bmatrix} -31 \\ 181 \\ 110 \end{bmatrix} - \begin{bmatrix} 50 \\ -75 \\ 61 \end{bmatrix} = \begin{bmatrix} -81 \\ 256 \\ 49 \end{bmatrix}$$

Magnitude of $\overrightarrow{LN} = \sqrt{(-81)^2 + 256^2 + 49^2} = \sqrt{74498}$

$$\overrightarrow{MO} = \mathbf{o} - \mathbf{m} = \begin{bmatrix} -94 \\ 37 \\ -2 \end{bmatrix} - \begin{bmatrix} 113 \\ 69 \\ 173 \end{bmatrix} = \begin{bmatrix} -207 \\ -32 \\ -175 \end{bmatrix}$$

Magnitude of $\overrightarrow{MN} = \sqrt{(-207)^2 + (-32)^2 + (-175)^2}$
$$= \sqrt{74498}$$

The diagonals of the quadrilateral are the same length.

Since the sides are all the same length and the diagonals are the same length, the quadrilateral is a square.

The area of the square is $193^2 = 37\,249$ square units.

7 Opposite sides of the quadrilateral are the same length and the diagonals are the same length. Therefore the quadrilateral is a rectangle.

8 a $\overrightarrow{PQ} = \mathbf{q} - \mathbf{p} = \begin{bmatrix} 12 \\ 21 \\ -34 \end{bmatrix} - \begin{bmatrix} -3 \\ 11 \\ -9 \end{bmatrix} = \begin{bmatrix} 15 \\ 10 \\ -25 \end{bmatrix} = 5\begin{bmatrix} 3 \\ 2 \\ -5 \end{bmatrix}$

$\overrightarrow{SR} = \mathbf{r} - \mathbf{s} = \begin{bmatrix} 13 \\ 0 \\ -14 \end{bmatrix} - \begin{bmatrix} 4 \\ -6 \\ 1 \end{bmatrix} = \begin{bmatrix} 9 \\ 6 \\ -15 \end{bmatrix} = 3\begin{bmatrix} 3 \\ 2 \\ -5 \end{bmatrix}$

$\overrightarrow{PQ}$ and $\overrightarrow{SR}$ have a common factor so are parallel.

b $5 : 3$

c Since $\overrightarrow{PQ}$ and $\overrightarrow{SR}$ are parallel but different lengths, PQRS is a trapezium.

9 a $4\overrightarrow{FG} = 3\overrightarrow{GH}$, so $\overrightarrow{FG}$ and $\overrightarrow{GH}$ share a common factor, so they are parallel. However, since they also share a common point G, they are collinear.

b $\begin{bmatrix} 23 \\ 34 \\ -20 \end{bmatrix}$

10 a $\begin{bmatrix} -1 \\ 5 \\ 7 \end{bmatrix}$

b ST and TU will be perpendicular if $ST^2 + TU^2 = SU^2$.

$\overrightarrow{TU} = \mathbf{u} - \mathbf{t} = \begin{bmatrix} 3 \\ 18 \\ 15 \end{bmatrix} - \begin{bmatrix} -3 \\ 14 \\ 17 \end{bmatrix} = \begin{bmatrix} 6 \\ 4 \\ -2 \end{bmatrix}$

$\overrightarrow{SU} = \mathbf{u} - \mathbf{s} = \begin{bmatrix} 3 \\ 18 \\ 15 \end{bmatrix} - \begin{bmatrix} -2 \\ 9 \\ 10 \end{bmatrix} = \begin{bmatrix} 5 \\ 9 \\ 5 \end{bmatrix}$

Magnitude of $\overrightarrow{ST} = \sqrt{(-1)^2 + 5^2 + 7^2} = \sqrt{75}$

Magnitude of $\overrightarrow{TU} = \sqrt{6^2 + 4^2 + (-2)^2} = \sqrt{56}$

Magnitude of $\overrightarrow{SU} = \sqrt{5^2 + 9^2 + 5^2} = \sqrt{131}$

Since $ST^2 + TU^2 = SU^2$, Pythagoras' theorem is satisfied and ST and TU are perpendicular.

c $\begin{bmatrix} 4 \\ 13 \\ 8 \end{bmatrix}$

d $\left(\dfrac{1}{2}, \dfrac{27}{2}, \dfrac{25}{2} \right)$

e The area of the triangle XST is $\dfrac{1}{4}$ of the area of the rectangle.

Area of STUV $= \sqrt{75} \times \sqrt{56} = 5\sqrt{3} \times 2\sqrt{14} = 10\sqrt{42}$

$\dfrac{1}{4} \times 10\sqrt{42} = \dfrac{5}{2}\sqrt{42}$ so $k = \dfrac{5}{2}$

11 a $\overrightarrow{EF} = \mathbf{f} - \mathbf{e} = \begin{bmatrix} 13 \\ 6 \\ -9 \end{bmatrix} - \begin{bmatrix} 9 \\ -2 \\ -1 \end{bmatrix} = \begin{bmatrix} 4 \\ 8 \\ -8 \end{bmatrix}$

$\overrightarrow{HG} = \mathbf{g} - \mathbf{h} = \begin{bmatrix} 5 \\ 20 \\ -8 \end{bmatrix} - \begin{bmatrix} -3 \\ 4 \\ 8 \end{bmatrix} = \begin{bmatrix} 8 \\ 16 \\ -16 \end{bmatrix} = 2\begin{bmatrix} 4 \\ 8 \\ -8 \end{bmatrix}$

Since $\overrightarrow{EF}$ and $\overrightarrow{HG}$ have a common factor, they are parallel.

b $\overrightarrow{EH} = \mathbf{h} - \mathbf{e} = \begin{bmatrix} -3 \\ 4 \\ 8 \end{bmatrix} - \begin{bmatrix} 9 \\ -2 \\ -1 \end{bmatrix} = \begin{bmatrix} -12 \\ 6 \\ 9 \end{bmatrix}$

Length of $\overrightarrow{EH} = \sqrt{(-12)^2 + 6^2 + 9^2} = 3\sqrt{29}$

$\overrightarrow{FG} = \mathbf{g} - \mathbf{f} = \begin{bmatrix} 5 \\ 20 \\ -8 \end{bmatrix} - \begin{bmatrix} 13 \\ 6 \\ -9 \end{bmatrix} = \begin{bmatrix} -8 \\ 14 \\ 1 \end{bmatrix}$

Length of $\overrightarrow{FG} = \sqrt{(-8)^2 + 14^2 + 1^2} = 3\sqrt{29}$

c 270 square units

11 Proof

Prior knowledge

1 Let $2n + 1$ be an odd number. Then $2n + 3$ and $2n + 5$ are consecutive odd numbers.

$$2n + 1 + 2n + 3 + 2n + 5 = 6n + 9$$
$$= 3(2n + 3)$$

which is a multiple of 3.

2 Let $2n$ be an even number and $2n + 1$ be the next number.

$$2n + 2n + 1 = 4n + 1$$

which is an odd number.

3 Let $2n$ be an even number and $2n + 1$ be the next number.

$$(2n + 1)^2 - (2n)^2 = 4n^2 + 4n + 1 - 4n^2$$
$$= 4n + 1$$
$$= (2n) + (2n + 1)$$

4 Let $2n$ be an even number and $2n + 2$ be the next even number.

$$(2n)^2 + (2n + 2)^2 = 4n^2 + 4n^2 + 8n + 4$$
$$= 8n^2 + 8n + 4$$
$$= 4(2n^2 + 2n + 1)$$

which is a multiple of 4.

5 $(2n - 1)^2 - (2n + 1)^2 = 4n^2 - 4n + 1 - (4n^2 + 4n + 1)$
$$= -8n$$

which is a multiple of 8.

Exercise 11.1A

1 Start by assuming that p is even.

So $p = 2n$ where n is a positive, whole number.

Then $p^2 = (2n)^2$
$$= 4n^2$$
$$= 2(2n^2)$$

$2(2n^2)$ is even; therefore p^2 is even.

This is a contradiction, so p must be odd.

2 Assume that $\sqrt{3}$ is rational.

Therefore, you can write $\sqrt{3}$ as $\dfrac{p}{q}$, where p and q are integers with no common factors other than 1.

If $\sqrt{3} = \dfrac{p}{q}$ then $3 = \dfrac{p^2}{q^2}$

Rearrange.

$$3q^2 = p^2$$

Therefore p^2 (and p) is a multiple of 3.

You can write $p = 3n$, so $p^2 = 9n^2$.

You have already shown that $3q^2 = p^2$, so now rewrite this.

$$3q^2 = 9n^2$$
$$q^2 = 3n^2$$

So q is also a multiple of 3.

If p and q are both multiples of 3, then they have a common factor of 3, which contradicts the assumption that they don't have a common factor other than 1.

So $\sqrt{3}$ is not rational – consequently, it must be irrational.

3 Assume that $\sqrt{9}$ is not rational.

Therefore $\sqrt{9}$ cannot be written as $\dfrac{p}{q}$, where p and q are positive integers.

If $p = 3$ and $q = 1$ then $\dfrac{p}{q} = \dfrac{3}{1}$

$$\left(\dfrac{p}{q}\right)^2 = \dfrac{p^2}{q^2}$$
$$\dfrac{p^2}{q^2} = \dfrac{9}{1} = 9$$

So $\sqrt{9} = \dfrac{3}{1}$ which contradicts the original assumption.

In this case proof by contradiction seems unnecessary as $\sqrt{9} = \pm 3$.

4 Assume that every non-zero real number has more than one reciprocal.

Let x be a non-zero real number and let both p and q be reciprocals of x where $p \notin q$.

So $px = 1$ and $qx = 1$

So $px = qx$

So $p = q$, which contradicts the original assumption.

5 Assume that $(x - a)^2 < 0$.

Let $n = x - a$ where $\{n \in \mathbb{R}\}$ as $\{a, x \in \mathbb{R}\}$.

So $(x - a)^2 = n^2$

So $\quad n^2 < 0$

But any real number squared is $\geqslant 0$.

So $\quad n^2 \geqslant 0$

So $(x - a)^2 \geqslant 0$, which contradicts the original assumption.

6 Assume that $x^3 \geqslant 0$ when $x < 0$ where $\{x \in \mathbb{R}\}$.

Let $n = x^3$ where $\{n \in \mathbb{R}\}$ as $\{x \in \mathbb{R}\}$.

So $n \geqslant 0$

So $x = \sqrt[3]{n}$

But if $x < 0$ then $\sqrt[3]{n} < 0$.

Then $\left(\sqrt[3]{n}\right)^3 < 0$.

So $x^3 < 0$, which contradicts the original assumption.

Exam-style questions 11   **page 214**

1 Assume that Elsa can sit next to Alf, who is already sitting next to Bukunmi. Then Elsa, Alf and Bukunmi are sitting in three adjacent seats. The remaining two seats, for Carlos and Dupika, are next to each other. But Carlos must not next to Dupika, so there is a contradiction. This means that the initial assumption, that Elsa can sit next to Alf, must be false. Hence it must be true that Elsa cannot sit next to Alf.

2 **a** Let $n = 2p$, where p is a positive, whole number.

   Then $n^2 = 4p^2$

   $2(2p^2)$ is even, therefore n^2 is even which was based on the assumption that n was even.

   b Assume that n is odd.

   So $n = 2p + 1$, where p is a positive, whole number.

   Then $n^2 = (2p + 1)^2$

   $\qquad = 4p^2 + 4p + 1$

   $\qquad = 2(2p^2 + 2p) + 1$

   $2(2p^2 + 2p)$ is even, so $2(2p^2 + 2p) + 1$ must be odd, therefore n^2 is odd.

   This is a contradiction, so n must be even.

   c In this case proof by deduction is much quicker than proof by contradiction.

3 **a** $\qquad (a + b)^2 = c^2 + (4)\left(\dfrac{ab}{2}\right)$

   $a^2 + 2ab + b^2 = c^2 + 2ab$

   $\qquad a^2 + b^2 = c^2$

   b Assumed that all the triangles have the same dimensions.

4 Let the angles in a quadrilateral be A, B, C and D.

   Assume the converse, that there is a quadrilateral that has two or more reflex angles. Let A and B both be reflex angles.

   Since A and B are reflex, $A > 180°$ and $B > 180°$.

   Hence $A + B > 180° + 180°$, so $A + B > 360°$.

   But the angles in a quadrilateral have a sum of $360°$, so $A + B + C + D = 360°$. For $A + B$ to be greater than $360°$, at least one of C or D must be negative. But an angle in a quadrilateral cannot be negative, so there is a contradiction. This means that the

initial assumption, that there is a quadrilateral that has two or more reflex angles, must be false. Hence it must be true that a quadrilateral has at most one reflex angle.

5 Assume that $\sqrt[3]{7}$ is rational.

   Therefore $\sqrt[3]{7} = \dfrac{p}{q}$ where p and q are integers with no common factors other than 1.

   Let $\sqrt[3]{7} = \dfrac{p}{q}$ then $7 = \dfrac{p^3}{q^3}$

   Rearrange.

   $$7q^3 = p^3$$

   Therefore p^3 (and p) is a multiple of 7.

   You can write $p = 7n$, so $p^3 = 343n^3$.

   You have already shown that $7q^3 = p^3$, so now rewrite this.

   $$7q^3 = 343n^3$$
   $$q^3 = 49n^3$$

   So q is also a multiple of 7.

   If p and q are both multiples of 7 then they have a common factor of 7, which contradicts the assumption that they don't have a common factor other than 1.

   So $\sqrt[3]{7}$ is not rational – consequently, it must be irrational.

6 Assume that neither of the shorter sides has an even length, i.e. that both a and b have odd lengths.

   Since a and b are positive odd integers, it is possible to write $a = 2m - 1$ and $b = 2n - 1$, where m and n are positive integers.

   Apply. Pythagoras' theorem, $a^2 + b^2 = c^2$.

   $\quad c^2 = (2m - 1)^2 + (2n - 1)^2$

   Expand.

   $\quad c^2 = 4m^2 - 4m + 1 + 4n^2 - 4n + 1$

   $\qquad = 4m^2 + 4n^2 - 4m - 4n + 2$

   $\qquad = 2(2m^2 + 2n^2 - 2m - 2n + 1)$

   Since m and n are positive integers, c^2 is an even integer and c is an even integer.

   Let $c = 2x$, where x is a positive integer.

   $c^2 = (2x)^2 = 4x^2$, where x^2 is a positive integer.

   However, $c^2 = 4(m^2 + n^2 - m - n + \frac{1}{2})$, where $m^2 + n^2 - m - n + \frac{1}{2}$ is not an integer.

   Therefore there is a contradiction. This means that the initial assumption, that both a and b have odd lengths, must be false. Hence it must be true that at least one of the shorter sides has an even length.

7 Assume that if x and y are positive real numbers, then $x + y < 2\sqrt{xy}$.

Since x and y are positive real numbers, squaring both sides will not change the inequality.

$$(x + y)^2 < 4xy$$
$$x^2 + 2xy + y^2 < 4xy$$
$$x^2 - 2xy + y^2 < 0$$
$$(x - y)^2 < 0$$

However, if $x = y$, then $(x - y)^2 = 0$ and if $x \neq y$, then $(x - y)^2 > 0$. Therefore there is a contradiction. This means that the initial assumption, that if x and y are positive real numbers then $x + y < 2\sqrt{xy}$, must be false. Hence it must be true that if x and y are positive real numbers, $x + y \geqslant 2\sqrt{xy}$.

8 As n is even, let $n = 2p$.

Then $n^2 + 3n + 2 = (2p)^2 + 3(2p) + 2$
$$= 4p^2 + 6p + 2$$
$$= 2(2p^2 + 3p + 1)$$

which is a multiple of 2 and so is an even number.

9 $f(x) = 2x + 5$

$g(x) = x^2 + 3$

$gf(x) = (2x + 5)^2 + 3$
$$= 4x^2 + 20x + 25 + 3$$
$$= 4x^2 + 20x + 28$$
$$= 4(x^2 + 5x + 7)$$

So $gf(x)$ is a multiple of 4 for all positive integer values of x.

10 Assume that if x and y are positive real numbers, then $x^2 + 4y^2 < 4xy$.

$$x^2 - 4xy + 4y^2 < 0$$
$$(x - 2y)^2 < 0$$

However, if $x = 2y$, then $(x - 2y)^2 = 0$ and if $x \neq 2y$, then $(x - 2y)^2 > 0$. Therefore there is a contradiction. This means that the initial assumption, that if x and y are positive real numbers then $x^2 + 4y^2 < 4xy$, must be false. Hence it must be true that if x and y are positive real numbers then $x^2 + 4y^2 \geqslant 4xy$.

11 Since SR = q, triangles PQS and PRS are isosceles, which means that $Q\hat{S}P = S\hat{P}Q$ and $R\hat{S}P = R\hat{P}S$

But this contradicts that $R\hat{S}P < Q\hat{S}P = S\hat{P}Q$.

So p, q, r is a right-angled triangle as it satisfies $p^2 + q^2 = r^2$.

12 Assume the converse, that there exist integers x and y for which $15x + 20y = 1$.

$15x$ and $20y$ both have a factor of 5. Hence $15x + 20y = 5(3x + 4y)$.

So if $15x + 20y = 1$, then $5(3x + 4y) = 1$, from which $3x + 4y = \frac{1}{5}$.

If x and y are integers, then $3x + 4y$ is also an integer, and cannot therefore be equal to $\frac{1}{5}$. Therefore there is a contradiction. This means that the initial assumption, that there exist integers x and y for which $15x + 20y = 1$, must be false. Hence it must be true that there exist no integers x and y for which $15x + 20y = 1$.

13 Assume the converse, that there exist integers x and y for which $3x - 9y = 4$.

$3x$ and $9y$ both have a factor of 3. Hence $3x - 9y = 3(x - 3y)$.

So if $3x - 9y = 4$, then $3(x - 3y) = 4$, from which $x - 3y = \frac{4}{3}$.

If x and y are integers, then $x - 3y$ is also an integer, and cannot therefore be equal to $\frac{4}{3}$. Therefore there is a contradiction. This means that the initial assumption, that there exist integers x and y, for which $3x - 9y = 4$, must be false. Hence it must be true that there exist no integers x and y for which $3x - 9y = 4$.

14 Gradient of radius $= \dfrac{b - d}{a - c}$

Gradient of tangent $= -\dfrac{a - c}{b - d}$

Equation of tangent is $y - d = -\dfrac{a - c}{b - d}(x - c)$
$$(y - d)(b - d) = -(a - c)(x - c)$$
$$(y - d)(b - d) + (x - c)(a - c) = 0$$

15 $y = \dfrac{x}{x^2 + A}$

$u = x$ so $\dfrac{du}{dx} = 1$

$v = x^2 + A$ so $\dfrac{dv}{dx} = 2x$

$$\frac{dy}{dx} = \frac{(x^2 + A)(1) - (x)(2x)}{(x^2 + A)^2}$$
$$\frac{dy}{dx} = \frac{A - x^2}{(x^2 + A)^2}$$

At a turning point, $\dfrac{dy}{dx} = 0$.

$$\frac{A - x^2}{(x^2 + A)^2} = 0$$
$$A - x^2 = 0$$
$$x^2 = A$$
$$x = \pm\sqrt{A}$$

16 Assume that a and b are odd square numbers and that they add up to c, which is also a square number.

Hence $a = m^2$, $b = n^2$ and $c = p^2$.

Since a is odd, so is m, and a can be written as $(2x + 1)^2$, where x is an integer.

Similarly, since b is odd, so is n, and b can be written as $(2y + 1)^2$, where y is an integer.

Hence $a + b = (2x + 1)^2 + (2y + 1)^2$
$$= 4x^2 + 4x + 1 + 4y^2 + 4y + 1$$
$$= 4(x^2 + x + y^2 + y) + 2$$

Since a and b are odd and $a + b = c$, c is even (as is p) and c can be written as $(2z)^2$, where z is an integer.

Hence $c = 4z^2$, so c is a multiple of 4.

However, $a + b = c = 4(x^2 + x + y^2 + y) + 2$, which is not a multiple of 4. There is a contradiction.

Therefore a and b cannot add up to c if a and b are odd square numbers and c is also a square number.

17 Let the dimensions of the cuboid be x cm, y cm and z cm.

$$V = xyz$$

The faces of the cuboid have areas of xy cm^2, xz cm^2 and yz cm^2.

$$xy \times xz \times yz = x^2y^2z^2$$
$$\sqrt{x^2y^2z^2} = xyz = V$$

18 Let the side of the square be x m.

The area of the square is x^2 m^2.

By Pythagoras' theorem, $x^2 + x^2 = d^2$
$$2x^2 = d^2$$
$$x^2 = \tfrac{1}{2}d^2$$

19 Equation of line is $y = \dfrac{r}{h}x$.

$$V = \pi \int_a^b y^2 \, dx$$

$$= \pi \int_0^h \left(\frac{r}{h}x\right)^2 dx$$

$$= \frac{\pi r^2}{h^2} \int_0^h x^2 \, dx$$

$$= \frac{\pi r^2}{h^2} \left[\tfrac{1}{3}x^3\right]_0^h$$

$$= \frac{\pi r^2}{h^2} \left(\left[\tfrac{1}{3}h^3\right] - [0]\right)$$

$$= \frac{\pi r^2 h^3}{3h^2}$$

$$= \tfrac{1}{3}\pi r^2 h$$

20 Assume the converse, that $\sin x + \cos x < 1$ for $0 \leqslant x \leqslant \dfrac{\pi}{2}$.

Note that for the domain $0 \leqslant x \leqslant \dfrac{\pi}{2}$, $\sin x > 0$ and $\cos x > 0$, so $\sin x + \cos x > 0$, and therefore $0 < \sin x + \cos x < 1$. This means that the square of $\sin x + \cos x$ will also be less than 1.

$$(\sin x + \cos x)^2 < 1$$

Expand

$$\sin^2 x + 2 \sin x \cos x + \cos^2 x < 1$$

From trigonometry it is known that $\sin^2 x + \cos^2 x = 1$.

So,
$$2 \sin x \cos x + 1 < 1$$
$$2 \sin x \cos x < 0$$
$$\sin x \cos x < 0$$

Since $\sin x > 0$ and $\cos x > 0$, this is a contradiction. This means that the initial assumption, that $\sin x + \cos x < 1$ for $0 \leqslant x \leqslant \dfrac{\pi}{2}$, must be false. Hence it must be true that $\sin x + \cos x \geqslant 1$ for $0 \leqslant x \leqslant \dfrac{\pi}{2}$.

21 Assume the converse, that if a, b and c are odd integers, then there is an integer solution for the equation $ax^2 + bx + c = 0$.

First, consider x being an even number.

If x is even, then ax^2 and bx are both even, as is $ax^2 + bx$.

Secondly, consider x being an odd number.

If x is odd, then ax^2 and bx are both odd and $ax^2 + bx$ is even.

In both cases, $ax^2 + bx$ is even and c is odd, so $ax^2 + bx + c$ is odd.

But if $ax^2 + bx + c$ is odd, then it cannot equal zero. Therefore there is a contradiction. This means that the initial assumption, that if a, b and c are odd integers, then there is an integer solution for the equation $ax^2 + bx + c = 0$, must be false. Hence it must be true that there exist no integer solutions for the equation $ax^2 + bx + c = 0$ if a, b and c are odd integers.

12 Probability

Prior knowledge   page 216

1 **a** 0.2

    b 0

    c 0.65

    d 0.1

    e 0.8

2 a

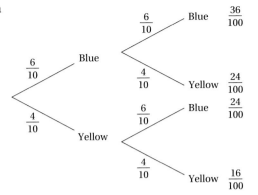

b 0.16

c 0.48

Exercise 12.1A   **page 221**

1 a $\frac{1}{20}$

   b $\frac{1}{5}$

   c $\frac{19}{20}$

2 a 0.6

   b 0.4

   c they are mutually exclusive

3 a $\frac{1}{13}$

   b $\frac{1}{4}$

   c $\frac{4}{13}$

   d $\frac{1}{52}$

   e $\frac{12}{13}$

   f $\frac{3}{13}$

4 0.4

5 The outcome 'dice 1 = 2, dice 2 = 2' is an outcome which satisfies both events. Therefore the events are not mutually exclusive.

6 a 0.6

   b 0.8

   c 0.4

   d 0.9

7 a 0.15

   b 0.45

   c 0.55

   d 0.25

   e 0.3

Exercise 12.2A   **page 225**

1 $\frac{1}{13}$

2 $\frac{3}{7}$

3 a 0.5

   b 0.5

4 a 0.7

   b $\frac{2}{3}$

   c 0.8

   d 0.4

5 a 0.3

   b 0.35

   c 0.4

Exercise 12.2B   **page 228**

1 a $\frac{5}{18}$ **b** $\frac{5}{18}$

   c $\frac{1}{6}$ **d** $\frac{5}{9}$

2 a

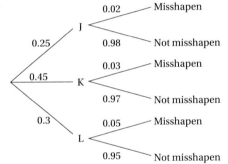

b 0.245

c 0.0335

d 0.597

3 a

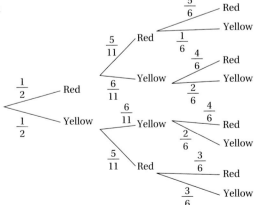

b $\frac{2}{3}$

c $\frac{10}{33}$

d $\frac{9}{11}$

e $\frac{25}{132}$

4 a

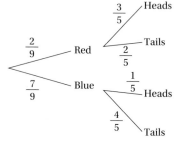

b $\frac{32}{45}$

c $\frac{7}{13}$

d $\frac{53}{81}$

Exercise 12.3A   page 233

1 a $\frac{1}{6}$ **b** Dice are six-sided, dice are fair

2 Probability of getting a job (up to and including age 24) = 1 − 0.1319 = 0.8681

This is assuming that England has the same unemployment rates as Australia.

The probability of Arthur getting a job in his lifetime will be greater than 0.8681, as Arthur will have more opportunities in life to acquire a job, assuming he is seeking a job.

3 The game will be fair if the probability of spinner 1 beating spinner 2 is equal to spinner 2 beating spinner 1. The same applies between spinner 1 and spinner 3, and between spinner 2 and spinner 3.

Probability of spinner 2 beating spinner 1

$$= \left(\frac{3}{6}\times\frac{1}{5}\right)+\left(\frac{2}{6}\times\frac{3}{5}\right)+\left(\frac{1}{6}\times\frac{4}{5}\right)=\frac{13}{30}$$

Probability of spinner 1 beating spinner 2

$$= \left(\frac{1}{5}\times\frac{3}{6}\right)+\left(\frac{1}{5}\times\frac{3}{6}\right)+\left(\frac{1}{5}\times1\right)=\frac{12}{30}$$

These probabilities are unequal; therefore the game is unfair. We do not agree that the game is fair.

Exam-style questions 12   page 234

1 a 0.625   **b** 0.65

  c 0.35   **d** 0.625

2 a i 0.065   **ii** 0.1275   **iii** 0.372 25

  b i 0.61

    ii 0.39

3 a 0.027

  b 0.7465

4 a i $s=1$ **ii** $q=9$ **iii** $p=15, r=7$

  b i $\frac{9}{16}$

    ii Number of students studying music and art = 9, therefore events are not independent

  c 0.103

5 a 9

  b i 0.91

    ii 0.418

  c Because a boy can do both sports.

  d $P(C \cap S) = 9 \div 100 = 0.09$

6 a i 0.3   **ii** 0.45   **iii** 0.75

  b 0.0081

  c 0.45

  d 0.2

7 a i 0.2   **ii** 0.44   **iii** 0.56

  b i 0.534

    ii 0.0220

8 a

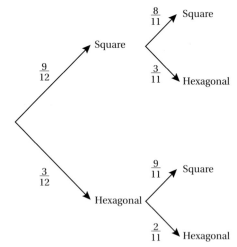

  b $\frac{1}{4}$

  c $\frac{8}{11}$

9 a

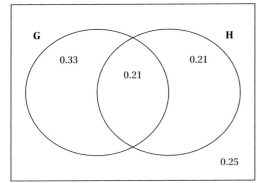

b P(G) = 0.54; P(H) = 0.42

c $\frac{33}{58}$

d the events are not independent.

13 Statistical distributions

Prior knowledge   **page 238**

1 a 3.75

b 2.12

c New mean = 3.32, new standard deviation = 2.24

Exercise 13.1A   **page 242**

1 a The distribution is symmetrical about the mean.

b The mode, median and mean are all the same.

c The total area below the curve is 1.

d The distribution is defined by two parameters: the mean and the standard deviation.

2 The standard deviations are equal, but the means are different, for the two distributions.

3 a 50%   **b** $\frac{17}{50}$

c 47.5%   **d** $\frac{779}{800}$

4 Curve A has a higher mean compared with curve B.

Curve A has a smaller standard deviation compared with curve B.

Exercise 13.2A   **page 244**

1 Fighter A ~ N(3, 4), Fighter B~N(2, 3^2)

2 Year 1 ~ N(0, 5), Year 2~N(−1, 2^2)

3 The 'N' and 'X' are in incorrect places; the mean should be 7 and the standard deviation should be 11. $X \sim N(7, 11^2)$

4 a Student A: mean = 5, variance = 49, standard deviation = 7

Student B: mean = 6, variance = 81, standard deviation = 9

b

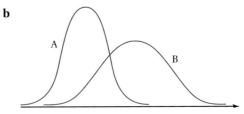

Exercise 13.2B   **page 246**

1 a

b

c

d

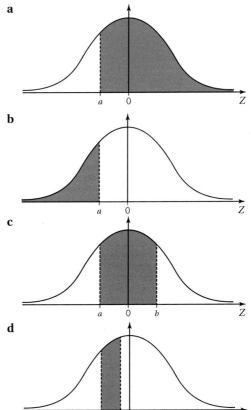

e

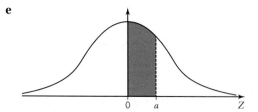

2 'N' and 'Z' are written in incorrect positions. The mean should equal 0. The standard deviation should equal 1.

$Z \sim N(0, 1^2)$

Exercise 13.2C   page 248

1 0.9192
2 0.8079
3 0.1446
4 0.0013
5 0.9066
6 0.1178
7 0.7934
8 0.0723

Exercise 13.3A   page 251

1 **a** 0.3849
   b 0.4878
   c 0.6247
   d 0.0228
2 **a** 0.6915
   b 0.1949
   c 0.8413
   d 0.1499
   e 0.5125
   f 0.0389
3 **a** 0.2376
   b 0.3334
   c 0.4284
4 **a** 0.0668
   b 0.0228
   c 0.9105
5 **a** 0.1057
   b 0.7887
   c 0.1057
   d 0.0829

6 **a** 0.382
   b 0.045
   c 0.05%

Exercise 13.3B   page 253

1 **a** **i** 0.5
     ii 0.0000
     iii 0.5
   b **i** 142.53 hours
     ii 146.22 hours
     iii 159.24 hours

2 Underweight: 491.14 g.
   Overweight: 508.06 g.

3 **a** 19.41
   b 17.02
4 **a** 53.26
   b 50.5
   c 38.6
   d 36.74
   e 6.74
   f 46.45
5 4480.75 hours
6 1.90 m
7 between 62.1 mph and 74.9 mph.
8 **a** 89
   b 56 and 104
   c 40

Exercise 13.3C   page 256

1 5.938
2 45.4 cm
3 $\sigma = -5$
   $\mu = 330$
4 $\sigma = 1.721$
   $\mu = 11.08$
5 **a** 0.0479
   b 0.1587
   c 0.7934
   d Stan's mean = 72.93 and standard deviation = 9.42.
6 $\sigma = 5.9382$
   $\mu = 135$

Exercise 13.4A   page 261

1 a 10

   b Variance = 5

      Standard deviation = 2.236

   c 0.005 127

2 a 0.6588

   b 0.6602

3 a 1

   b 0.9999

   c 0

   d 0.055 73

4 a You may use the normal distribution as an approximation for the binomial $B(n, p)$ (where n is the number of trials, each having probability p of success) when n is large and p is close to 0.5.

   b 0.2173

   c 0.2146

   d 1.24%

5 a 0.1645

   b 0.1641

   c 0.27% error.

   d Worse agreement with a 5.4% error between the two values.

6 a $P(\text{any one correct}) = \dfrac{1}{5}$

   b Therefore the pass mark must be 13 to the nearest whole mark.

Exam-style questions 13   page 263

1 a 0.0062

   b 0.1587

   c 0.8351

2 a 0.6965

   b 163.4 cm

3 0.212

4 a 0.8183

   b 0.0345

   c 0.3183

   d 1256 days

5 $212.6 - \mu = 0.44\sigma$ (1)

   $211.8 - \mu = -1.175\sigma$ (2)

$\sigma = 0.4954$

$\mu = 212.38$

32% are rejected.

6 a 0.3792

   b 3.7676

   c 0.0228

7 6.8788

8 a 0.2335

   b 277.41 g

   c $\sigma = 21.186$

      $\mu = 255.15$

9 a 14.49

   b 0.000 017 31

   c The normal distribution is not a suitable model.

      $100 + 3\sigma = 143.47$, so 8.37 p.m. for latest arrival.

10 136.7

14 Statistical hypothesis testing

Prior knowledge   page 265

1 a and b

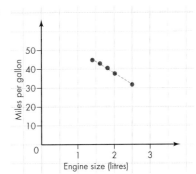

   c Negative correlation

   d As the engine size increases, the number of miles per gallon decreases.

Exercise 14.1A   page 268

1 All values above 0 up to and including 1

2 a Slight positive correlation

   b Almost perfect negative correlation

   c Perfect negative correlation

   d Very slight positive correlation

3 −0.85 strongest, −0.05 weakest

4 a i

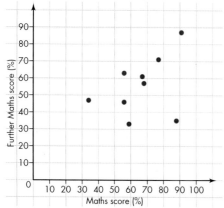

ii The scatter graph shows a positive correlation, albeit not very strong. PMCC = 0.392485 supports this.

b i

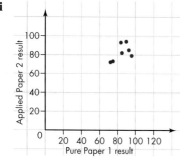

ii The scatter graph shows a positive correlation, albeit not very strong. PMCC = 0.519396 supports this.

c i

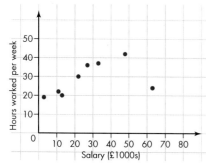

ii The scatter graph shows quite a strong positive correlation. PMCC = 0.77931 supports this.

5 No correlation between maths and art results.

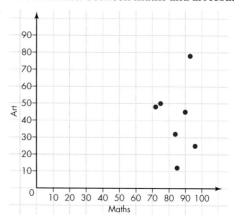

Exercise 14.1B   **page 273**

1 a H_0: $\rho = 0$

H_1: $\rho \neq 0$

b The critical value for a 2-tail test at the 5% significance level for $n = 20$ is 0.444. Since $0.444 < 0.695$, reject H_0. There is not enough evidence to suggest that there is a positive correlation between the practical and written papers.

c The conclusion would be the same, as the critical region would be 0.378.

2 a H_0: $\rho = 0$

H_1: $\rho \neq 0$

b The critical value for a 2-tail test at the 5% significance level for $n = 15$ is 0.441. Since $0.441 < 0.896$, reject H_0. There is evidence to suggest that there is a correlation between height and weight.

3 a

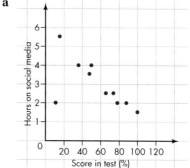

b H_0: $\rho = 0$

H_1: $\rho < 0$

c the critical region for a sample size of 10 and significance level of 2.5% is $x < -0.632$.

Since $-0.647 < -0.632$, reject H_0. There is evidence to suggest that there is a correlation between test scores and time spent on social media.

d By omitting the outlier at 2 hours and score of 11, the PMCC is -0.974 and the critical value is -0.666. The results are still evidence to reject H_0.

Exercise 14.2A   page 278

1 a 0.4207

b there is not enough evidence to suggest there is a decrease in pin weights.

2 a H_0: The population mean is unchanged, i.e. $\mu = 750\,g$.

H_1: The population mean is different, i.e. $\mu \neq 750\,g$.

b 2-tail test, as she is unsure whether her guinea pigs are of this breed, not specifying above or below the mean.

c There is evidence to suggest her guinea pigs are the correct breed.

3 the evidence suggests the mean volume is set to 8 litres.

4 there is sufficient evidence to suggest the journey time has in fact increased.

5 there is not sufficient evidence to suggest the bulb lifetime is lower.

6 there is sufficient evidence to suggest the restaurant would accept the samphire.

7 a 0.8064

b 0.2023

8 There is not enough evidence to suggest that the mass is decreasing.

Exercise 14.3A   page 282

1 a $g = ah^c$

Take logarithms.

$$\log g = \log ah^c$$
$$= \log a + \log h^c$$
$$= \log a + c\log h$$

b This is the equation of a straight-line graph with gradient c and intercept $\log a$.

c
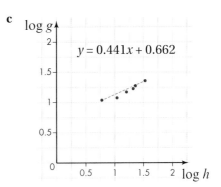

a is approximately 4.59.

c is approximately 0.44.

2 a $k = mn^s$

Take logarithms.

$$\log k = \log mn^s$$
$$= \log m + \log n^s$$
$$= \log m + s\log n$$

b This is the equation of a straight-line graph with gradient $\log n$ and y-intercept $\log m$.

c
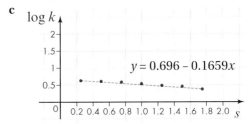

m is approximately 4.97.

n is approximately 0.683.

3 a $y = kx^n$ is an appropriate model as the log plot produces a linear graph.

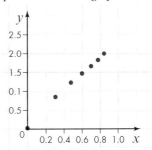

b k is approximately 1.21

n is approximately 2.3

c approximately 224.

4 **a** $y = na^x$ would be a suitable model.

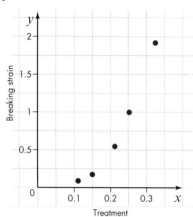

Treatment

b $\log y = \log n + x \log a$

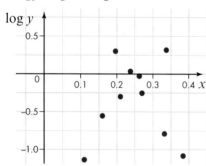

n is approximately 0.0162

a is approximately 4.30×10^6

Approximate relation is $y = 0.0162(4.30 \times 10^6)^x$

Exam-style questions 14   page 284

1 **a**

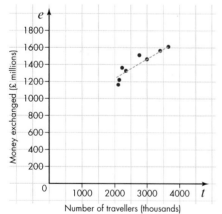

Number of travellers (thousands)

As the number of travellers increases, the amount of money exchanged increases.

b The value of r agrees with the findings on the scatter diagram – a positive correlation.

2 **a** 0.036 could be rounded to 0, so no correlation.

b

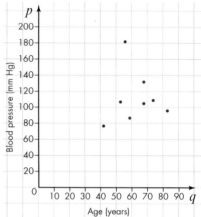

Age (years)

There is no trend between the variables, so no correlation.

3 **a** $\log J = \log a + h \log b$

b Find the intercept of the line to find the value of a.

c Find the gradient of the line to find the value of b.

4 **a** $H_0: \rho = 0$

$H_1: \rho \neq 0$

b The critical value for a 2-tail test at the 5% significance level for $n = 10$ is -0.6319.

Since $-0.6319 > -0.859$, reject H_0. There is enough evidence to suggest that there is correlation between speed and fuel consumption.

c The conclusion would be the same, as the critical region would be -0.5494.

5 **a**

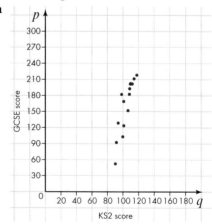

KS2 score

b $H_0: \rho = 0$

$H_1: \rho > 0$

The teacher believes there is a positive relationship.

c The critical region for a sample size of 15 and significance level of 5% is $x > 0.4409$.

Since $0.4409 < 0.878$, reject H_0. There is evidence to suggest that there is correlation between the students' KS2 and GCSE scores.

6 **a** $y = kx^n$ is an appropriate model as the log plot produces a linear graph.

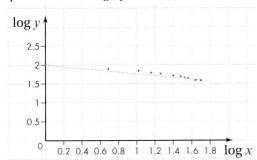

b k is approximately 100.

n is approximately −0.2.

c Using extrapolation, x is approximately 56; it would take about 56 minutes.

7 **a** 712

b 1007

c Evidence suggests that the mean score has increased.

8 C

9 **a** $H_0: \rho = 0$

b $H_1: \rho > 0$

c reject H_0, as evidence suggests that students who are good at mathematics are also good at physics.

10 $H_0: \mu = 68$

$H_1: \mu < 68$, where μ is the population mean journey time.

Let X be the distribution of cyclists' times.

$X \sim N(68, \frac{9}{100})$

Using the tables, $P(Z < k) = 0.05$

$k = -1.645$

So the critical value is $68 + \left(\frac{-1.645 \times 3}{\sqrt{100}} \right) = 67.51$

So reject the null hypothesis for values less than 67.51.

Since $67.51 > 65.3$, the null hypothesis is rejected. There is sufficient evidence to suggest the journey time has, in fact, decreased.

11 H_0: The population mean is unchanged, i.e. $\mu = 3.4$ kg.

H_1: The population mean has increased, i.e. $\mu > 3.4$ kg.

1-tail test at 1% significance level.

The distribution of sample means, $\bar{X}$, is $N(\mu, \frac{\sigma^2}{n})$.

According to the null hypothesis, $\mu = 3.4$, and it is known that $\sigma = 0.8$ and $n = 75$.

So this distribution is $N(3.4, \frac{0.8^2}{75})$.

Using the tables, $P(Z < k) = 0.99$

$k = 2.326$

So the critical value is $3.4 + \left(\frac{2.326 \times 0.8}{\sqrt{75}} \right) = 3.615$

So reject the null hypothesis for values above 3.615.

Since $3.55 < 3.615$, the null hypothesis is accepted. There is not sufficient evidence to suggest that the mass of an adult male cat is increasing.

15 Kinematics

Prior knowledge **page 287**

1 **a** $5.6\,\mathrm{m\,s^{-1}}$

b $10.4\,\mathrm{m\,s^{-1}}$

2 $90\,\mathrm{km\,h^{-1}}$

3 **a** $\begin{bmatrix} 60 \\ -25 \end{bmatrix}$ N

b $65\,\mathrm{N}$

22.6° below horizontal

4 **a** $15t^2 + \frac{3}{2}t^{\frac{1}{2}}$

b $\frac{1}{15}(3t - 2)^5 + c$

Exercise 15.1A **page 290**

1 **a** $36\,\mathrm{m}$

b $6\,\mathrm{s}$

2 **a** $0.9\,\mathrm{m\,s^{-2}}$

b $15\,\mathrm{m\,s^{-1}} = (\frac{15}{1000} \times 3600)\,\mathrm{m\,h^{-1}} = 54\,\mathrm{km\,h^{-1}}$

$54\,\mathrm{km\,h^{-1}} \approx \frac{54}{1.6}\,\mathrm{mph} = 33.75\,\mathrm{mph}$, which is greater than the speed limit of 30 mph.

3 $9.20\,\mathrm{m}$

4 $u = 39.2\,\mathrm{m\,s^{-1}}, a = -9.8\,\mathrm{m\,s^{-2}}$

a $t = 2\,\mathrm{s}$

Substitute into $v = u + at$.

$v = 39.2 - 9.8 \times 2 = 19.6\,\mathrm{m\,s^{-1}}$

b $78.4\,\mathrm{m}$

c $8\,\mathrm{s}$

5 **a** $0.467\,\mathrm{s}$ or $2.19\,\mathrm{s}$

b $3.74\,\mathrm{s}$

6 **a** $v = \dfrac{\mathrm{d}r}{\mathrm{d}t} = 30t - \dfrac{5}{2}t^{\frac{3}{2}}$

$a = \dfrac{\mathrm{d}^2 r}{\mathrm{d}t^3} = 30 - \dfrac{15}{4}t^{\frac{1}{2}}$

b $202.5\,\mathrm{m\,s^{-1}}$

c For maximum velocity, $a = 0$.

$$30 - \frac{15}{4}t^{\frac{1}{2}} = 0$$

$$120 - 15\sqrt{t} = 0$$

$$15\sqrt{t} = 120$$

$$\sqrt{t} = 8$$

$$t = 64\,\mathrm{s}$$

When $t = 64$,

$$v = 30(64) - \frac{5}{2}(64)^{\frac{3}{2}} = 1920 - 1280 = 640\,\mathrm{m\,s^{-1}}$$

7 a $v = 9 - 3t^2$

$r = 9t - t^3 + 2$

b When $v = 0$, $9 - 3t^2 = 0$

$$3t^2 = 9$$

$$t^2 = 3$$

$$t = \sqrt{3}\,\mathrm{s}$$

c When $t = 2$, $r = 9(2) - (2)^3 + 2$

$$r = 18 - 8 + 2$$

$$r = 12\,\mathrm{m}$$

Exercise 15.2A page 296

1 a $(9\mathbf{i} - 12\mathbf{j})\,\mathrm{km}$

b $15\,\mathrm{km}$

c $[(-7\mathbf{i} + 9\mathbf{j}) + t(3\mathbf{i} - 4\mathbf{j})]\,\mathrm{km}$

d $(20\mathbf{i} - 27\mathbf{j})\,\mathrm{km}$

2 $a = 1$

$b = -2$

3 a $\begin{bmatrix} -80 \\ 40 \end{bmatrix}\,\mathrm{km\,h^{-1}}$

b $\begin{bmatrix} -106 \\ 1 \end{bmatrix}\,\mathrm{km}$

4 a $\mathbf{r}_A = (8 + 3t)\mathbf{i} + (1 - t)\mathbf{j}$

$\mathbf{r}_B = (12 + t)\mathbf{i} + (-11 + 5t)\mathbf{j}$

b $t = 2$

$(14\mathbf{i} - \mathbf{j})\,\mathrm{m}$

c $19.0\,\mathrm{m}$

5 a $\mathbf{r}_G = \begin{bmatrix} 13 \\ -5 \end{bmatrix} + t\begin{bmatrix} -1 \\ 6 \end{bmatrix} = \begin{bmatrix} 13 - t \\ -5 + 6t \end{bmatrix}$

$\mathbf{r}_H = \begin{bmatrix} 2 \\ 17 \end{bmatrix} + t\begin{bmatrix} 3 \\ -2 \end{bmatrix} = \begin{bmatrix} 2 + 3t \\ 17 - 2t \end{bmatrix}$

$\mathbf{r}_G - \mathbf{r}_H = \begin{bmatrix} 13 - t \\ -5 + 6t \end{bmatrix} - \begin{bmatrix} 2 + 3t \\ 17 - 2t \end{bmatrix} = \begin{bmatrix} 11 - 4t \\ -22 + 8t \end{bmatrix}$

At 8:30 a.m., $t = 1$

$$\mathbf{r}_G - \mathbf{r}_H = \begin{bmatrix} 11 - 4 \times 1 \\ -22 + 8 \times 1 \end{bmatrix} = \begin{bmatrix} 7 \\ -14 \end{bmatrix}\,\mathrm{km}$$

Distance $= \sqrt{7^2 + (-14)^2} = 15.7\,\mathrm{km}$

b $11.2\,\mathrm{km}$

c 10:15 a.m.

6 a $(3 - 2t)\mathbf{i} + (-7 + t)\mathbf{j}$

b i 3:30 p.m.

ii 5:20 p.m.

iii 9:00 p.m.

c Coefficients of $\mathbf{i}$ and $\mathbf{j}$ are equal and opposite.

$$3 - 2t = -(-7 + t)$$

$$3 - 2t = 7 - t$$

$$t = -4$$

Since t cannot be negative, C will never be southeast of D.

7 a A is faster.

b $d^2 = 2t^2 - 28t + 148$

c 9:50 p.m.

7.07 km

8 a 7:15 p.m.

$(2\mathbf{i} + 8\mathbf{j})\,\mathrm{km}$

b 7.55 km

Exercise 15.3A page 300

1 a i $(16\mathbf{i} - 3\mathbf{j})\,\mathrm{m\,s^{-1}}$

ii $5\,\mathrm{m\,s^{-1}}$

b i $(42\mathbf{i} + 18\mathbf{j})\,\mathrm{m}$

ii $45.7\,\mathrm{m}$

2 a $(-5\mathbf{i} + 2\mathbf{j})\,\mathrm{m\,s^{-1}}$

b $(-25\mathbf{i} + 10\mathbf{j})\,\mathrm{m}$

c $(-2.8\mathbf{i} + 3.6\mathbf{j})\,\mathrm{m\,s^{-2}}$

3 $\mathbf{u} = \begin{bmatrix} 7 \\ 3 \end{bmatrix}\,\mathrm{m\,s^{-1}}, t = 4\,\mathrm{s}$

Substitute into $\mathbf{F} = m\mathbf{a}$.

$$\mathbf{a} = \begin{bmatrix} -10 \\ 12 \end{bmatrix} \div 2 = \begin{bmatrix} -5 \\ 6 \end{bmatrix}\,\mathrm{m\,s^{-2}}$$

Substitute into $\mathbf{s} = \mathbf{u}t + \frac{1}{2}\mathbf{a}t^2$.

$$\mathbf{s} = \begin{bmatrix} 7 \\ 3 \end{bmatrix} \times 4 + \frac{1}{2} \times \begin{bmatrix} -5 \\ 6 \end{bmatrix} \times 4^2$$

$$=\begin{bmatrix}28\\12\end{bmatrix}+\begin{bmatrix}-40\\48\end{bmatrix}=\begin{bmatrix}-12\\60\end{bmatrix}\text{m}$$

This is the displacement from $\begin{bmatrix}10\\1\end{bmatrix}$ m, so

$$\text{position}=\begin{bmatrix}10\\1\end{bmatrix}+\begin{bmatrix}-12\\60\end{bmatrix}=\begin{bmatrix}-2\\61\end{bmatrix}\text{m}$$

4 a $5\,\text{m s}^{-1}$

b i $(64\mathbf{i}+12\mathbf{j})$ m

ii $(23\mathbf{i}+6.5\mathbf{j})$ m

5 a Substitute into $\mathbf{s}=\mathbf{u}t+\dfrac{1}{2}\mathbf{a}t^2$.

$$\mathbf{a}=\frac{\mathbf{s}-\mathbf{u}t}{\frac{1}{2}t^2}=\frac{(15\mathbf{i}-6\mathbf{j})-3(-\mathbf{i}+7\mathbf{j})}{\frac{1}{2}\times3^2}=\frac{18\mathbf{i}-27\mathbf{j}}{\frac{1}{2}\times3^2}$$

$$=(4\mathbf{i}-6\mathbf{j})\,\text{m s}^{-2}$$

Magnitude $=\sqrt{4^2+\left(-6\right)^2}=\sqrt{52}=2\sqrt{13}\ \text{m s}^{-2}$

b Substitute into $\mathbf{v}=\mathbf{u}+\mathbf{a}t$.

$$\mathbf{v}=(-\mathbf{i}+7\mathbf{j})+(4\mathbf{i}-6\mathbf{j})\times3=(11\mathbf{i}-11\mathbf{j})\,\text{m s}^{-1}$$

$$\text{Speed}=\sqrt{11^2+\left(-11\right)^2}=\sqrt{242}=11\sqrt{2}\ \text{m s}^{-1}$$

6 Firstly, Stephen has not explained where the expressions $-0.1\mathbf{i}+T\times0.5\mathbf{i}$ and $0.3\mathbf{j}+T\times0.2\mathbf{j}$ have come from. He has used $\mathbf{v}=\mathbf{u}+\mathbf{a}t$ but should have stated this.

Secondly, his answer does not work. If $T=0.625$, then $-0.1\mathbf{i}+T\times0.5\mathbf{i}=\dfrac{17}{80}\mathbf{i}$ and $0.3\mathbf{j}+T\times0.2\mathbf{j}=\dfrac{17}{40}\mathbf{j}$, so Stephen has made the coefficient of $\mathbf{j}$ twice as large as the coefficient of $\mathbf{i}$, whereas it needs to be the other way round. This can be seen by putting the coefficient of $\mathbf{i}$ equal to $2k$ and the coefficient of $\mathbf{j}$ equal to k as shown below:

$$-0.1\mathbf{i}+T\times0.5\mathbf{i}=(0.5T-0.1)\mathbf{i}=2k\mathbf{i}$$

$$0.3\mathbf{j}+T\times0.2\mathbf{j}=(0.2T+0.3)\mathbf{j}=k\mathbf{j}$$

Hence $k=0.2T+0.3$, from which

$$0.5T-0.1=2(0.2T+0.3)$$

$$0.5T-0.1=0.4T+0.6$$

$$0.1T=0.7$$

$$T=7\,\text{s}$$

Now, $-0.1\mathbf{i}+7\times0.5\mathbf{i}=3.4\mathbf{i}$ and $0.3\mathbf{j}+7\times0.2\mathbf{j}=1.7\mathbf{j}$

So $(3.4\mathbf{i}+1.7\mathbf{j})=1.7(2\mathbf{i}+\mathbf{j})$

Therefore the cyclist is travelling parallel to $(2\mathbf{i}+\mathbf{j})$.

7 a Substitute into $\mathbf{v}=\mathbf{u}+\mathbf{a}t$.

$$\mathbf{v}=\begin{bmatrix}6\\2\end{bmatrix}+\begin{bmatrix}-4\\1\end{bmatrix}t$$

$$\mathbf{v}=\begin{bmatrix}6\\2\end{bmatrix}+\frac{8}{3}\begin{bmatrix}-4\\1\end{bmatrix}=\begin{bmatrix}-\frac{14}{3}\\\frac{14}{3}\end{bmatrix}=\frac{14}{3}\begin{bmatrix}-1\\1\end{bmatrix}$$

b i $5\,\text{s}$

ii $12\,\text{s}$

c Assume that $\begin{bmatrix}6\\2\end{bmatrix}+\begin{bmatrix}-4\\1\end{bmatrix}t=k\begin{bmatrix}-4\\1\end{bmatrix}$

i: $6-4t=-4k$

j: $2+t=k$

Hence $6-4t=-4(2+t)$

$$6-4t=-8-4t$$

$$14=0$$

which is a contradiction, so the particle will never be moving parallel to $\begin{bmatrix}-4\\1\end{bmatrix}$.

8 $40\mathbf{j}$ m

1 A

2 a $\begin{bmatrix}2\\9\end{bmatrix}\text{m s}^{-1}$

b $2\sqrt{5}\,\text{m s}^{-2}$

3 a $\mathbf{v}=\dfrac{\text{d}\mathbf{r}}{\text{d}t}=[(3t^2-24t+p)\mathbf{i}+(42t+q-6t^2)\mathbf{j}]$

When $t=2$,

$$\mathbf{v}=[(3(2)^2-24(2)+p)\mathbf{i}+(42(2)+q-6(2)^2)\mathbf{j}]$$

$$=[(p-36)\mathbf{i}+(q+60)\mathbf{j}]$$

$$p-36=-15$$

$$p=21$$

b -72

c i $3\,\text{s}$ or $4\,\text{s}$

ii $1\,\text{s}$ or $7\,\text{s}$

4 $(24\mathbf{i}+28\mathbf{j})$ N

5 a $\mathbf{v}=\dfrac{\text{d}\mathbf{r}}{\text{d}t}=\dfrac{1}{6}\begin{bmatrix}18t-12\\6t^2\end{bmatrix}=\begin{bmatrix}3t-2\\t^2\end{bmatrix}\text{m s}^{-1}$

$$\mathbf{a}=\dfrac{\text{d}\mathbf{v}}{\text{d}t}=\begin{bmatrix}3\\2t\end{bmatrix}\text{m s}^{-2}$$

When $t=4$, $\mathbf{a}=\begin{bmatrix}3\\2\times4\end{bmatrix}=\begin{bmatrix}3\\8\end{bmatrix}\text{m s}^{-2}$

b $\begin{bmatrix}40\\96\end{bmatrix}\text{m s}^{-1}$

1 $\mathbf{a}=\dfrac{\mathbf{F}}{m}=\dfrac{20t\mathbf{i}-15\mathbf{j}}{2.5}=(8t\mathbf{i}-6\mathbf{j})\,\text{m s}^{-2}$

$$\mathbf{v}=\int\mathbf{a}\,\text{d}t=\int(8t\mathbf{i}-6\mathbf{j})\text{d}t=(4t^2+C)\mathbf{i}+(-6t+D)\mathbf{j}$$

When $t=0$, $C\mathbf{i}+D\mathbf{j}=-6\mathbf{i}+34\mathbf{j}$

So $C=-6$, $D=34$

$$\mathbf{v}=[(4t^2-6)\mathbf{i}+(-6t+34)\mathbf{j}]\,\text{m s}^{-1}$$

When $t = 3$, $\mathbf{v} = [(4(3)^2 - 6)\mathbf{i} + (-6(3) + 34)\mathbf{j}]$

$\qquad = (30\mathbf{i} + 16\mathbf{j})\,\text{m s}^{-1}$

Speed $= \sqrt{30^2 + 16^2} = 34\,\text{m s}^{-1}$

2 $\quad \mathbf{r} = \int \mathbf{v}\,dt = \begin{bmatrix} \dfrac{1}{3}t^3 - 2t + C \\ 14t - \dfrac{3}{2}t^2 + D \end{bmatrix}$

When $t = 0$, $\begin{bmatrix} C \\ D \end{bmatrix} = \begin{bmatrix} 4 \\ -7 \end{bmatrix}$

$\mathbf{r} = \begin{bmatrix} \dfrac{1}{3}t^3 - 2t + 4 \\ 14t - \dfrac{3}{2}t^2 - 7 \end{bmatrix}\,\text{m}$

When $t = 4$, $\mathbf{r} = \begin{bmatrix} \dfrac{1}{3}(4)^3 - 2(4) + 4 \\ 14(4) - \dfrac{3}{2}(4)^2 - 7 \end{bmatrix} = \begin{bmatrix} \dfrac{52}{3} \\ 25 \end{bmatrix}\,\text{m}$

Distance $= \sqrt{\left(\dfrac{52}{3}\right)^2 + 25^2} = 30.4\,\text{m}$

3 $\quad \dfrac{1}{10}\,[\mathbf{i} + (21 + 4\ln 2)\mathbf{j}]\,\text{m}$

4 $\quad$ **a** $\quad 8.2\,\text{m s}^{-2}$

$\quad$ **b** $\quad \mathbf{r} = [(8t + 3)\mathbf{i} + (\tfrac{1}{3}(t^2 + 9)^{\frac{3}{2}} - 9)\mathbf{j}]$

5 $\quad$ **a** $\quad 4\,\text{s or } 6\,\text{s}$

$\quad$ **b** $\quad t = 4$, $\mathbf{v} = \begin{bmatrix} 6 \\ 4 \end{bmatrix}\,\text{m s}^{-1}$

$\qquad t = 6$, $\mathbf{v} = \begin{bmatrix} 24 \\ 16 \end{bmatrix}\,\text{m s}^{-1}$

6 $\quad$ **a** $\quad$ Chloe has not used the information that X initially has a velocity of $-\mathbf{i}\,\text{m s}^{-1}$.

$\qquad$ She has also found the displacement but added its $\mathbf{i}$ and $\mathbf{j}$ components to find the distance.

$\quad$ **b** $\quad \mathbf{v} = \int \mathbf{a}\,dt = (2t^2 + C)\mathbf{i} + (8t + D)\mathbf{j}$

$\qquad$ When $t = 0$, $C\mathbf{i} + D\mathbf{j} = -\mathbf{i}$

$\qquad$ So $C = -1$, $D = 0$

$\qquad\qquad \mathbf{v} = [(2t^2 - 1)\mathbf{i} + 8t\mathbf{j}]\,\text{m s}^{-1}$

$\qquad \mathbf{r} = \int \mathbf{v}\,dt = [(\tfrac{2}{3}t^3 - t + C_1)\mathbf{i} + (4t^2 + D_1)\mathbf{j}]\,\text{m}$

$\qquad$ When $t = 0$, $C_1\mathbf{i} + D_1\mathbf{j} = 0$

$\qquad$ So $C_1 = 0$, $D_1 = 0$

$\qquad\qquad \mathbf{r} = [(\tfrac{2}{3}t^3 - t)\mathbf{i} + 4t^2\mathbf{j}]\,\text{m}$

When $t = 3$, $\mathbf{r} = [(\tfrac{2}{3}(3)^3 - (3))\mathbf{i} + 4(3)^2\mathbf{j}]$

$\qquad = (15\mathbf{i} + 36\mathbf{j})\,\text{m}$

Distance $= \sqrt{15^2 + 36^2} = 39\,\text{m}$

7 $\quad \mathbf{r} = [(\tfrac{1}{2}t^3 - \tfrac{5}{2}t^2 + 4t - 3)\mathbf{i} + (-\tfrac{1}{6}t^3 + \tfrac{9}{2}t^2 - 4t + 2)\mathbf{j}]\,\text{m}$

8 $\quad$ **a** $\quad p = 2$, $q = 3$

$\quad$ **b** $\quad \mathbf{v}_P = [(5 + 2t)\mathbf{i} + (3t^2 - 5)\mathbf{j}]\,\text{m s}^{-1}$

$\qquad \mathbf{r}_P = \int \mathbf{v}\,dt = [(5t + t^2 + C)\mathbf{i} + (t^3 - 5t + D)\mathbf{j}]$

$\qquad$ When $t = 0$, $C\mathbf{i} + D\mathbf{j} = 4\mathbf{j}$

$\qquad$ So $C = 0$, $D = 4$

$\qquad\qquad \mathbf{r}_P = [(5t + t^2)\mathbf{i} + (t^3 - 5t + 4)\mathbf{j}]\,\text{m}$

$\qquad \mathbf{r}_Q = [(8\mathbf{i} + 12\mathbf{j}) + t(3\mathbf{i} - 5\mathbf{j})] = [(8 + 3t)\mathbf{i} + (12 - 5t)\mathbf{j}]\,\text{m}$

$\qquad$ P and Q will collide if their position vectors are the same at the same time.

$\qquad$ Equating the coefficients of $\mathbf{i}$:

$\qquad\qquad 5t + t^2 = 8 + 3t$

$\qquad\qquad t^2 + 2t - 8 = 0$

$\qquad\qquad (t + 4)(t - 2) = 0$

$\qquad\qquad\qquad t = -4 \text{ or } 2$

$\qquad$ Since t cannot be negative, $t = 2\,\text{s}$.

$\qquad$ Equating the coefficients of $\mathbf{j}$:

$\qquad\qquad t^3 - 5t + 4 = 12 - 5t$

$\qquad\qquad\qquad t^3 = 8$

$\qquad\qquad\qquad t = 2$

$\qquad$ Hence the particles collide when $t = 2$.

$\quad$ **c** $\quad (14\mathbf{i} + 2\mathbf{j})\,\text{m}$

1 $\quad$ **a** $\quad 54.2\,\text{m}$

$\quad$ **b** $\quad 27.8\,\text{m}$

$\quad$ **c** $\quad 18.9\,\text{m s}^{-1}$

2 $\quad 19.3\,\text{m}$

$\quad 69.0\,\text{m}$

3 $\quad$ **a** $\quad$ Vertically:

$\qquad u = 24.5\sin\alpha\,\text{m s}^{-1}$

$\qquad a = -9.8\,\text{m s}^{-2}$

$\qquad s = 0\,\text{m}$

$\qquad t = 4\,\text{s}$

$\qquad$ Substitute into $s = ut + \dfrac{1}{2}at^2$.

$\qquad\qquad 0 = 24.5 \times 4\sin\alpha + \dfrac{1}{2} \times -9.8 \times 4^2$

$\qquad\qquad 78.4 = 98\sin\alpha$

$\qquad\qquad \sin\alpha = \dfrac{78.4}{98} = \dfrac{4}{5}$

$\quad$ **b** $\quad 58.8\,\text{m}$

4 a 5 s

b 127 m

c 42.7 m s^{-1}

d 53.4° below the horizontal

5 Vertically:

$u = -20 \sin 10° \, \text{m s}^{-1}$

$a = -9.8 \, \text{m s}^{-2}$

$s = -50 \, \text{m}$

Substitute into $s = ut + \dfrac{1}{2}at^2$.

$-50 = (-20 \sin 10°)t + \dfrac{1}{2} \times -9.8 \times t^2$

$4.9t^2 + (20 \sin 10°)t - 50 = 0$

$t = \dfrac{-20 \sin 10 \pm \sqrt{(20 \sin 10)^2 - 4 \times 4.9 \times -50}}{2 \times 4.9}$

$t = -3.57 \, \text{s or } 2.86 \, \text{s}$

Since t cannot be negative, $t = 2.86 \, \text{s}$.

Horizontally:

$u = 20 \cos 10° \, \text{m s}^{-1}$

$a = 0 \, \text{m s}^{-2}$

$s = x \, \text{m}$

$t = 2.86 \, \text{s}$

Substitute into $s = ut + \dfrac{1}{2}at^2$.

$x = 20 \times 2.86 \cos 10° + 0 = 56.3 \, \text{m}$

6 a 44.1 m.

b 19.6 m s^{-1}

7 a 13 m s^{-1}

b 5.68 m s^{-1}

c Direction $= \tan^{-1}\left(\dfrac{27}{5}\right) = \arctan 0.54$

d 0.714 s

1.73 s

8 a $\dfrac{8}{15}$

b 51 m s^{-1}

Exercise 15.5B   **page 316**

1 a Vertically:

$u = U \sin 30° = \dfrac{1}{2}U \, \text{m s}^{-1}$

$a = -g \, \text{m s}^{-2}$

$s = 0 \, \text{m}$

Substitute into $s = ut + \dfrac{1}{2}at^2$.

$0 = \dfrac{1}{2}Ut + \dfrac{1}{2} \times -g \times t^2$

$0 = \dfrac{1}{2}t(U - gt)$

So $t = 0$ (when the golf ball is initially on the ground)

or $U - gt = 0$

$U = gt$

$t = \dfrac{U}{g}$

b Horizontally:

$u = U \cos 30° = \dfrac{\sqrt{3}}{2}U \, \text{m s}^{-1}$

$a = 0 \, \text{m s}^{-2}$

$s = x \, \text{m}$

Substitute into $s = ut + \dfrac{1}{2}at^2$.

$x = \dfrac{\sqrt{3}}{2}U \times \dfrac{U}{g} + 0$

$= \dfrac{U^2\sqrt{3}}{2g} \, \text{m}$

c Vertically:

$u = \dfrac{1}{2}U \, \text{m s}^{-1}$

$a = -g \, \text{m s}^{-2}$

$v = 0 \, \text{m s}^{-1}$

Substitute into $v^2 = u^2 + 2as$

$s = \dfrac{v^2 - u^2}{2a}$

$= \dfrac{0^2 - \left(\dfrac{1}{2}U\right)^2}{2 \times -g}$

$= \dfrac{U^2}{8g} \, \text{m}$

2 For Anna, $\theta = 135°$ would be launching the particle backwards. Her range is correct but negative. She should have found the smallest value of 2θ for which $\cos 2\theta = 0$, which would have given an acute value of θ. If $2\theta = 90°$, then $\theta = 45°$ and the maximum range is $\dfrac{U^2}{g} \, \text{m}$.

For Johann, although the graph of $\sin \theta$ has a maximum value for 90°, the graph of $\sin 2\theta$ has a maximum value for 45°, from which the maximum range is $\dfrac{U^2}{g} \, \text{m}$.

3 a Horizontally:

$u = 40 \cos \theta \, \text{m s}^{-1}$

$s = 60 \, \text{m}$

$t = T \, \text{s}$

$a = 0 \, \text{m s}^{-2}$

Substitute into $s = ut + \dfrac{1}{2}at^2$.

$60 = 40T \cos \theta + 0$

$\dfrac{3}{2} = T \cos \theta$

$T = \dfrac{3}{2} \sec \theta$

b Vertically:

$u = 40\sin\theta\,\text{m s}^{-1}$

$s = 0\,\text{m}$

$t = T\,\text{s}$

$a = -g\,\text{m s}^{-2}$

Substitute into $s = ut + \frac{1}{2}at^2$.

$0 = 40T\sin\theta + \frac{1}{2}\times -g\times T^2$

$0 = 40T\sin\theta - \frac{1}{2}gT^2$

$0 = T(40\sin\theta - \frac{1}{2}gT)$

So $T = 0$ (when the cannonball is fired initially)

or $40\sin\theta = \frac{1}{2}gT$

$80\sin\theta = gT$

c $\quad 80\sin\theta = g\times\frac{3}{2}\sec\theta$

$80\sin\theta\cos\theta = \frac{3}{2}g$

$160\sin\theta\cos\theta = 3g$

$80(2\sin\theta\cos\theta) = 3g$

$80\sin 2\theta = 3g$

$\sin 2\theta = \dfrac{3g}{80}$

4 Horizontally:

Let the angle $= \theta$

$u = V\cos\theta\,\text{m s}^{-1}$

$a = 0\,\text{m s}^{-2}$

$s = x\,\text{m}$

Substitute into $s = ut + \frac{1}{2}at^2$.

$x = Vt\cos\theta + 0$

$t = \dfrac{x}{V\cos\theta}$

Vertically:

$u = V\sin\theta\,\text{m s}^{-1}$

$a = -g\,\text{m s}^{-2}$

$s = y\,\text{m}$

$t = \dfrac{x}{V\cos\theta}\,\text{s}$

Substitute into $s = ut + \frac{1}{2}at^2$.

$y = V\sin\theta\times\dfrac{x}{V\cos\theta} + \dfrac{1}{2}\times -g\times\left(\dfrac{x}{V\cos\theta}\right)^2$

$= x\tan\theta - \dfrac{gx^2}{2V^2\cos^2\theta}$

$= x\tan\theta - \dfrac{gx^2\sec^2\theta}{2V^2}$

$= x\tan\theta - \dfrac{gx^2\left(1+\tan^2\theta\right)}{2V^2}$

Given that $\tan\theta = 2$,

$y = 2x - \dfrac{gx^2\left(1+2^2\right)}{2V^2}$

$= 2x - \dfrac{5gx^2}{2V^2}$

$= \dfrac{2x\times 2V^2 - 5gx^2}{2V^2}$

$= \dfrac{4xV^2 - 5gx^2}{2V^2}$

$= \dfrac{x\left(4V^2 - 5gx\right)}{2V^2}$

5 a Vertically:

$u = V\sin\theta = \dfrac{3}{5}V\,\text{m s}^{-1}$

$a = -g\,\text{m s}^{-2}$

$s = 0\,\text{m}$

$t = T\,\text{s}$

Substitute into $s = ut + \frac{1}{2}at^2$.

$0 = \dfrac{3}{5}VT - \dfrac{1}{2}gT^2$

$0 = T(\dfrac{3}{5}V - \dfrac{1}{2}gT)$

So $T = 0$ (initially) or $\dfrac{3}{5}V = \dfrac{1}{2}gT$

$$T = \dfrac{6V}{5g}\,\text{s}$$

If $\sin\theta = \dfrac{3}{5}$, then $\cos\theta = \dfrac{4}{5}$

Horizontally:

$u = V\cos\theta = \dfrac{4}{5}V\,\text{m s}^{-1}$

$a = 0\,\text{m s}^{-2}$

$s = x\,\text{m}$

$t = \dfrac{6V}{5g}\,\text{s}$

Substitute into $s = ut + \frac{1}{2}at^2$.

$x = \dfrac{4}{5}V\times\dfrac{6V}{5g} + 0$

$= \dfrac{24V^2}{25g}\,\text{m}$

b Vertically:

$u = V\sin\alpha\,\text{m s}^{-1}$

$a = -g\,\text{m s}^{-2}$

$s = 0\,\text{m}$

$t = T\,\text{s}$

Substitute into $s = ut + \frac{1}{2}at^2$.

$0 = VT\sin\alpha - \frac{1}{2}gT^2$

$0 = T(V\sin\alpha - \frac{1}{2}gT)$

So $T = 0$ (initially) or $V\sin\alpha = \frac{1}{2}gT$

$$T = \frac{2V\sin\alpha}{g}\text{ s}$$

Horizontally:

$u = V\cos\alpha\text{ m s}^{-1}$

$a = 0\text{ m s}^{-2}$

$s = x\text{ m}$

$t = \frac{2V\sin\alpha}{g}\text{ s}$

Substitute into $s = ut + \frac{1}{2}at^2$.

$x = V\cos\alpha \times \dfrac{2V\sin\alpha}{g} + 0$

$= \dfrac{2V^2\sin\alpha\cos\alpha}{g} = \dfrac{V^2\sin 2\alpha}{g}\text{ m}$

If the particle is projected at $(90° - \alpha)$ instead,

$x = \dfrac{V^2\sin 2(90-\alpha)}{g} = \dfrac{V^2\sin(180-2\alpha)}{g}$

Apply the compound angle formula for $\sin(A - B)$.

$x = \dfrac{V^2(\sin 180°\cos 2\alpha - \sin 2\alpha\cos 180°)}{g}$

$= \dfrac{V^2(0\cos 2\alpha - \sin 2\alpha \times -1)}{g}$

$= \dfrac{V^2\sin 2\alpha}{g}$, which is the same range as for α.

6 Start from the formula $y = x\tan\theta - \dfrac{gx^2(1+\tan^2\theta)}{2U^2}$
(from **question 4**).

For particle A, $\theta = 0$, $U = 14\text{ m s}^{-1}$

$y = x\tan 0 - \dfrac{gx^2(1+\tan^2 0)}{2 \times 14^2} = -\dfrac{9.8x^2}{2 \times 14^2} = -\dfrac{x^2}{40}$

For particle B, $\theta = 45°$, $U = 28\text{ m s}^{-1}$

$y = x\tan 45° - \dfrac{gx^2(1+\tan^2 45°)}{2 \times 28^2} = x - \dfrac{9.8 \times 2x^2}{2 \times 28^2} = x - \dfrac{x^2}{80}$

Since A is 60 m vertically above B,

$x - \dfrac{x^2}{80} = -\dfrac{x^2}{40} + 60$

$\dfrac{x^2}{80} + x - 60 = 0$

$x^2 + 80x - 4800 = 0$

$(x + 120)(x - 40) = 0$

$x = -120$ or 40

Since the displacement is positive, $x = 40\text{ m}$.

1 **a** $(-3\mathbf{i} + 5\mathbf{j})\text{ m s}^{-2}$

   b 50 m s^{-1}

2 **a** $54.0°$ below horizontal

   b 28.7 m s^{-1}

   c 25.6 m

3 54 kmh^{-1}

4 2:15 p.m.

5 **a** $\begin{bmatrix} 1.5 \\ 6.5 \end{bmatrix}\text{ m s}^{-1}$

   b 1.58 N

6 **a** $(4.2\mathbf{i} + 8.75\mathbf{j})\text{ m}$

   b 0.361 m s^{-2}

   c 0.875 m s^{-2}

7 **a** $\mathbf{r}_W = [(-6\mathbf{i} + 63\mathbf{j}) + t(8\mathbf{i} - 5\mathbf{j})]$
      $= [(-6 + 8t)\mathbf{i} + (63 - 5t)\mathbf{j}]\text{ km}$

   $\mathbf{r}_B = [(60\mathbf{i} - 27\mathbf{j}) + t(-3\mathbf{i} + 10\mathbf{j})]$
      $= [(60 - 3t)\mathbf{i} + (-27 + 10t)\mathbf{j}]\text{ km}$

   $\mathbf{r}_W - \mathbf{r}_B = [(-66 + 11t)\mathbf{i} + (90 - 15t)\mathbf{j}]\text{ km}$

   Putting the coefficient of $\mathbf{i}$ equal to zero:

   $-66 + 11t = 0$

   $11t = 66$

   $t = 6$

   6 hours after 3:30 a.m. is 9:30 a.m.

   Note that putting the coefficient of $\mathbf{j}$ equal to zero gives the same answer:

   $90 - 15t = 0$

   $15t = 90$

   $t = 6$

   When $t = 6$, $\mathbf{r}_W = [(-6 + 8 \times 6)\mathbf{i} + (63 - 5 \times 6)\mathbf{j}]$
      $= (42\mathbf{i} + 33\mathbf{j})\text{ km}$

   b HMS *Barron* alters course at 7:30 a.m., when $t = 4$.

   $\mathbf{r}_B = [(60 - 3 \times 4)\mathbf{i} + (-27 + 10 \times 4)\mathbf{j}] = (48\mathbf{i} + 13\mathbf{j})\text{ km}$

   Position of HMS *Barron* at 9:30 a.m.
   is $(48\mathbf{i} + 13\mathbf{j}) + 2(-2\mathbf{i} + 10\mathbf{j}) = (44\mathbf{i} + 33\mathbf{j})\text{ km}$

   c 2 km

8 **a** 20.4 m

   b Vertically:

   $v = 0\text{ m s}^{-1}$, $a = -9.8\text{ m s}^{-2}$, $t = \dfrac{100}{49}\text{ s}$

   Substitute into $v = u + at$.

   $u = v - at$

   $u = 0 - (-9.8) \times \dfrac{100}{49} = 20\text{ m s}^{-1}$

Let the angle be θ, so $\tan\theta = \dfrac{3}{4}$.

So $\sin\theta = \dfrac{3}{5}$ and $\cos\theta = \dfrac{4}{5}$.

The initial velocity is $20 \div \dfrac{3}{5} = \dfrac{100}{3}\,\text{ms}^{-1}$

Horizontally:

$s = x\,\text{m}$

$u = \dfrac{100}{3}\cos\theta = \dfrac{100}{3} \times \dfrac{4}{5} = \dfrac{80}{3}\,\text{ms}^{-1}$

$t = \dfrac{200}{49}\,\text{s}$

$a = 0\,\text{ms}^{-2}$

Substitute into $s = ut + \dfrac{1}{2}at^2$.

$x = \dfrac{80}{3} \times \dfrac{200}{49} + 0 = 108.8\,\text{m}$

The range of the particle is just under $109\,\text{m}$.

9 a $53.1°$

b $25\,\text{ms}^{-1}$

10 a 5:29 p.m.

b $d^2 = (-18 + 3t)^2 + (29 - 6t)^2$

$= 324 - 108t + 9t^2 + 841 - 348t + 36t^2$

$= 45t^2 - 456t + 1165$

$\dfrac{\text{d}}{\text{dt}}(d^2) = 90t - 456 = 0$

$t = \dfrac{456}{90}\,\text{h} = \dfrac{76}{15}\,\text{h} = 5\,\text{h}\,4\,\text{min}$

$5\,\text{h}\,4\,\text{min}$ after 1:49 p.m. is 6:53 p.m.

c $3.13\,\text{km}$

11 a $4\,\text{s}$

b $\mathbf{r} = [4t^{\frac{5}{2}}\mathbf{i} + (\frac{4}{3}t^{\frac{3}{2}} + 12t)\mathbf{j}]\,\text{m}$

12 a Let the time taken to hit the basket be $T\,\text{s}$.

Horizontally:

$u = U\cos\theta, a = 0\,\text{ms}^{-2}, s = 20\,\text{m}, t = T\,\text{s}$

Substitute into $s = ut + \dfrac{1}{2}at^2$.

$20 = UT\cos\theta$

$T = \dfrac{20}{U\cos\theta}$

Vertically:

$u = U\sin\theta, a = -g\,\text{ms}^{-2}, s = 4\,\text{m}, t = T\,\text{s}$

Substitute into $s = ut + \dfrac{1}{2}at^2$.

$4 = U\sin\theta \times \dfrac{20}{U\cos\theta} - \dfrac{1}{2}g \times \left(\dfrac{20}{U\cos\theta}\right)^2$

$= 20\tan\theta - \dfrac{1}{2}g \times \dfrac{400}{U^2\cos^2\theta}$

$= 20\tan\theta - \dfrac{200g\sec^2\theta}{U^2}$

$= 20\tan\theta - \dfrac{200g(1 + \tan^2\theta)}{U^2}$

Given that $U^2 = 25g$,

$4 = 20\tan\theta - \dfrac{200g(1 + \tan^2\theta)}{25g}$

$= 20\tan\theta - 8(1 + \tan^2\theta)$

$= 20\tan\theta - 8 - 8\tan^2\theta$

$8\tan^2\theta - 20\tan\theta + 12 = 0$

$2\tan^2\theta - 5\tan\theta + 3 = 0$

b $56.3°$ or $45°$

16 Forces

Prior knowledge page 321

1 a $28.1\,\text{cm}$

  b $23.1°$

2 $25.5\,\text{m}$

3 a $44.1\,\text{N}$

  b $2\,\text{ms}^{-2}$

4 The acceleration is $2.45\,\text{ms}^{-2}$.

The tension is $36.8\,\text{N}$

Exercise 16.1A page 327

1 a

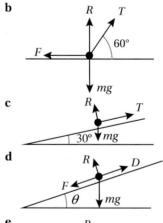

b

c

d

e

f

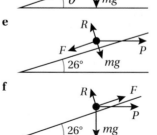

g

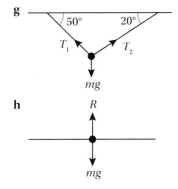

h

R

mg

2 a $2T\sin\theta - mg = 0$

b $T = \dfrac{mg}{2\sin\theta}$

3 a $T\cos 19° - F = 0$

b $T\sin 19° + R - W = 0$

4 a $T_B\cos 65° - T_A\cos 25° = 0$

b $T_A\sin 25° + T_B\sin 65° - mg = 0$

5 a R($\nearrow$)

$F - mg\sin\theta = 0$

R($\nwarrow$)

$R - mg\cos\theta = 0$

b R($\nearrow$)

$X\cos\theta - 9g\sin\theta - F = 0$

R($\nwarrow$)

$R - X\sin\theta - 9g\cos\theta = 0$

c R($\nearrow$)

$T\cos 15° - 4g\sin 27° - F = 0$

R($\nwarrow$)

$R + T\sin 15° - 4g\cos 27° = 0$

6 a Mary has not included X.

Correct answer: $R = 3g\cos 24° + X\sin 24°$

b Mary has written $X\cos 24°$ rather than X.

Correct answer: $F + X = 3g\sin 24°$

c Mary has written $3g\cos 24°$ rather than $3g\sin 24°$.

Correct answer: $X\cos 24° = F + 3g\sin 24°$

d Mary has written $X\cos 24°$ rather than $X\sin 24°$.

Correct answer: $R = X\sin 24° + 3g\cos 24°$

e Mary has written $X\cos 64°$ rather than $X\cos 40°$.

Correct answer: $X\cos 40° = F + 3g\sin 24°$

Exercise 16.1B   **page 331**

1 a $43.3\,$N

b $23.0\,$N

2 $2.24\,$kg

3 R($\nearrow$)

$2400 - 800g\sin 15° - F = 0$

$\qquad\qquad F = 2400 - 800g\sin 15° = 371\,$N

4 Both solutions are correct. Victrix's solution requires resolving twice whereas Colin's solution requires resolving once. Victrix's solution is possibly simpler since it just requires horizontal and vertical resolving. By resolving perpendicular to the tension, Colin's solution does not require the tension to be found.

5 a All three forces are in the horizontal plane, whereas weight acts vertically and perpendicular to all the three forces.

b $10.9\,$N

6 $28.9\,$N

7 friction is acting up the slope with a magnitude of $6.34\,$N

8 $8.43\,$N and $12.0\,$N

Exercise 16.2A   **page 334**

1 a $17\,$N

b $28.1°$

2 a $44.4\,$N

b $239°$

3 a i: $(9 + 7\cos 50°)\,$N

j: $(7\sin 50°)\,$N

Magnitude $= \sqrt{(9 + 7\cos 50°)^2 + (7\sin 50°)^2}$

$= 14.5\,$N

b Angle $= \tan^{-1}\left(\dfrac{7\sin 50°}{9 + 7\cos 50°}\right) = 21.7°$

4 Magnitude $= 10.8\,$N

Angle $= 37.4°$ to O_x

5 a Hita has drawn out the diagram with the forces tip to tail.

She has applied the cosine rule to the $3\,$N and $5\,$N forces to obtain the resultant of these two as $7\,$N.

She has then applied the sine rule to obtain the angle opposite the $5\,$N force as $38.2°$.

Using alternate angles, she calculates the angle opposite the overall resultant force as $(90 + 38.2) = 128.2°$.

She then applies the cosine rule again to the $4\,$N and $7\,$N forces to obtain the overall resultant force of $9.98\,$N.

She then applies the sine rule again to obtain the angle opposite the 4 N force as 18.4°.

She then adds 38.2° and 18.4° to obtain the angle 56.6°. The direction of the resultant force is 56.6° to the horizontal.

b **i**: $(3 + 5\cos 60°)$ N

j: $(4 + 5\sin 60°)$ N

Magnitude $= \sqrt{(3 + 5\cos 60°)^2 + (4 + 5\sin 60°)^2}$

$= 9.98$ N

Angle $= \tan^{-1}\left(\dfrac{4 + 5\sin 60°}{3 + 5\cos 60°}\right) = 56.6°$

6 $\sin\theta = 0.28 = \dfrac{7}{25}$

so $\cos\theta = \dfrac{\sqrt{25^2 - 7^2}}{25} = \dfrac{24}{25}$

i: $(7\sin\theta) = (7 \times \dfrac{7}{25}) = 1.96\,\text{m s}^{-1}$

j: $(15 + 7\cos\theta) = (15 + 7 \times \dfrac{24}{25}) = 21.72\,\text{m s}^{-1}$

Speed $= \sqrt{1.96^2 + 21.72^2} = 21.8\,\text{m s}^{-1}$

7 **a** 13.1 N

b $21.6°$ to O_x

8 **a** **i**: $(3 + 3\cos 40° - 5\cos 40° - 5\cos 50°) = -1.746$ N

j: $(3\sin 40° + 5\sin 50° - 5\sin 40°)$
$\qquad\qquad = (5\sin 50° - 2\sin 40°) = 2.545$ N

Angle $= \tan^{-1}\left(\dfrac{2.545}{1.746}\right) = 55.5°$

Hence the resultant force acts at approximately $124°$ to O_x.

b Magnitude

$= \sqrt{(-1.746)^2 + 2.545^2}$

$= 3.09$ N

Exercise 16.3A   page 337

1 0.532

2 1520 N

3 $R(\rightarrow)$

$T\cos 20° - F = 0$

$\qquad F = T\cos 20°$

$R(\uparrow)$

$R + T\sin 20° - 5g = 0$

$\qquad R = 5g - T\sin 20°$

Substitute into $F = \mu R$.

$T\cos 20° = \dfrac{1}{4}(5g - T\sin 20°)$

$4T\cos 20° = 5g - T\sin 20°$

$4T\cos 20° + T\sin 20° = 5g$

$T(4\cos 20° + \sin 20°) = 5g$

$T = \dfrac{5g}{4\cos 20° + \sin 20°} = 11.9$ N

4 $16.7°$

5 8.69 N

6 149 kg

7 **a** $R(\nearrow)$

$4g - F - 17g\sin\alpha = 0$

$\qquad\qquad F = 4g - 17g\sin\alpha$

$R(\nwarrow)$

$R - 17g\cos\alpha = 0$

$\qquad R = 17g\cos\alpha$

Substitute into $F = \mu R$.

$4g - 17g\sin\alpha = \mu(17g\cos\alpha)$

From $F = \mu R$, $\mu = \dfrac{F}{R}$, so

$\mu = \dfrac{4g - 17g\sin\alpha}{17g\cos\alpha} = \dfrac{4 - 17\sin\alpha}{17\cos\alpha}$

b $\tan\alpha = \dfrac{13}{84}$

so $\sin\alpha = \dfrac{13}{\sqrt{13^2 + 84^2}} = \dfrac{13}{85}$

and $\cos\alpha = \dfrac{84}{85}$

$\mu = \dfrac{4 - 17 \times \dfrac{13}{85}}{17 \times \dfrac{84}{85}} = \dfrac{340 - 221}{17 \times 84} = \dfrac{17 \times 20 - 17 \times 13}{17 \times 84}$

$\qquad = \dfrac{17 \times 7}{17 \times 84} = \dfrac{1}{12}$

c 25.5 kg

8 71.1 N

Exercise 16.3B   page 341

1 $\dfrac{1}{2}$

2 **a** 3 s

b $4.2\,\text{m s}^{-1}$

3 **a** 0.206

b $1.01\,\text{m s}^{-2}$

4 **a** $2.35\,\text{m s}^{-2}$

b 1.45 m

5 **a** $2.25\,\text{N}$

   b R($\nwarrow$)

$$0.5g\sin 40° - F = 0.5a$$

$$0.5g\sin 40° - 2.25 = 0.5a$$

$$0.5a = 0.900$$

$$a = 1.80\,\text{m s}^{-2}$$

   Since a is positive, motion will occur.

6 **a** $13.5\,\text{N}$

   b $1.07\,\text{m s}^{-2}$

Exercise 16.4A   page 350

1 **a** $0.24\,\text{m s}^{-2}$

   b $5.04\,\text{m s}^{-2}$

   c Whilst coupled, $s = 75\,\text{m}$, $u = 0\,\text{m s}^{-1}$, $a = 0.24\,\text{m s}^{-2}$

   Substitute into $v^2 = u^2 + 2as$.

$$v^2 = 0 + 2 \times 0.24 \times 75$$

$$= 36$$

$$v = 6\,\text{m s}^{-1}$$

   R($\nearrow$) for the horsebox whilst uncoupled:

$$-1776 = 600a$$

$$a = -2.96\,\text{m s}^{-2}$$

$$v = 0\,\text{m s}^{-1},\ u = 6\,\text{m s}^{-1},\ a = -2.96\,\text{m s}^{-2}$$

   Substitute into $v^2 = u^2 + 2as$.

$$s = \frac{v^2 - u^2}{2a}$$

$$= \frac{0 - 6^2}{2 \times -2.96} = 6.08\,\text{m}$$

   Hence the horsebox travels just over $6\,\text{m}$ after the uncoupling.

2 **a** The lighter particle does not collide with the heavier particle

   b $17\,\text{cm}$ apart.

3 **a** $0.75\,\text{m s}^{-2}$

   b $T\cos 12° - 400 = 500 \times 0.75$

$$T\cos 12° = 775$$

$$T = \frac{775}{\cos 12°} = 792\,\text{N}$$

   Hence the tension in the tow-bar is $790\,\text{N}$, correct to 2 significant figures.

4 **a** $0.6g\,\text{N}$

   b $4.9\,\text{m s}^{-2}$

   c $15.7\,\text{N}$

   d $8\,\text{m}$

5 **a** $0.148\,\text{m s}^{-2}$

   b $990\,\text{N}$

   c Whilst coupled, $t = 8\,\text{s}$, $a = 0.148\,\text{m s}^{-2}$, $u = 0\,\text{m s}^{-1}$

   Substitute into $v = u + at$.

$$v = 0 + 0.148 \times 8 = 1.19\,\text{m s}^{-1}$$

   R($\nearrow$) for the combine harvester whilst uncoupled:

$$-250 - 400g\sin 10° = 400a$$

$$a = \frac{-250 - 400g\sin 10°}{400}$$

$$= -2.33\,\text{m s}^{-2}$$

$$v = 0\,\text{m s}^{-1},\ a = -2.33\,\text{m s}^{-2},\ u = 1.19\,\text{m s}^{-1}$$

   Substitute into $v = u + at$:

$$t = \frac{v - u}{a}$$

$$= \frac{0 - 1.19}{-2.33} = 0.509\,\text{s}$$

   Hence the combine harvester continues to move for just over half a second before it comes to instantaneous rest.

6 **a** $4.08\,\text{m s}^{-2}$

   b $4.57\,\text{N}$

   c $18.4\,\text{m}$

   d $30.6\,\text{m}$

7 **a** R($\nwarrow$) for A:

$$R - 1.8g\cos 30° = 0$$

$$R = 1.8g\cos 30°$$

   Substitute into $F = \mu R$.

$$F = \frac{\sqrt{3}}{9} \times 1.8g\cos 30° = 0.3g$$

   R($\nearrow$) for A:

$$T - F - 1.8g\sin 30° = 1.8a$$

$$T - 0.3g - 0.9g = 1.8a$$

$$T - 1.2g = 1.8a \qquad ①$$

   R($\downarrow$) for B:

$$5.2g - T = 5.2a \qquad ②$$

   Add ① and ②.

$$4g = 7a$$

$$a = \frac{4}{7}g\,\text{m s}^{-2}$$

   b $21.8\,\text{N}$

   c $3.71\,\text{s}$

1 $\frac{5}{12}$

2 **a** Assume the rope is light and inextensible, and the peg bag is a particle.

b 8.62 N and 8.59 N.

3 $G = 4\,\text{N}$

4 **a** 20.8 N

b 53.3°

5 **a** $0.8\,\text{m}\,\text{s}^{-2}$

b The rope is inextensible.

c 1888 N

d $12\,\text{m}\,\text{s}^{-1}$

e 15 s

6 3.03 kg

7 **a** $0.75\,\text{m}\,\text{s}^{-2}$

b 1012 N

c $1.38\,\text{m}\,\text{s}^{-2}$

d $-0.5\,\text{m}\,\text{s}^{-2}$

e 36 m

8 $0.36\,\text{g}\,\text{m}\,\text{s}^{-1}$

9 $R(\nearrow)$

$$T\cos\theta + F - Mg\sin\theta = 0$$
$$F = Mg\sin\theta - T\cos\theta$$

$R(\nwarrow)$

$$R + T\sin\theta - Mg\cos\theta = 0$$
$$R = Mg\cos\theta - T\sin\theta$$

Substitute into $F = \mu R$.

$$Mg\sin\theta - T\cos\theta = \mu(Mg\cos\theta - T\sin\theta)$$
$$= \mu Mg\cos\theta - \mu T\sin\theta$$
$$\mu T\sin\theta - T\cos\theta = \mu Mg\cos\theta - Mg\sin\theta$$
$$T(\mu\sin\theta - \cos\theta) = \mu Mg\cos\theta - Mg\sin\theta$$
$$T = \frac{\mu Mg\cos\theta - Mg\sin\theta}{\mu\sin\theta - \cos\theta}$$
$$= \frac{Mg(\mu\cos\theta - \sin\theta)}{\mu\sin\theta - \cos\theta}$$

10 **a** C and D are modelled as particles so that all their mass is concentrated at a single point.

b $3.528\,\text{m}\,\text{s}^{-2}$ and 18.8 N

c 2.4 s

11 **a** $R(\uparrow)$ for P:

$$R - 2.1g = 0$$
$$R = 2.1g$$
$$F = \mu R = \mu \times 2.1g$$

$R(\rightarrow)$ for P:

$$T - F = 2.1a$$
$$T - \mu \times 2.1g = 2.1a \qquad \text{①}$$

$R(\downarrow)$ for Q:

$$1.5g - T = 1.5a \qquad \text{②}$$

Add ① and ②.

$$1.5g - \mu \times 2.1g = 3.6a$$
$$5g - 7g\mu = 12a$$
$$g(5 - 7\mu) = 12a$$
$$a = \frac{(5 - 7\mu)g}{12}$$

b Since $a > 0$, $\frac{(5 - 7\mu)g}{12} > 0$

$$5 - 7\mu > 0$$
$$5 > 7\mu$$
$$7\mu < 5$$
$$\mu < \frac{5}{7}$$

Whilst taut, $u = 0\,\text{m}\,\text{s}^{-1}$, $a = \frac{(5 - 7\mu)g}{12}\,\text{m}\,\text{s}^{-2}$, $s = 1\,\text{m}$

Substitute into $v^2 = u^2 + 2as$.

$$v^2 = 0 + 2 \times \frac{(5 - 7\mu)g}{12} \times 1$$
$$= \frac{(5 - 7\mu)g}{6}$$

$R(\rightarrow)$ for P whilst slack:

$$-\mu \times 2.1g = 2.1a$$
$$a = -\mu g$$

$u^2 = \frac{(5 - 7\mu)g}{6}$, $v = 0\,\text{m}\,\text{s}^{-1}$, $a = -\mu g\,\text{m}\,\text{s}^{-2}$, $s < 1.5\,\text{m}$

Substitute into $v^2 = u^2 + 2as$.

$$s = \frac{v^2 - u^2}{2a} < 1.5$$
$$\frac{0 - \frac{(5 - 7\mu)g}{6}}{2 \times -\mu g} < 1.5$$
$$-\frac{(5 - 7\mu)g}{6} > -3\mu g$$
$$-g(5 - 7\mu) > -18\mu g$$
$$-5g + 7g\mu > -18\mu g$$
$$25g\mu > 5g$$
$$\mu > \frac{1}{5}$$

Hence $\frac{1}{5} < \mu < \frac{5}{7}$

12 a 75 N

b 21 N

13 Let the friction force be F N and the reaction force be R N.

$R(\nearrow)$

$5g \sin \alpha - X \cos \alpha - F = 0$

$R(\nwarrow)$

$R - X \sin \alpha - 5g \cos \alpha = 0$

On the point of sliding, $F = \mu R$

$5g \sin \alpha - X \cos \alpha = \mu(X \sin \alpha + 5g \cos \alpha)$

$\qquad\qquad\qquad = \mu X \sin \alpha + 5g\mu \cos \alpha$

$5g \sin \alpha - 5g\mu \cos \alpha = \mu X \sin \alpha + X \cos \alpha$

$\qquad 5g \tan \alpha - 5g\mu = \mu X \tan \alpha + X$

$\qquad X(1 + \mu \tan \alpha) = 5g(\tan \alpha - \mu)$

$$X = \frac{5g(\tan \alpha - \mu)}{1 + \mu \tan \alpha}$$

14 a $5.7°$

b 0.251

17 Moments

Prior knowledge page 355

1 Magnitude $= 8.60$ N

Direction $= 54.5°$ to the horizontal

2 1.225 N

3 Horizontal component $= 5.54$ N

Vertical component $= 7.09$ N

4 0.4

Exercise 17.1A page 359

1 a i 30 N m   **ii** 42 N m

   b i 7 N   **ii** 3.75 N

2 3 N m anticlockwise

3 a 27 N m clockwise

   b 35 N m clockwise

   c 45 N m clockwise

   d 57 N m clockwise

4 a 7 N m anticlockwise

   b 49 N m clockwise

   c 7 N m anticlockwise

   d 30.9 N m anticlockwise

5 13 N m anticlockwise

6 Both methods are valid. Both methods lead to an answer of the form $Fd \sin \theta$. Sophie's method is more efficient in this example.

7 a 14.0 N m anticlockwise

   b The turning moment will still be anticlockwise, but smaller (11.0 N).

Exercise 17.1B page 363

1 a 2

   b 8 N m anticlockwise

2 1.2 m

3 a 82 N m anticlockwise

   b i 41 N

   vertically downwards.

   ii 39 N

4 a $3 + (-1) + 1 = 0$

   Since net force $= 0$, forces are in equilibrium.

   b 9

   c $\dfrac{26}{9}$

5 The value of U is correct but the working is incorrect.

She has resolved incorrectly with cos and sin.

She has then also evaluated the compound angle formulae incorrectly.

   a Resolving vertically:

   $T \cos \alpha = U \cos \alpha$

   $\qquad T = U$

   $\qquad q = U$

   b Taking moments about O:

   $T \times 8 \sin \alpha = U \times 3 \cos \alpha$

   $q \times 8 \sin \alpha = q \times 3 \cos \alpha$

   $\qquad \tan \alpha = \dfrac{3}{8}$

   $\qquad \alpha = \tan^{-1}\left(\dfrac{3}{8}\right) = 20.6°$

   c Resolving horizontally:

   $V = T \sin \alpha + U \sin \alpha$

   $\quad = q \sin \alpha + q \sin \alpha$

   $\quad = 2q \sin \alpha$

   $\quad = 2q \times \dfrac{3}{\sqrt{73}} = \dfrac{6q\sqrt{73}}{73}$

page 370

1 a

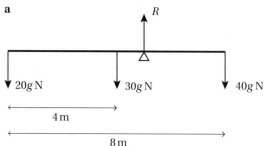

b

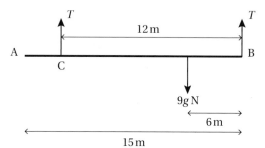

c

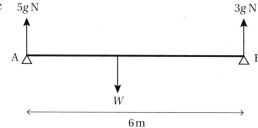

d

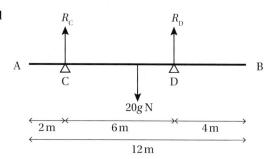

e

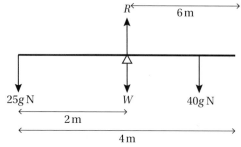

f

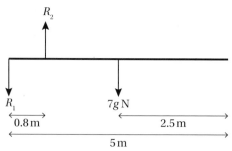

2 a The log is rigid and straight.

b $R_D = 588\,\text{N}$

$R_C = 784\,\text{N}$

3 a $2.1\,\text{m}$

b $510\,\text{N}$

4 $172\,\text{N}$

5 a Let JL be $x\,$m.

Moments about L:

$40g(12 - x) = 24gx + 16g(x - 6)$

$480 - 40x = 24x + 16x - 96$

$576 = 80x$

$x = 7.2\,\text{m}$

b Resolving vertically:

$R_L = 24g + 16g + 40g = 80g = 784\,\text{N}$

c The sack is a particle with its mass concentrated at a single point. The beam is a rigid rod with its weight acting at its centre point.

6 a $\dfrac{1}{15}$

b $147\,\text{N}$

7 a $588\,\text{N}$

b $4.8\,\text{m}$ from A.

c Resolving vertically:

$R_A + R_B = 64g + mg$

$R_A + 5R_A = 64g + mg$

$6R_A = 64g + mg$

$R_A = \dfrac{1}{6}(64g + mg)$

Moments about B:

$mg \times 1.2 + R_A \times 4.8 = 64g \times 1.8$

$mg \times 1.2 + \dfrac{1}{6}(64g + mg) \times 4.8 = 64g \times 1.8$

$7.2m + 307.2 + 4.8m = 691.2$

$12m = 384$

$m = 32$

d The boulders are particles with their masses concentrated at a single point so that the perpendicular distances are exact.

8 a 294 N

  b 6 m

1 $\frac{1}{3} d$ m

2 Let the distance of the centre of mass from A be x m.

$R_C = 0$

Moments about D:

$4g \times 0.7 = 10g(1.4 - x)$

$\qquad 2.8 = 14 - 10x$

$\qquad 10x = 11.2$

$\qquad\quad x = 1.12$

Distance of the centre of mass from A is 1.12 m.

3 a 0 N

  b 7.5 kg

4 a $R_Q = 78.4$ N

  $R_P = 118$ N

  b 60 kg

5 a 98 N

  b Let the mass of the added load be M kg.

  $R_U = 0$

  Moments about V:

  $(M + 5)g \times 3 = 10g \times 2 + 70g \times 6$

  $\qquad 3M + 15 = 20 + 420$

  $\qquad\quad 3M = 425$

  $\qquad\quad\; M = \frac{425}{3}$

  The mass of the load that has been added is $\frac{425}{3}$ kg, and a = 425.

6 $M = 54$ kg

  Centre of mass is 3 m from A

1 a Mg N

  b $0.4Mg$ N

  c $0.4Mg$ N

  d 38.7°

2 5.49 m

3 Let the reaction at the ground be R_G, the friction at the ground be F_G and the reaction at the wall be R_W.

Resolving horizontally:

$R_W = F_G$

Resolving vertically:

$36g + 45g = R_G$

$\qquad R_G = 81g$

Moments about G:

$45g \times 2 \sin 30° + 36g \times 3 \sin 30° = R_W \times 6 \cos 30°$

$\qquad\qquad\quad 198g \sin 30° = 6R_W \cos 30°$

$\qquad\qquad\quad 33g \tan 30° = R_W$

$\qquad\qquad\qquad\quad R_W = 11g\sqrt{3}$

$F_G = \mu R_G$

$\mu = \dfrac{F_G}{R_G} = \dfrac{R_W}{81g} = \dfrac{11g\sqrt{3}}{81g} = \dfrac{11\sqrt{3}}{81}$

4 Let the reaction at the ground be R_C, the friction at the ground be F_C, the reaction at the wall be R_D and the friction at the wall be F_D.

$F_D = \frac{1}{6} R_D$

Resolving horizontally:

$F_C = R_D = 6F_D$

Resolving vertically:

$F_D + R_C = 54g$

$\qquad F_C = 6(54g - R_C) = 324g - 6R_C$

Moments about D:

$54g \times a \cos \alpha + F_C \times 2a \sin \alpha = R_C \times 2a \cos \alpha$

$\qquad\qquad\quad 54g + 2F_C \tan \alpha = 2R_C$

$\qquad\qquad\quad 27g + F_C \tan \alpha = R_C$

Given that $\sin \alpha = \frac{4}{5}$, $\tan \alpha = \frac{4}{3}$.

$\qquad\qquad 27g + \frac{4}{3}F_C = R_C$

$27g + \frac{4}{3}(324g - 6R_C) = R_C$

$\qquad 27g + 432g - 8R_C = R_C$

$\qquad\qquad\qquad 459g = 9R_C$

$\qquad\qquad\qquad\quad R_C = 51g$ N

$\qquad\qquad\quad F_C = 324g - 6 \times 51g$

$\qquad\qquad\qquad\; = 324g - 306g = 18g$ N

$\qquad \mu_C = \dfrac{F_C}{R_C} = \dfrac{18g}{51g} = \dfrac{6}{17}$

5 2.5 m

6 $\frac{1}{3}$

page 381

1 **a** 79.9 N

   b 33.8 N

   c 45.2 N

   d 56.4 N

   53.2° to the horizontal

2 **a** 45.3 N

   b 45.3 N

   60° to the horizontal

3 **a** Let the friction at the wall be FN, the reaction at the wall be RN and the tension be TN.

   From the right-angled triangle, $\cos \alpha = \frac{12}{13}$, from which $\sin \alpha = \frac{5}{13}$.

   Moments about A:

   $Mg \times 4a = T \sin \alpha \times 12a$

   $T \sin \alpha = \frac{1}{3} Mg$

   $T \times \frac{5}{13} = \frac{1}{3} Mg$

   $T = \frac{13}{15} Mg$

   b Resolving vertically:

   $F + T \sin \alpha = Mg$

   $F + \frac{1}{3} Mg = Mg$

   $F = \frac{2}{3} Mg$

   Resolving horizontally:

   $R = T \cos \alpha$

   $= \frac{13}{15} Mg \times \frac{12}{13} = \frac{4}{5} Mg$

   $F \leqslant \mu R$

   $\mu \geqslant \frac{F}{R}$

   $\mu \geqslant \dfrac{\frac{2}{3} Mg}{\frac{4}{5} Mg}$

   $\mu \geqslant \frac{5}{6}$

4 Let the components of the reaction at A be XN and YN.

   Moments about A:

   $6g \times 5a = 2F \times 2a + F \times 10a$

   $14F = 30g$

   $F = 21$ N

Resolving vertically:

$Y + F = 6g$

$Y = 6g - 21 = 37.8$ N

Resolving horizontally:

$X = 2F = 42$ N

Direction $= \arctan^{-1}\left(\dfrac{37.8}{42} \right) = \arctan 0.9$ to the horizontal

5 **a** 33.6°

   b 0.829

6 Let the components of the reaction at A be XN and YN and the thrust be TN. Let AB be xm.

   a Moments about A:

   $0.6g \times \frac{1}{2} x = T \cos 75° \times \frac{3}{4} x$

   $T \cos 75° = 0.4g$

   $T = \dfrac{0.4g}{\cos 75°} = 15.1$ N

   b 14.8 N

7 **a** $T_B = 19.6$ N

   $T_A = 13.4$ N

   b $x = 2.30$

page 383

1 **a** 931 N

   b 4 m

2 **a** The broomstick can be modelled as a uniform rod and the broom head as a particle.

   b 0.175 m

3 **a** $3.25d$

   b Plank is unlikely to be rigid and straight.

   Parcel is unlikely to have mass acting at centre.

4 10.5 m

5 **a** Light and inextensible

   b Moments about D:

   $W \times 5 + 28 \times 11 = T_C \times 7$

   $T_C = \frac{5}{7} W + 44$

   c 84 N

6 **a** R_A needs to act vertically downwards rather than upwards. Romain has also misinterpreted 2300 grams as a weight of 2300g.

   b Moments about A:

   $R_B \times 0.5 = 2.3g \times 2.4$

   $R_B = 11.04g = 108$ N

Resolving vertically:

$$R_A + 2.3g = R_B$$

$$R_A = 11.04g - 2.3g = 8.74g$$

$$= 85.7\,N$$

7 Clockwise moments:

$$10 \times 3 = 30\,N\,m$$

Anticlockwise moments:

$$4 \times 1 + 3 \times 11 + 2 \times 6 = 49\,N\,m$$

Overall turning moment $= 49 - 30$

$$= 19\,N\,m\ anticlockwise$$

8 a 4500 N

  b 4.8 m

9 2.75 m

10 a 19.6 N

  b The beam is a rod and the cable is a light inextensible string.

  c Resolving vertically:

$$R_A + T\sin\beta = 2.5g$$

$$T\sin\beta = 2.5g - 2g = 0.5g$$

$$0.28T = 0.5g$$

$$T = 17.5\,N$$

11 The mass of the rod is 30 kg.

Centre of mass is 2 m from A.

12 a From Nevita's diagram, if the system is at rest in equilibrium under the action of three non-parallel forces, then all three forces must act through a single point. Hence the vertical distance below the centre of mass is given

by $x\tan 40°$ from the left-hand triangle and $(7 - x)\tan 60°$ from the right-hand triangle.

  b Resolving horizontally:

$$T_1\cos 40° = T_2\cos 60°$$

$$T_1 = \frac{T_2\cos 60°}{\cos 40°}$$

Resolving vertically:

$$T_1\sin 40° + T_2\sin 60° = 3g$$

$$\frac{T_2\cos 60°}{\cos 40°}\sin 40° + T_2\sin 60° = 3g$$

$$T_2(\cos 60°\tan 40° + \sin 60°) = 3g$$

$$T_2 = \frac{3g}{\cos 60°\tan 40° + \sin 60°} = 22.9\,N$$

  c Nevita's method is more efficient for finding the centre of mass because it does not require finding the tensions first.

13 a $R_G = \dfrac{Mg}{1 + \mu_W\mu_G}$

  b $R_G = \dfrac{Mg}{4 - 4\mu_G\tan\theta}$

  c Hence $\dfrac{Mg}{1 + \mu_W\mu_G} = \dfrac{Mg}{4 - 4\mu_G\tan\theta}$

$$1 + \mu_W\mu_G = 4 - 4\mu_G\tan\theta$$

$$\mu_W\mu_G = 3 - 4\mu_G\tan\theta$$

$$\mu_W = \frac{3 - 4\mu_G\tan\theta}{\mu_G}$$

14 $\dfrac{7}{2}a$ m.

1 Algebra and functions 1: Functions   page 387

1 a Given $f(x) = \dfrac{x}{x^2 - 4} - \dfrac{1}{x + 2}$ to rewrite as asked, factorise the denominator $(x^2 - 4)$ by using the difference of two squares:

$$\dfrac{x}{x^2 - 4} - \dfrac{1}{x + 2}$$

$$= \dfrac{x}{(x - 2)(x + 2)} - \dfrac{1}{(x + 2)}$$

Find the common denominator for the two fractions which is $(x - 2)(x + 2)$ so

$$\dfrac{1}{(x + 2)} = \dfrac{(x - 2)}{(x - 2)(x + 2)}$$

therefore $\dfrac{x}{(x - 2)(x + 2)} - \dfrac{1}{(x + 2)}$

$$= \dfrac{x}{(x - 2)(x + 2)} - \dfrac{(x - 2)}{(x - 2)(x + 2)}$$

$$= \dfrac{x - (x - 2)}{(x - 2)(x + 2)}$$

$$= \dfrac{2}{(x - 2)(x + 2)}$$

$$= \dfrac{2}{(x^2 - 4)}$$

b $\{y : y > 0\}$.

c $f^{-1} : x \mapsto \sqrt{4 + \dfrac{2}{x}}$

$\{x : x > 0\}$

d $gf(3) = -\ln 5$

2 a $f^{-1}(x) = \dfrac{3 + x}{x}$

b $g^{-1}(x) = \dfrac{4}{x}$

c $x = 0$ or 1

3 a $f^{-1}(x) = \dfrac{1 - 5x}{4x + 3}$

b Domain $\left\{ x \in R, \ x \neq -\dfrac{3}{4} \right\}$, range $\left\{ x \in R, \ f^{-1}(x) \neq -\dfrac{5}{4} \right\}$

c $x = -1$

4 a

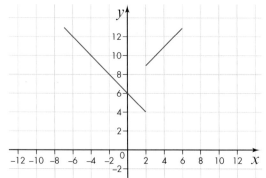

b $x = \dfrac{5}{2}$

5 a

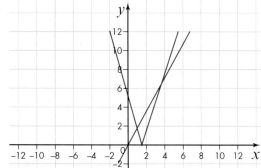

b $x > 5$
and
$x < 1$

c

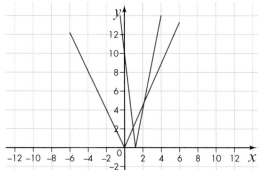

d $x > \dfrac{5}{2}$
or
$x < \dfrac{5}{4}$

6 a $f^{-1}(x) = \ln x$

b $fg(x) = e^{2x^2 - 3}$

c $x = -1.6$ or 1.6

7 a $f(x) = \dfrac{3x + 11}{(x - 2)(x + 2)}$ with domain $\{x \in R, x \neq 2\}$

b $f^{-1}(x) = \dfrac{3 \pm \sqrt{16x^2 + 44x + 9}}{2x}$

8 $x = 0.511$

9 a $fg(x) = 2 \sin x$

$gf(x) = \sin 2x$

b

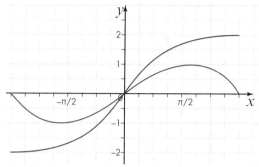

The transformation that maps $g(x)$ to $fg(x)$ is a vertical stretch of a scale factor of 2.

The transformation that maps $g(x)$ to $gf(x)$ is a horizontal stretch of a scale factor of $\frac{1}{2}$.

c $g^{-1}(x) = \arcsin x$ with domain $\{x \in R, -1 \leqslant x \leqslant 1\}$

d

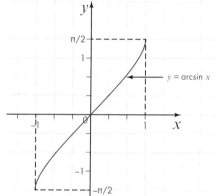

$y = \arcsin x$

10 a $f^{-1}(x) = \sqrt[3]{x - 17}$

b $x = -2.650$ or 2.650

2 Algebra and functions 2: Partial fractions   page 389

1 a $\dfrac{1}{x+2} + \dfrac{2}{x-1}$

b $-\dfrac{3}{2} - \dfrac{9}{4}x - \dfrac{15}{8}x^2$

valid for $\{x: -1 < x < 1\}$

c $\ln 28$

2 a $f(-2) -16 + 44 - 20 - 8 = 0$, so $(x+2)$ is a factor of $f(x)$.

$$
\begin{array}{r}
2x^2 + 7x - 4 \\
x+2\,\overline{\smash{\big)}\,2x^3 + 11x^2 + 10x - 8} \\
\underline{2x^3 + \ 4x^2} \\
7x^2 + 10x \\
\underline{7x^2 + 14x} \\
-4x - 8 \\
\underline{-4x - 8} \\
0
\end{array}
$$

$2x^2 + 7x - 4 = (x+4)(2x-1)$

$\dfrac{5}{2x^3 + 11x^2 + 10x - 8} = \dfrac{A}{x+2} + \dfrac{B}{x+4} + \dfrac{C}{2x-1}$

$5 = A(x+4)(2x-1) + B(x+2)(2x-1) + C(x+2)(x+4)$

$x = -4$

$B = \dfrac{5}{18}$

$x = -2$

$A = -\dfrac{1}{2}$

$x = \dfrac{1}{2}$

$C = \dfrac{4}{9}$

$\dfrac{5}{2x^3 + 11x^2 + 10x - 8}$

$= -\dfrac{1}{2(x+2)} + \dfrac{5}{18(x+4)} + \dfrac{4}{9(2x-1)}$

b $\dfrac{10}{49}$

3 a $\dfrac{7}{4(x+2)} + \dfrac{5}{4(x-2)}$

b there are no real solutions.

4 a $\dfrac{14}{(3x-2)} - \dfrac{14}{(3x+2)}$

b $\dfrac{14}{3} \ln\left(\dfrac{14}{11}\right)$

5 a $\dfrac{3}{x-3} - \dfrac{8}{x-7}$

b Hence means that you can use the result from part (a), so

$\dfrac{3}{x-3} = 3\left[-\dfrac{1}{3}\left(1 - \dfrac{x}{3}\right)^{-1}\right]$

$= -1\left(1 + (-1)\left(-\dfrac{x}{3}\right) + \dfrac{(-1)(-2)}{2}\left(-\dfrac{x}{3}\right)^2 + \cdots\right)$

$= -1\left(1 + \dfrac{x}{3} + \dfrac{x^2}{9} + \cdots\right)$

$= -1 - \dfrac{x}{3} - \dfrac{x^2}{9} + \cdots$

Similarly,

$$\frac{8}{x-7} = 8\left[-\frac{1}{7}\left(1-\frac{x}{7}\right)^{-1}\right]$$

$$= -\frac{8}{7}\left(1+(-1)\left(-\frac{x}{7}\right)+\frac{(-1)(-2)}{2}\left(-\frac{x}{7}\right)^2+\cdots\right)$$

$$= -\frac{8}{7}\left(1+\frac{x}{7}+\frac{x^2}{49}+\cdots\right)$$

$$= -\frac{8}{7}-\frac{8x}{49}-\frac{8x^2}{343}+\cdots$$

For the coefficient of x^2

$$-\frac{x^2}{9}-\frac{8x^2}{343} = \frac{-343+72}{3087}x^2$$

Coefficient of x^2 is $-\frac{271}{3087}$.

6 $\frac{1}{4(x+2)}+\frac{11}{4(x-2)}-\frac{2}{x-3}$

7 a $\frac{3}{(x-2)}+\frac{16}{(x-2)^2}+\frac{13}{(x-2)^3}$

b $-\frac{3}{(x-2)^2}-\frac{32}{(x-2)^3}-\frac{39}{(x-2)^4}$

8 a $\frac{4x^2+7x-3}{(x-1)(x-2)^2} = \frac{A}{x-1}+\frac{B}{x-2}+\frac{C}{x-2}$

$4x^2+7x-3 = A(x-2)^2+B(x-1)(x-2)+$
$\qquad C(x-1)\,x-2)$

$x=1$
$4+7-3=A$
$A=8$
$x=0$
$-35=2B+2C$ ⓵
$x=-1$
$-78=6B+6C$ ⓶
⓵ × 3
$-105=6B+6C$ ⓷

Equations ⓵ and ⓷ are contradictory and so cannot be solved.

Consequently, $\frac{4x^2+7x-3}{(x-1)(x-2)^2}$ cannot be split into

partial fractions in the form $\frac{A}{x-1}+\frac{B}{x-2}+\frac{C}{x-2}$.

b $\frac{8}{x-1}-\frac{4}{x-2}+\frac{27}{(x-2)^2}$

9 $-\frac{4}{9}$

10 a $\frac{8}{(x+1)}+\frac{9}{(x+1)^2}$

b $17-26x+35x^2-44x^3+\cdots$

3 Coordinate geometry: Parametric equations page 390

1 a $-\frac{4}{3}\sin t$

b $y = -\frac{2\sqrt{3}}{3}x+\frac{5}{2}$

c $y = 1-\frac{2x^2}{9}$

2 a A is $(3,0)$
$(\pi^2+3,0)$

b $\left(\frac{\pi^2}{4}+3,1\right)$

c 2π

3 a

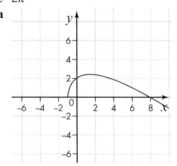

b $2x-y-16=0$.

4 a

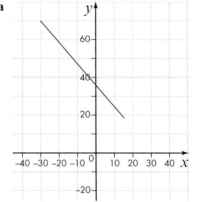

b 57 units

c $\sqrt{130}$

d $322°$

5 a $\left(\frac{68}{13},0\right)$

b $y = \frac{68-13x}{4x+1}$

6 $x = 5\cos t+1$
$y = 3\sin t-2$
This is an ellipse with centre $(1,-2)$
For lengths of minor and major axes, the centre is not relevant.
$x = 5\cos t$

When $x = 0$, $t = \dfrac{\pi}{2}, \dfrac{3\pi}{2}$

$y = 3, -3$

$y = 3\sin t$

When $y = 0$, $t = 0, \pi$

$x = 5, -5$

So the major axis is 10 units and the minor axis is 6 units. Consequently, the ellipse will not fit inside a rectangle that measures 11 units by 5.5 units.

7 $\left(0, -\dfrac{29}{2}\right)$

8 $(x + 5)^2 + (y - 7)^2 = 1$

A circle centre $(-5, 7)$, radius 1 unit.

9 $\dfrac{44}{5}$ units^2

10 Area = 8.299 units^2

$y^2 = t$

Substituting

$x = \left(y^2\right)^3 - 2$

$x = y^6 - 2$

$(x + 2)^{\frac{1}{6}} = y$

Area under curve = $\int (x + 2)^{\frac{1}{6}} \, \mathrm{d}x$ between 5 and -2.

Area under curve = $\left[\dfrac{6(x + 2)^{\frac{7}{6}}}{7}\right]$ between 5 and -2.

Area under curve = 8.299 units^2

11 a $\left(\dfrac{x}{4}\right)^2 + \left(\dfrac{y}{5}\right)^2 = 1$ (ellipse)

b $\left(\dfrac{x - 1}{3}\right)^2 + \left(\dfrac{y + 4}{7}\right)^2 = 1$ (ellipse)

c $x^2 + y^2 = 1$ (Circle)

12 a $y = 1 - 2x^2$

b $y = \dfrac{3}{2}x^2 - 3$

c $y^2 = \dfrac{x^2}{1 + x^2}$

13 $x = \cos t$, $x^2 = \cos^2 t$ ①

$y = \sin 2t = 2\sin t \cos t$

$y = 2\sin t \, (x)$

$\sin t = \dfrac{y}{2x}$, $\sin^2 t = \dfrac{y^2}{4x^2}$ ②

Add ① + ②

$\dfrac{y^2}{4x^2} + x^2 = \sin^2 t + \cos^2 t = 1$

$y^2 + 4x^4 = 4x^2$

$y^2 = 4x^2 (1 - x^2)$

$y = 2x\sqrt{1 - x^2}$

4 Sequences and series
page 391

1 13 years old

2 a $a_{n+1} - a_n = \dfrac{2(n + 1) + 3}{(n + 1) + 1} - \dfrac{2n + 3}{n + 1} = \dfrac{2n + 5}{n + 2} - \dfrac{2n + 3}{n + 1}$

$\qquad = \dfrac{(2n + 5)(n + 1) - (2n + 3)(n + 2)}{(n + 2)(n + 1)}$

$\qquad = \dfrac{2n^2 + 7n + 5 - 2n^2 - 7n - 6}{(n + 2)(n + 1)}$

$\qquad = \dfrac{-1}{(n + 2)(n + 1)}$.

This is always negative which shows that a_{n+1} is always less than a_n, so the sequence is decreasing.

b You can write $a_n = \dfrac{2(n + 1) + 1}{n + 1} = 2 + \dfrac{1}{n + 1}$. As n increases, $\dfrac{1}{n + 1} \to 0$ and $2 + \dfrac{1}{n + 1}$ approaches a limit of 2.

3 14 terms

4 22 months

5 a £256

b 1 398 100

c 5 592 404

d The numbers become large very quickly – you will run out of people. After 10 stages the number of postcards is about one-tenth of the population of England.

6 a $1 + 3 + 5 + \cdots$ (n terms) is an arithmetic sequence with $a = 1$ and $d = 2$

$S_n = \dfrac{n}{2}\{2a + (n - 1)d\} = \dfrac{n}{2}\{2 + 2(n - 1)\}$

$\qquad = \dfrac{n}{2}(2 + 2n - 2) = \dfrac{n}{2} \times 2n = n^2$

b $\dfrac{3}{2}n(n + 1)$

7 a $n(78 - 2n)$

b If $n = 39$, the sum is 0, and if the sum is negative, then $n > 39$ as all subsequent terms are negative.

8 18 months

9 a $1 - x + x^2 - x^3$

b $\dfrac{1}{2} - \dfrac{1}{4}x + \dfrac{1}{8}x^2 - \dfrac{1}{16}x^3$

c $\dfrac{1}{2} - \dfrac{3}{4}x + \dfrac{7}{8}x^2 - \dfrac{15}{16}x^3$

10 a $2 + \dfrac{1}{12}x^2 - \dfrac{1}{288}x^4 + \dfrac{5}{20736}x^6$

The expansion is valid if $-\sqrt{8} < x < \sqrt{8}$

b If $8 + x^2 = 9$ then $x^2 = 1$ and the estimate is

$2 + \dfrac{1}{12} - \dfrac{1}{288} + \dfrac{5}{20736} = 2.0801022...$

On a calculator, $\sqrt[3]{9} = 2.0800838...$ which to
4 d.p. is 2.0801 and the same as the estimate.

11 a Write $S = a + ar + ar^2 + ... + ar^{n-2} + ar^{n-1}$
Multiply by r: $rS = ar + ar^2 + ... + ar^{n-1} + ar^n$
Subtract one from the other and all terms
except two cancel out: $S - rS = a - ar^n$
Factorise: $S(1 - r) = a(1 - r^n)$ and rearrange:

$S = \dfrac{a(1 - r^n)}{1 - r}$

b 28 terms

12 a 2078

b A geometric sequence increases by a constant
proportion from one term to the next. This
is more reasonable for a population than an
arithmetic sequence which increases by a
constant amount.

13 $\dfrac{80}{69}$

14 a $1 - \dfrac{1}{2}x - \dfrac{1}{8}x^2$

b $1 - \dfrac{3}{2}x + \dfrac{11}{8}x^2$

15 a Write $S = a + (a + d) + (a + 2d) + ... + (a + (n - 1)d)$.
Then write the series in reverse:
$S = (a + (n - 1)d) + (a + (n - 2)d) ... + (a + d) + a$
Add corresponding terms and each total is
$2a + (n - 1)d$
Therefore $2S = \{2a + (n - 1)d\} \times n$, so

$S = \dfrac{n}{2}\{2a + (n - 1)d\}$

b $d = 3$
$a = 25$

5 Trigonometry   page 394

1 2.417 radians

2 $\sin\theta \approx \theta$ and $\cos\theta \approx 1 - \dfrac{1}{2}\theta^2$ so

$\dfrac{\sin^2 2x}{\cos 0.5x - 1} \approx \dfrac{(2x)^2}{1 - \frac{1}{2}(0.5x)^2 - 1} = \dfrac{4x^2}{-\frac{1}{2} \times \frac{x^2}{4}}$

$= \dfrac{4x^2}{-\frac{1}{8}x^2} = -32$

3

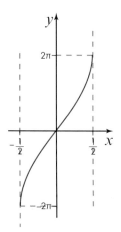

The domain is $-0.5 \leq x \leq 0.5$ and the range is
$-2\pi \leq y \leq 2\pi$

4 (0.427, 0.828)

5 a The x-coordinate of A is $6\cos\theta$. The angle
between AB and the vertical is θ so the
horizontal length of AB is $4\sin\theta$. Add these to
get the x-coordinate.

b $7.2111\cos(\theta - 0.5880)$

c $\theta = 1.393$ or 6.067

6 a

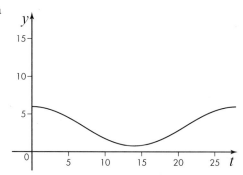

b before 07:20 and after 20:40.

7 a $\sin\left(x + \dfrac{\pi}{4}\right) = \sin x \cos\dfrac{\pi}{4} + \cos x \sin\dfrac{\pi}{4}$

$= \sin x \times \dfrac{1}{\sqrt{2}} + \cos x \times \dfrac{1}{\sqrt{2}}$

so $\sqrt{2}\sin\left(x + \dfrac{\pi}{4}\right) = \sin x + \cos x$

b $\dfrac{\sqrt{3}+1}{2\sqrt{2}}$

8 a $\sin(x + y) + \sin(x - y) = \sin x \cos y + \cos x \sin y + \sin x \cos y - \cos x \sin y = 2\sin x \cos y$

b In part (a), write $x + y = a$ and $x - y = b$.

Add the equations to get $2x = a + b$. So $x = \dfrac{a+b}{2}$.

Subtract the equations to get $2y = a - b$ so

$y = \dfrac{a-b}{2}$

Substitute into (a) to get

$\sin a + \sin b = 2\sin\dfrac{a+b}{2}\cos\dfrac{a-b}{2}$

c $2\cos\dfrac{a+b}{2}\cos\dfrac{a-b}{2}$

9 $\cos 2x = 2\cos^2 x - 1$ and so $\cos 2(2x) = 2\cos^2 2x - 1$
$= 2(2\cos^2 x - 1)^2 - 1$
Square the bracket: $\cos 4x = 2(4\cos^4 x - 4\cos^2 x + 1) - 1$
$= 8\cos^4 x - 8\cos^2 x + 2 - 1 = 8\cos^4 x - 8\cos^2 x + 1$

10 $x = 1.162$ or 3.329

11 a $\dfrac{\sqrt{3}+1}{2\sqrt{2}}$

b $\tan 75° = \tan(45° + 30°) = \dfrac{\tan 45° + \tan 30°}{1 - \tan 45° \tan 30°} =$

$\dfrac{1 + \frac{1}{\sqrt{3}}}{1 - 1 \times \frac{1}{\sqrt{3}}} = \dfrac{\sqrt{3}+1}{\sqrt{3}-1} = \dfrac{(\sqrt{3}+1)(\sqrt{3}+1)}{(\sqrt{3}-1)(\sqrt{3}+1)} =$

$\dfrac{3 + 2\sqrt{3} + 1}{3 - 1} = \dfrac{4 + 2\sqrt{3}}{2} = 2 + \sqrt{3}$

12 $\sin 3\theta = \sin(2\theta + \theta) = \sin 2\theta \cos\theta + \cos 2\theta \sin\theta$ so

$\dfrac{\sin 3\theta}{\sin\theta} = \dfrac{\sin 2\theta \cos\theta}{\sin\theta} + \cos 2\theta$

Now $\sin 2\theta = 2\sin\theta\cos\theta$ so

$\dfrac{\sin 2\theta\cos\theta}{\sin\theta} = \dfrac{2\sin\theta\cos\theta\cos\theta}{\sin\theta} = 2\cos^2\theta$ and

therefore

$\dfrac{\sin 3\theta}{\sin\theta} = \cos 2\theta + 2\cos^2\theta$

13 a $x = 15°, 75°, 195°, 255°$

b $-2 < a < 2$

14 $\sec 2x = \dfrac{1}{\cos 2x}$. Now $\cos 2x = 2\cos^2 x - 1$, so

$\sec 2x = \dfrac{1}{2\cos^2 x - 1}$.

Divide the numerator and the denominator by $\cos^2 x$:

$\sec 2x = \dfrac{\frac{1}{\cos^2 x}}{2 - \frac{1}{\cos^2 x}}$ or $\sec 2x = \dfrac{\sec^2 x}{2 - \sec^2 x}$

15 $\tan 3x = \tan(x + 2x) = \dfrac{\tan x + \tan 2x}{1 - \tan x \tan 2x}$

$= \dfrac{\tan x + \frac{2\tan x}{1 - \tan^2 x}}{1 - \tan x \frac{2\tan x}{1 - \tan^2 x}} = \dfrac{\tan x(1 - \tan^2 x) + 2\tan x}{1 - \tan^2 x - 2\tan^2 x}$

$= \dfrac{3\tan x - \tan^3 x}{1 - 3\tan^2 x}$

6 Differentiation   page 395

1 a 9 years

b 177 animals per year

2 At a stationary point, $\dfrac{dy}{dx} = 0$; $\dfrac{dy}{dx} = 4x^3 - 24x^2$.

If $4x^3 - 24x^2 = 0$, then $x^3 - 6x^2 = 0$

So $x^2(x - 6) = 0$ and $x = 0$ or 6. Now $\dfrac{d^2 y}{dx^2} = 12x^2 - 48x$.

If $x = 0$, $\dfrac{d^2 y}{dx^2} = 0$, so this could be a point of inflection. Check values close to 0.

x	-1	0	1
y	$1 + 8 + 20 = 29$	20	$1 - 8 + 20 = 13$

This shows that $(0, 20)$ is a point of inflection.

If $x = 6$ then, $\dfrac{d^2 y}{dx^2} = 12 \times 36 - 48 \times 6 = 144$, which is positive so this is a minimum point.
$y = 6^4 - 8 \times 6^3 + 20 = -412$, so the coordinates are $(6, -412)$.

3 a $\left(\dfrac{\pi}{6}, \dfrac{1}{4}\right), \left(\dfrac{\pi}{2}, 0\right)$ and $\left(\dfrac{5\pi}{6}, \dfrac{1}{4}\right)$.

b

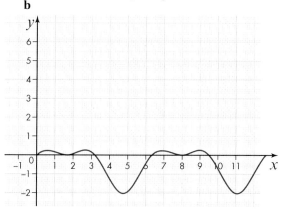

4 $-\dfrac{4x}{3}\left(1 - 2x^2\right)^{-\frac{2}{3}}$

5 The only stationary point is $(5, 125)$.

6 a $-4x\sin 2x^2$

b $-4\cos 2x\sin 2x$

7 a $y = e^{ax} - e^{-ax}$; $\dfrac{dy}{dx} = ae^{ax} + ae^{-ax}$;

$\dfrac{d^2 y}{dx^2} = a \times ae^{ax} + a \times -ae^{-ax}$

$\dfrac{d^2 y}{dx^2} = a^2 e^{ax} - a^2 e^{-ax} = a^2\left(e^{ax} - e^{-ax}\right) = a^2 y$

b $y = \cos ax - \sin ax;$ $\dfrac{dy}{dx} = -a\sin ax - a\cos ax;$

$$\dfrac{d^2y}{dx^2} = -a \times a\cos ax - a \times -a\sin ax$$

$$\dfrac{d^2y}{dx^2} = -a^2\cos ax + a^2\sin ax = -a^2(\cos ax - \sin ax),\text{ so}$$

in this case $\dfrac{d^2y}{dx^2} = -a^2y.$

8 $y = \ln(1 + 2x),$ so $\dfrac{dy}{dx} = \dfrac{1}{(1+2x)} \times 2 = \dfrac{2}{1+2x}.$ You can

write this as $\dfrac{dy}{dx} = 2(1+2x)^{-1},$

so $\dfrac{d^2y}{dx^2} = 2 \times -1 \times (1+2x)^{-2} \times 2 = -\dfrac{4}{(1+2x)^2}$

and then $\left(\dfrac{dy}{dx}\right)^2 = \left(\dfrac{2}{1+2x}\right)^2 = \dfrac{4}{(1+2x)^2} = -\dfrac{d^2y}{dx^2}$ or

$\dfrac{d^2y}{dx^2} = -\left(\dfrac{dy}{dx}\right)^2.$

9 a $f(1) = 2^{\sqrt{1}} = 2$ and $f(4) = 2^{\sqrt{4}} = 2^2 = 4$

b $f(x) = 2^{\sqrt{x}} = e^{(\ln 2)\sqrt{x}}.$ Use the chain rule to differentiate this. If $y = e^u$ and $u = (\ln 2)\sqrt{x},$

then $\dfrac{dy}{du} = e^u$ and $\dfrac{du}{dx} = \dfrac{d}{dx}\left((\ln 2)x^{\frac{1}{2}}\right) =$

$\ln 2 \times \dfrac{1}{2}x^{-\frac{1}{2}}.$ $f'(x) = \dfrac{dy}{du} \times \dfrac{du}{dx};$

$f'(x) = e^{(\ln 2)\sqrt{x}}\ln 2 \times \dfrac{1}{2}x^{-\frac{1}{2}} = \dfrac{\ln 2}{2\sqrt{x}} \times 2^{\sqrt{x}};$

$f'(1) = \dfrac{\ln 2}{2\sqrt{1}} \times 2^{\sqrt{1}} = \ln 2$ and $f'(4) = \dfrac{\ln 2}{2\sqrt{4}} \times 2^{\sqrt{4}}$

$= \dfrac{1}{4} \times 2^2 = \ln 2$

Since the gradients are the same at both points, the tangents are parallel.

10 a $-2\cos x \sin x$

b $-\sin 2x$

c They are identical because the double angle formula states that

$\cos 2x = 2\cos^2 x - 1,$ and so $\dfrac{1}{2}\cos 2x = \cos^2 x - \dfrac{1}{2}.$

Because $\dfrac{1}{2}\cos 2x$ and $\cos^2 x$ differ by a constant, they will have the same derivative.

11 The derivative of $\sin 2x$ is $\lim\limits_{\delta x \to 0} \dfrac{\sin 2(x + \delta x) - \sin 2x}{\delta x}.$

Now $\sin 2(x + \delta x) = \sin(2x + 2\delta x)$

$= \sin 2x \cos 2\delta x + \cos 2x \sin 2\delta x.$

Use the small angle approximations $\sin 2\delta x \approx 2\delta x$

and $\cos 2\delta x \approx 1 - \dfrac{1}{2}(2\delta x)^2 = 1 - 2(\delta x)^2.$

Then $\sin 2(x + \delta x) - \sin 2x$

$= \sin 2x \times \{1 - 2(\delta x)^2\} + \cos 2x \times 2\delta x - \sin 2x$

$= -2\sin 2x \times (\delta x)^2 + 2\cos 2x \times \delta x.$

So $\dfrac{\sin 2(x + \delta x) - \sin 2x}{\delta x} = -2\sin 2x \times \delta x + 2\cos 2x.$

As $\delta x \to 0$, the first term $\to 0$, and so

$\lim\limits_{\delta x \to 0} \dfrac{\sin 2(x + \delta x) - \sin 2x}{\delta x} = 2\cos 2x$ and this is the derivative.

12 $(0.857, -1)$

13 $y = \ln\dfrac{2x^2}{1+x^2} = \ln 2x^2 - \ln(1 + x^2)$

$\dfrac{dy}{dx} = \dfrac{1}{2x^2} \times 4x - \dfrac{1}{1+x^2} \times 2x = \dfrac{2}{x} - \dfrac{2x}{1+x^2}$

$= \dfrac{2(1+x^2) - 2x^2}{x(1+x^2)} = \dfrac{2}{x(1+x^2)}$

14 $y = (\cos x)^2,$ so

$\dfrac{dy}{dx} = 2\cos x \times -\sin x = -2\sin x \cos x = -\sin 2x.$

Then $\dfrac{d^2y}{dx^2} = -\cos 2x \times 2 = -2\cos 2x.$

Now $\cos 2x = 2\cos^2 x - 1 = 2y - 1$ so

$\dfrac{d^2y}{dx^2} = -2(2y - 1) = -4y + 2$ and therefore

$\dfrac{d^2y}{dx^2} + 4y = 2$

15 a $c = \dfrac{1}{3}$

b $\left(-\dfrac{1}{3}, -\dfrac{1}{27}\right)$

7 Further differentiation page 397

1 $y = 1.5x - 7$

2 $\dfrac{dx}{dt} = 1 + \dfrac{1}{t^2}$ and $\dfrac{dy}{dt} = 1 - \dfrac{4}{t^2}.$

Therefore $\dfrac{dy}{dx} = \dfrac{1 - \dfrac{4}{t^2}}{1 + \dfrac{1}{t^2}} = \dfrac{t^2 - 4}{t^2 + 1}.$

At a stationary point, $\dfrac{dy}{dx} = 0$ and so $t^2 = 4; t$ is positive, so $t = 2.$

Then $x = 2 - \dfrac{1}{2} = 1.5$ and $y = 2 + \dfrac{4}{2} = 4,$ so the

coordinates are $(1.5, 4).$

If $t = 1,$ the point is $(0, 5)$ and if $t = 3,$ the point is

$\left(2\frac{2}{3}, 4\frac{1}{3}\right);$ these are both above $(1.5, 4)$ so it is a minimum point.

3 $-\dfrac{4}{9}$ and $\dfrac{4}{9}$

4 $y = \sec x$, so $\dfrac{dy}{dx} = \sec x \tan x$. Use the product rule:

$$\dfrac{d^2y}{dx^2} = \sec x \tan x \times \tan x + \sec x \times \sec^2 x$$

$$= \sec x \tan^2 x + \sec^3 x$$

Substitute: $\tan^2 x = \sec^2 x - 1$;

$$\dfrac{d^2y}{dx^2} = \sec x\left(\sec^2 x - 1\right) + \sec^3 x = 2\sec^3 x - \sec x$$

5 **a** $1.84\,\mathrm{m\,s^{-2}}$

   b $13.5\,\mathrm{m\,s^{-1}}$

6 **a** $\left(1, \dfrac{e}{2}\right)$

   b The values of y are increasing as you move from left to right, so this is a point of inflection.

7 $(2, 1.107)$ and $(-2, -1.107)$

8 **a** $\dfrac{x}{\sqrt{x^2+3}}$.

   b $y^2 = x^2 + 3$, so $2y\dfrac{dy}{dx} = 2x$

     Rearrange: $\dfrac{dy}{dx} = \dfrac{x}{y} = \dfrac{x}{\sqrt{x^2+3}}$.

   c $x^2 = y^2 - 3$; $x = \left(y^2 - 3\right)^{\frac{1}{2}}$

     $\dfrac{dx}{dy} = \dfrac{1}{2}\left(y^2 - 3\right)^{-\frac{1}{2}} \times 2y = \dfrac{y}{\sqrt{y^2-3}}$

     Hence $\dfrac{dy}{dx} = \dfrac{\sqrt{y^2-3}}{y} = \dfrac{\sqrt{x^2}}{\sqrt{x^2+3}} = \dfrac{x}{\sqrt{x^2+3}}$,

     as before.

9 $5e^{3x}\sin 2x + 12e^{3x}\cos 2x$

10 **a** $-\dfrac{1}{4}$ and $-\dfrac{1}{4}$.

    b $\left(-\sqrt{12}, \sqrt{12}\right)$

11 **a** $10.7\,\mathrm{cm\,s^{-1}}$

    b $0.39\,\mathrm{s}$

12 If $x = \dfrac{2t}{1+t^2}$, then $\dfrac{dx}{dt} = \dfrac{2\left(1+t^2\right) - 2t \times 2t}{\left(1+t^2\right)^2}$

$$= \dfrac{2 + 2t^2 - 4t^2}{\left(1+t^2\right)^2} = \dfrac{2 - 2t^2}{\left(1+t^2\right)^2}.$$

If $y = \dfrac{1-t^2}{1+t^2}$, then $\dfrac{dy}{dt} = \dfrac{-2t\left(1+t^2\right) - \left(1-t^2\right) \times 2t}{\left(1+t^2\right)^2} =$

$$\dfrac{-2t - 2t^3 - 2t + 2t^3}{\left(1+t^2\right)^2} = \dfrac{-4t}{\left(1+t^2\right)^2}.$$

Then $\dfrac{dy}{dx} = \dfrac{dy}{dt} \div \dfrac{dx}{dt} = \dfrac{-4t}{2 - 2t^2} = \dfrac{-2t}{1-t^2}$. But

$\dfrac{x}{y} = \dfrac{2t}{1-t^2}$, so $\dfrac{dy}{dx} = -\dfrac{x}{y}$.

13 1.9365

14 $x^x\left(1 + \ln x\right)$

15 **a** Differentiate $\dfrac{x^2}{9} + \dfrac{y^2}{4} = 1$ to get $\dfrac{2x}{9} + \dfrac{2y}{4}\dfrac{dy}{dx} = 0$.

    Rearrange: $\dfrac{2y}{4}\dfrac{dy}{dx} = -\dfrac{2x}{9}$; $\dfrac{dy}{dx} = -\dfrac{4x}{9y}$.

   b The area is $AD \times AB = 2x \times 2y = 4xy$

   c $\dfrac{d}{dx}$ (area) $= 4y + 4x\dfrac{dy}{dx}$

     If the area is a maximum, this will be 0, so

     $4y + 4x\dfrac{dy}{dx} = 0$ or $\dfrac{dy}{dx} = -\dfrac{y}{x}$.

     But, from part (a), $\dfrac{dy}{dx} = -\dfrac{4x}{9y}$, so the area is a

     maximum when $= -\dfrac{4x}{9y} = -\dfrac{y}{x}$.

     Therefore $y^2 = \dfrac{4}{9}x^2$ or taking the positive root,

     $y = \dfrac{2}{3}x$

     Substitute into the equation of the curve:

     $\dfrac{x^2}{9} + \dfrac{1}{4}\left(\dfrac{2}{3}x\right)^2 = 1$; $\dfrac{x^2}{9} + \dfrac{x^2}{9} = 1$; $\dfrac{2x^2}{9} = 1$; $x^2 = \dfrac{9}{2}$.

     So $x = \dfrac{3}{\sqrt{2}}$. Then $y = \dfrac{2}{3} \times \dfrac{3}{\sqrt{2}} = \dfrac{2}{\sqrt{2}}$ and the

     maximum area is $4 \times \dfrac{3}{\sqrt{2}} \times \dfrac{2}{\sqrt{2}} = 12$ square units.

8 Integration   page 400

1 36

2 $-\dfrac{1.25}{\left(4x-1\right)^2} + c$

3 **a** $y = e^{-x^2 + c}$ or $y = ae^{-x^2}$ where $a = e^c$

b

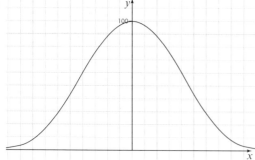

4 17.75

5 $x \ln 2x - x + c.$

6 **a** $\dfrac{1}{\ln 2}\left(1 - 2^{-a}\right)$

   b $\dfrac{1}{\ln 2}$

7 $\dfrac{2}{5}(x+4)^{\frac{5}{2}} - \dfrac{8}{3}(x+4)^{\frac{3}{2}} + c$

8 $-\dfrac{1}{4}x\cos 2x + \dfrac{1}{8}\sin 2x + c$

9 **a** $\dfrac{1}{2}\arctan\dfrac{x}{2} + c$

   b $\dfrac{1}{2}\ln\left(x^2 + 4\right) + c$

   c $x - 2\arctan\dfrac{x}{2} + c$

10 $\dfrac{1}{2}\pi$

11 $-\dfrac{2}{5}e^{-0.5x}\sin x - \dfrac{4}{5}e^{-0.5x}\cos x + c$

12 $\displaystyle\int_0^{\frac{\pi}{2}} \tan 0.5x \, dx = \int_0^{\frac{\pi}{2}} \dfrac{\sin 0.5x}{\cos 0.5x} dx.$ If $u = \cos 0.5x$ then

$\dfrac{du}{dx} = -0.5\sin 0.5x$

$\displaystyle\int_0^{\frac{\pi}{2}} \dfrac{\sin 0.5x}{\cos 0.5x} dx = \int_0^{\frac{\pi}{2}} \dfrac{\sin 0.5x}{u}\dfrac{dx}{du} du =$

$\displaystyle\int_0^{\frac{\pi}{2}} \dfrac{(\sin 0.5x)}{u}\dfrac{1}{(-0.5\sin 0.5x)} du = \int_0^{\frac{\pi}{2}} \dfrac{-2}{u} du = \left[-2\ln|u|\right]_0^{\frac{\pi}{2}}$

$= \left[-2\ln(\cos 0.5x)\right]_0^{\frac{\pi}{2}} = \left[-2\ln\cos\dfrac{\pi}{4}\right] - \left[-2\ln 1\right] =$

$-2\ln\dfrac{1}{\sqrt{2}} - 0 = -2\left(\ln 1 - \ln\sqrt{2}\right) = 2\ln 2^{\frac{1}{2}}$

$= 2 \times \dfrac{1}{2}\ln 2 = \ln 2$

13 **a** $x = 57.1e^{0.08t}$

   b If steps are taken to reduce infection, the differential equation will change. The rate will also reduce when a large proportion of the population of the city are infected.

14 $2\ln|x + 1| - \ln|x + 3| + c = \ln\dfrac{(x+1)^2}{|x+3|} + c$

15 **a** The radius of the disc is y and the volume is $\pi y^2 \delta x$.
   However, $x^2 + y^2 = r^2$ and so $y^2 = r^2 - x^2$, and the volume of the disc is $\pi(r^2 - x^2)\delta x$.

   b The volume of the sphere is $\displaystyle\lim_{\delta x \to 0}\sum \pi\left(r^2 - x^2\right)\delta x$

$= \displaystyle\int_{-r}^{r} \pi\left(r^2 - x^2\right) dx$

$= \left[\pi\left(r^2 x - \dfrac{1}{3}x^3\right)\right]_{-r}^{r} = \left[\pi\left(r^3 - \dfrac{1}{3}r^3\right)\right] -$

$\left[\pi\left(-r^3 + \dfrac{1}{3}r^3\right)\right] = 2\pi r^3 - \dfrac{2}{3}\pi r^3 = \dfrac{4}{3}\pi r^3$

9 Numerical methods   page 402

1 **a** $x = 0.2, y = 1.380$
   $x = 0.6, y = 2.627$
   $x = 0.8, y = 3.624$

   b 2.507

2 **a** Given $f(x) = x^3 - 2x^2 - 3$
   $x = 2 \Rightarrow y = 2^3 - 2\times 2^2 - 3 = -3$
   $x = 3 \Rightarrow y = 3^3 - 2\times 3^2 - 3 = +6$
   $f(x)$ is a continuous function and changes sign between $x = 2$ and $x = 3$. Therefore the equation $f(x) = 0$ has a root in the interval $[2, 3]$.

   b Given $f(x) = 0$
   $\Rightarrow x^3 - 2x^2 - 3 = 0$
   $\Rightarrow x^3 = 2x^2 + 3$
   $\Rightarrow x^2 = \dfrac{2x^2 + 3}{x} = \dfrac{2x^2}{x} + \dfrac{3}{x}$, with $x \neq 0$
   $\Rightarrow x^2 = 2x + \dfrac{3}{x}$
   Taking the square root of both sides of this equation,
   $\Rightarrow x = \sqrt{2x + \dfrac{3}{x}}$

   c $x_2 = 2.345$
   $x_3 = 2.443$
   $x_4 = 2.473$

d To prove that the root is 2.486 to 3 decimal places, you need to show a change of sign of the function for values on either side of $x = 2.486$. Start with $x = 2.4855$ (take the x-value to four decimal places to ensure accuracy for the final answer of x to three decimal places):

f(2.4855) = −0.00072

f(2.4865) = +0.0079

f(x) is a continuous function and changes sign between $x = 2.4855$ and $x = 2.4865$. Therefore the equation f(x) = 0 has a root in the interval [2.4855, 2.4865]. The root is 2.486 correct to 3 decimal places.

3 a

x	0	0.5	1.0	1.5	2.0
y	1	$e^{-\frac{1}{4}}$	e^{-1}	$e^{-\frac{9}{4}}$	e^{-4}

b 0.8806

4 a $x = 5$, $x^2 - 4x - 7 = -2$

$x = 6$, $x^2 - 4x - 7 = 5$

There is a change of sign so there is a root in the interval $5 < x < 6$.

b 5.317

c 0.007%

5 a 0.190

b 0.190

c Both iteration formulae converge to the same level of accuracy on the same root.

6 a $\ln\sqrt{2}$

b 0.3590108264

c 0.349758334

d 0.919%

7 a f(−1) = −0.6321205588

f(0) = 1

There is a change of sign so there is a root in the interval $-1 < x < 0$.

b −0.567

c

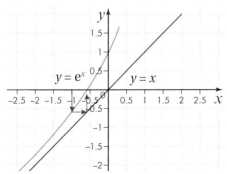

8 a

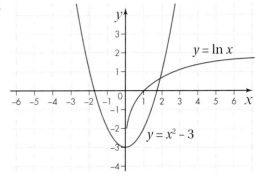

b $x^2 - \ln x - 3 = 0$

$x \to 0$, $x^2 - \ln x - 3 \to \infty$

$x = 1$, $x^2 - \ln x - 3 = -2$

There is a change of sign, so there is a root in the interval $0 < x < 1$.

$x = 1$, $x^2 - \ln x - 3 = -2$

$x = 2$, $x^2 - \ln x - 3 = 0.3068528194$

There is a change of sign, so there is a root in the interval $1 < x < 2$.

c i 0.0499

ii 1.9092

d Both iteration formulae converge to a root but the formula in (c)(i) converges to the root in the interval $0 < x < 1$ whereas the formula in (c)(ii) converges to the root in the interval $1 < x < 2$.

9 a

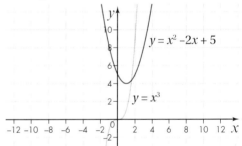

b $x = 0$, $x^3 - x^2 + 2x - 5 = -5$

$x = 2$, $x^3 - x^2 + 2x - 5 = 3$

There is a change of sign so there is a root in the interval $0 < x < 2$.

c 1.64

10 2.5649

10 Three-dimensional vectors
page 404

1 **a** All three offices and the centre of the circle are of the form $a\mathbf{i} + b\mathbf{j} + 12\mathbf{k}$, so lie in the same plane.

For first office, $d^2 = (7-7)^2 + (9-4)^2 = 0 + 25 = 25$, $d = 5$

For second office, $d^2 = (11-7)^2 + (1-4)^2 = 16 + 9 = 25$, $d = 5$

For third office, $d^2 = (3-7)^2 + (1-4)^2 = 16 + 9 = 25$, $d = 5$

All three offices are equidistant from $(7\mathbf{i} + 4\mathbf{j} + 12\mathbf{k})$, and hence $(7\mathbf{i} + 4\mathbf{j} + 12\mathbf{k})$ is the centre of the circle.

   b 13

2 **a** $p = 2$

$q = 22$

   b $\overrightarrow{BD} = \begin{bmatrix} 20 \\ 11 \\ 25 \end{bmatrix} - \begin{bmatrix} 8 \\ 2 \\ 22 \end{bmatrix} = \begin{bmatrix} 12 \\ 9 \\ 3 \end{bmatrix}$

$\overrightarrow{BC} = \begin{bmatrix} 4 \\ 14 \\ 2 \end{bmatrix} - \begin{bmatrix} 8 \\ 2 \\ 22 \end{bmatrix} = \begin{bmatrix} -4 \\ 12 \\ -20 \end{bmatrix}$

$\overrightarrow{CD} = \begin{bmatrix} 20 \\ 11 \\ 25 \end{bmatrix} - \begin{bmatrix} 4 \\ 14 \\ 2 \end{bmatrix} = \begin{bmatrix} 16 \\ -3 \\ 23 \end{bmatrix}$

$BD^2 = 12^2 + 9^2 + 3^2 = 234$

$BC^2 = (-4)^2 + 12^2 + (-20)^2 = 560$

$CD^2 = 16^2 + (-3)^2 + 23^2 = 794$

Since $234 + 560 = 794$, $BD^2 + BC^2 = CD^2$

Pythagoras' theorem is satisfied. $\angle$CBD is a right angle and BD is perpendicular to AC.

3 **a** $\overrightarrow{WY} = (116.5\mathbf{i} + 42\mathbf{j} + 45.5\mathbf{k}) - (8\mathbf{i} + 3\mathbf{j} + 9.5\mathbf{k}) = (108.5\mathbf{i} + 39\mathbf{j} + 36\mathbf{k})$

$WY^2 = 108.5^2 + 39^2 + 36^2 = 14589.25$

$\overrightarrow{XZ} = (102.5\mathbf{i} + 24\mathbf{j} - 17.5\mathbf{k}) - (22\mathbf{i} + 21\mathbf{j} + 72.5\mathbf{k}) = (80.5\mathbf{i} + 3\mathbf{j} - 90\mathbf{k})$

$XZ^2 = 80.5^2 + 3^2 + (-90)^2 = 14589.25$

Hence $WY = XZ$

   b $\overrightarrow{WX} = (22\mathbf{i} + 21\mathbf{j} + 72.5\mathbf{k}) - (8\mathbf{i} + 3\mathbf{j} + 9.5\mathbf{k}) = (14\mathbf{i} + 18\mathbf{j} + 63\mathbf{k})$

$WX^2 = 14^2 + 18^2 + 63^2 = 4489$, $WX = 67$

$\overrightarrow{WZ} = (102.5\mathbf{i} + 24\mathbf{j} - 17.5\mathbf{k}) - (8\mathbf{i} + 3\mathbf{j} + 9.5\mathbf{k}) = (94.5\mathbf{i} + 21\mathbf{j} - 27\mathbf{k})$

$WZ^2 = 94.5^2 + 21^2 + (-27)^2 = 10100.25$, $WZ = 100.5$

67 is $\frac{2}{3}$ of 100.5, so $WX = \frac{2}{3} WZ$.

   c Rectangle

4 **a** $\overrightarrow{XM} = \begin{bmatrix} 4 \\ -1 \\ 2 \end{bmatrix}$

   b $\overrightarrow{OZ} = \begin{bmatrix} -4 \\ 9 \\ -1 \end{bmatrix}$

5 **a** $\overrightarrow{OT} = \frac{1}{2}((-63\mathbf{i} + 102\mathbf{j} - 26\mathbf{k}) + (25\mathbf{i} - 52\mathbf{j} + 30\mathbf{k})) = (-19\mathbf{i} + 25\mathbf{j} + 2\mathbf{k})$

   b -42

   c 8649

6 It is sufficient to show that the diagonals are perpendicular and share a midpoint but are of different lengths.

Diagonals are SU and TV.

Let midpoint be X.

Midpoint of $SU = \frac{1}{2}((9\mathbf{i} + \mathbf{j} - 9\mathbf{k}) + (-11\mathbf{i} + 14\mathbf{j} + 20\mathbf{k})) = (-\mathbf{i} + \frac{15}{2}\mathbf{j} + \frac{11}{2}\mathbf{k})$

Midpoint of $TV = \frac{1}{2}((2\mathbf{i} + 11\mathbf{j} + 6\mathbf{k}) + (-4\mathbf{i} + 4\mathbf{j} + 5\mathbf{k})) = (-\mathbf{i} + \frac{15}{2}\mathbf{j} + \frac{11}{2}\mathbf{k})$

Diagonals share a midpoint.

$\overrightarrow{SU} = (-11\mathbf{i} + 14\mathbf{j} + 20\mathbf{k}) - (9\mathbf{i} + \mathbf{j} - 9\mathbf{k}) = (-20\mathbf{i} + 13\mathbf{j} + 29\mathbf{k})$

$SU^2 = (-20)^2 + 13^2 + 29^2 = 1410$

$\overrightarrow{TV} = (-4\mathbf{i} + 4\mathbf{j} + 5\mathbf{k}) - (2\mathbf{i} + 11\mathbf{j} + 6\mathbf{k}) = (-6\mathbf{i} - 7\mathbf{j} - \mathbf{k})$

$(TV)^2 = (-6)^2 + (-7)^2 + (-1)^2 = 86$

Diagonals are different lengths.

$(SX)^2 = \frac{1}{4}(SU)^2 = 352.5$

$(TX)^2 = \frac{1}{4}(TV)^2 = 21.5$

$\overrightarrow{ST} = (2\mathbf{i} + 11\mathbf{j} + 6\mathbf{k}) - (9\mathbf{i} + \mathbf{j} - 9\mathbf{k}) = (-7\mathbf{i} + 10\mathbf{j} + 15\mathbf{k})$

$(ST)^2 = (-7)^2 + 10^2 + 15^2 = 374$

Since $(SX)^2 + (TX)^2 = (ST)^2$, diagonals are perpendicular

$STUV$ is a rhombus

7 **a** $\overrightarrow{OE} = (23\mathbf{i} + 19\mathbf{j} + 5\mathbf{k})$

   b $\overrightarrow{BE} = (23\mathbf{i} + 19\mathbf{j} + 5\mathbf{k}) - (27\mathbf{i} + 15\mathbf{j} + 3\mathbf{k}) = (-4\mathbf{i} + 4\mathbf{j} + 2\mathbf{k}) = 2(-2\mathbf{i} + 2\mathbf{j} + \mathbf{k})$

$\overrightarrow{ED} = (9\mathbf{i} + 33\mathbf{j} + 12\mathbf{k}) - (23\mathbf{i} + 19\mathbf{j} + 5\mathbf{k}) = (-14\mathbf{i} + 14\mathbf{j} + 7\mathbf{k}) = 7(-2\mathbf{i} + 2\mathbf{j} + \mathbf{k})$

$\overrightarrow{BE} : \overrightarrow{ED} = 2 : 7$

   c Given that E is the midpoint of the diagonal AC but not BD, it is sufficient to show that AE and BE are perpendicular.

$\overrightarrow{AE} = (23\mathbf{i} + 19\mathbf{j} + 5\mathbf{k}) - (17\mathbf{i} + 16\mathbf{j} - \mathbf{k}) = (6\mathbf{i} + 3\mathbf{j} + 6\mathbf{k})$

$\overrightarrow{AB} = (27\mathbf{i} + 15\mathbf{j} + 3\mathbf{k}) - (17\mathbf{i} + 16\mathbf{j} - \mathbf{k}) = (10\mathbf{i} - \mathbf{j} + 4\mathbf{k})$

$(AE)^2 = 6^2 + 3^2 + 6^2 = 81$

$(AB)^2 = 10^2 + (-1)^2 + 4^2 = 117$

$(BE)^2 = (-4)^2 + 4^2 + 2^2 = 36$

Since $(AE)^2 + (BE)^2 = (AB)^2$, AE and BE are perpendicular.

d 243

8 a $\overrightarrow{PS} = \begin{bmatrix} 15 \\ -4 \\ 7 \end{bmatrix} - \begin{bmatrix} 5 \\ 6 \\ 2 \end{bmatrix} = \begin{bmatrix} 10 \\ -10 \\ 5 \end{bmatrix} = 5\begin{bmatrix} 2 \\ -2 \\ 1 \end{bmatrix}$

$\overrightarrow{QR} = \begin{bmatrix} 13 \\ 10 \\ 12 \end{bmatrix} - \begin{bmatrix} 9 \\ 14 \\ 10 \end{bmatrix} = \begin{bmatrix} 4 \\ -4 \\ 2 \end{bmatrix} = 2\begin{bmatrix} 2 \\ -2 \\ 1 \end{bmatrix}$

Since $\overrightarrow{PS}$ and $\overrightarrow{QR}$ are both multiples of $\begin{bmatrix} 2 \\ -2 \\ 1 \end{bmatrix}$, PS and QR are parallel.

b 5:2

c For $\angle QPS$ to be 90°, $(PQ)^2 + (PS)^2 = (QS)^2$

$\overrightarrow{PQ} = \begin{bmatrix} 9 \\ 14 \\ 10 \end{bmatrix} - \begin{bmatrix} 5 \\ 6 \\ 2 \end{bmatrix} = \begin{bmatrix} 4 \\ 8 \\ 8 \end{bmatrix}$

$\overrightarrow{QS} = \begin{bmatrix} 15 \\ -4 \\ 7 \end{bmatrix} - \begin{bmatrix} 9 \\ 14 \\ 10 \end{bmatrix} = \begin{bmatrix} 6 \\ -18 \\ -3 \end{bmatrix}$

$(PQ)^2 = 4^2 + 8^2 + 8^2 = 144$

$(PS)^2 = 10^2 + (-10)^2 + 5^2 = 225$

$(QS)^2 = 6^2 + (-18)^2 + (-3)^2 = 369$

Since $144 + 225 = 369$, $(PQ)^2 + (PS)^2 = (QS)^2$ and $\angle QPS = 90°$.

d 126

9 a $\overrightarrow{OD} = \begin{bmatrix} 29 \\ -13 \\ 13 \end{bmatrix}$

b $\overrightarrow{OE} = \begin{bmatrix} 21 \\ -8 \\ 16 \end{bmatrix}$

c Kite

d 294

10 a $\overrightarrow{OM} = \frac{1}{2}(\overrightarrow{OA} + \overrightarrow{OB})$

$\overrightarrow{OM} = \frac{1}{2}((-2.6\mathbf{i} - 8\mathbf{j} - 13\mathbf{k}) + (16.6\mathbf{i} + 4\mathbf{j} + 19\mathbf{k})) = (7\mathbf{i} - 2\mathbf{j} + 3\mathbf{k})$

$\overrightarrow{MC} = (47\mathbf{i} - 26\mathbf{j} - 12\mathbf{k}) - (7\mathbf{i} - 2\mathbf{j} + 3\mathbf{k}) = (40\mathbf{i} - 24\mathbf{j} - 15\mathbf{k})$

$\overrightarrow{AM} = (7\mathbf{i} - 2\mathbf{j} + 3\mathbf{k}) - (-2.6\mathbf{i} - 8\mathbf{j} - 13\mathbf{k}) = (9.6\mathbf{i} + 6\mathbf{j} + 16\mathbf{k})$

$\overrightarrow{AC} = (47\mathbf{i} - 26\mathbf{j} - 12\mathbf{k}) - (-2.6\mathbf{i} - 8\mathbf{j} - 13\mathbf{k})$
$= (49.6\mathbf{i} - 18\mathbf{j} + \mathbf{k})$

$(MC)^2 = 40^2 + (-24)^2 + (-15)^2 = 2401$

$(AM)^2 = 9.6^2 + 6^2 + 16^2 = 384.16$

$(AC)^2 = 49.6^2 + (-18)^2 + 1^2 = 2785.16$

Since $(MC)^2 + (AM)^2 = (AC)^2$, MC is perpendicular to AB.

b $\overrightarrow{OD} = -33\mathbf{i} + 22\mathbf{j} + 18\mathbf{k}$

11 Proof　page 405

1 Let $n = 2m$, an even number.

$(n - 1)^3 = (2m - 1)^3$

$(2m - 1)^3 = (2m - 1)(4m^2 - 4m + 1)$

$(2m - 1)(4m^2 - 4m + 1)$

$= 8m^3 - 8m^2 + 2m - 4m^2 + 4m - 1$

$8m^3 - 8m^2 + 2m - 4m^2 + 4m - 1$

$= 8m^3 - 12m^2 + 6m - 1$

$8m^3 - 12m^2 + 6m - 1 = 2(4m^3 - 6m^2 + 3m) - 1$

Since $2(4m^3 - 6m^2 + 3m)$ is even,

$2(4m^3 - 6m^2 + 3m) - 1$ is an odd number.

2 Assume the converse, that there exist integers x and y, for which $7x + 28y = 11$. $7x$ and $28y$ both have a factor of 7. Hence $7x + 28y = 7(x + 4y)$. So if $7x + 28y = 11$, then $7(x + 4y) = 11$, from which $x + 4y = \frac{11}{7}$. If x and y are integers, then $x + 4y$ is also an integer, and cannot therefore be equal to $\frac{11}{7}$. Therefore there is a contradiction. This means that the initial assumption, that there exist integers x and y, for which $7x + 28y = 11$, must be false. Hence it must be true that there exist no integers x and y for which $7x + 28y = 11$.

3 Assume that every non-zero real number has more than one reciprocal.

Let x be a non-zero real number and let both p and q be reciprocals of x where $p \neq q$.

So $px = 1$ and $qx = 1$.

So $px = qx$

So $p = q$ which contradicts the original assumption.

4 $ax^2 - bx - c = 0$

$a\left(x^2 - \dfrac{bx}{a} - \dfrac{c}{a}\right) = 0$

$\left(x - \dfrac{b}{2a}\right)^2 - \dfrac{b^2}{4a^2} - \dfrac{c}{a} = 0$

$\left(x - \dfrac{b}{2a}\right)^2 = \dfrac{b^2 + 4ac}{4a^2}$

$x = \dfrac{b \pm \sqrt{b^2 + 4ac}}{2a}$

Since a, b, $c \in \mathbb{R}$ and a, b, $c > 0$, $b^2 + 4ac > 0$ and, consequently, quadratic equations in this form will always have two distinct real roots.

5 Let $n = 2p + 1$ where p is a positive, whole number.
Then $n^2 + 2n + 1 = (2p + 1)^2 + 2(2p + 1) + 1$
$= 4p^2 + 4p + 1 + 4p + 2 + 1$
$= 4p^2 + 8p + 4$
$= 4(p^2 + 2p + 1)$
$4(p^2 + 2p + 1)$ is even.

6 Assume that $\sqrt[3]{11}$ is rational.

Therefore $\sqrt[3]{11} = \dfrac{p}{q}$ where p and q are integers with no common factors other than 1.

If $\sqrt[3]{11} = \dfrac{p}{q}$ then $11 = \dfrac{p^3}{q^3}$

Rearranging gives:
$11q^3 = p^3$
Therefore p^3 (and p) is a multiple of 11.
You can say that $p = 11n$, so $p^3 = 1331n^3$
You have already shown that $11q^3 = p^3$, which can now be rewritten as:
$11q^3 = 1331n^3$
$q^3 = 121n^3$
So q is also a multiple of 11.
If p and q are both multiples of 11, then they have a common factor of 11, which contradicts the assumption that they don't have a common factor other than 1.

So $\sqrt[3]{11}$ is not rational, and, consequently, it must be irrational.

7 Assume that if x and y are positive real numbers, then $x + y < \sqrt{xy}$.
Since x and y are positive real numbers, squaring both sides will not change the inequality.
$(x + y)^2 < xy$
$x^2 + 2xy + y^2 < xy$
$x^2 + xy + y^2 < 0$
$\left(x + \dfrac{y}{2}\right)^2 - \dfrac{y^2}{4} + y^2 < 0$
$\left(x + \dfrac{y}{2}\right)^2 + \dfrac{3y^2}{4} < 0$

However, if $x = y$ or $x \neq y$, then $\left(x + \dfrac{y}{2}\right)^2 + \dfrac{3y^2}{4} \geq 0$.

Therefore there is a contradiction. This means that the initial assumption, that if x and y are positive real numbers then $x + y < \sqrt{xy}$, must be false. Hence it must be true that if x and y are positive real numbers, $x + y \geq \sqrt{xy}$.

8 Assume that $2\sqrt{2}$ is rational.

Therefore, you can write $2\sqrt{2}$ as $2\left(\dfrac{p}{q}\right)$ where p and q are integers with no common factors other than 1.

If $2\sqrt{2} = 2\left(\dfrac{p}{q}\right)$ then $2 = \dfrac{p^2}{q^2}$

Rearranging gives:
$2q^2 = p^2$.
Therefore p^2 (and p) is a multiple of 2.
You can say that $p = 2n$, so $p^2 = 4n^2$.
You have already shown that $2q^2 = p^2$, which can now be rewritten as
$2q^2 = 4n^2$
$q^2 = 2n^2$
So q is also a multiple of 2.
If p and q are both multiples of 2 then they have a common factor of 2, which contradicts the assumption that they don't have a common factor other than 1.

So $\sqrt{2}$ is not rational – consequently $2\sqrt{2}$ is not rational, so it must be irrational.

9 Let's say that $p^2 - 4q - 2 \neq 0$ is false and so $p^2 - 4q - 2 = 0$.
So integers p, $q \in \mathbb{Z}$ exist for which $p^2 - 4q - 2 = 0$.
Rearranging gives $p^2 - 4q = 2$
$p^2 = 4q + 2$
$p^2 = 2(2q + 1)$
So p^2 is even.
So we can say that $p = 2r$.
Substituting gives $4r^2 = 4q + 2$
$4r^2 - 4q = 2$
$2(2r^2 - 2q) = 1$

But 1 is not an even number, so, consequently, there is a contradiction. Therefore the original assumption that $p^2 - 4q - 2 = 0$ is wrong and so the statement that $p^2 - 4q - 2 \neq 0$ where p, $q \in \mathbb{Z}$ is true.

10 Let's say that both a and b are odd.
Let $a = 2n + 1$ and $b = 2m + 1$
$a^2 + b^2 = (2n + 1)^2 + (2m + 1)^2$
$a^2 + b^2 = 4n^2 + 4n + 1 + 4m^2 + 4m + 1$
$a^2 + b^2 = 2(2n^2 + 2n + 2m^2 + 2m + 1)$

Which doesn't have an even power of 2, so, consequently, $2(2n^2 + 2n + 2m^2 + 2m + 1) = c^2$ does not have an integer solution for c. Consequently both a and b cannot be odd so one of them must be even.

12 Probability   page 406

1 0.74
2 a 0.99
   b 0.36
   c 0.35
   d 0.36

3 **a** 0.172

   b 0.3246

   c 0.4554

4 **a**

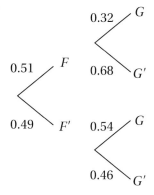

   b 0.4278

   c 0.394

5 Probability of getting accused of gun crime (before and including up to age 39) = 0.234

   This is assuming that England has the same rate of gun crime in 30 to 39 year olds as America.

   The probability of John Paul getting accused of gun crime is greater than 0.234, as John Paul will have more opportunities in life to acquire a gun licence, assuming he is seeking to acquire a gun.

6 **i** 0.28413

   ii 0.25416

7 **a** 47

   b **i** 0.511

      ii 0.708

   c Because a girl can do both disciplines.

   d If S and D are independent, then $P(S \cap D) = P(S) \times P(D)$

   $$\frac{40}{50} \times \frac{30}{50} = 0.48$$

   But $P(S \cap D) = \frac{23}{50} = 0.46 \neq 0.48$ so they are not independent

8 **a** **i** 0.267

      ii 0.162

      iii 0.467

      iv 0.81

   b **i** Since $P(A \cap B) \neq 0$ so A and B are not mutually exclusive.

      ii Since $P(A|B) \neq P(A)$, A and B are NOT independent.

9 **a** $P(F \cup N) = P(F) + P(N) - P(F \cap N)$, assuming they are not mutually exclusive.

   b 0.0264

   c 0.434

   d 0.87

10 **a**

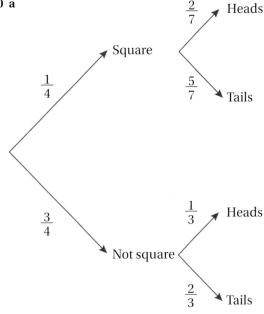

   b $\frac{19}{28}$

   c $\frac{2}{9}$

13 Statistical distributions page 408

1 **a** 0.1666

   b 0.3979

   c $m = 46.7$

2 **a** 0.142

   b 0.241

   c 0.0.0342

   d Evidence suggests height and hand span are positively correlated / linked, so the assumption of independence is not sensible.

3 **a** Symmetrical (about the mean μ)

   Mode = mean = median

   Horizontal axis asymptotic to curve

   Distribution is 'bell shaped'

   99.97% of data lies within 3 standard deviations of the mean.

   Any 3 sensible properties

   b 0.683

4 0.17

5 a

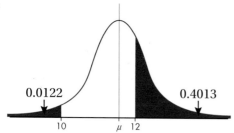

b i $P\left(Z < \dfrac{10 - \mu}{\sigma}\right) = 0.0122 \Rightarrow 10 - \mu = -2.25\sigma$

$P\left(Z > \dfrac{12 - \mu}{\sigma}\right) = 0.4013 \Rightarrow 12 - \mu = 0.25\sigma$

$2 = 2.5\sigma \Rightarrow \sigma = 0.8$

ii 11.8

c 0.9848

6 a 72.95

b 0.392

7 a 0.1056

b 0.6826

8 70.3 days

9 69.9%

10 a 0.1923

b 8.46

14 Statistical hypothesis testing page 410

1 a

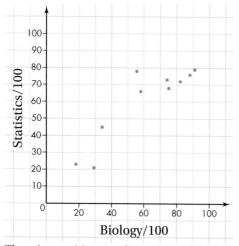

There is a positive correlation, so as the biology scores improve, the statistics scores also improve.

b $H_0: p = 0$

$H_1: p > 0$

where p is a positive correlation between the two variables.

c The critical value for $n = 10$ at the 1% significance level for a one-tailed test is found from tables to be 0.7155.

0.897 > 0.7155 so reject the null hypothesis as there is evidence to suggest that as students increase their scores in one subject, their scores in the other subject also increase.

2 a

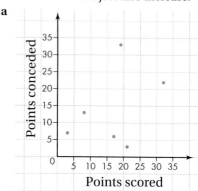

Very slight positive correlation.

b $H_0: p = 0$

$H_1: p > 0$

where p is a positive correlation between the two variables.

c The critical value for $n = 6$ at the 10% significance level for a one-tailed test is found from tables to be 0.6084.

0.377 < 0.7293 so accept the null hypothesis as there is not enough evidence to suggest that the new training routine is improving performance.

3 a

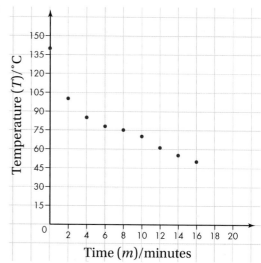

b To test whether the relationship is of the form $T = ka^m$, take logarithms of both sides to give

$\log T = \log k + m \log a$ and plot the graph of $\log T$ against m.

As the graph is a straight line, the relationship is of the form $T = ka^m$.

Time $(m)/$ minutes	Temperature $(T)/°C$	$\log T$
0	140	2.146128
2	100	2
4	85	1.929419
6	78	1.892095
8	75	1.875061
10	70	1.845098
12	61	1.78533
14	55	1.740363
16	50	1.69897

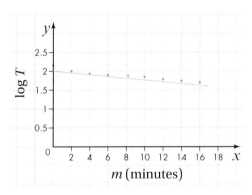

m (minutes)

c $a = 0.946$

$k = 118.2$

4 a $\log_{10} y = -0.31 \log_{10} x + \log_{10} 3.16$

b $y = 3.16 x^{-0.31}$

5 a

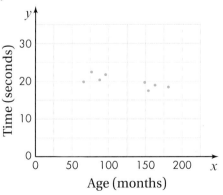

Age (months)

This shows a negative correlation, which implies that as age increases, the slide time decreases. There is no data between 100 and 150 months.

b $H_0: p = 0$

$H_1: p \neq 0$

where p is a correlation between the two variables.

c The critical value for $n = 8$ at the 5% significance level for a two-tailed test is found from tables to be 0.7067.

$-0.754 > -0.7067$ so reject the null hypothesis as there is evidence to suggest a connection between slide time and age.

d Need to redo the test to include all age groups and a larger sample.

6 Null hypothesis, $H_0: p = 60$. The speed is normal. Alternate hypothesis, $H_1: p > 60$. The speed has increased.

1-tailed test at the 10% significance level.

The distribution of sample means, $\bar{X}$, is $N(\mu, \frac{\sigma^2}{n})$.

According to the null hypothesis, $\mu = 60$ and it is known that $\sigma = 5.2$ and $n = 10$.

So this distribution is $N\left(60, \frac{5.2^2}{10}\right)$.

The sample mean is 66.3 mph.

The probability of the mean, $\bar{X}$, of a randomly chosen sample being greater than the sample mean is given by:

$$P(\bar{X} \geqslant 66.3) = 1 - \Phi\left(\frac{66.3 - 60}{\frac{5.2}{\sqrt{10}}}\right)$$

$$P(\bar{X} \geqslant 66.3) = 1 - (\Phi 3.831) = 0.1128$$

Since $0.1128 > 10\%$, the required significance level, the null hypothesis is accepted, as there is not enough evidence.

7 H_0: The population mean is unchanged.

$\mu = 8$

H_1: The population mean has increased.

$\mu > 8$

Calculate the mean and standard deviation:

$$\bar{x} = \frac{\Sigma x}{n} = \frac{9200}{1000} = 9.2$$

$$s = \sqrt{\frac{\Sigma x^2 - n\bar{x}^2}{n}} = \sqrt{\left(\frac{86023 - 1000 \times 9.2^2}{n}\right)} = 1.176$$

Now standardise the z-value corresponding to $\bar{x} = 9.2$ calculated using $\mu = 8$ and by approximating σ by $s = 1.176$

$$z = \frac{\bar{x} - \mu_0}{\frac{\sigma}{\sqrt{n}}} = \frac{(9.2 - 8)}{\frac{1.176}{\sqrt{1000}}} = 32.268$$

The critical region value is $z = 2.236$ when the significance level is 1%.

32.268 > 2.236, therefore the null hypothesis is rejected. The evidence suggests the electrician's claims are true – the lifetime of a TV has increased. You have made the assumption that the TV lives are normally distributed.

8 the evidence suggests that the mean volume is set to 5 litres.

9 there is not sufficient evidence to suggest that the battery life is any better.

10 $H_0: \rho = 0$. There is no correlation between the amount of sleep a teenager gets and their IQ. $H_1: \rho \neq 0$. There is a correlation between the amount of sleep a teenager gets and their IQ. Significance level is 5%.

The critical value for a two-tailed test at the 5% significance level for $n = 50$ is 0.2787.

Since 0.2743 < 0.2787, reject H_1. There is not enough evidence to suggest that there is correlation between hours sleeping and IQ.

15 Kinematics   page 412

1 1.47 s and 4.53 s

2 **a** 1.76 s

   b 14.6 m s^{-1}

3 **a** $(8\mathbf{i} + 9\mathbf{j})$ m

   b $(-5\mathbf{i} - 2\mathbf{j})$ m s^{-1}

   c 4 s

   d 15.8 m

4 **a** 28 m s^{-1}

   b 17.5 m

   c 40 m

5 **a** $\frac{4}{7}$ s

   b $\mathbf{a} = (24\mathbf{i} + 14t\mathbf{j})$ m s^{-2}

   When $t = \frac{1}{2}$, $\mathbf{a} = (24\mathbf{i} + 7\mathbf{j})$ m s^{-2}

   $a = \sqrt{24^2 + 7^2} = 25$ m s^{-2}

6 **a** $(16\mathbf{i} - 12\mathbf{j})$ km h^{-1}

   b Let T hours be the time after 15:40
   $\mathbf{r}_U = (-5\mathbf{i} + 80\mathbf{j}) + T(16\mathbf{i} - 12\mathbf{j})$
   $\mathbf{r}_U = (16T - 5)\mathbf{i} + (80 - 12T)\mathbf{j}$
   $\mathbf{r}_V = (46 + 2(T - 1))\mathbf{i} + (18(T - 1) - 7)\mathbf{j}$
   $\mathbf{r}_V - \mathbf{r}_U = (51 + 2(T - 1) - 16T)\mathbf{i} + (18(T - 1) + 12T - 87)\mathbf{j}$
   i: $51 + 2(T - 1) - 16T = 0$
   $51 + 2T - 2 - 16T = 0$
   $14T = 49$
   $T = 3.5$
   j: $18(T - 1) + 12T - 87 = 0$
   $18T - 18 + 12T - 87 = 0$

$30T = 105$
$T = 3.5$
Hence the ships will collide when $T = 3.5$, at 19:10.
$\mathbf{r}_U = [(16(3.5) - 5)\mathbf{i} + (80 - 12(3.5))\mathbf{j}]$
   $= (51\mathbf{i} + 38\mathbf{j})$ km

   c 2.5 km

7 **a** $\left(\frac{48}{g}\mathbf{i} + \left(\frac{28}{g} + 8\right)\mathbf{j}\right)$ m and $\left(\frac{168}{g}\mathbf{i} + \left(\frac{28}{g} + 8\right)\mathbf{j}\right)$ m

   b 52°

8 **a** $(-\mathbf{i} + 3\mathbf{j})$ m s^{-2}

   b $(6\mathbf{i} - 7\mathbf{j})$ m s^{-1}

   c $\frac{7}{3}$ s

   d 30.1 m

9 **a**

   b 447 m

10 **a** 36 m

   b 9.83 m

11 **a** $[(5t - 9)\mathbf{i} + (1 + 2t)\mathbf{j}]$ km

   b 1.5 hours

   c $d^2 = 10\left(\frac{3}{2}\right)^2 - 30\left(\frac{3}{2}\right) + 25 = \frac{5}{2}$

   $d = \sqrt{\frac{5}{2}} = 1.58$ km

12 **a** Vertically, $u = 20\sin\beta$ m s^{-1}, $a = -g$ m s^{-2}, $s = 0$ m,
   let $t = T$ s

   $s = ut + \frac{1}{2}at^2$

   $0 = (20\sin\beta)T + \frac{1}{2} \times -gT^2$

   $0 = T(20\sin\beta - \frac{1}{2}gT)$

   $T = \frac{40\sin\beta}{g}$

   Horizontally, $u = 20\cos\beta$ m s^{-1}, $a = 0$ m s^{-2},
   $t = \frac{40\sin\beta}{g}$ s, $s = d$ m

   $s = ut + \frac{1}{2}at^2$

$$d = (20\cos\beta)\frac{40\sin\beta}{g} + 0$$

$$d = \frac{800\sin\beta\cos\beta}{g}$$

$$d = \frac{400 \times 2\sin\beta\cos\beta}{g} = \frac{400\sin2\beta}{g}$$

b $\frac{400}{g}$ m

13 a $(27\mathbf{i} + 13\mathbf{j})$ m

b $(83\mathbf{i} + 45\mathbf{j})$ m

16 Forces   page 414

1 Let the angle be θ
$R(\nearrow)$: $X\cos\theta - F - 0.6g\sin\theta = 0$
$R(\nwarrow)$: $R - X\sin\theta - 0.6g\cos\theta = 0$
$F = \mu R$

$X\cos\theta - 0.6g\sin\theta = \frac{1}{17}(X\sin\theta + 0.6g\cos\theta)$

$X - 0.6g\tan\theta = \frac{1}{17}(X\tan\theta + 0.6g)$

$17X - 10.2g \times \frac{1}{3} = \frac{1}{3}X + 0.6g$

$\frac{50}{3}X = 4g$

$X = \frac{6}{25}g$ N

2 0.227

3 a $9\,\mathrm{m\,s^{-2}}$

b $054.6°$

4 $7.25\,\mathrm{m}$

5 a $R(\nearrow)$: $T - F - 30g\sin14 = 0$
$R(\nwarrow)$: $R - 30g\cos14 = 0$
$F = \mu R$
$T - 30g\sin14 = 0.05(30g\cos14)$
$T = 85.4\,\mathrm{N}$

b Light inextensible string. Tension is constant.

6 a $1200\,\mathrm{kg}$

b $85.1\,\mathrm{N}$

c $4.8\,\mathrm{s}$

7 a $R(\swarrow)$: $Mg\sin\theta - F = Ma$
$R(\nwarrow)$: $R - Mg\cos\theta = 0$
$F = \mu R$
$Mg\sin\theta - Ma = \mu Mg\cos\theta$
$g\sin\theta - a = \mu g\cos\theta$
$a = g\sin\theta - \mu g\cos\theta$
$a = g(\sin\theta - \mu\cos\theta)$

b $4.52\,\mathrm{s}$

8 a $R(\searrow)_Q$: $5g\sin60 - F - T = 5a$
$R(\nearrow)_Q$: $R - 5g\cos60 = 0$
$F = \mu R$
$5g\sin60 - \mu5g\cos60 - T = 5a$
$R(\nearrow)_P$: $T - F - 3g\sin60 = 3a$
$R(\nwarrow)_P$: $R - 3g\cos60 = 0$
$T - \mu3g\cos60 - 3g\sin60 = 3a$
Adding: $2g\sin60 - \mu8g\cos60 = 8a$
$g\sqrt{3} - 4\mu g = 8a$
$a = \frac{(\sqrt{3} - 4\mu)g}{8}\,\mathrm{m\,s^{-2}}$

b $1\frac{1}{9}$ m

9 a $R(\nearrow)$: $P\cos2\alpha - F - 4g\sin\alpha = 0$
$R(\nwarrow)$: $R - P\sin2\alpha - 4g\cos\alpha = 0$
$F = \mu R$

$P\cos2\alpha - 4g\sin\alpha = \frac{1}{8}(P\sin2\alpha + 4g\cos\alpha)$

$8P\cos2\alpha - 32g\sin\alpha = P\sin2\alpha + 4g\cos\alpha$
$8P\cos2\alpha - P\sin2\alpha = 32g\sin\alpha + 4g\cos\alpha$
$P(8\cos2\alpha - \sin2\alpha) = 32g\sin\alpha + 4g\cos\alpha$

$P = \frac{4g(\cos\alpha + 8\sin\alpha)}{8\cos2\alpha - \sin2\alpha}$

b Given that $\sin\alpha = \frac{1}{2}$, $\cos\alpha = \frac{\sqrt{3}}{2}$

$\sin2\alpha = 2\sin\alpha\cos\alpha = 2(\frac{1}{2})(\frac{\sqrt{3}}{2}) = \frac{\sqrt{3}}{2}$

$\cos2\alpha = \cos^2\alpha - \sin^2\alpha = (\frac{\sqrt{3}}{2})^2 - (\frac{1}{2})^2 = \frac{1}{2}$

$P = \frac{4g\left(\frac{\sqrt{3}}{2} + 8 \times \frac{1}{2}\right)}{8 \times \frac{1}{2} - \frac{\sqrt{3}}{2}} = 4g\frac{8 + \sqrt{3}}{8 - \sqrt{3}}$

10 a $x = 22.1°$

b $M = 0.669\,\mathrm{kg}$

11 $5.14\,\mathrm{m\,s^{-1}}$

12 a $R_Y(\downarrow)$: $4g - T = 4a$
Let the angle of the slope be α
$R(\nearrow)_X$: $T - F - 3g\sin\alpha = 3a$
$R(\nwarrow)_X$: $R - 3g\cos\alpha = 0$
$T - \mu3g\cos\alpha - 3g\sin\alpha = 3a$
Adding: $4g - \mu3g\cos\alpha - 3g\sin\alpha = 7a$

$4g - \mu3g \times \frac{4}{5} - 3g \times \frac{3}{5} = 7a$

$\frac{11}{5}g - \frac{12}{5}\mu g = 7a$

Since $a > 0$

$\frac{11}{5}g - \frac{12}{5}\mu g > 0$

$11 - 12\mu > 0$

$\mu < \frac{11}{12}$

b Force on pulley $= \sqrt{(T + T\sin\alpha)^2 + (T\cos\alpha)^2}$

Force $= \sqrt{\left(T + \dfrac{3}{5}T\right)^2 + \left(\dfrac{4}{5}T\right)^2}$

Force $= \sqrt{\dfrac{80T^2}{25}} = \dfrac{4\sqrt{5}T}{5}$

Alternative method.

Force exerted on pulley $= 2T\cos\left(45 - \dfrac{1}{2}\alpha\right)$

Force $= 2T[\cos 45\cos\dfrac{1}{2}\alpha + \sin 45\sin\dfrac{1}{2}\alpha]$

$\sin\alpha = \dfrac{3}{5}$

$2\sin\dfrac{1}{2}\alpha\cos\dfrac{1}{2}\alpha = \dfrac{3}{5}$

$2\sin\dfrac{1}{2}\alpha\sqrt{1 - \sin^2\dfrac{1}{2}\alpha} = \dfrac{3}{5}$

$4\sin^2\dfrac{1}{2}\alpha(1 - \sin^2\dfrac{1}{2}\alpha) = \dfrac{9}{25}$

$\sin^4\dfrac{1}{2}\alpha - \sin^2\dfrac{1}{2}\alpha + \dfrac{9}{100} = 0$

$\sin^2\dfrac{1}{2}\alpha = \dfrac{1 \pm \sqrt{(-1)^2 - 4 \times 1 \times \dfrac{9}{100}}}{2 \times 1}$

$\sin^2\dfrac{1}{2}\alpha = \dfrac{1}{10}$ or $\dfrac{9}{10}$

$\cos^2\dfrac{1}{2}\alpha = \dfrac{9}{10}$ or $\dfrac{1}{10}$

$\sin\dfrac{1}{2}\alpha = \dfrac{\sqrt{10}}{10}$ or $\dfrac{3\sqrt{10}}{10}$

$\cos\dfrac{1}{2}\alpha = \dfrac{3\sqrt{10}}{10}$ or $\dfrac{\sqrt{10}}{10}$

Force $= 2T[\dfrac{\sqrt{2}}{2} \times \dfrac{3\sqrt{10}}{10} + \dfrac{\sqrt{2}}{2} \times \dfrac{\sqrt{10}}{10}]$

$= 2T[\dfrac{3\sqrt{20}}{20} + \dfrac{\sqrt{20}}{20}]$

$= 2T[\dfrac{\sqrt{20}}{5}]$

$= 2T[\dfrac{2\sqrt{5}}{5}]$

$= \dfrac{4}{5}\sqrt{5}T$

17 Moments   page 417

1 **a** 157 N

   b 26.7 kg

   c As a rod which is one-dimensional, rigid object.

2 27 N m anticlockwise

3 $x = \dfrac{1}{2}$

4 **a** 42.4 N

   b 107 N

   c 4.23 m

5 **a** Let the reaction force at the peg be R.
   Moments about P: $3\sqrt{3}g \times 3.5\cos 30 + 2\sqrt{3}g \times$
   $7\cos 30 - R \times \dfrac{49}{8} = 0$

   $\dfrac{49}{8}R = 3\sqrt{3}g \times 3.5 \times \dfrac{\sqrt{3}}{2} + 2\sqrt{3}g \times 7 \times \dfrac{\sqrt{3}}{2}$

   $\dfrac{49}{8}R = 3\sqrt{3}g \times \dfrac{7}{2} \times \dfrac{\sqrt{3}}{2} + 2\sqrt{3}g \times 7 \times \dfrac{\sqrt{3}}{2}$

   $\dfrac{49}{8}R = \dfrac{63}{4}g + 21g = \dfrac{147}{4}g$

   $R = \dfrac{147}{4}g \times \dfrac{8}{49} = 6g \text{ N}$

   b Let reaction force at P be R_p
   $R(\rightarrow)\ F - R\sin 30 = 0$
   $F = 6g\sin 30 = 3g$
   $R(\downarrow)\ 3\sqrt{3}g + 2\sqrt{3}g - R\cos 30 - R_p = 0$
   $R_p = 5\sqrt{3}g - 6g\cos 30$
   $R_p = 5\sqrt{3}g - 3\sqrt{3}g = 2\sqrt{3}g$
   $F = \mu R_p$

   $\mu = \dfrac{F}{R_p} = \dfrac{3g}{2\sqrt{3}g} = \dfrac{\sqrt{3}}{2}$

6 **a** 255 N

   b 11 kg

7 **a** $M = 30$ kg

   b $x = 1.6$ m

8 **a** $\dfrac{26g}{5}$ N

   b i 61.2 N

   ii 39.8°

9 $R(\rightarrow)\ F_X - R_Z = 0$
   $R(\downarrow)\ Mg - F_Z - R_X = 0$
   Moments about X:
   $Mg \times 2a\cos\alpha - R_Z \times 6a\sin\alpha - F_Z \times 6a\cos\alpha = 0$
   $Mg - 3R_Z\tan\alpha - 3F_Z = 0$
   $F_Z + R_X - 3R_Z\tan\alpha - 3F_Z = 0$
   $R_X - 3R_Z\tan\alpha - 2F_Z = 0$

   $\dfrac{F_X}{\mu_X} - 3F_X\tan\alpha - 2\mu_Z R_Z = 0$

   $\dfrac{F_X}{\mu_X} - 3F_X\tan\alpha - 2\mu_Z F_X = 0$

   $\dfrac{1}{\mu_X} - 3\tan\alpha - 2\mu_Z = 0$

   $1 - 3\mu_X\tan\alpha - 2\mu_X\mu_Z = 0$

   $1 - 2\mu_X\mu_Z = 3\mu_X\tan\alpha$

   $\tan\alpha = \dfrac{1 - 2\mu_X\mu_Z}{3\mu_X}$

10 225 kg

FORMULAE

Formulae that students are expected to know for A Level Mathematics are given below and should be learnt by heart.

Pure Mathematics

Quadratic Equations

$ax^2 + bx + c = 0$ has roots $\dfrac{-b \pm \sqrt{b^2 - 4ac}}{2a}$

Laws of Indices

$a^x a^y \equiv a^{x+y}$

$a^2 \div a^y \equiv a^{x-y}$

$(a^x)^y \equiv a^{xy}$

Laws of Logarithms

$x = a^n \Leftrightarrow n = \log_a x$ for $a > 0$ and $x > 0$

$\log_a x + \log_a y \equiv \log_a(xy)$

$\log_a x - \log_a y \equiv \log_a\left(\dfrac{x}{y}\right)$

$k\log_a x \equiv \log_a(x^k)$

Coordinate Geometry

A straight line graph, gradient m passing through (x_1, y_1) has equation $y - y_1 = m(x - x_1)$

Straight lines with gradients m_1 and m_2 are perpendicular when $m_1 m_2 = -1$

Sequences

General term of an arithmetic progression:

$u_n = a + (n - 1)d$

General term of a geometric progression:

$u_n = ar^{n-1}$

Trigonometry

In the triangle ABC

Sine rule: $\dfrac{a}{\sin A} = \dfrac{b}{\sin B} = \dfrac{c}{\sin C}$

Cosine rule: $a^2 = b^2 + c^2 - 2bc\cos A$

Area $= \dfrac{1}{2}ab\sin C$

$\cos^2 A + \sin^2 A \equiv 1$

$\sec^2 A \equiv 1 + \tan^2 A$

$\operatorname{cosec}^2 A \equiv 1 + \cot^2 A$

$\sin 2A \equiv 2\sin A\cos A$

$\cos 2A \equiv \cos^2 A - \sin^2 A$

$\tan 2A \equiv \dfrac{2\tan A}{1 - \tan^2 A}$

Mensuration

Circumference and area of circle, radius r and diameter d:

$C = 2\pi r = \pi d \qquad A = \pi r^2$

Pythagoras' theorem:

In any right-angled triangle where a, b and c are the lengths of the sides and c is the hypotenuse, $c^2 = a^2 + b^2$

Area of a trapezium $= \dfrac{1}{2}(a + b)h$, where a and b are the lengths of the parallel sides and h is their perpendicular separation.

Volume of a prism = area of cross section × length

For a circle of radius r, where an angle at the centre of θ radians subtends an arc of length s and encloses an associated sector of area A:

$s = r\theta \quad A = \dfrac{1}{2}r^2\theta$

Calculus and Differential Equations

Differentiation

Function	Derivative
x^n	nx^{n-1}
$\sin kx$	$k\cos kx$
$\cos kx$	$-k\sin kx$
e^{kx}	ke^{kx}
$\ln x$	$\dfrac{1}{x}$
$f(x) + g(x)$	$f'(x) + g'(x)$
$f(x)g(x)$	$f'(x)g(x) + f(x)g'(x)$
$f(g(x))$	$f'(g(x))g'(x)$

Integration

Function	Integral
x^n	$\dfrac{1}{n+1}x^{n+1} + c,\ n \neq -1$
$\cos kx$	$\dfrac{1}{k}\sin kx + c$
$\sin kx$	$-\dfrac{1}{k}\cos kx + c$
e^{kx}	$\dfrac{1}{k}e^{kx} + c$
$\dfrac{1}{x}$	$\ln|x| + c,\ x \neq 0$
$f'(x) + g'(x)$	$f(x) + g(x) + c$
$f'(g(x))g'(x)$	$f(g(x)) + c$

Area under a curve $= \displaystyle\int_a^b y\,dx\,(y \geqslant 0)$

Vectors

$|x\mathbf{i} + y\mathbf{j} + z\mathbf{k}| = \sqrt{(x^2 + y^2 + z^2)}$

Statistics

The mean of a set of data: $\bar{x} = \dfrac{\sum x}{n} = \dfrac{\sum fx}{\sum f}$

The standard Normal variable:

$$Z = \dfrac{X - \mu}{\sigma} \text{ where } X \sim N(\mu,\ \sigma^2)$$

Mechanics

Forces and Equilibrium

Weight $=$ mass $\times$ g

Friction: $F \leqslant \mu R$

Newton's second law in the form: $F = ma$

Kinematics

For motion in a straight line with variable acceleration:

$$v = \dfrac{dr}{dt} \qquad a = \dfrac{dv}{dt} = \dfrac{d^2r}{dt^2}$$

$$r = \int v\,dt \qquad v = \int a\,dt$$

Glossary

1-tail test A test of the region under one of the tails (sides) of a statistical distribution.

2-tail test A test of the region under both of the tails (sides) of a statistical distribution.

absolute value See **modulus**.

acceleration The rate of change of velocity. If the displacement is x then the acceleration is $\dfrac{d^2x}{dt^2}$.

addition formula An identity that expresses a trigonometric function of the sum or difference of two angles in terms of trigonometric functions of the individual angles. An example is $\cos(A + B) = \cos A \cos B - \sin A \sin B$.

addition rule A formula you can use for two events that links the probability of the union and the intersection: $P(A \cup B) = P(A) + P(B) - P(A \cap B)$.

alternative hypothesis A hypothesis which the researcher tries to prove.

amplitude The measure of change over a period of a periodic variable or function. For example, the trigonometric function sine has a period of 2π. Its amplitude over a period is 1. See also periodic function.

anticlockwise Turning in the opposite direction to the hands of a clock.

arc Part of the circumference of a circle. The word can also refer to part of any curve.

arccosine Usually written as arccos or $\cos^{-1}$, this is the inverse of the cosine function. If $\cos\theta = x$ then $\arccos x = \theta$. The principal values are $0 \leqslant \arccos x \leqslant \pi$.

arcsine Usually written arcsin or $\sin^{-1}$, this is the inverse of the sine function. If $\sin\theta = x$ then $\arcsin x = \theta$. The principal values are $-\dfrac{\pi}{2} \leqslant \arcsin x \leqslant \dfrac{\pi}{2}$.

arctangent Usually written as arctan or $\tan^{-1}$, this is the inverse of the tangent function. If $\tan\theta = x$ then $\arctan x = \theta$. The principal values are $-\dfrac{\pi}{2} \leqslant \arctan x \leqslant \dfrac{\pi}{2}$.

arithmetic sequence A sequence of numbers in which each term differs by a constant amount (the common difference) from the one before.

asymptote A straight line that a curve approaches but never meets.

bearing An angle measured clockwise from north.

binomial distribution The number of successes, x, in n repeated trials with a constant probability, p.

binomial expansion If n is a positive integer then
$$(a + b)^n = a^n + na^{n-1}b + \frac{n(n-1)}{2}a^{n-2}b^2 + \ldots + b^n.$$
More generally, for any n,
$$(1 + x)^n = 1 + nx + \frac{n(n-1)}{2}x^2 + \frac{n(n-1)(n-2)}{3!}x^2 + \ldots$$
is a convergent series if $|x| < 1$.

bivariate data Data that has two variables.

Cartesian coordinates Coordinate system invented by René Descartes (1596–1650) that allows each point to be specified uniquely by a pair of coordinates, for example (1, 2). The coordinates are the directed distances from two perpendicular axes, for example x and y, to the point.

Cartesian equation An equation in which the variables are the Cartesian coordinates of a point on a line, curve or surface.

chain rule If y is a function of u and u is a function of x then $\dfrac{dy}{dx} = \dfrac{dy}{du} \times \dfrac{du}{dx}$.

clockwise Turning in the same direction as the hands of a clock.

cobweb diagram A graphical representation of an iterative formula where $g'(x) < 0$.

coefficient A number written in front of a variable in an algebraic term; for example, in $8x$, 8 is the coefficient of x.

coefficient of friction The value of μ in the inequality $F \leqslant \mu R$.

column vector A vector written in the form $\begin{bmatrix} a \\ b \end{bmatrix}$ where a and b are the coefficients of $\mathbf{i}$ and $\mathbf{j}$.

component The amount of a force acting in a particular direction.

composite function The application of one function more than once, or the combination of two or more functions, makes a composite function.

conditional probability The calculation of the probabilities of dependent events, looking at the probability of an event given that another event occurs first: $P(A \mid B) = \dfrac{P(A \cap B)}{P(B)}$.

continuity correction A continuity correction is an adjustment that is made when a discrete distribution is approximated by a continuous distribution.

continuous distribution Data that can take values within a range.

convergent sequence   A sequence for which the nth term gets closer and closer to some limit as n increases.

correlation   A positive correlation means that if one variable gets bigger, the other variable tends to get bigger. A negative correlation means that if one variable gets bigger, the other variable tends to get smaller.

correlation coefficient   A numerical measure of the strength of the linear association between two variables.

cosecant   The reciprocal of the sine of an angle. It is usually abbreviated to cosec or csc and so $\operatorname{cosec}\theta = \dfrac{1}{\sin\theta}$.

cotangent   The reciprocal of the tangent of an angle. It is usually abbreviated to cot and so $\cot\theta = \dfrac{1}{\tan\theta}$.

critical region   The area of the sampling distribution of a statistic that will lead to the rejection of the hypothesis tested when that hypothesis is true.

decreasing sequence   A sequence in which each term is less than the previous one.

degree   The unit of measure of an angle. For example, a right angle is 90 degrees.

dependent events   Events that will influence one another.

derivative   The result of differentiating a function.

differential equation   An equation containing derivatives of a function. Solving a differential equation involves integration.

discontinuity   A point where there is a break in a function. For example, $f(x) = \dfrac{1}{x}$ has a discontinuity at $x = 0$. The line $x = 0$ is an asymptote.

discrete probability distribution   The probability distribution of a discrete random variable.

displacement   The distance, in a specified direction, of a moving point from its starting point.

divergent sequence   A sequence for which the sum increases without limit as n increases.

domain   The set of values of the variable for which a function is defined. For example, the largest possible domain of the function $\sqrt{x+2}$ is $x \geqslant 2$.

double angle formula   An identity that expresses a trigonometric function of double an angle in terms of trigonometric functions of the single angle. An example is $\sin 2A = 2\sin A \cos A$.

elementary function   Functions including polynomials, rational functions, radical expressions, exponential functions, logarithms, trigonometric functions and inverse trigonometric functions.

equilibrium   When forces all balance and, as a result, the sum of the forces in any direction is zero.

friction   The force which opposes motion when a surface is rough.

function   A special mapping such that every member of the domain maps to exactly one member of the range.

general solution   The general solution of a differential equation is an equation containing an arbitrary constant. It will correspond to a family of curves. For example, the general solution of the differential equation $\dfrac{\mathrm{d}y}{\mathrm{d}x} = x$ is $y = \frac{1}{2}x^2 + c$, which corresponds to a set of parabolas.

geometric sequence   A sequence of numbers in which each term differs by a constant multiple (the common ratio) from the one before.

histogram   A diagram consisting of rectangles with area proportional to the frequency of a variable and width equal to the class interval.

hypothesis test   A formal process to determine whether to reject a null hypothesis, based on sample data.

i j k notation   A way of representing a vector in three dimensions, where $\mathbf{i}$ is one unit along the x-axis, $\mathbf{j}$ is one unit along the y-axis and $\mathbf{k}$ is one unit along the z-axis.

identity   An equation that cannot be solved because it is true for all values of the variables. An example is $\sin^2 x + \cos^2 x = 1$. Identities are sometimes written using the $\equiv$ sign: $\sin^2 x + \cos^2 x \equiv 1$.

image   The resulting position after the transformation of an object.

implicit equation   An equation which gives a relationship that the dependent and independent variables must satisfy, rather than defining the dependent variable directly (explicitly). For example, the circle with the implicit equation $x^2 + y^2 = 100$.

improper fraction   A numeric fraction where the numerator is greater than or equal to the denominator, or an algebraic fraction where the degree of the numerator is greater or equal to the degree of the denominator.

increasing sequence   A sequence in which each term is greater than the previous one.

independent events   Two events, *A* and *B*, are independent if the fact that *A* occurs does not affect the probability of *B* occurring.

inflection points   The points between the mean, μ, and the standard deviation, σ– that is, the inflection points are $\mu - \sigma$ and $\mu + \sigma$.

integrand   The function being integrated.

integration by parts   A method of integrating a product of two functions, u and $\dfrac{du}{dx}$, using the rule $\int u \dfrac{du}{dx} dx = uv - \int v \dfrac{du}{dx} dx$. The aim is to change a difficult integration into an easier one.

integration by substitution   Changing the independent variable in an integration using the rule $\int f(x) dx = \int f(x) \dfrac{dx}{du} du$. The aim is to make $f(x) \dfrac{dx}{du}$ a function of u that is easier to integrate than the original integrand.

inverse function   A function that maps members of the range back to members of the domain.

irreducible form   The form of a fraction where the numerator and the denominator have no common factors other than 1.

iteration   The repetition of a process or action.

limiting equilibrium   When a particle is in limiting equilibrium, it is either moving or on the point of moving, and $F = \mu R$.

magnitude   The size (or length) of a vector.

many-to-one function   A function in which some members of the domain map to the same member of the range.

mapping   Transforms one set of numbers into a different set of numbers.

maximum point   A point on a graph where the gradient is zero, which is higher than the points either side of it. Can be a global maximum where it is the highest point anywhere on the graph, or a local maximum where it is the highest point in that part of the graph.

minimum point   A point on a graph where the gradient is zero, which is lower than the points either side of it. Can be a global minimum where it is the lowest point anywhere on the graph, or a local minimum where it is the lowest point in that part of the graph.

mixed number   A number that comprises an integer part and a proper fraction.

modelling   A model allows you to simplify a real event or situation.

modulus   The modulus of a real number is the magnitude of the number. It is shown by two vertical lines. For example, $|3.5| = |{-3.5}| = 3.5$.

mutually exclusive events   Two or more events that cannot occur at once.

non-standardised variables   A special case of the normal distribution is the standard normal distribution. Any normal distribution can be represented by standard normal distribution. A non-standardised variable is one which has not been manipulated for the standard normal distribution. See also standard normal distribution.

non-uniform   A rod is said to be uniform if its mass is not evenly distributed through the rod. As a result, its centre of mass may not be assumed to be positioned at the centre of the rod.

normal distribution   Normal distributions are defined by two parameters, the mean (μ) and the standard deviation (σ). If a variable X has a normal distribution then you can write this using mathematical notation: $X \sim N(\mu, \sigma^2)$.

null hypothesis   A hypothesis which the researcher tries to disprove, reject or nullify.

one-to-one function   A function in which one member of the domain maps to exactly one member of the range.

order (of a periodic sequence)   The length of the cycle of recurring values.

oscillating sequence   A sequence in which the terms are successively greater and then less than the one before.

parallel   Two vectors are parallel if they point in the same direction. They do not need to have the same magnitude.

parametric equation   An equation that involves an independent variable in terms of which both coordinates of a point on a graph can be expressed.

partial fraction   One of two or more fractions into which a more complex fraction can be split.

particular solution   A solution of a differential equation satisfies a particular condition. For example, it might pass through a particular point.

period   The interval over which a periodic function repeats. For example, the trigonometric function sine has a period of 2π.

periodic function   A function that repeats in uniform periods. The trigonometric functions are examples of periodic functions.

periodic sequence   A sequence in which the terms cycle round the same set of numbers as n increases.

point of inflection   A point on a graph, other than a turning point, at which the gradient of the curve is zero.

position vector   The relative displacement of a point from the origin (for example, a point with coordinates $(-2, 3)$ has the position vector $(-2\mathbf{i} + 3\mathbf{j})$)

principal values   The chosen representative values in the range of a function that could have more than one value. For example, $\arcsin 0.5$ could have many values but the principal value is $\frac{\pi}{6}$.

probability   The measure of the chance that a particular event will occur as a result of an experiment.

product moment correlation coefficient (PMCC)   A numerical measure of the strength of a correlation between two variables. It ranges from -1 to 1 and can have any value in this range.

product rule (differentiation)   A rule for differentiating the product of two functions. If $y = uv$, where u and v are functions of x, then $\frac{dy}{dx} = u\frac{dv}{dx} + v\frac{du}{dx}$.

product rule (probability)   If the formula for conditional probability is rearranged you get an important probability law called the product rule.

projectile   An object propelled through the air such that its subsequent motion takes place in two dimensions rather than one.

proof by contradiction   Method of proof that begins by assuming that the opposite of what you are trying to prove is true, followed by logical steps that lead to a contradiction.

proper fraction   A numeric fraction where the numerator is less than the denominator, or an algebraic fraction where the degree of the numerator is less than the degree of the denominator.

quadrant   One of the four sections between a pair of axes crossing at the origin, each with both positive and negative values.

quotient rule   A rule for differentiating the quotient of two functions. If $y = \frac{u}{v}$ where u and v are functions of x, then $\frac{dy}{dx} = \frac{v\frac{du}{dx} - u\frac{dv}{dx}}{v^2}$.

radian   A unit of measurement for angles. π radians = $180°$ and one radian is approximately $57.3°$.

range   The set of values that a function can take. For example, the range of $\sin\theta$ is $-1 \leqslant \sin\theta \leqslant 1$.

recurrence relation   An equation that gives each term in a sequence in terms of the previous one. If the first term is known then subsequent terms can be calculated.

resolving   The process of summing the forces in a particular direction.

resultant force   The sum of the forces in a particular direction.

rod   A rigid one-dimensional object. A plank is modelled as a rod when calculating moments, so it is assumed to remain straight even when it is loaded.

rough   If a surface is rough, it will produce a resistive frictional force.

sample space   The range of values of a random variable.

secant   The reciprocal of the cosine of an angle. It is usually abbreviated to sec and so $\sec\theta = \frac{1}{\cos\theta}$.

second derivative   The result of differentiating a function twice. It is written as $\frac{d^2y}{dx^2}$ or $f''(x)$.

sector   The part of a circle bounded by an arc and two radii.

self inverse   When $f(x) = f^{-1}(x)$.

sense   The sense of a turning moment is whether it is clockwise or anticlockwise.

separating the variables   A method of solving a differential equation by rewriting it so that each side of the equation can be directly integrated with respect to one of the variables. For example, the differential equation $\frac{dy}{dx} = 2xy$ can be rewritten as $\int\frac{1}{y}dy = \int 2x\,dx$.

sequence   An ordered set of numbers, usually subject to a rule describing them.

series   The sum of the terms of an arithmetic or geometric sequence.

set   A collection of objects or elements.

set notation   A way of listing the members of a set.

sigma notation   The sign Σ (sigma) is used to indicate a sum. Often a subscript is used to show the terms being added. For example $\sum_{i=1}^{5} 3i^2 = 3 + 12 + 27 + 48 + 75$.

significance level   The null hypothesis is rejected if the p-value is less than the significance level.

simplify   To make an equation or expression easier to work with or understand by combining like terms or cancelling.

staircase diagram A graphical representation of an iterative formula where $g'(x) > 0$.

standard deviation A numerical value used to indicate how widely individuals in a group vary, equal to the square root of the variance.

standard normal distribution A normal distribution with a mean of 0 and a standard deviation of 1.

standardising The process to manipulate a non-standardised normally distributed variable to the standard normal distribution using:

$$z = \frac{\text{value} - \text{mean}}{\text{standard deviation}}$$

$$= \frac{x - \mu}{\sigma}.$$

stationary point A point on a curve where $\frac{dy}{dx} = 0$. It can be a maximum point, a minimum point or a point of inflection.

sum to infinity If a geometric series has a common ratio with a modulus less than 1, the sum of n terms approaches a limit as n increases. This is called the sum to infinity.

tilting A rod is said to be on the point of tilting when it is about to lose contact with a support (or when a string that it is suspended from is about to go slack).

turning moment The rotation that takes place when a force is applied to an object (which is fixed or supported at a point) along an axis that does not pass through the point. If the moments are in equilibrium, then the object will not rotate.

uniform A rod is said to be uniform if its mass is evenly distributed through the rod. As a result, its centre of mass is positioned at the centre of the rod.

unit vector A vector with a magnitude of 1.

vector A quantity with a size and a direction.

velocity The rate of change of displacement.

velocity vector A vector for which the magnitude is the speed and the direction is the direction in which the object is moving.

INDEX

Notes

Notes

Notes

Notes

Notes

Notes

Notes

Acknowledgements

The publishers wish to thank the following for permission to reproduce photographs. Every effort has been made to trace copyright holders and to obtain their permission for the use of copyright materials. The publishers will gladly receive any information enabling them to rectify any error or omission at the first opportunity.

Cover & p1 Paladin12/Shutterstock

p1 Phovoir/Shutterstock, p31 PORTRAIT IMAGES ASIA BY NONWARIT/Shutterstock, p45 Tungphoto/Shutterstock, p61 Andrew Park/Shutterstock, p80 AkeSak/Shutterstock, p89 Brian S/Shutterstock, p90 ToySupachai/Shutterstock, p107 Tyler Olson/Shutterstock, p121 Fouad A. Saad/Shutterstock, p143 MilanB/Shutterstock, p171 Es sarawuth/Shutterstock, p198 REUTERS/Alamy Stock Photo, p209 Caron Badkin/Shutterstock, p216 Monkey Business Images/Shutterstock, p238 ESB Professional/Shutterstock, p265 bunyarit/Shutterstock, p287 XiXinXing/Shutterstock, p321 Ivonne Wierink/Shutterstock, p355 PhotosIndia.com LLC/Alamy Stock Photo.